Advances in powder metallurgy

Related titles:

High-energy ball milling: mechanochemical processing of nanopowders
(ISBN 978-1-84569-531-6)

Sintering of advanced materials
(ISBN 978-1-84569-562-0)

Fundamentals of metallurgy
(ISBN 978-1-85573-927-7)

Details of these books and a complete list of titles from Woodhead Publishing can be obtained by:

- visiting our web site at www.woodheadpublishing.com
- contacting Customer Services (e-mail: sales@woodheadpublishing.com; fax: +44 (0) 1223 832819; tel.: +44 (0) 1223 499140 ext. 130; address: Woodhead Publishing Limited, 80, High Street, Sawston, Cambridge CB22 3HJ, UK)
- in North America, contacting our US office (e-mail: usmarketing@woodheadpublishing.com; tel.: (215) 928 9112; address: Woodhead Publishing, 1518 Walnut Street, Suite 1100, Philadelphia, PA 19102-3406, USA

If you would like e-versions of our content, please visit our online platform: www.woodheadpublishingonline.com. Please recommend it to your librarian so that everyone in your institution can benefit from the wealth of content on the site.

We are always happy to receive suggestions for new books from potential editors. To enquire about contributing to our Materials series, please send your name, contact address and details of the topic/s you are interested in to francis.dodds@woodheadpublishing.com. We look forward to hearing from you.

The team responsible for publishing this book:
Commissioning Editor: Nell Holden
Publications Coordinator: Emily Cole
Project Editor: Diana Paulding
Editorial and Production Manager: Mary Campbell
Production Editor: Richard Fairclough
Copyeditor: Marilyn Grant
Proofreader: Tony Phillips
Cover Designer: Terry Callanan

Woodhead Publishing Series in Metals and Surface Engineering: Number 60

Advances in powder metallurgy

Properties, processing and applications

Edited by

Isaac Chang and Yuyuan Zhao

Oxford Cambridge Philadelphia New Delhi

Published by Woodhead Publishing Limited,
80 High Street, Sawston, Cambridge CB22 3HJ, UK
www.woodheadpublishing.com
www.woodheadpublishingonline.com

Woodhead Publishing, 1518 Walnut Street, Suite 1100, Philadelphia,
PA 19102-3406, USA

Woodhead Publishing India Private Limited, 303, Vardaan House, 7/28 Ansari Road,
Daryaganj, New Delhi – 110002, India
www.woodheadpublishingindia.com

First published 2013, Woodhead Publishing Limited

British Library Cataloguing in Publication Data
A catalogue record for this book is available from the British Library.

Library of Congress Control Number: 2013939414

ISBN 978-0-85709-420-9 (print)
ISBN 978-0-85709-890-0 (online)
ISSN 2052-5559 Woodhead Publishing Series in Metals and Surface Engineering (print)
ISSN 2052-5567 Woodhead Publishing Series in Metals and Surface Engineering (online)

The publisher's policy is to use permanent paper from mills that operate a sustainable forestry policy, and which has been manufactured from pulp which is processed using acid-free and elemental chlorine-free practices. Furthermore, the publisher ensures that the text paper and cover board used have met acceptable environmental accreditation standards.

Cover photo courtesy of MIBA Sinter Austria GmbH, A-4655 Vorchdorf, Austria
Typeset by Replika Press Pvt Ltd, India
Printed by Lightning Source

Contents

Contributor contact details

(* = main contact)

Editors

Dr Isaac Chang
School of Metallurgy and Materials
College of Physical Sciences and Engineering
University of Birmingham
Birmingham, B15 2TT, UK

Email: i.t.chang@bham.ac.uk

Dr Yuyuan Zhao
School of Engineering
University of Liverpool
Brownlow Hill
Liverpool, L69 3GH, UK

Email: Y.Y.Zhao@liverpool.ac.uk

Chapter 1

Dr John Dunkley
Atomising Systems Limited
371 Coleford Road
Sheffield, S9 5NF

Email: jjd@atomising.co.uk

Chapter 2

Professor George Z. Chen
Department of Chemical and Environmental Engineering and Energy and Sustainability Research Division
Faculty of Engineering
University of Nottingham
Nottingham, NG7 2RD, UK

Email: george.chen@nottingham.ac.uk

Chapter 3

Professor Challapalli Suryanarayana*
Department of Mechanical and Aerospace Engineering
University of Central Florida
Orlando
FL 32816-2450, USA

Email: Challapalli.Suryanarayana@ucf.edu

Dr Eugene Ivanov
Tosoh SMD, Inc.
Grove City
Ohio
OH, 43123-1895, USA

Email: Eugene.Ivanov@tosoh.com

Chapter 4

Dr Isaac Chang
School of Metallurgy and Materials
College of Physical Sciences and Engineering
University of Birmingham
Birmingham, B15 2TT, UK

Email: i.t.chang@bham.ac.uk

Chapter 5

Professor Arash Simchi,* Amirali Nojoomi
Department of Materials Science and Engineering
Sharif University of Technology
Azadi St.
Tehran, 11155-11365, Iran

Email: arashsimchi@gmail.com

Chapter 6

Professor Iain Todd*
Department of Materials Science and Engineering
The University of Sheffield
Sir Robert Hadfield Building
Mappin Street
Sheffield, S1 3JD, UK

Email: i.todd@sheffield.ac.uk

Dr Alfred T. Sidambe
Department of Materials Science and Engineering
The University of Sheffield
Sir Robert Hadfield Building
Mappin Street
Sheffield, S1 3JD, UK

Chapter 7

Professor Herbert Danninger and Dr Christian Gierl-Mayer
Vienna University of Technology
Vienna, Austria

Email: hdanning@mail.zserv.tuwien.ac.at; cgierl@mail.tuwien.ac.at

Chapter 8

Professor F H (Sam) Froes
5208 Ridge Drive NE
Tacoma, Wa 98422, USA

Email: ssfroes@comcast.net

Chapter 9

Professor Nuria Llorca-Isern
Dept. Ciencia dels Materials i Enginyeria Metal·lurgica
Facultat de Química
Universitat de Barcelona
Marti Franques, 1
08028 Barcelona, Spain

Email: nullorca@ub.edu

Chapter 10

Dr Russell Goodall
Department of Materials Science & Engineering
The University of Sheffield
Sir Robert Hadfield Building,
Mappin Street
Sheffield, S1 3JD, UK

Email: r.goodall@sheffield.ac.uk

Chapter 11

Professor Herbert Danninger, Dr Christian Gierl-Mayer and Dr Susanne Strobl
Vienna University of Technology
Vienna, Austria

Email: hdanning@mail.zserv.tuwien.ac.at; cgierl@mail.tuwien.ac.at; sstrobl@mail.tuwien.ac.at

Chapter 12

Professor Dinesh Agrawal
Pennsylvania State University
USA

Email: dxa4@psu.edu

Chapter 13

Dr Cem Selcuk
Brunel Innovation Centre
c/o TWI Ltd.
Abington Hall
Granta Park
Great Abington
Cambridge, CB21 6AL, UK

Email: cem.selcuk2@twi.co.uk

Chapter 14

Dr Gyu M. Lee*
Department of Industrial Engineering
Pusan National University
Busan, South Korea

Email: glee@pusan.ac.kr

Dr Seong Jin Park
Department of Industrial Engineering
POSTECH
San 31
Hyoja-Dong
Pohang
Kyungbuk 790-784, South Korea

Email: sjpark87@postech.ac.kr

Chapter 15

Dr Cem Selcuk
Brunel Innovation Centre
c/o Abington Hall
Granta Park
Great Abington
Cambridge, CB21 6AL, UK

Email: cem.selcuk@brunel.ac.uk

Chapter 16

Professor Nikhilesh Chawla
Fulton Professor of Materials Science and Engineering
Arizona State University
USA

Email: nchawla@asu.edu

Chapter 17

Dr P Ramakrishnan
Professor Emeritus, IIT Bombay
402, OPAL, Powai-Vihar Complex
Adishan Karahcarya Marg
Powai, Mumbai-400076, India

Email: prrmpm@yahoo.com

Chapter 18

Dr Martin Bram*
Institute of Energy and Climate Research
(IEK-1: Materials Synthesis and Processing)
Forschungszentrum Jülich GmbH
Wilhelm-Johnen-Strasse
52425 Jülich, Germany

Email: m.bram@fz-juelich.de

Dr Thomas Ebel
Department Powdertechnology
Institute of Materials Research
Helmholtz-Zentrum Geesthacht
Max-Planck-Straße 1
21502 Geesthacht, Germany

E-mail: thomas.ebel@hzg.de

Dr Ana Paula Cysne Barbosa
LabPlasma - Laboratório de Processamento de Materiais por Plasma
Universidade Federal do Rio Grande do Norte
Natal - RN
CEP 59078-970, Brasil

E-mail: anacysne@yahoo.com

Dr Nihan Tuncer
Department of Material Science and Engineering
Anadolu University
Iki Eylul Campus
26555 Eskisehir, Turkey

E-mail: nihant@anadolu.edu.tr

Martin Wolff
Department Powdertechnology
Institute of Materials Research
Helmholtz-Zentrum Geesthacht
Max-Planck-Straße 1
21502 Geesthacht, Germany

E-mail: martin.wolff@hzg.de

Chapter 19

Professor Janusz Konstanty
UGH University of Science and Technology
Poland

Email: konstant@uci.agh.edu.pl

Woodhead Publishing Series in Metals and Surface Engineering

1 **Nickel and chromium plating**
J.K. Dennis and T. E. Such

2 **Microbiologically influenced corrosion handbook**
S. Borenstein

3 **Surface engineering casebook**
Edited by J.S. Burnell-Gray and P.K. Datta

4 **Duplex stainless steels**
Edited by R. Gunn

5 **Engineering coatings**
S. Grainger and J. Blunt

6 **Developments in marine corrosion**
Edited by J.P. Blitz and C.B. Little

7 **Fundamental and applied aspects of chemically modified surfaces**
J.P. Blitz and C.B. Little

8 **Paint and surface coatings**
Edited by R. Lambourne and T. A. Strivens

9 **Surfacing: core research from TWI**
TWI

10 **Recommended values of thermophysical properties for selected commercial alloys**
K.C. Mills

11 **Corrosion of austenitic stainless steels**
Edited by H.S. Katal and B. Raj

12 **Fundamentals of metallurgy**
Edited by S. Seetharaman

13 **Energy absorption of structures and materials**
G. Lu and T. X. Yu

14 **The Hatfield memorial lectures: developments in iron and steel processing**
Edited by P.R. Beely

15 **Laser shock peening**
K. Ding and L. Ye

16 **Structural shear joints**
G. T. Hahn, C.A. Rubin and K. A. Iyer

17 **Direct strip casting of metals and alloys**
M. Ferry

18 **Surface coatings for protection against wear**
Edited by B.G. Mellor

19 **Handbook of gold exploration and evaluation**
E. MacDonald

20 **The cold spray materials deposition process**
Edited by V.K. Champagne

21 **The SGTE casebook: thermodynamics at work: Second Edition**
Edited by K. Hack

22 **Belt conveying of minerals**
E.D. Yardley and L.R. Stace

23 **Techniques for corrosion monitoring**
Edited by L. Yang

24 **Creep-resistant steels**
Edited by F. Abe

25 **Developments in high temperature corrosion and protection of materials**
Edited by W. Gao

26 **Mineral wool: production and properties**
B. Sirok and B. Blagojevic

27 **High-performance organic coatings**
Edited by A.S. Khana

28 **Hydrometallurgy: principles and applications**
T. Havlik

29 **Corrosion control in the aerospace industry**
Edited by S. Benavides

30 **Multiaxial notch fatigue**
L. Susmel

31 **Titanium alloys**
W. Sha and S. Malinox

32 **Advances in marine antifouling coatings and technologies**
Edited by C. Hellio and D.M. Yebra

33 **Maraging steels**
W. Sha and W. Gao

34 **Surface engineering of light alloys**
Edited by H. Dong

35 **Sintering of advanced materials**
Edited by Z.Z. Fang

36 **Managing wastes from aluminium smelter plants**
B. Mazumber and B.K. Mishra

37 **Fundamentals of aluminium metallurgy**
Edited by R. Lumley

38 **Electroless copper and nickel-phosphorus plating**
W. Sha and X. Wu

39 **Thermal barrier coatings**
Edited by H. Xu and H. Guo

40 **Nanostructured metals and alloys**
Edited by S.H. Wang

41 **Corrosion of magnesium alloys**
Edited by G.L. Song

42 **Shape memory and superelastic alloys**
Edited by Y. Yamauchi and I. Ohkata

43 **Superplasticity and grain boundaries in ultrafine-grained materials**
A.L. Zhilyaev and A.I Pshenichnyuk

44 **Superplastic forming of advanced metallic materials**
Edited by G. Guiliano

45 **Nanocoatings and ultra-thin films**
Edited by A. S.H. Makhlouf and I. Tiginyanu

46 **Stress corrosion cracking**
Edited by V. S. Raja and T. Shoji

47 **Tribocorrosion of passive metals and coatings**
Edited by D. Landolt and S. Mischler

48 **Metalworking fluids (MWFs) for cutting and grinding**
Edited by V.P Astakhov and S. Joksch

49 **Corrosion protection and control using nanomaterials**
Edited by V.S. Saji and R. Cook

50 **Laser surface modification of alloys for corrosion and erosion resistance**
Edited by C.T. Kowk

51 **Gaseous hydrogen embrittlement of materials in energy technologies Volume 1: the problem, its characterisation and effects on particular alloy classes**
Edited by R. P. Gangloff and B. P. Somerday

52 **Gaseous hydrogen embrittlement of materials in energy technologies Volume 2: mechanisms, modelling and future developments**
Edited by R. P. Gangloff and B. P. Somerday

53 **Advances in wrought magnesium alloys**
Edited by C. Bettles

54 **Handbook of metal injection molding**
Edited by D. Heaney

55 **Microstructure evolution in metal forming processes**
Edited by J. Lin and D. Balint

56 **Phase transformations in steels Volume 1: fundamentals and diffusion-controlled transformations**
Edited by E. Pereloma and D. V. Edmonds

57 **Phase transformations in steels Volume 2: diffusionless transformations, high strength steels, modelling and advanced analytical techniques**
Edited by E. Pereloma and D. V. Edmonds

58 **Corrosion prevention of magnesium alloys**
Edited by G.L. Song

59 **Fundamentals of magnesium alloy metallurgy**
Edited by M. Pekguleryuz, K. Kainer and A. Kaya

60 **Advances in powder metallurgy**
Edited by I. Chang and Y. Zhao

61 **Rare earth-based corrosion inhibitors**
Edited by M. Forsyth and B. Hinton

62 **Thermochemical surface engineering of steels**
Edited by M. Somers and E. Mittemeijer

Part I

Forming and shaping of metal powders

1 Advances in atomisation techniques for the formation of metal powders

J. J. DUNKLEY, Atomising Systems Ltd, UK

DOI: 10.1533/9780857098900.1.3

Abstract: This chapter is a review of recent developments (since 2000) and aims to provide the reader with an introduction to a range of atomisation techniques developed in the last century, in order to understand the latest developments. The basic principles of gas, water and centrifugal atomisation are covered, followed by discussion of problems and advances in each area. The chapter will not attempt to cover all the classical techniques in detail. A previous review[1] or major book[2] can be consulted for this.

Key words: atomisation, metal powders.

1.1 Introduction

1.1.1 Atomisation

This may be defined as breaking liquid into droplets. Items like fire sprinklers, crop sprayers, aerosols and so on utilise atomisation methods for cold liquids. There is a very extensive literature on this surveyed in Yule and Dunkley[2] and Nasr *et al.*[3] and much of it is very useful for those interested in melt atomisation. Clearly, the atomisation of melts, especially at temperatures as high as molten steels, is a very different proposition in practical terms, but some techniques used for cold liquids are also applied to metals. These include gas and air atomisation, which are referred to as 'two fluid' atomisation methods (along with water atomisation, which is almost unique to melt atomisation), ultrasonic and centrifugal atomisation.

1.1.2 Powders

It is generally accepted that in order to be described as a powder, a granular solid should have a particle size below about 1 mm (1000 μm). For powder metallurgy (PM) uses, we would describe as coarse powders, those with significant proportions above about 150 μm (100 mesh). A good example of coarse powder would be the powders used in the hot isostatic pressing (HIP) of shapes and billets. These at one time could range up to 850 μm for high speed steel (HSS) billet production, but most producers now tend to use <600 or 500 μm and much finer sizes are in use for more special alloys. The classical PM (cold compact and sinter) industry has generally used moderately fine powders, nominally <150 μm, but often containing as much as 10–20% >150 μm but below 180, 212, or 250 μm. The mass median size (D_{50}) of these is around 60 μm (Fig. 1.1). The mass median is the particle size below which 50% by mass of the particle size distribution is found.

Recent developments such as metal injection moulding (MIM) have led to a demand for much finer powders, with median sizes around 10 μm. We refer to these as fine powders. For such PM materials as diamond tools and tungsten carbide, it is quite normal to utilise powders with a median size of around 1 μm. These could perhaps be described as 'superfine' but there is no accepted terminology. This is also true of 'nanopowders' where the median particle size is sub-micron, but whether a 0.5 μm powder is sensibly classified as a nanopowder is debatable. When sizes get down to 10–100 nm (0.01–0.1 μm), the 'nanopowder' classification is clearly reasonable.

The question arises: 'What sizes of powder can be made by atomisation?' The answer is that for industrially applied atomisation methods for metallic powders, median particle sizes from shot (>1 mm) down to 10 μm can be made and some producers offer powders as fine as 2 μm, while not divulging just how small a yield of this size range is produced prior to classification. Median particle sizes below 10 μm are not difficult to make in lower surface tension metals, like tin or lead. Sub-micron or nanopowders are not made by atomisation (Fig. 1.2).

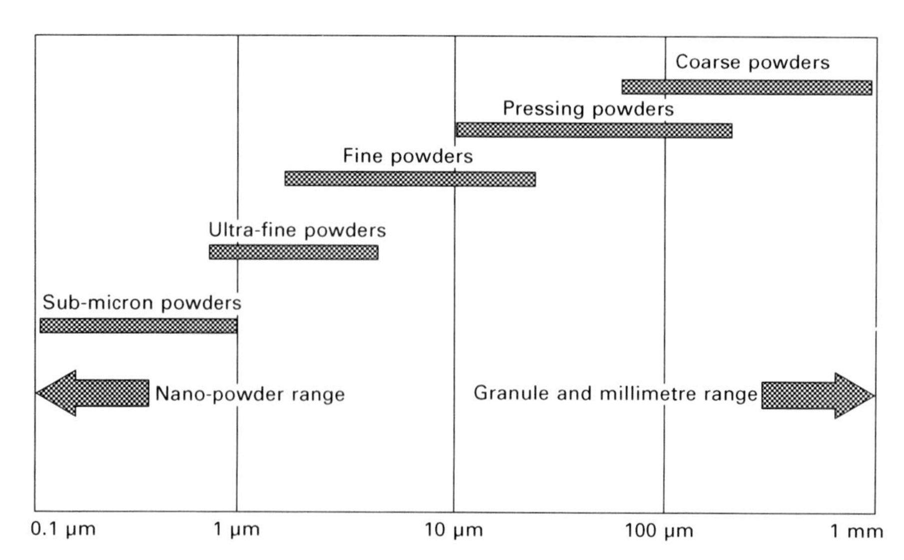

1.1 Approximate descriptive size ranges of particles.

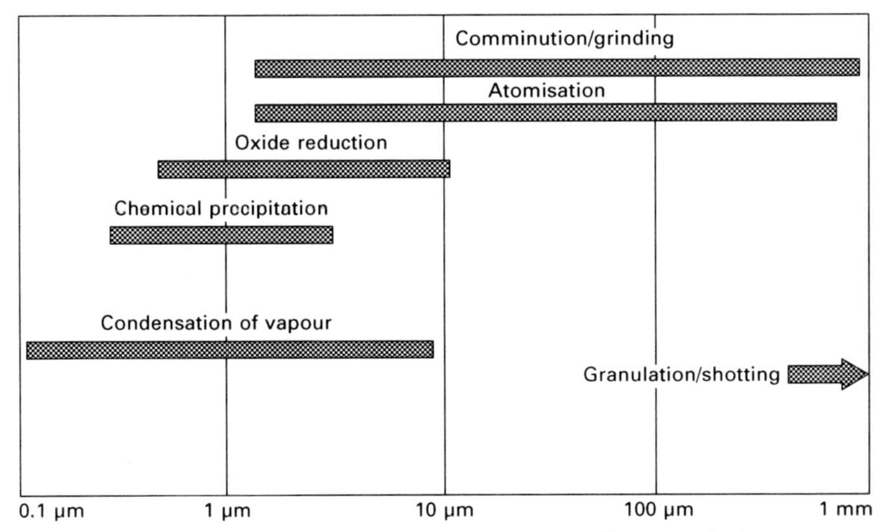

1.2 Approximate size ranges produced by different techniques.

1.1.3 Particle size distributions

It is vital to appreciate that all atomisation (indeed all production) processes do not make a monosize powder, but often make a quite wide distribution of particle sizes about a median value. It has been found that atomised powders mostly conform to a log-normal statistical distribution. This distribution is completely described by just two parameters: the mass-median size (D_{50}) and the standard deviation (σ) which is 1.0 for a monosize powder and rises to about 2.3 for typical gas or water atomised powders. The standard deviation (σ) can be readily calculated as D_{84}/D_{50} or D_{50}/D_{16}. Thus, yield in a narrow range specification, which is often demanded for industrial applications, is improved when σ is reduced. This is of considerable economic importance. Figure 1.3 shows a cumulative undersize plot on log-normal graph paper, which is generally close to a straight line.

1.2 Atomisation techniques

1.2.1 Gas/air atomisation

In this technique, a jet or jets of gas or air is used to break up the molten metal and to freeze it to a powder. There are two basic types of atomising nozzle: closed or close-coupled and open or free-fall.

Basic gas atomiser types

In the open or free-fall nozzle a stream of metal falls a significant distance, often 50–150 mm, from a ceramic nozzle before it is impacted by jets of gas

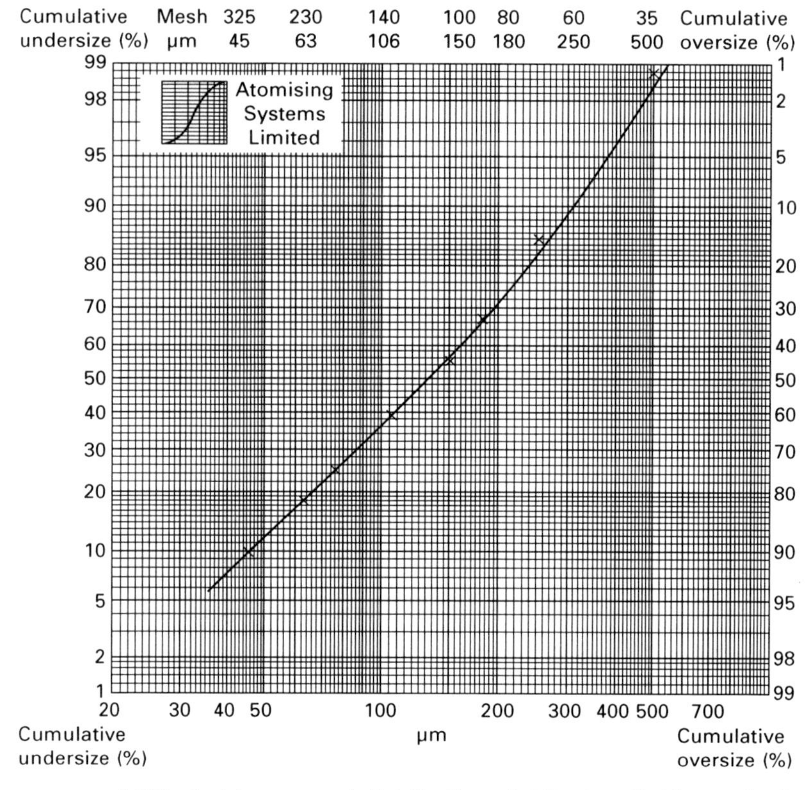

1.3 Typical log-normal distribution plot for two-fluid atomisation.

or air. In close-coupled atomisation, the gas or air jet strikes the melt as it emerges from the nozzle. It is generally accepted that free-fall nozzles are rather less efficient for making fine powder, as the gas jets have normally travelled 50–150 mm before hitting the melt and have thus decayed in velocity compared to the close-coupled kind, where the gas jet may only travel 10–30 mm before impacting the melt. It is also generally acknowledged that free-fall nozzle systems, where the gas and metal nozzles are well separated, is more reliable and easier to set up than a close-coupled nozzle, where the gas can both chill the melt nozzle and create either negative or positive pressures at its outlet, which affect melt flow rates.

Normally gas/metal ratios for heavy metals (Fe, Ni, Cu, etc) are in the range 0.5–4.0 $m^3\ kg^{-1}$ at normal temperature and pressure. The lower end of this range, used in large, multi-ton atomisers for HSS production, gives coarse powders with median sizes of 150–250 μm. The upper end of this range, used in smaller units for special alloys, and with close-coupled nozzles, might allow particle sizes of as fine as 30–40 μm for steels.

Particle size distributions are normally close to log-normal and σ is normally in the range 2.0–2.3, although operation at very slow pouring speeds, such as <1 kg min^{-1} (which implies using a nozzle of about 1 mm diameter) can allow σ to be reduced to as low as 1.7–1.8 in some cases. Powders made by gas atomisation are conventionally assumed to be spherical. This is not by any means always the case. In particular the air atomisation of metals such as Zn and Al leads to an oxide film that interferes with spherodisation, and finer powders often exhibit 'satelliting.' (Fig. 1.4)[5]

The productivity of gas atomisation is somewhat limited. It is unusual to operate a single nozzle at more than 50 kg min^{-1} (3 t h^{-1}). This is partly due to heat transfer limitations at this level as, for steel, about 1.25 MW of heat must be removed and large equipment is needed. Thus a plant producing much above 10 000 t per year is unusual.

1.2.2 Water atomisation

In this technique, a falling stream of melt is struck by water jets. Obviously, cooling in this technique is much more rapid than cooling in gas atomisation and owing to the presence of oxygen in the water, oxidation takes place by the reaction:

$$M + H_2O = MO + H_2$$

This can only occur when the free energy of formation of the metal oxide is higher than that of water, which is true for many industrially important metals, including Fe, Cr, Mn and Si. Oxidation of Ni and Co is marginal, and Cu and Ag are easily reducible by hydrogen, meaning that water-atomised

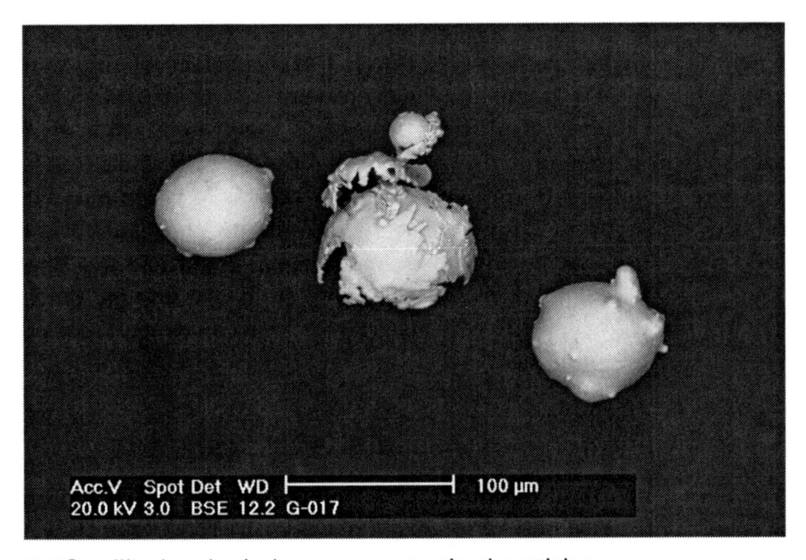

1.4 Satellited and misshapen gas atomised particles.

Cu and Ag powders can have very low oxygen contents, <0.05%. For the more reactive metals, the oxygen content of the powder depends on reaction kinetics and the properties of the oxide film (protective or not). Thus pure iron powder can be about 0.3% oxygen, while FeMn with a fusible and non-stoichimetric oxide can reach 3% and FeSi, with a high melting and very protective oxide, can be well below 0.1%.

The other effect of oxide formation is on particle shape. As for gas/air atomisation, if a strong oxide film forms that is solid at the melting temperature of the metal, it often interferes with spherodisation and an irregular powder results, as it does for Cr-containing stainless steels. If a fusible oxide is formed, as for NiCrBSi, then a more spherical powder results. Particle shape can also be affected by the rapid cooling rate. This is apparent only for metals that melt below around 700°C, as above this temperature heat transfer is limited by film boiling. For melting points in the range 200–400°C (e.g. Sn, Pb) cooling is so fast that the resulting powder is often in the form of ligaments that have frozen before finishing break-up. In such cases surprisingly coarse powder can result with densities of only 10% of solid. These effects are discussed by Dunkley.[6]

Because of its excellent quenching properties, very coarse powders, or even shot with median sizes of up to several millimetres, can be made. The use of pressures in the 100–200 bar range typically results in powders in the 50–100 μm range (D_{50}). Higher pressures, up to the level where the spray breaks the sound barrier (~500 bar, 50 MPa) result in finer powders, down to 30 μm or less. Typically a doubling of pressure nearly halves the particle size (Fig. 1.5) with

$$D = K P^{-0.8}$$

where D is median particle size (D_{50}), K is a constant relating to apparatus and metal and P is atomising water pressure.

There are reports of ultra-high pressure water atomisation by Japanese authors where pressures of 1–2000 bar (100–200 MPa) are used to make powders as fine as 10 μm, but details are sketchy. The productivity of this process ranges from a few kilograms per minute to at least 500 kg min^{-1} (30 t h^{-1}). This is achievable with little change in particle size distribution. Thus huge outputs of over 250 000 t yr^{-1} from one line are quite feasible and designs for over 1 MT yr^{-1} and 180 t h^{-1} have been made for processing copper smelting intermediates.

1.2.3 Centrifugal atomisation

The above two-fluid atomisation techniques (water and gas atomisation) account for the vast bulk of atomised powders in production. However, they both produce rather wide particle size distributions with σ being 2.0–2.3 in

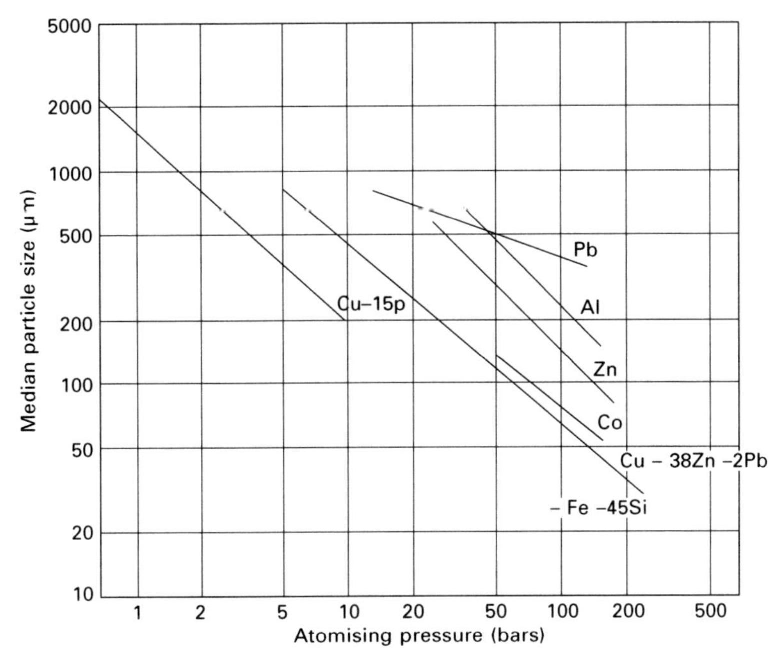

1.5 Water atomised particle size versus pressure for common metals.

most cases. Centrifugal atomisation is atomisation by spinning the melt at high speed (Fig. 1.6). It is used on an industrial scale for several special applications, particularly for lower melting alloys where the problems of erosion of the spinning disk or cup are less difficult. Thus Pb- and Sn-based solders, zinc and some aluminium are made in this way.[7]

The great advantage of this technique is that σ can be greatly reduced, from 2.0–2.3 for two fluid methods to as low as 1.3–1.5 in many cases. This dramatically increases yield in narrow size ranges, such as the 25–45 μm and 20–38 μm ranges used in electronic solders (Fig. 1.7).

1.3 Problems and advances in gas atomisation

As noted above, gas atomisation suffers from some limitations. In particular the production of finer particle sizes is desirable for some applications like MIM, as well as narrower size ranges. The high energy cost associated with the large amount of gas used or compressed is also a burden. Cleanliness is vital for HIP applications and satelliting is undesirable for many uses.

1.3.1 Finer powders

Sometimes these limitations are combined. For example, to make MIM powders it is sometimes necessary to raise operating pressures from the

1.6 Centrifugal cup atomisation of steel at 30 t h^{-1}.

1.7 Industrial centrifugal atomisation system for electronic solder powders.

normal 20 bars to over 50 bars, and to use gas/metal ratios as high as 4–5 normal m^3 kg^{-1}. This raises costs substantially, while yields of less than 25 μm remain poor, perhaps 30–60%. At the same time, very fine powders tend to agglomerate and satelliting can become severe, reducing packing density and MIM properties, as well as increasing apparent particle size. One way of reducing gas consumption and improving yields is to utilise hot gas.[8]

For a fixed nozzle geometry, it has been found that gas consumption falls inversely with the square root of absolute temperature, while the median size falls at a similar rate. An anti-satellite system, that works to prevent collisions between the hot spray and cold fines, can also greatly improve particle shape by reducing satellites.[5]

1.3.2 Clean powder production

The hot isostatic pressing of high integrity, high alloy components, such as are used in the nuclear, aerospace, oil-drilling and chemical industries, demands defect-free powders. Defects in this context can include a number of problems:

- non-metallic inclusions
- run-to-run alloy cross contamination
- oxide films
- foreign bodies (e.g. organics).

Inclusions are critical for fatigue applications and for rotating aerospace parts. The drastic measure of screening all powder less than 50 or 60 μm has been adopted to control their maximum size in some cases. For economic reasons, it is now preferable to take precautions against them by optimising melt practice, where major advances have been made in past decades, so that nearly 100% of the powder can be utilised. Melting itself is difficult to carry out in normal tilt-pouring induction furnaces without significant inclusions being picked up, especially when pouring through air into a slag-covered tundish. In aerospace applications, vacuum melting and pouring is often used, which improves matters greatly, but at a huge cost in slower production and

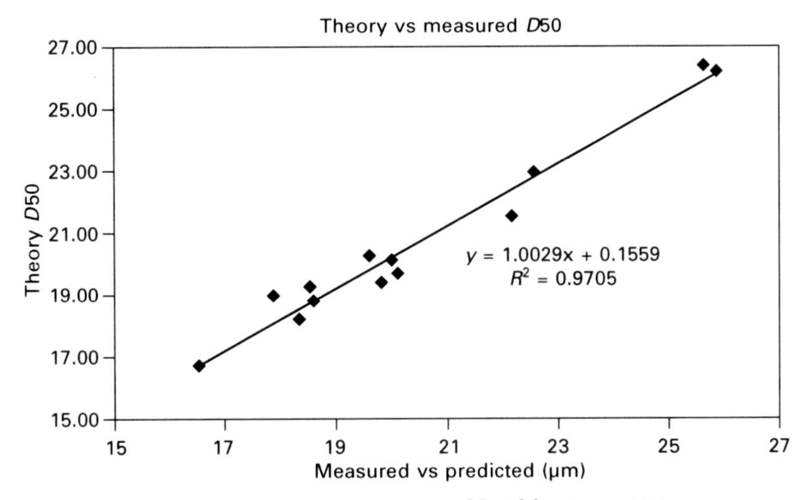

1.8 Plot of data using equation $D = k_6 M^{0.5} T^{-0.6}$, where M is mass flow and T is temperature from Aderhold and Dunkley.[8]

increased system cost. For really clean powders, it has been found best to use very large tundishes, sometimes over 5 t in capacity, to allow a long period of holding during which inclusions can float out of the melt. Bottom pouring via slidegates is used in advanced systems and produces excellent quality powders in terms of non-metallic inclusions.[9]

Run-to-run contamination is always a concern in any plant not dedicated to a single product. For HIP applications, where quite coarse (*ca.* 100 μm) powders are used, a single particle of, for example a ferritic stainless steel in an austenitic stainless steel, will not diffuse away in processing and can have serious results in terms of corrosion or fatigue performance. Thus it is necessary to design gas atomisers which can be quickly and thoroughly cleaned. This is very demanding, as to dissipate the megawatt or more of heat to cool the powder requires a large surface area in the system, all of which must be cleaned. Techniques borrowed from the food, chemical and pharmaceutical industries, especially clean in place (CIP) methods are needed. Of course, if even this is not good enough, then one can resort to wash heats where an entire melt is processed and rejected to clean the system. Obviously this is an expensive and unattractive option, unless changes in grade are very infrequent, or the customer will pay very high prices.

Sieving machines are very hard to clean and may best be dedicated and/or swapped when grades are changed. Filters are even more of a cleaning problem and cyclones are much more cleanable. Luckily cyclones can capture down to 2–5 μm powders, so for HIP applications, the fines removed in filters before exhausting the gas to atmosphere or re-compressing, can be rejected or remelted, as they are <1% of the good product.

Oxide films are generally less of a common problem. They should be eliminated by adequate purging of the system, careful sealing and control of the system pressure (slight positive pressure). It is not too difficult to use oxygen meters to monitor system oxygen levels before pouring, but they are of limited value during operation (except on recycled gas) as the metal will get into the system to sub-ppm levels and powder fines are very unpopular inside the sensors.

Foreign body contamination with such substances as wood, plastic and mineral dust in HIP powders is highly undesirable, leading to defects in HIP products. Plain bad housekeeping is obviously the first suspect and all transfer operations, containers and so on need to be examined to ensure they are scrupulously clean and sealed at all times. For aerospace applications, some have resorted to containing all powder transfer work in a clean room with filtered air. The next common source of contamination is seal materials such as gaskets, elastomeric valve seats or seals and so on. Eternal vigilance is the only way to ensure quality in this respect, but plant design must be carefully considered.

1.3.3 Satelliting thermal spray powders

Satellites, which are superfine (e.g. 2–10 μm) powders stuck to the surface of much coarser (e.g. 40–100 μm) powders (see Fig 1.4), are a concern in thermal spraying for two reasons. First, they adversely affect the flow of the powder through the powder feeder to the spray system and second, they interact differently with the flame or plasma, owing to locally smaller cross-sections that lead to local overheating and perhaps oxidation. While there is little, if any, published data, it is quite clear in the industry that smooth, satellite-free spheres are the ideal. The Hall flow rate of the powder is often used as a good indicator of quality of shape and, of course, of flowability in feeders. Satellites are formed inside the atomiser, more specifically in the area close to the atomising nozzle where the spray is extremely hot and particles can weld together (Fig. 1.9).[5]

Fine particles are sucked up into this area as the expanding gas jets entrain about ten times their volume of ambient gas. Obviously, the finer the powder size distribution overall, the higher will be the amount of superfines and the higher the concentration of such particles in the atomisation zone. Thus satelliting tends to get worse as finer powders are targeted. The thermal spray industry uses a wide range of particle sizes, from as coarse as –200+75 μm for plasma transferred arc (PTA) welding, to 10–30 μm for some plasma or high velocity oxy-fuel spraying (HVOF) processes. Overall, the trend in the industry has been for the processes using finer powders to have grown more rapidly, so satelliting is becoming more of an issue.

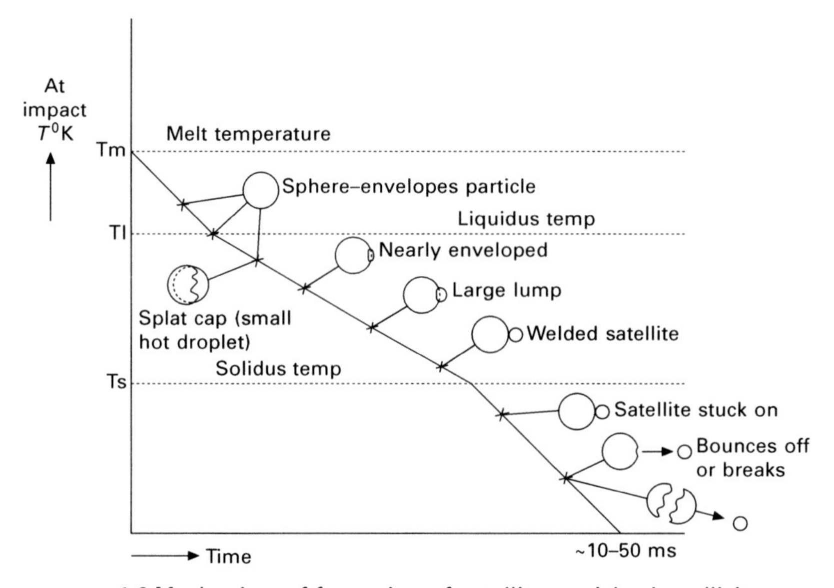

1.9 Mechanism of formation of satellite particles in collisions.

Techniques have been developed which do permit a major reduction in satellite particles.[5]

1.3.4 Process economics

The costs of gas atomisation include capital costs, labour costs and consumable (including energy) costs. All of these are then increased by the yield loss factor. If we have a yield of x%, cost has been increased by a factor of $100/x$. Given that yields can range from as low as 20% to as high as 90–98%, this is of huge importance. A larger scale of operation tends to reduce capital costs and follows a rough square root law. Thus a doubling of melt capacity or output/hour will only result in a ~40–50% increase in capital cost. Labour costs are almost directly reduced by hourly capacity. However, utilisation is a key factor. The capital cost per kilogram produced will be lower at constant output in a smaller, cheaper plant. If twice as much powder can be sold, as can be made on a small plant, doubling the capacity of the small plant will reduce capital costs per kilogram.

Consumable costs in gas atomisation are highly significant. Smaller plants, producing less than 1000 t per year, commonly use nitrogen on a total loss basis. Bulk liquid nitrogen prices vary widely, but are currently around €0.15 m^{-3} in the UK. Thus if we use several cu.m/kg our costs are as high as €0.5 kg^{-1} for gas alone. To reduce these costs it is very attractive to use hot gas atomisation.[8] It is possible to halve gas consumption by heating the gas to just 200–300°C without degrading particle size. The energy costs of doing this only account for about 10% of the resulting cost savings, making it a very attractive investment.

It pays to monitor yield with great care given its vital importance. Yield can be affected as much outside as inside the atomiser. Matters such as dust losses to filters, tundish skull losses, dust losses from sieving, and especially inefficient sieving during very difficult separations such as –50+25 μm, need careful consideration. It is also vital, in cases where yield is low, to discuss with users the precise limits on over- and under-size particles and to take full advantage of any latitude allowed. If the yield is <50% an allowance of a few percent on over- or under-size can affect costs by as much as 10%.

1.4 Problems and advances in water atomisation

Finer powders: Water atomisation, as mentioned above, typically operates at pressures of 100–500 bar and produces powders with D_{50} values from 100 down to about 30 μm. The velocity of the water resulting from such pressures ranges from 140–316 m s^{-1}, assuming no friction losses. The water jets at high pressures typically consist of about 99 vol% gas and less than 1 vol% water droplets when they actually hit the melt. As velocities rise towards 300 m s^{-1}, the speed of sound in air/nitrogen, the behaviour of the gas becomes more and more important. Even though the 99 vol% gas equates to only 10 wt%, so that it does not absorb a lot of momentum from the water, it is found that the behaviour of the gas when water jets collide can lead to local pressure rises that disturb atomisation very badly. Thus totally different designs are needed at pressures above this range. Some Japanese authors have apparently managed to operate at extreme pressures of over 1000 bar (100 MPa) and to produce MIM grade powders. However few details have been published.

High production rates: It is already commonplace to operate water atomisers for steel at 250 or 500 kg min^{-1}. As this equates to a production rate of 15–30 t h^{-1}, or at 6000 h per year utilisation, to 90–180 kt per year, there are few requirements in the PM industry for higher outputs. In smelters, where outputs of >100 t h^{-1} are encountered, it is possible that envisage systems that operate at such rates, simply by using multiple tundish outlets, or by scaling up the nozzles. Annual outputs of 0.5 Mt per year to over 1 Mt per year are certainly feasible. To reduce costs, it is normal to reduce the amount of high pressure water used to much less than the flow needed to extract all the heat from the melt without boiling. This simple measure can reduce pump energy demand and thus pump capital cost by half.

1.5 Centrifugal atomisation

Centrifugal atomisation, as mentioned above, has the great advantage of producing narrow size range powders with σ of 1.3–1.5. It has some other attractions. It can operate continuously, in an inert gas atmosphere and it uses very little energy compared with gas and even water atomisation. However, the equipment size is dictated by the distance needed for the powder/droplets to fly before they freeze and can be very large, as the radius of the vessel is thus dictated. Cleanability is obviously not great with large equipment (it can range from 2–10 m diameter). Thus capital costs can be fairly high, although there are no large pumps or compressors needed and the cost is just for simple fabrication work. Thus centrifugal atomisation (Fig. 1.10) finds applications where:

- The finished product has a narrow size range.
- One product is desired in large volumes without specification changes.

1.5.1 Application to electronic solders

Electronic solders are used to make the solder pastes that are used in an electronic device assembly on a very large scale (thousands of tons per year). Specifications are extremely demanding *viz*:

1.10 Centrifugal atomisation of steel at 70 kg min^{-1} to ~100 μm powder.

- perfect sphericity and freedom from satellite particles (>98% perfect spheres)
- extremely low oxygen content, <100 ppm
- very narrow powder size distributions, e.g. Type III at 25–45 μm, Type IV at 20–38 μm.

In the early days some producers used gas atomisation to make these products and found yields as low as 20%, as well as problems with oxygen pick-up and satellite formation. Ultrasonic atomisation provided a solution for the yield, raising it to ~50% for Type III, but it proved difficult to make the finer Type IV powder size at high yields and unit outputs were very low, around 10–30 kg h^{-1}.

Centrifugal atomisation, with rotation speeds of over 50 000 rpm and a tiny disk (~30 mm diameter) has proved able to make the main grades with yields as high as 70% and at atomising rates of 200 kg h^{-1} and this is now the preferred method for this product. Plants have quite small diameters, only about 2.5–3 m, and the latest units incorporate lean phase pneumatic conveying in inert gas directly to a classifier to separate the fines before the product is discharged for final screening at 38 or 45 μm.

1.5.2 Other non-ferrous powders

Zinc powder for alkaline batteries is produced using rotation speeds of 2000–10,000 rpm to make ~100–300 μm powders. Outputs of 1–3t h^{-1} are achieved using large chambers, *ca* 5–10 m in diameter. The final specifications vary, but typically ask for the maximum/minimum size ratio to be about 3 or 4:1. Air atomisation is still used by some producers, but yields of ~70% are a major cost problem, while centrifugal atomisation can reach >95% yields. Energy costs to drive the rotor are a tiny fraction of those needed for a compressor, so the economics are attractive. Particle shape can be controlled by atmospheric oxygen content with apparent powder densities from 1.0–4.0 g ml^{-1} depending on oxygen. It is not difficult to imagine other fields of application, especially coarse Al powders where an absence of explosible fines would be most desirable. The production of filter bronzes could also make sense, where extremely tight cuts are used.

1.6 Other atomisation techniques

Space does not permit a discussion here of other minor techniques, such as ultrasonic atomisation of solder powders, and vibrating orifice atomisation of solders for BGA applications. See Yule and Dunkley[2] for further information.

1.7 Conclusion

Atomisation breaks liquid metal into droplets, which then cool to form powder and can be processed further. Recent developments in powdered metal processes have led to demands for finer grained powders and atomisation techniques can be used to create powders up to (but not including) the nanoscale. Different techniques such as gas/air, water and centrifugal atomisation have been discussed. Potential problems such as inclusions, contamination and oxide films can occur when processing metal powder and this chapter has discussed some of the ways to mitigate these problems. Finally, the economics of powder production and some of the scores of applications for atomised metal powders have been touched upon.

1.8 References

1 Dunkley, J. J., *Atomization*, ASM Handbook, Vol 7 Powder Metal Technologies and Applications, ASM International, 1998, 35–52.
2 Yule, A. J. and Dunkley, J. J., *Atomization of Melts*. Clarendon Press, Oxford, 1994.
3 Nasr, G. G., Yule, A. J. and Bendig, L, *Industrial Sprays and Atomization*. Springer Verlag, London, 2002.
4 Unal A, 'Effect of processing variables on particle size in gas atomisation of rapidly solidified aluminium powders'. *Materials Science and Technology*, 1987, **3**, 1029–39.
5 Dunkley, J. J. and Telford, B., 'Control of "satellite" particles in gas atomisation'. *Advances in Powder Metallurgy and Particulate Materials*, 2002, **3**, 103–10.

6 Dunkley, J. J., 'The production of metal powders by water atomisation'. *Powder Metallurgy International*, 1978, **10**, 39–41.
7 Dunkley, J. J. and Aderhold, D., 'Centrifugal atomisation of metal powders'. *Advances in Powder Metallurgy and Particulate Materials*, 2007, **2**, 26–31.
8 Aderhold, D, and Dunkley, J. J., 'Experience of industrial-scale hot inert gas atomisation'. *Advances in Powder Metallurgy and Particulate Materials*, 2008, **2**, 1–7.
9 Sundin S, Nordberg L-O, Cocco M and Walve R, Innovations in high performance gas-atomized PM steels. *Proceedings of the 2008 World PM Congress*, Session 51–1, Washington, MPIF, 2008.

2
Forming metal powders by electrolysis

G. Z. CHEN, University of Nottingham, UK

DOI: 10.1533/9780857098900.1.19

Abstract: Fossil fuels will eventually be replaced by renewables. Currently, the most feasible and efficient way of utilising renewable energy is to convert it to electricity. In response to this change, fossil energy-based pyrometallurgical processes will inevitably shift to electricity driven processes. This chapter considers the feasibility of direct conversion of mineral to metal powder using a new electrochemical method, the FFC Cambridge process (Fray, Farthing and Chen). The discussion will be on the background of electrometallurgy and powder metallurgy, the principles of the new process and its application for metal powder production, and the direct route from oxide precursors to alloyed powders.

Key words: alloys, electrolysis, metal oxides, metals, molten salts, powders.

2.1 Background of electrometallurgy and powder metallurgy

The predictable exhaustion of fossil resources in the near future[1] challenges the current metallurgical industry which is largely based on carbothermic reduction of minerals to their respective metals.[2] Obviously, without the cheap supply of carbon (e.g. coal and natural gas), production of metals will become expensive or even unaffordable. To maintain the sustainability of metal supply, alternative processes have to be developed that can utilise renewables. In fact, in the carbothermic reduction of metal oxide-based minerals, the carbon functions in two ways: it provides the electrons needed to reduce the minerals to the metals and reacts with oxygen to supply energy (heat) to enable the reduction. Fortunately, these two roles of carbon can be undertaken by electrons which are commonly the direct product of conversion of renewables to electricity. In other words, in the renewable era most, if not all, carbothermic reduction processes will and must be replaced by direct or indirect electrolytic processes.

However, the truth is that metal production via electrolysis has long been established and plays important roles in the extraction and/or refining of some important metals, such as aluminium, nickel and copper. In these industrialised electrolytic processes, a common feature is the cathodic deposition of the metal from an electrolyte containing the metal ion. The differences are found in the anode reaction. In the electrolytic extraction of aluminium, the anodic process is the discharge (or oxidation) of the oxide ion, O^{2-}. The net process is the decomposition of the feed alumina, Al_2O_3, into aluminium metal and oxygen, which in turn reacts with the carbon (anode) to produce CO_2 at elevated working temperatures (e.g. 900°C). The process does not change the composition of the electrolyte. For nickel extraction, the feed mineral is nickel sulfide, NiS, which is attached to the anode. During electrolysis, the solid sulfide is anodically decomposed into solid sulfur and the nickel ion, Ni^{2+}, which enters the electrolyte and is then transported to the cathode to be reduced to nickel metal. Again, the process does not incur changes in the electrolyte. Copper is primarily extracted from copper sulfide, first by converting sulfide to oxide (roast) followed by carbothermic reduction. However, the product still contains too many impurities and needs refining, which is achieved by electrolysis using impure copper as the anode. During electrolysis, the impure copper is anodically oxidised to the copper ion, Cu^{2+}, leaving most of the impurities, particularly the non-metallic ones, on the anode. The Cu^{2+} ion then enters the electrolyte, is transported to and is reduced to purer copper at the cathode. In principle, the process should also only separate copper from the non-metallic impurities without altering the electrolyte.

It should be noted that all electrolytic processes occur in a liquid electrolyte which can be an aqueous solution containing an appropriate supporting electrolyte, such as $H_2SO_4 + CuSO_4$ for copper refining, and $H_2SO_4 + NiSO_4$ for nickel extraction, or a molten salt at elevated temperatures, for example, the molten mixture of Na_3AlF_6 (cryolite), AlF_3 and Al_2O_3 for aluminium extraction. There are two other types of electrolytes that are more commonly used in laboratories, that is organic electrolyte composed of an organic solvent and an appropriate inorganic or organic salt, and ionic liquids which are basically room temperature molten salts composed of relatively large cations and anions which are often organic.

In comparison with more commonly used pyrometallurgy (particularly via carbothermic reduction), electrometallurgy is advantageous in several aspects. It has a relatively simpler device requirement, has greater process efficiency and a lower environmental impact. The process operation and product quality are also easier to control. One disadvantage is that all electrolytic processes require a higher energy input and hence a greater cost which is offset by the fast increasing cost of using carbon-based fossil resources. The other disadvantage is that all existing electrolytic processes are specific for a particular metal. This situation may change because a more generic technology, that is the FFC Cambridge process (Fray, Farthing and Chen), emerged in recent years and can in principle be applied to the extraction of most pure metals, as well as to direct synthesis of alloys.[3,4] This new process is especially attractive because it can be applied to produce powders of either a pure metal or an alloy, as will be introduced in more detail in the next section.

Metal or alloy powders can be used for various purposes, but they are mostly applied in powder metallurgy, particularly for the manufacture of metallic products with complicated or near-net shapes.[5,6] There are a number of conventional manufacturing methods for producing metallic products, such as melting and casting, and machining. The former requires the metal to be melted at elevated temperatures and hence is energy intensive. The latter produces a large quantity of waste (e.g. turnings and spent lubricant). In contrast, powder metallurgy produces specially shaped products directly from the powder of either pure metal or alloy through sintering, often pressurised, at temperatures much lower than that used to melt the metal. It is therefore simple in processing equipment and operation, effective in product quality control, and low in energy input, and produces less waste. In addition, powder metallurgy is also preferred to traditional melting and casting technologies to produce and process metal alloys containing metals of very different reactivities, densities and melting points. These advantages of powder metallurgy can only be realised on the basis of the low cost availability of the various metal and alloy powders.

Metal powders are industrially produced by two main techniques. For metals with low melting points such as aluminium, the so-called atomisation process is usually chosen.[7] The principle of atomisation is to force a liquid metal together with a stream of gas through a small orifice that turns the liquid into small droplets which, upon cooling, solidify into fine particles. The direct product from atomisation usually features spheres, as shown in Fig. 2.1(a).[8] The particle sizes of metal powders produced by atomisation vary, ranging from tens of micrometres to sub-micrometres, although products can be made more uniform by process control and/or post-process sieving.

Atomisation is possibly the most applied method for producing metal powders. However, for metals with high melting points, other techniques are needed. For example, chromium powders find many applications, including high-performance non-ferrous alloys,[9] sputter target material for plasma or

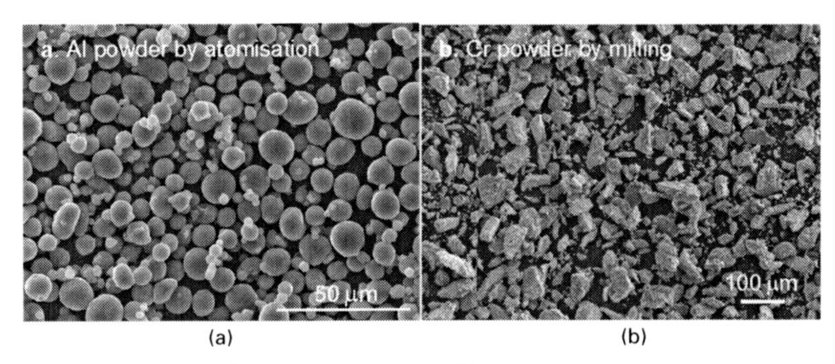

2.1 (a) Aluminium[8] and (b) chromium[17] powders produced by atomisation and milling, respectively.

spray coating,[10] and cermets, which are metal–ceramic composites with high electric conductivity, thermal stability and corrosion (oxidation) resistance.[5,6] Chromium has a melting temperature of 1857°C. Thus, chromium powder is very difficult, if not impossible, to produce by the atomisation method. At present, commercial production of chromium powder involves two main steps: (1) metallothermic (e.g. Al or Ca) or electrolytic (extraction) reduction, and (2) mechanical milling,[2, 11] which both contribute significantly to the high market price of the material (US$6–10 per kg). The chromium products of these extraction methods are brittle, particularly the electrolytic product which is contaminated by hydrogen[2,11] and can be crushed and milled into powder. However, the milling process not only has a long operation time, but also involves the risk of contamination of the powder product by constituents in air and by the mill tools.[12] In addition, the milled powder has a wide range of particle sizes and shapes and very fine materials are usually screened out and rejected. Figure 2.1(b) shows a typical example of the commercial chromium powder prepared by milling. It can be seen that the powder particles have a great number of irregular shapes and very many different sizes, ranging from sub-micrometres to a few tens of micrometres.

There are, of course, other methods for production of metal powders,[5,6] but they are all a follow-on step to the extraction of the metal from the mineral. It will obviously be advantageous if the metal powder can be produced directly from the mineral, which is a more direct process and also saves energy. In the following sections, the recent development of the FFC Cambridge process is reviewed for making metal powders directly from the mineral precursors, for example oxides and sulfides, using molten salts as the electrolytes.

2.2 Principle and main technological prospects for the FFC Cambridge process

The FFC Cambridge process was proposed in late 1990s, named after the inventors, Derek J. Fray, Tom W. Farthing and George Z. Chen who were then working at the University of Cambridge.[13,14] To date, most reported studies have revealed that the energy consumption of the FFC Cambridge process is generally lower than that of other existing technologies, although variations can be significant between different metal oxides.[4,15–20] For example, according to published literature, the electrolysis energy consumption in making titanium by the FFC Cambridge process is below 20 kWh kg^{-1} Ti in laboratory,[16] whilst the overall energy consumption is over 50 kWh kg^{-1} Ti in the Kroll process which is currently the dominant industrial method for extraction of titanium and zirconium. Similar findings are reported for other metals and alloys in the form of either a porous body or powder,[4,15–20] making this new process a potential commercial alternative to existing carbon-based technologies.

Briefly, the FFC Cambridge process can be explained using the solid state electroreduction of SiO_2 to Si as an example.[18,19] It starts by making the oxide into a cathode which is then placed in the molten salt, typically $CaCl_2$ at 900°C. After inserting the anode, for example a graphite rod, electrolysis is carried out between the cathode and the anode at a voltage that is sufficiently higher than the decomposition voltage of the oxide at the cathode. In this process, driven by the negative potential applied on the cathode, the oxide is reduced to silicon which remains on the cathode, whilst the oxygen in SiO_2 is ionised, enters the molten salts, is transported by diffusion and convection to and is discharged at the anode. The electrode and overall cell reactions can be summarised below using a carbon-based anode.

Oxide cathode: $SiO_2 + 4\ e = Si + 2\ O^{2-}$ [2.1]

Carbon anode: $x\ O^{2-} + C = CO_x + 2x\ e\ (x = 1 \text{ and } 2)$ [2.2]

Cell: $x\ SiO_2 + 2\ C = x\ Si + 2\ CO_x$ [2.3]

In reaction [2.2], CO_x refers to both CO and CO_2 as observed in past studies of the FFC Cambridge process and other molten salt-based electrochemical processes, such as the aluminium extraction mentioned in the previous section. The ratio of the two gases depends particularly on the temperature and the cell voltage. Obviously, this anode reaction consumes the anode and emits greenhouse gases, although it is supposed to lower the cell voltage. To avoid frequent replacement of the anode and a carbon impact on the environment, effort has long been made to develop a so-called inert anode. This is typically made from materials that are electrochemically and chemically inert in molten salts, but conductive to electrons or the oxide ion, O^{2-}. A few recent studies have shown promising results using polycrystalline SnO_2 (often doped with Cu and other elements), a composite of $CaRuO_3$ and $CaTiO_3$, and oxide ion conducting membranes for the FFC Cambridge process.[21–25] If this inert anode is used, the cathode reaction will remain the same, but the anode and cell reactions will change as follows:

Inert anode: $2O^{2-} = O_2 + 4\ e$ [2.4]

Cell: $SiO_2 = Si + O_2$ [2.5]

Reactions [2.1] to [2.5] represent the overall electrode and cell reactions in the FFC Cambridge process, but the practice is more complicated, particularly on the cathode as explained in the following sections.

It is worth mentioning that direct electrochemical reduction of a solid compound attached to a cathode has been known for years, for example AgCl to Ag in the commonly used reference electrode, and $PbSO_4$ to Pb in the well-known lead acid battery. However, before proposal of the FFC Cambridge process, no previous work had succeeded to utilise this electrochemistry for extraction of metals, particularly reactive metals and semi-metals like titanium and silicon.

Cell designs are a core technological issue in the research and development of the FFC Cambridge process. A number of cell designs have been proposed, as shown in Fig. 2.2, and tested successfully in the laboratory for proof of the concept (I),[3] fundamental research (II to VI),[3,4,17–19,26–30] near-net-shape manufacturing (VII),[30] and scale up (VIII), whilst type IX was specifically proposed for metal powder production.[31] Several other technological issues are discussed below in connection with the configuration of the cell.

The FFC Cambridge process has been mostly researched using metal oxides as the feedstock to construct the cathode,[4] whilst sulfides are emerging as an emission-free option when used together with a graphite anode (see later discussion).[32–34] Because components of air can attack metals at elevated temperatures (e.g. formation of oxides and nitrides), the process needs to be operated with appropriate protection. The electrolytic extraction of aluminium in industry is protected by a solidified surface layer (salt crust) above the molten salt. The 'salt crust approach' is in principle applicable to the FFC Cambridge process, but it is not as convenient as using an inert gas and not necessarily cheaper in laboratory. Argon is often chosen, although nitrogen can be used if it does not react with the metal produced. Vacuum should not be applied for protection to avoid evaporation of the molten salt.

Another technological issue is the selection of molten salts. The FFC Cambridge process needs an electrolyte that can dissolve and transport the anions released at the cathode, e.g. O^{2-} or S^{2-}, to the anode. There are only a few pure molten chloride salts that can dissolve and transport the oxide ion, including $CaCl_2$, $BaCl_2$ and LiCl. Of these three, $CaCl_2$ is the cheapest with the lowest toxicity and hence is widely used. However, one of its main disadvantages is a relatively high melting temperature (762°C) and quite a large electronic conductivity when the applied cell voltage is sufficiently high to force a cathode potential close to that for the reduction of the Ca^{2+} ion.[29] In a two electrode cell, Ca^{2+} ion reduction may occur at an applied voltage that decreases from the decomposition voltage of $CaCl_2$ to that of CaO (3.099 to 2.520 V at 900°C) with an increasing concentration of the O^{2-} ion in the molten salt. The other common chloride salts, such as NaCl and KCl, have relatively poor solubility for the O^{2-} ion, and are thus not suitable to be used alone in the FFC Cambridge process.[35] Of course, two or more of these chloride salts can be mixed to achieve some desired properties, such as a lower working liquid temperature to save energy, and/or a lower solubility of the O^{2-} ion to reduce the electronic conductivity of the electrolyte. These changes should be maintained at a certain level so that the reaction kinetics at the electrodes and ion transport through the electrolyte are not compromised too much. More recently, some fluoride

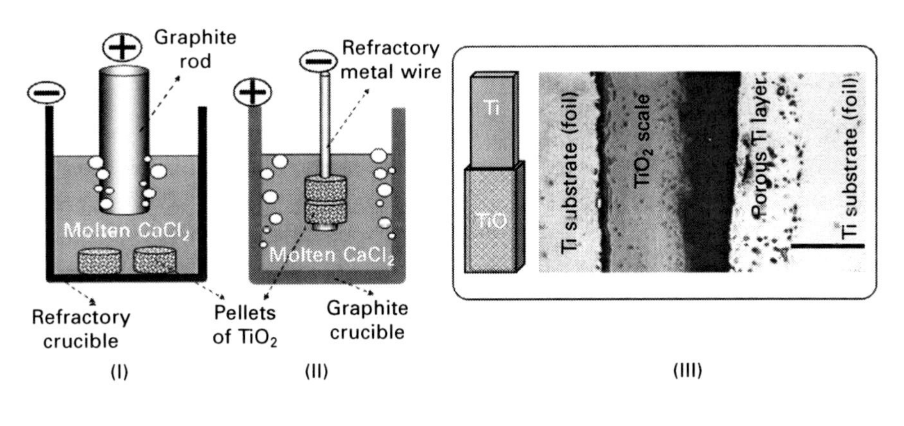

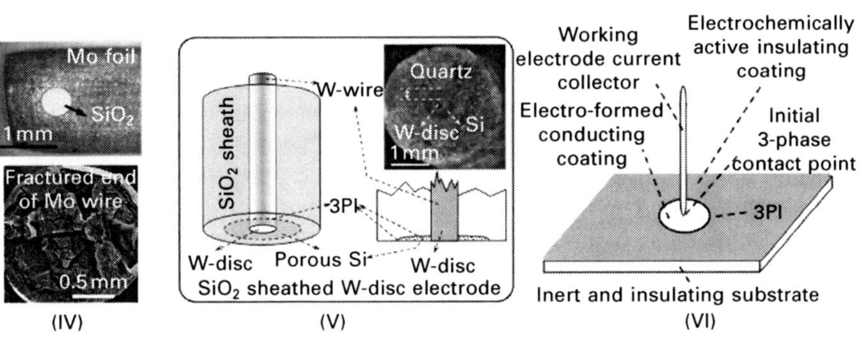

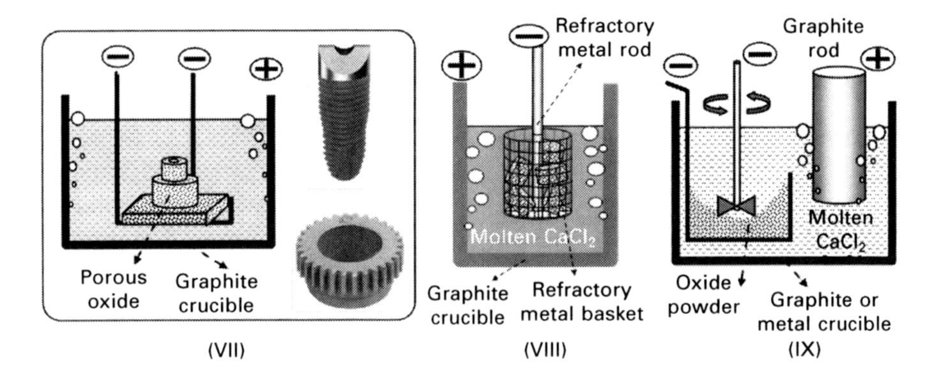

2.2 Cell designs with a cathode of type: (I) pellets-in-crucible,[3] (II) suspended pellet(s),[3] (III) a thin coating (image: cross sections of TiO_2 coated Ti foil before and after electroreduction),[26] (IV) powder-in-cavity (top: SiO_2 powder loaded in the through-hole of a Mo foil; bottom: cracks in the fractured end of Mo wire)[27–29] (V) a disc-with-oxide-sheath,[19] (VI) a contact-pin,[18,30] (VII) porous-oxide-preforms (for near-net-shape manufacturing of engineering components as shown by the images),[30] (VIII) pellets-in-basket,[3,15] (IX) stirred powder-in-ladle.[31]

salts were applied to support the FFC Cambridge process,[36] although the toxicity of these salts may cause environmental concerns if used at large scales.

As will be discussed below, all electrode reactions are driven by the electrode potential or more strictly, the overpotential. Thus, control of the electrode potential is important to ensure high process efficiency and satisfactory product quality. In common electrochemical practices, electrode potential control is achieved using a reference electrode. For molten salts research, the most often used reference electrode is the Ag/AgCl couple contained in a closed thin wall tube of glass, Mullite or quartz.[37–39] However, these container materials are unstable for the long term operation of the FFC Cambridge process in which the molten salt can become rich in O^{2-} ions which can attack SiO_2-based materials.[39] The FFC Cambridge process has so far been researched and developed in a two-electrode cell, in most cases, to demonstrate its electrochemical and engineering feasibility. However, it can be anticipated that optimisation of the process for commercial purposes will require further improvement in both process efficiency and product quality, depending on the ability to control the electrode potential. Particularly, in order to produce metal powders with the desired properties, such as particle size and elemental composition, accurate potential control can become a crucial technical issue, calling for effort to develop more reliable and long lasting reference electrodes.

2.3 Production of metal powders by the FFC Cambridge process

It is possible to produce metal powders by operating the FFC Cambridge process under appropriate conditions when the feedstock is a metal oxide or sulfide powder and is converted to a porous pellet or other preforms for attachment to the cathode (current collector). Electrolysis temperature and time are the two most important factors. It is generally observed in powder metallurgy that sintering a metal powder becomes significant when the temperature reaches or is above two-thirds of the melting point (in Kelvin) of the metal.[40] The effectiveness of sintering improves with the heating time. To make metal powders by the FFC Cambridge process, sintering should and can be avoided or minimised by applying a temperature below two-thirds of the metal's melting point for a sufficiently short time so that the reduction is completed. The third control parameter is the cell voltage. Because the electroreduction speed increases with the cell voltage, a higher cell voltage means a shorter time for complete reduction and hence less sintering of the metal powder produced at the cathode. Nevertheless, the cell voltage should still be lower than the decomposition voltage of the molten salt(s) to avoid contamination and unwanted by-products.

When using a metal oxide powder as the feedstock, the basic change in the FFC Cambridge process is the removal of large oxide ions, leaving the small metal atoms on the cathode. Consequently, both the mass and the volume of the material left on the cathode would in most cases be reduced (unless the molar volume of the metal is larger than that of the metal oxide). In all successful cases using the FFC Cambridge process to extract metals from their oxide precursors, the molar volume of the metal is always smaller than that of the oxide. Otherwise, as in the case of MgO and Mg, the electroreduction would be unable to continue.[16] Therefore, if sintering is avoided, the particle size of the produced metal powder will be smaller than that of the metal oxide powder. For example, the molar volume of tantalum is $V_{Ta} = 10.868$ cm^3 mol^{-1}, whilst that of its oxide, Ta_2O_5 is $V_{Ta_2O_5} = 53.889$ cm^3 mol^{-1}. This change in molar volume corresponds to a dimension reduction of $(2V_{Ta}/V_{Ta_2O_5})^{1/3} = 0.739$. In other words, a Ta_2O_5 powder with a particle size of 300 nm can be reduced to a Ta powder 200 nm in size. This feature of the FFC Cambridge process is useful because smaller Ta particles give greater specific surface areas and larger capacitance.[41]

According to the literature, it is interesting to note that the body shapes of the oxide or sulfide precursors (e.g. porous cylindrical pellets) are retained in metallised products in FFC Cambridge process, presenting an opportunity for near-net-shape manufacturing, see Fig. 2.2 part VI.[42] However, at particulate levels, there is no direct relationship between the feedstock and the product. As shown in Fig. 2.3, the electrolytically produced metal powders can be lightly interconnected nodules (Fig. 2.3(b) and (d)) (Ti, Ta, Nb, Zr, Ni, Mo, W),[3,32,34,41–43] more regularly shaped crystallites, such as cuboids (Fig. 2.3(f)) (Cr, Fe)[17,44] and even micro- and nano-wires (Fig. 2.3h) (Si, Cu).[45] The causes of such differences in particle morphology are not fully understood, although it is possible to start from the discussion below to develop understanding further.

The cathode reactions in the FFC Cambridge process can be basically explained by the models of charge transfer at the compound/metal/molten salt three phase interline (3PI) (boundary), particularly if the compound, such as oxide and sulfide, has poor conductivity. Whilst details of the 3PI models can be found in the literature,[17,18,30,46] Fig. 3.4 illustrates the concept using a spherical and insulating metal oxide particle as an example. The process starts at the very first 3PI (which could be a point), converting the compound (oxide) there to metal. The newly formed metal with the oxide and molten salt next to it then forms new 3PIs. When these processes continue, the 3PIs move along the surface of the particle until all the surface of the particle has been metallised. Afterwards, because the metallised surface layer is porous due to the molar volume of the metal being smaller than that of the oxide, molten salt can access the interior of the particle, oxygen ions can move out through the molten salt contained in the pores of the metallised surface and,

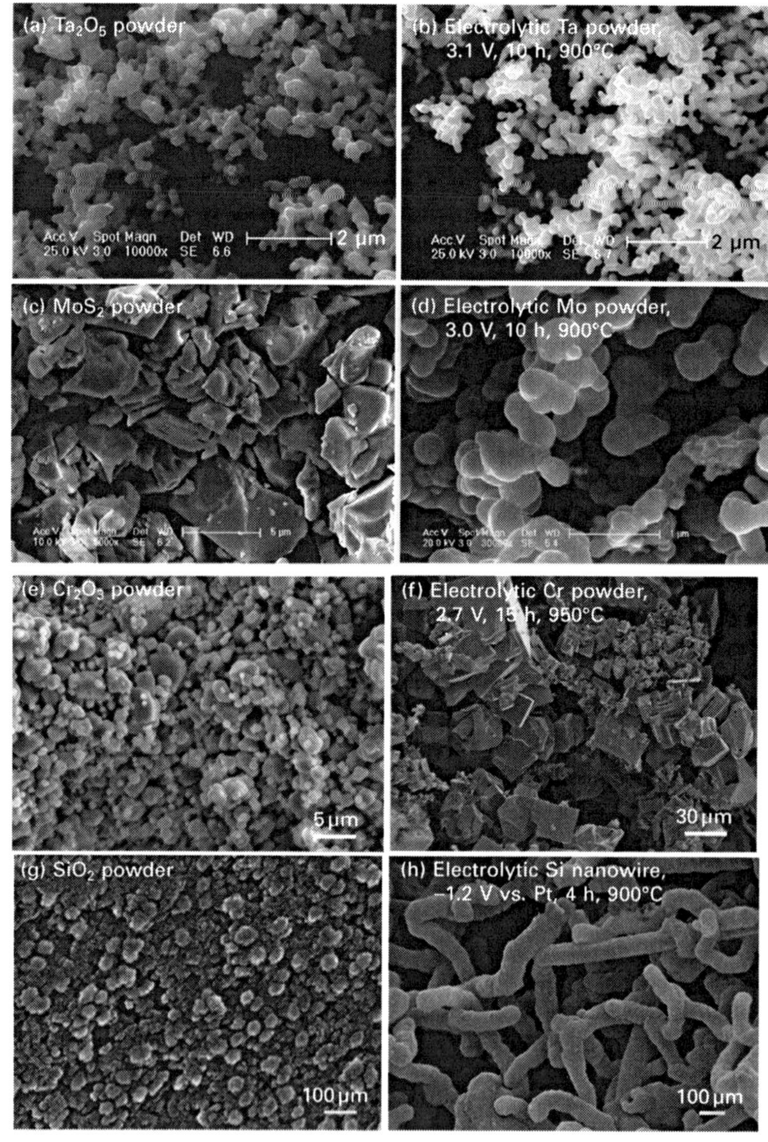

2.3 SEM images of the commercial powders of (a) Ta_2O_5,[41] (c) MoS_2,[44] (e) Cr_2O_3,[17] and (g) SiO_2,[45] and their electroreduction products (powders), (b) Ta,[41] (d) Mo,[44] (f) Cr[17] and (h) Si,[45] prepared under the indicated conditions. Molten $CaCl_2$ was used in all these cases. Figures 2.3(g) and (h) by permission of The Royal Society of Chemistry.

the reduction continues, accompanied by the 3PIs moving along the depth direction into the interior of the oxide particle.

It is worth pointing out that the boundary connecting three different phases

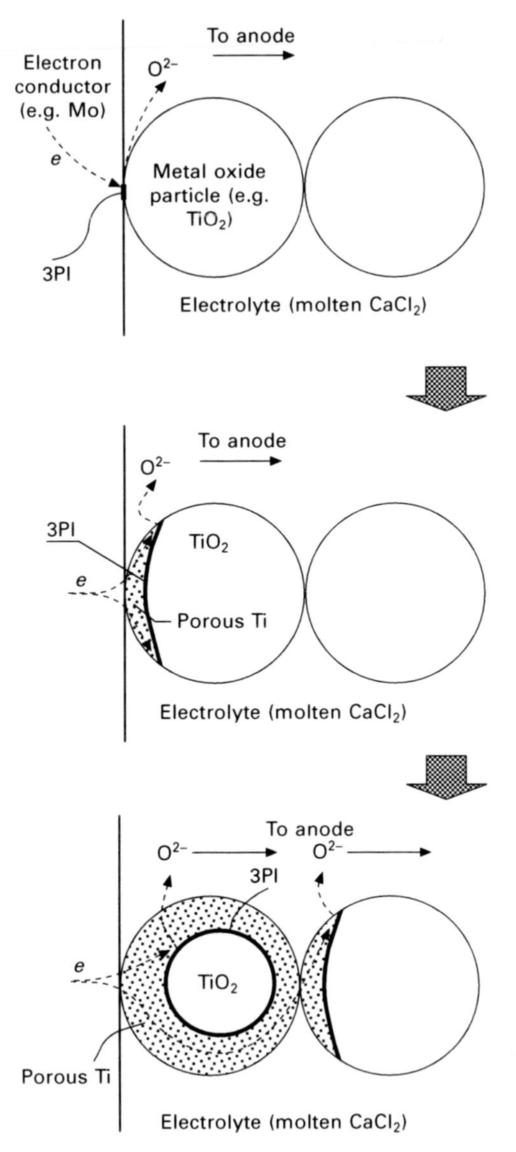

2.4 Schematic illustration of the propagation of the three phase interlines (3PIs)[30] during the electroreduction of two connected spherical particles of a metal oxide (TiO_2 in this figure) whose molar volume is larger than that of the metal.[16] The extra thick lines indicate the 3PIs.

can only be a line or point, although charge transfer reactions involve both electrons and ions in close proximity to the geometric 3PI. The emphasis here, similar to the use of the concept of the 'two phase interface' for conventional electron and ion transfer reactions involving the electrode and the electrolyte, is on the concept of the 'three phase interline' which helps to focus on the geometric location of the reactions.[46]

Qualitatively, electroreduction will be quicker to complete on smaller metal oxide particles, but the actual time or rate of reduction should also depend on the nature of the metal oxide. For example, at the same applied cathode potential and temperature in molten $CaCl_2$, it was observed that NiO or Fe_2O_3 could be reduced much quicker than TiO_2[20,47]. To explain this phenomenon requires understanding of the reduction mechanism and kinetics which are interdependent.

Electrochemical reactions are largely driven by the applied electrode potential. The comparison of the rates between different electrode reduction reactions needs to be made on the basis of how much the applied potential, E, is more negative than the standard reduction potential of each oxide, $E°$. The difference is defined as the overpotential, $\eta = |E° - E|$. The value of $E°$ can be derived from the Gibbs free energy of formation, $\Delta G°$ for each compound at a given temperature, for example, at 900°C, it is 1.960 V (vs. Ca/Ca^{2+}) for NiO/Ni, 1.089 V for TiO_2/TiO, and 0.421 V for TiO/Ti. Note that the thermodynamically calculated potential for TiO_2/Ti is 0.755 V, which seems to imply that the direct reduction of TiO_2 to Ti is easier than from TiO to Ti. However, because the reduction from TiO_2 to TiO occurs at a more positive potential, formation of TiO is inevitable during the reduction of TiO_2, in agreement with experimental findings.[16,26] Thus, to obtain Ti metal from TiO_2, the potential applied needs to be at or less positive than 0.421 V for TiO/Ti.

Besides the applied potential (or overpotential), it is commonly observed that the electroreduction of a metal oxide often includes intermediate steps where, in the case of electroreduction of TiO_2 in molten $CaCl_2$, various calcium titanate (or perovskite) phases are formed.[26,28,31] These intermediate phases may be formed either electrochemically or chemically and exert a detrimental effect on process efficiency.[28] This is because the inclusion of Ca expands the local volume in the cathode and blocks nearby pores that are needed to transport the O^{2-} ions. A couple of recent reports show that the influence of the perovskites may be minimised by properly engineering the oxide precursors, such as perovskitisation of the oxide precursors,[28] or making the precursor highly porous using NH_4HCO_3 as a fugitive porogenic agent.[16]

However, in making reactive metal powders, the FFC Cambridge process suffers from the same problem as the other metal powder production techniques. Because of the relatively large specific surface area of a powder, upon exposing it to air or water, natural oxidation of the reactive metal occurs readily on the surface of the metal particles. The added challenge to the FFC Cambridge process is to separate the metal powder from the solidified molten salt. Washing with water is effective because of the large solubility of alkali and alkaline earth chlorides. However, for reactive metals such as Ti and Zr, oxidation of the surface occurs spontaneously in water. The formed oxide layer may be only a few nanometres thick, but the overall oxygen content can increase significantly with decreasing particle size. For example, for Ti spherical particles 2 μm in diameter with a 3 nm thick surface oxide layer, the oxygen content can be about 3900 ppm, which agrees very well with a recent study of the Ti product from the FFC Cambridge process.[16]

To minimise the influence of water, attempts were made in the past to use a non-aqueous solvent to wash rare earth metal powders produced by the FFC Cambridge process. (In fact, it is impossible to use water for rare earth metals because they react with each other.) The tests were fairly successful in the laboratory, but require further development at larger scales to assess the effectiveness, costs and environmental impact. Nevertheless, it is worth mentioning that inorganic chloride salts are usually poorly soluble in an organic solvent, whilst in many cases the residual water in the solvent might have produced interesting effects. In fact, even under strictly dry conditions (e.g. in a working glovebox), dissolving alkali and alkaline earth chlorides into an organic solvent may be assisted by using a chelating ligand, such as crown ethers whose ring sizes offer good selectivity towards different cations. These organic systems are an area that is worthy of further investigation, particularly in association with production of reactive metal powders by the FFC Cambridge process.

On the other hand, in principle, separation of the solidified molten salt from the metal powder may be achieved through distillation under vacuum, which is a standard step in the Kroll process. In addition to increased energy consumption, the application of vacuum distillation to the FFC Cambridge process also requires technical modification considering the boiling temperature of the molten salt used. At present, $CaCl_2$ is commonly studied in laboratories for the FFC Cambridge process, whilst in the Kroll process, $MgCl_2$ by-product must be removed. The boiling temperature is 1412°C for $MgCl_2$, but it is higher than 1600°C for $CaCl_2$. To compromise, it is technically possible to apply a higher vacuum to distil $CaCl_2$ by balancing energy consumption in terms of heating and vacuum.

The other potential approach is to replace molten $CaCl_2$ by another molten salt with a lower boiling temperature. In the current literature, molten LiCl is one of the few systems studied. LiCl has a boiling temperature below 1400°C and is thus easier to distil. Molten LiCl with added Li_2O was successfully used to accommodate the electroreduction of uranium oxides to uranium metal.[48] However, different results were reported for the electroreduction of TiO_2 in molten LiCl,[49,50] calling for further investigations to explore the potential benefit to the separation of solidified salt from the reactive metal product by vacuum distillation. Considering the electrochemical aspects, the standard potential of the TiO/Ti couple is 56.2 mV vs. Li/Li^+ at 900°C. Such a small difference between TiO/Ti and Li/Li^+ makes it a practical challenge to reduce TiO_2 to Ti in molten LiCl. However, to make high quality powders of those less reactive metals, molten LiCl remains a candidate worthy of investigation.

In more recent years, the FFC Cambridge process has been applied to prepare metal powders from their sulfide precursors, with results promising an emission free, low energy and fast metal powder production method whilst still using a carbon anode. Metal sulfides that were successfully tested include CuS, MoS_2 and WS_2.[32–34] There are similarities and differences between electroreduction of metal oxide and metal sulfide. The sulfide cathode process is very similar to reaction [2.1], just replacing the oxide ion, O^{2-}, by the sulfide ion, S^{2-}. The products of electroreduction of metal sulfides are also nodular in shape as exemplified in Fig. 2.3(d). However, on a carbon anode, the discharge of the O^{2-} ion makes CO and CO_2, whilst discharge of S^{2-} ion produces only sulfur vapour, S_2, which condenses to solid readily after moving from the electrolyser to a collector at ambient temperature. In other words, the carbon anode behaves as an inert anode in the discharge of S^{2-}, making the anode reaction similar to reaction [2.4] by replacing O with S. Figure 2.5(a) shows photographs of a carbon anode before and after use in a 60 h sulfide electrolysis experiment, confirming the absence of any noticeable erosive or corrosive attack on the carbon, even at microscopic levels.[34]

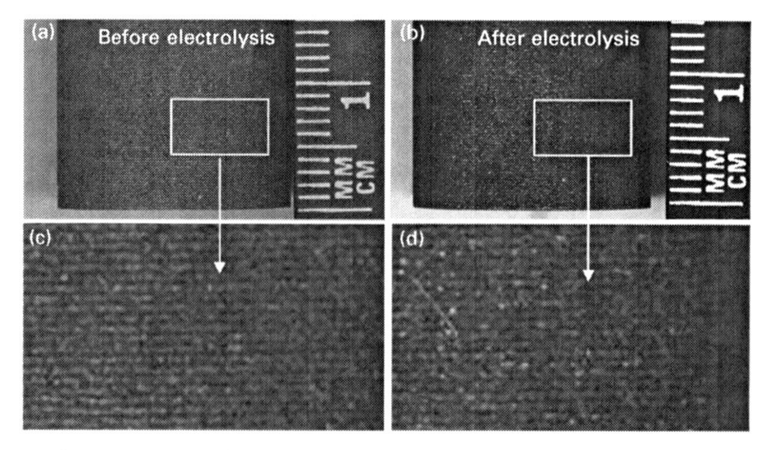

2.5 Photographs of a graphite rod anode before (a) and after (b) electrolysis of WS_2 in molten equimolar NaCl–KCl at 2.7 V and 700°C for 60 h.[34] The two boxed regions in (a) and (b) are enlarged in (c) and (d) respectively to show that machining marks remain almost unchanged, proving the inertness of the graphite towards the anodic discharge of sulfide ion.

2.4 Direct route from oxide precursors to alloyed powders

In conventional alloy making processes, there are at least two main steps: the composing metals of an alloy are extracted separately from their minerals and often refined to a sufficient purity, and the metals are then melted together, probably repeatedly to ensure uniformity, to form the alloy. A special advantage of the FFC Cambridge process is its capability to produce alloys, either a porous body or a powder, directly from the mixed oxide precursor without melting. This approach is energy saving in terms of both thermodynamics and practice. Thermodynamically, each of the composing elements of a stable alloy is in a lower energy state than the pure form of the element. Thus, mixing the component metals by melting, energy is released as heat, which may in theory compensate for the energy input for heating in a continuous process. However, this heat of alloy formation is largely wasted in batch operations. In practice, melting and casting, particularly if repeated, are also energy, time and labour consuming operations. Thus, by avoiding the melting step, the FFC Cambridge process should intrinsically consume less energy than conventional melting-casting techniques for production of alloys.

It is worth pointing out that one of the reasons for repeated melting and casting operations in the conventional method is to overcome the inhomogeneity in the product containing component metals with a large difference in density and melting temperature. For example, the well known shape memory NiTi alloy, also called Nitinol commercially, finds many applications. However, making this alloy is not straightforward because the two metals have very different melting temperatures, densities and molar volumes, as compared in Table 2.1. Melting is achieved by an electric arc under high vacuum in the water-cooled copper hearth. Particularly, the large difference in density leads to segregation of nickel from titanium in the molten mixture owing to the effect of gravity. Without melting any metal, the NiTi alloy in the porous body or powders, can be readily prepared by the FFC Cambridge process from porous pellets of mixed NiO and TiO_2 powders.[20,51] Figure 2.6 compares the X-ray diffraction (XRD) patterns and scanning electron microscopy (SEM) images of the products of electrolysis of an equimolar mixture of TiO_2 and NiO powders at different voltages.[20] The differences

Table 2.1 Basic properties of nickel and titanium[a]

Property	FM	Density (g cm^{-3})	mp (°C)	Molar volume (cm^3)
Nickel	58.693	8.908	1455	6.59
Titanium	47.867	4.507	1668	10.64

[a]Information source: http://www.webelements.com

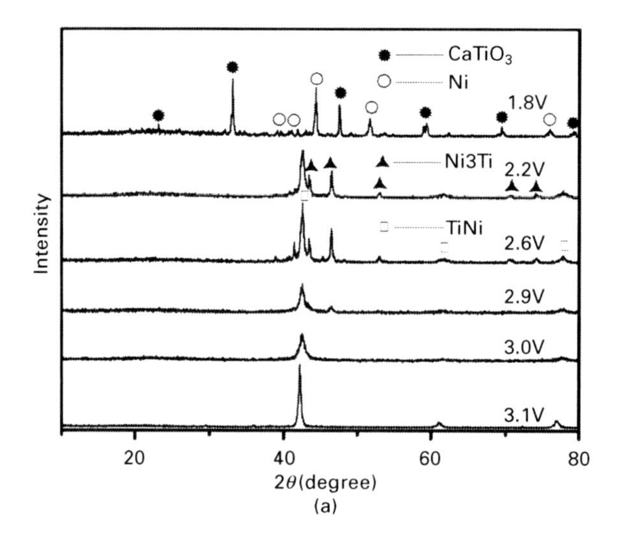

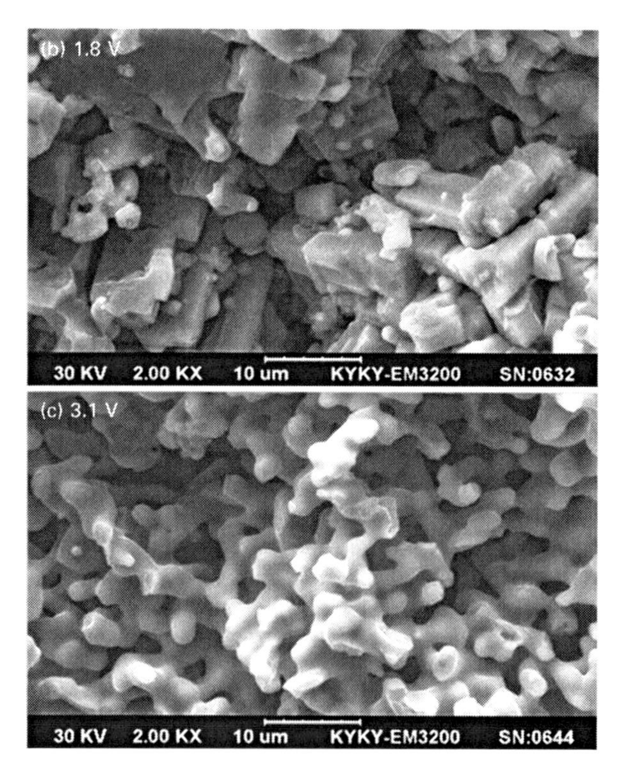

2.6 (a) XRD patterns and (b,c) SEM images of the products of electroreduction of equimolar mixture of TiO_2 and NiO in molten $CaCl_2$ at 900°C at the indicated cell voltages for 12 h.[20] Note that at 1.8 V, the product contained small nickel metal particles distributed amongst the much larger perovskite crystallites.

between these results are indicative of the process starting with the formation of Ni metal particles which then alloys with Ti from the slower reduction of TiO_2.

The FFC Cambridge process also works very well for making ternary alloy powders. The first example is a Ni_2MnGa powder which is a ferromagnetic shape memory alloy that plays an important role in highly efficient sound devices.[52] By mixing the three oxide powders in the designated metal ratio, pressing the mixture to a cylindrical pellet and sintering at 540°C for 2 h, electrolysis was very successful at 900°C and 3.0 V in molten $CaCl_2$ for about a day. The reduced pellet retained its body shape, but can be readily manually ground to a powder. Table 2.2 shows the elemental compositions measured by energy dispersive X-ray spectroscopy (EDX) on two randomly selected locations in the electrolytic powder sample, which matched expectation very well. The XRD pattern of the powder exhibited all the diffraction peaks, in line with those on the standard pattern for Ni_2MnGa.

Another example of ternary alloy powder is related to so-called giant magnetostrictive materials which usually have the composition $Tb_xDy_{1-x}Fe_2$ ($0 \leq x \leq 1$).[53] In an initial attempt, the $TbFe_2$ intermetallic compound was prepared by electroreduction of porous pellets of mixed Fe_2O_3 and Tb_4O_7 powders in which Tb:Fe = 1:2. In molten $CaCl_2$ at 900°C, the electroreduction proceeded to completion to a phase-pure product ($TbFe_2$) within 10 h at 3.1 V. More importantly, the study revealed very high recovery of the expensive Tb, reaching over 97%. The process consumed energy lower than 14 kWh $kg^{-1}TbFe_2$ with the product containing less than 1300 ppm oxygen. These achievements were claimed to be at least partly a result of pre-compounding the Fe_2O_3 and Tb_3O_4 powders at a sintering temperature of 1200°C, which led to the formation of co-oxide phases, such as $Tb_3Fe_5O_{12}$ and $TbFeO_3$, which were well interfused into a porous structure, as shown in Fig. 2.7(a). An important step adopted in this study was the use of an organic solvent, dimethylsulfoxide (DMSO), to wash the electrolysis product with a magnetic stirrer in a sealed conical flask, which avoided oxidation and washing loss of the magnetic product. A photograph taken during the washing process is displayed in Fig. 2.7(b). Another interesting finding of this study was that the electrolytic product which was composed of 5–10 μm nodules, see Fig. 2.7(c), could be pulverised to smaller particles by application of sonication to assist washing in DMSO, as shown in Fig. 2.7(d). The pre-compounding approach used in the making of $TbFe_2$ is a simple but highly effective way

Table 2.2 EDX analyses of electrolytic Ni_2MnGa alloy powders[52]

Location	Ni (at.%)	Mn (at.%)	Ga (at.%)	χ^2
1	49.62	24.67	25.71	1.73
2	49.18	27.56	23.26	1.53

2.7 SEM images of the interiors of the Tb_4O_7–Fe_2O_3 (Tb:Fe = 1:2) pellet: (a) sintered in air at 1200°C for 2 h, and (b) its as-electrolysed product at 2.6 V and 900°C for 12 h.[27] (c) shows alignment along the magnetic lines of the particles of the collected $TbFe_2$ powder on the magnetic stirrer during washing in DMSO in a sealed conical flask. The SEM image of the $TbFe_2$ powder from washing in DMSO and magnetic collection is shown in (d). The elemental composition was confirmed by inductively coupled plasma – atomic absorption spectroscopy (ICP–AAS) analysis.

of controlling uniformity in the composition of an alloy produced by the FFC Cambridge process. This is largely because the distribution of the alloying elements in the compounded precursor is uniform at the atomic level within a short distance. Thus, once the oxygen atoms are removed at appropriate potentials, these closely arranged alloying atoms can react quickly to form the desired alloys or intermetallics.

The importance of the pre-compounding step can be also appreciated when producing $LaNi_{5-x}M_x$ hydrogen storage alloys (powders).[54, 55] La_2O_3 is a hygroscopic oxide and can react spontaneously with molten $CaCl_2$, which challenges the usual method of preparing the oxide precursor, and maintain its integrity in the molten salt during electrolysis. By pre-compounding (sintering) the La_2O_3 powder mixed well with the NiO powder at elevated temperatures in air, formation of several co-oxide phases may be achieved, including $LaNiO_3$ (< 1000°C), $La_4Ni_3O_{10}$ and $La_3Ni_2O_7$ (1100–1250°C), and/or La_2NiO_4 (> 1300°C). These co-oxide phases are all very stable in air and molten $CaCl_2$, ensuring success in the follow-on electrolysis. The pre-compounding step can also be applied to make $LaNi_{5-x}M_x$ hydrogen storage alloys (powders) with three or more components so that their charge/discharge performance can be improved. The as-produced electrolytic $LaNi_{5-x}M_x$ powders possessed fairly reversible activity in alkaline electrolyte for hydrogen storage, as evidenced in Fig. 2.8(a) by the cyclic voltammograms of the electrolytic powder of $LaNi_4Co$.[55] Figure 2.8(b) compares the discharging capacity of

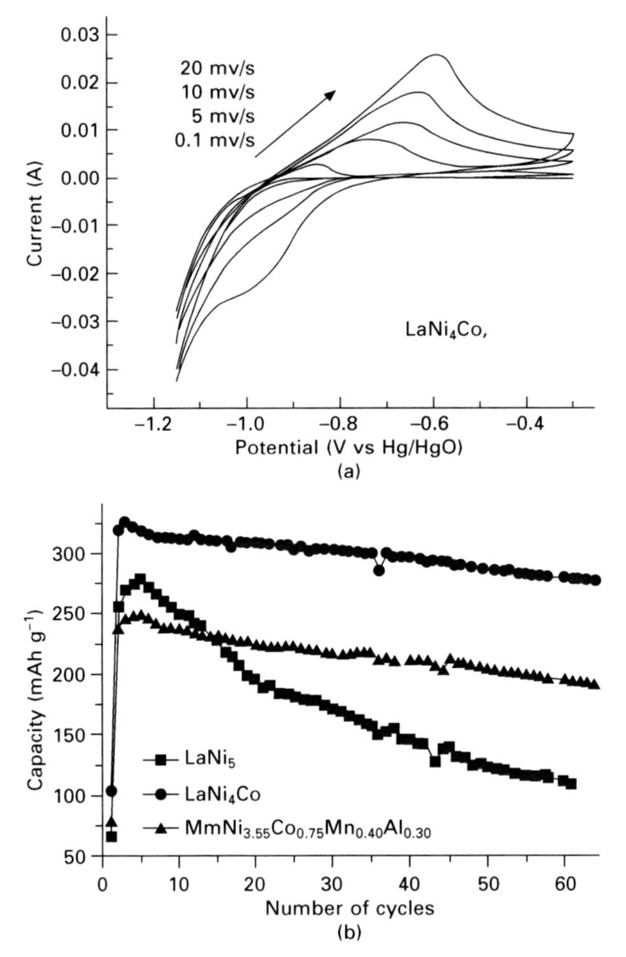

2.8 (a) Cyclic voltammograms[55] and (b) discharge capacity variation versus the number of charge/discharge cycles (60 mA g^{-1}, 1.45 V/0.90 V)[54] of the indicated electrolytic $LaNi_5$-type powders measured in 6.0 M KOH at 25°C. (c) A schematic comparison of the conventional process and the FFC Cambridge process for making $LaNi_{5-x}M_x$ powders; superscripts indicate original references in Zhu *et al.*[54]

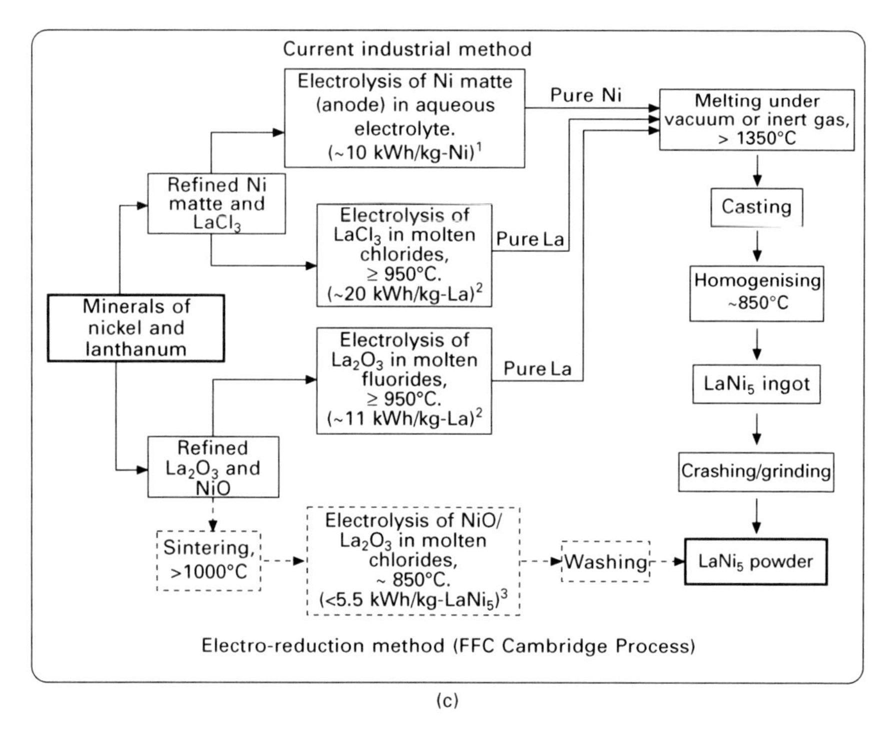

2.8 Continued

the electrolytic $LaNi_5$ powder with two multi-component $LaNi_{5-x}M_x$ samples prepared by electrolysis of the pre-compounded precursors. Note that the theoretical electrochemical hydrogen storage capacity for charging $LaNi_5$ to $LaNi_5H_6$ is 371.9 mAh g^{-1}, which is not too far from the maximum discharging capacities measured for the electrolytic $LaNi_5$ and $LaNi_4Co$ powders. Fig. 2.8(c) illustrates the flowcharts of the conventional and the FFC Cambridge processes for making the $LaNi_5$ powders, showing clearly the advantages of the FFC approach: simple, direct and energy efficient.[54,55]

2.5 Conclusions and future trends

Since the publication of the original patent in 1999, the FFC Cambridge process has attracted great attention in both the industry and academia. While its commercial development is beyond the scope of this chapter, the prospects for using this new process to produce metal powders, particularly pre-alloyed powders, are introduced in terms of the principles and the technological advantages compared with existing industrial processes. In particular, research on the FFC Cambridge process has so far succeeded in production of most tested pure metal, binary and ternary alloy powders with accurate control of the elemental composition and low energy consumption. Many of these developments offer great versatility and flexibility to meet application needs. On the other hand, this chapter also offers analyses of the technical challenges ahead of the optimisation and commercialisation of the FFC Cambridge process, and possible solutions. It is obvious that the space allocated here is insufficient to give a more comprehensive review of the literature on this new process for making metal powders. Nevertheless, it is hoped that this chapter may stimulate and be of some assistance to readers with interests in making wider investigation and further development of the FFC Cambridge process for production of metal powders, particularly in demonstrating its great technological potential in the post-fossil era.

2.6 Acknowledgement

The author is grateful to all past and present co-workers whose names appear in the reference list for their valuable contributions to the research that has formed the basis of this chapter.

2.7 References

1 N. Armaroli and V. Balzani, *Energy Environ. Sci.* 2011, **4**, 3193.
2 F. Habashi, *Handbook of Extractive Metallurgy*, Vols. I-IV, Wiley-VCH, New York, 1997.
3 G. Z. Chen, D. J. Fray and T. M. Farthing, *Nature*, 2000, **407**, 361.
4 D. H. Wang, X. B. Jin and G. Z. Chen, *Annu. Rep. Prog. Chem., Sect. C: Phys. Chem.*, 2008, **104**, 189.
5 G. Dowson, *Powder Metallurgy – The Process and Its Products,* Adam Higher, New York, 1990.
6 W. Schatt and K.-P. Wieters, *Powder Metallurgy – Processing and Materials*, European Powder Metallurgy Association, Shrewsbury, 1997.
7 A.J. Yule and J.J. Dunkley, *Atomization of Melts – For Powder Production and Spray Deposition*, Clarendon Press, Oxford, 1994.
8 http://yijiastar.en.ec21.com/High_Purity_Aluminum_Powder--5211867_5211970.html (viewed on 14 June 2012).
9 G. H. Gessinger, *Powder Metallurgy of Superalloys*, Butterworth and Co, London, 1984.
10 http://www.exotech.com/Products/ChromiumPowder.aspx (viewed on 14 June 2012).
11 A.H. Sully and E.A. Brandes, *Chromium*, 2nd edition, Butterworth and Co, London, 1967.
12 W. Abdul-Razzaq and M.S. Seehra, *Phys. Status Solidi A*, 2002, **193**, 94.
13 D. J. Fray, T. W. Farthing and Z. Chen, *Patent*, WO9964638.
14 http://www.metalysis.com/index.php (viewed on 14 June 2012).
15 C. Schwandt, G. R. Doughty, D. J. Fray, *Key Eng. Mater*, 2010, **436**, 13.
16 W. Li W, X. B. Jin, F. L. Huang, G. Z. Chen, *Angew. Chem. Int. Ed.*, 2010, **49**, 3203.

17 G. Z. Chen, E. Gordo, D. J. Fray, *Metall. Mater. Trans. B*, 2004, **35**, 223.
18 T. Nohira, K. Yasuda, Y. Ito, *Nat. Mater.*, 2003, **2**, 397.
19 X. B. Jin, P. Gao, D. H. Wang, X. H. Hu, G. Z. Chen, *Angew. Chem. Int. Ed.*, 2004, **43**, 733.
20 Y. Zhu, M. Ma, D. H. Wang, K. Jiang, X. H. Hu, X. B. Jin, G. Z. Chen, *Chin. Sci. Bull.*, 2006, **51**, 2535.
21 K. T. Kilby, S. Q. Jiao, D. J. Fray, *Electrochim. Acta*, 2010, **55**, 7126.
22 S. Q. Jiao, L. L. Zhang, H. M. Zhu, D. J. Fray, *Electrochim. Acta*, 2010, **55**, 7016.
23 A. Krishnan, X. G. Lu, U. B. Pal, *Metall. Mater. Trans. B*, 2005, **36**, 463.
24 U. B. Pal, A. C. Powell, *JOM*, 2007, **59**, 44.
25 C. Y. Chen, X. G. Lu, *Acta Metall. Sinica*, 2008, **44**, 145.
26 G. Z. Chen, D. J. Fray, *J. Electrochem. Soc.*, 2002, **149**, E455.
27 G. H. Qiu, M. Ma, D. H. Wang, X. B. Jin, X. H. Hu and G. Z. Chen, *J. Electrochem. Soc.*, 2005, **152**, E328.
28 K. Jiang, X. H. Hu, M. Ma, D. H. Wang , G. H. and Qiu, X. B. Jin, G. Z. Chen, *Angew. Chem. Int. Ed.*, 2006, **45**, 428–32.
29 G. H. Qiu, K. Jiang, M. Ma, D. H. Wang, X. B. Jin, G. Z. Chen, *Z. Naturforsch.*, 2007, **62a**, 292.
30 Y. Deng , D. H. Wang, W. Xiao, X. B. Jin, X. H. Hu, G. Z. Chen, *J. Phys. Chem. B*, 2005, **109**, 14043.
31 G. Z. Chen, D. J. Fray, *Light Metals 2004*, 2004, 881.
32 G. M. Li, D. H. Wang, X. B. Jin, G. Z. Chen, *Electrochem. Commun.*, 2007, **9**, 1951.
33 X. L. Ge, X. D. Wang, S. Seetharaman, *Electrochim. Acta*, 2009, **54**, 4397.
34 T. Wang, H. P. Gao, X. B. Jin, H. L. Chen, J. J. Peng, G. Z. Chen, *Electrochem. Commun.*, 2011, **13**, 1492.
35 D. J. Fray, *JOM*, 2001, **53**, 26.
36 M. Gibilaro, J. Pivato, L. Cassayre, L. Massot, P. Chamelot, P. Taxil, *Electrochim. Acta*, 2011, **56**, 5410.
37 J. O'M. Bockris, G. J. Hills, D. Inman, L. Young, *J. Sci. Instrum.*, 1956, **33**, 438.
38 K. Yasuda, T. Nohira, R. Hagiwara, Y. H. Ogata, *J. Electrochem. Soc.*, 2007, **154**, E95.
39 P. Gao, X.B. Jin, D.H. Wang, X.H. Hu, G.Z. Chen, *J. Electroanal. Chem.*, 2005, **579**, 321.
40 W. Horobin (ed), *How It Works – Science and Technology*, 3rd edition, Marshall Cavendish, New York, 2003, p. 105.
41 T. Wu, X. B. Jin, W. Xiao, X. H. Hu, D. H. Wang, G. Z. Chen, *Chem. Mater.*, 2007, **19**, 153.
42 J. J. Peng, K. Jiang, W. Xiao, D. H. Wang, X. B. Jin, G. Z. Chen, *Chem. Mater.*, 2008, **20**, 7274.
43 T. Wu, X. B. Jin, W. Xiao, C. Liu, D. H. Wang, G. Z. Chen, *Phys. Chem. Chem. Phys.*, 2008, **10**, 1809.
44 G. M. Li, D. H. Wang, Z. Chen, *J. Mater. Sci. Technol.*, 2009, **25(6)**, 767.
45 J. Y. Yang, S. G. Lu, S. R. Kan, X. J. Zhang, J. Du, *Chem. Commun.*, 2009, 3273.
46 W. Xiao, X. B. Jin, Y. Deng, D. H. Wang and G. Z. Chen, *Chem. Eur. J.*, 2007, **13**, 604.

47 M. Ma, D. H. Wang, X. H. Hu, X. B. Jin, G. Z. Chen, *Chem. Eur. J.*, 2006, **12**, 5075.
48 J.-M. Hur, S. M. Jeong, H. Lee, *Electrochem. Commun.*, 2010, **12**, 706.
49 J.-M. Hur, S.-C. Lee, S.-M. Jeong, C.-S. Seo, *Chem. Lett.*, 2007, **1028**, 36.
50 K. Jiang, X. H. Hu, H. J. Sun, D. H. Wang, X. B. Jin, Y. Y. Ren, G. Z. Chen, *Chem. Mater.*, 2004, **16**, 4324.
51 B. Jackson, M. Jackson, D. Dye, D. Inman, R. Dashwood, *J. Electrochem. Soc.*, 2008, **155**, E171.
52 A. J. Muir Wood, R. C. Copcutt, G. Z. Chen, D. J. Fray, *Adv. Eng. Mater.*, 2003, **5**, 650.
53 G. H. Qiu, D. H. Wang, M. Ma, X. B. Jin, G. Z. Chen, *J. Electroanal. Chem.*, 2006, **589**, 139.
54 Y. Zhu, D. H. Wang, M. Ma, X. H. Hu, X. B. Jin, G. Z. Chen, *Chem. Commun.*, **2515**, 2007.

3 Mechanochemical synthesis of nanocrystalline metal powders

C. SURYANARAYANA, University of Central Florida, USA, and E. IVANOV, Tosoh SMD, Inc., USA

DOI: 10.1533/9780857098900.1.42

Abstract: This chapter introduces the novel method of mechanochemical synthesis as an effective method for synthesizing metal powders in the nanocrystalline state. After introducing the basic principles of the process, process parameters that affect the constitution and microstructure of the processed powders are discussed. The mechanisms of alloying and grain refinement are also described. Methods for achieving the smallest possible grain size are highlighted. Current problems associated with the consolidation of powders to bulk shape are described. The ubiquitous problem of powder contamination during milling and solutions to eliminate or minimize this are also emphasized.

Key words: mechanochemical synthesis, minimum grain size, nanocrystalline powders, powder consolidation, powder contamination.

3.1 Introduction

Development of novel materials and improvement of the properties of existing materials has been the preoccupation of materials scientists for several decades. The second half of the last century was witness to the development of novel materials such as metallic glasses (including bulk metallic glasses), high-temperature superconductors, and quasicrystalline alloys, to name a few. Several new techniques for synthesizing these novel materials have also been developed and improved upon including rapid solidification processing, mechanical alloying, ion implantation, plasma processing, physical and chemical vapor deposition methods, and others. With the help of these techniques, it has been possible to produce a variety of monolthic and composite materials with vastly improved properties. A common underlying theme in all these processes is to bring the material into a highly energetic condition by increasing either the temperature or pressure, or by irradiation, or storing of mechanical energy. The material is then 'quenched' to retain the high-temperature/high-pressure phase at either lower temperatures or atmospheric pressure and allowed to transform slowly into more stable phases. Using this approach, it has been possible to achieve both constitutional and microstructural changes in the materials. The advent of new characterization techniques during this time period has also greatly aided understanding of structure–property correlations in these novel materials (Suryanarayana, 1999).

Powders are ubiquitous and they can be synthesized in many different ways starting from the vapor, liquid, or solid states of matter (German, 1992). Atomization and rapid solidification processing are the common methods for producing powders starting from the liquid state. While the atomization process is used to produce large amounts of steel and other powders, the method of reduction of metal oxides has been used to produce thousands of tons of Mo and W powders. In fact, practically all of Mo and W powders are produced this way. Occasionally, hydrogen decrepitation is also used to make metal powders. Vapor deposition methods are utilized to deposit thin films, or synthesize small quantities of powders. But this is an expensive process and the productivity is also very low. Further, the large surface to volume ratio of these products necessitates handling them in ultra-high vacuum conditions to prevent their contamination. Additionally, production of powders from the vapor or liquid states involves vaporization or melting of the materials, involving achieving very high temperatures. However, these methods may not be suitable for materials with very high melting temperatures. This is because, first, a suitable solid container may not be available to withstand the high temperatures and second, the chances of the material becoming easily oxidized and thus introducing impurities and contamination into the material are very high. Consequently, powder production starting from the solid-state condition has been the most common method, especially for high-melting point materials. Of these solid-state processing methods, techniques that could be grouped under the broad category of mechanochemical processing have been the most popular for producing both monolithic and composite powders in a variety of alloy systems.

3.2 Mechanochemical processing

Mechanochemical processing (MCP) is the term applied to the powder process in which chemical reactions, structural changes and phase transformations are activated by the application of mechanical energy. MCP methods have been used since time immemorial. Use of mechanical energy to grind down various materials dates back to the beginning of human history. Research activity in the field of MCP has a long history with the first publication dating back to 1892 (Carrey Lea, 1892). It was shown that the halides of gold, silver, platinum and mercury decomposed to halogen and the metal during fine grinding in a mortar. This study clearly established that chemical changes could be brought about, not only by heating but also by mechanical action. But the use of mechanically activated processes, however, dates back to the early history of mankind, when fires were initiated by rubbing flints against one another. Ostwald coined the term 'mechanochemical' in 1891

(Ostwald, 1919). Heinicke's much later definition that 'mechanochemistry is a branch of chemistry which is concerned with chemical and physico-chemical transformations of substances in all states of aggregation produced by the effect of mechanical energy' has been widely accepted (Heinicke, 1984). While the scientific basis underlying MCP was investigated from the very beginning, applications of MCP products were slow to come about, mostly because of limitations on the productivity of MCP reactors, purity of the products and the economics of the process. The general phenomenon of MCP has been a popular research topic in Germany (Tiessen *et al.*, 1966; Heinicke, 1984), former USSR and Eastern Europe (Boldyrev, 2006). The topic of MCP of materials has been reviewed periodically (Avvakumov, 1986; Tkacova, 1989; Gutman, 1994, 1998; Juhász and Opoczky, 1990; Ivanov, 1993; McCormick, 1995, 1997; Ivanov and Suryanarayana, 2000; Avvakumov *et al.*, 2001; Suryanarayana, 2001, 2004; Takacs, 2002; Boldyrev, 2006; Baláž, 2008; Sopika-Lizer, 2010). These researchers concentrated their efforts on both the fundamental principles of MCP and the potential applications for materials produced by MCP.

It is now accepted that the MCP technique embraces three different processes, mechanical alloying, mechanical milling, and reaction milling. All these three processes involve cold welding, fracturing and rewelding of powder particles during repeated collisions with grinding balls in a high-energy milling device. But depending on the actual process, other features may be present.

3.2.1 Mechanical alloying

Mechanical alloying (MA) is the generic term used to denote processing of metal powders in high-energy ball mills. But, more specifically, MA describes the process where mixtures of powders (of different metals or alloys/compounds) are milled together. Thus, if powders of pure metals A and B are milled together to synthesize a solid solution (either equilibrium or supersaturated), intermetallic, or an amorphous phase, the process is referred to as MA. Material transfer is involved in this process to obtain a homogeneous alloy. The present explosion of activity in this area can be traced to the original synthesis of oxide dispersion strengthened (ODS) nickel-based superalloys by John Benjamin in 1966 at the INCO laboratories (Benjamin, 1992). In these alloys, two different strengthening mechanisms are combined, the intermediate temperature strength due to formation of the γ' precipitates and the high-temperature strength due to dispersion hardening by the addition of insoluble oxide particles. This approach has been subsequently extended to Fe-based ODS alloys as well.

3.2.2 Mechanical milling

When powders with uniform (often stoichiometric) composition, such as pure metals, intermetallics, or prealloyed powders, are milled in a high-energy ball mill, and material transfer is *not* required for homogenization, the process has been termed mechanical milling (MM). It may be noted that when a mixture of two intermetallics is processed and alloying occurs, this will be referred to as MA since material transfer is involved. But if a pure metal or an intermetallic is processed only to reduce particle (or grain) size and increase the surface area, it will be referred to as MM, since material transfer is not involved. If a phase transformation occurs in a single-phase material under mechanical action, this will also fall under the category of MM. The destruction of long-range order in intermetallics to produce either a disordered intermetallic (solid solution) or an amorphous phase has been referred to as mechanical disordering (MD). The advantage of MM/MD over MA is that since the powders are already alloyed and only a reduction in particle size and/or other transformations need to be induced mechanically, the time required for processing is short. For example, MM requires half the time required for MA to achieve the same effect.

3.2.3 Reaction milling

Reaction (or reactive) milling (RM) is the milling process accompanied by a solid–gas or solid-state reaction and was pioneered by Jangg (Jangg 1989). In this process the initial powder mixture reacts to produce fine dispersions of oxides, carbides, and so on in a metal matrix. The dispersion of carbides, for example, is achieved by adding lamp-black or graphite during aluminum milling. Milling metal powders in the presence of reactive solids/liquids/gases (enabling a chemical reaction to take place) is now regularly employed to synthesize metal oxides, nitrides, carbides and carbonyls (Suryanarayana, 2004). These oxides, carbides, borides or nitrides are incorporated into the alloy matrix providing additional strength and high-temperature stability. Some exchange reactions can also occur leading to the formation of composites, reduction of oxides, chlorides, and so on to pure metals.

Some other terms used in the literature for mechanical processing of metal powders include cryomilling (in which the milling operation is carried out at cryogenic (very low) temperatures and/or milling of materials is done in a cryogenic medium such as liquid nitrogen) (Luton *et al.*, 1989; Witkin and Lavernia, 2006), mechanically activated annealing (M2A), double mechanical alloying (dMA) and mechanically activated self-propagating high-temperature synthesis (MASHS) (Suryanarayana, 2004).

3.2.4 Exchange reactions

The category of exchange reactions in MCP has received a lot of attention in recent years (McCormick, 1997; Suryanarayana, 2001, 2004). This emanated from reports in 1989 that MCP could be used to induce a wide variety of solid–solid and even liquid–solid chemical reactions. For example, it was shown that CuO could be reduced to pure metal Cu by ball milling CuO at room temperature with a more reactive metal like Ca (Schaffer and McCormick, 1989). Milling together of CuO and ZnO with Ca has resulted in the direct formation of β'-brass (McCormick *et al.*, 1989). These reports virtually led to an explosion of research activity in this area.

In all the above processes, microstructural changes take place leading to refinement in the powder particle size and grain size. Thus, depending on the type of processing used and the composition of the powder mix, a variety of microstructural and constitutional changes can occur in the milled powders.

3.2.5 Attributes of mechanochemical processing (MCP)

MCP is normally a dry, high-energy ball milling technique and has been employed to produce a variety of commercially useful and scientifically interesting materials. The formation of an amorphous phase by mechanical grinding of a Y-Co intermetallic compound in 1981 (Ermakov *et al.*, 1981) and its formation in the Ni-Nb system by ball milling of blended elemental powder mixtures (Koch *et al.*, 1983) brought about the recognition that this technique is a potential non-equilibrium processing technique. Beginning in the mid-1980s, a number of investigations have been carried out to synthesize a variety of equilibrium and non-equilibrium phases including supersaturated solid solutions, crystalline and quasicrystalline intermediate phases, and amorphous alloys. Additionally, it has been recognized that powder mixtures can be mechanically activated to induce chemical reactions, at room temperature or at least at much lower temperatures than normally required, to produce pure metals, nanocomposites and a variety of commercially useful materials. Efforts have also been under way since the early 1990s to understand the process fundamentals of MA through modeling studies. Because of all these special attributes, this simple but effective processing technique has been applied to metals, ceramics, polymers and composite materials. The attributes of mechanochemical processing are listed below. However, in the present chapter, the focus will be on the synthesis of nanocrystalline metal particles.

1. Production of fine dispersion of second phase (usually oxide) particles
2. Extension of solid solubility limits
3. Refinement of grain sizes down to nanometer range

4. Synthesis of novel crystalline and quasicrystalline phases
5. Development of amorphous (glassy) phases
6. Disordering of ordered intermetallics
7. Possibility of alloying of difficult to alloy elements/metals
8. Inducement of chemical (displacement) reactions at low temperatures for (a) Mineral and Waste processing, (b) Metals refining, (c) Combustion reactions, and (d) Production of discrete ultrafine particles
9. Scalable process

Nanocrystalline materials are single- or multi-phase polycrystalline solids with a grain size of the order of a few nanometers (1 nm = 10^{-9} m = 10 Å), typically 1–100 nm in at least one dimension. Since the grain sizes are so small, a significant volume of the microstructure in nanocrystalline materials is composed of interfaces, mainly grain boundaries. That is, a large volume fraction of the atoms resides in the grain boundaries. Consequently, nanocrystalline materials exhibit properties that are significantly different from, and often an improvement on, their conventional coarse-grained polycrystalline counterparts. Compared to the material with a more conventional grain size, that is, larger than a few micrometers, nanocrystalline materials show increased strength, high hardness, extremely high diffusion rates and consequently reduced sintering times for powder compaction, and improved deformation characteristics. Several excellent reviews are available giving details on different aspects of processings, properties, and applications of these materials (Gleiter, 1989; Suryanarayana, 1995a, 2005).

3.3 The process

3.3.1 Processing

MCP involves loading the blended elemental or pre-alloyed powders together with the grinding medium into a vial and subjecting them to heavy deformation. During this process, the powder particles are repeatedly flattened, cold welded, fractured and rewelded. The processes of cold welding and fracturing, their kinetics and predominance at any stage depend mostly on the deformation characteristics of the starting powders. Sometimes, about 1–2 wt% of a process control agent (PCA) (also referred to as lubricant or surfactant, which is usually an organic compound) is added to the powder mixture during milling. Amongst the several PCAs used, stearic acid is the most common. The PCA adsorbs on the surfaces of the powder particles and minimizes excessive cold welding of powder particles amongst themselves and/or to the milling container and the grinding medium; this inhibits agglomeration of powder particles. The PCA is most commonly added whenever ductile materials are milled.

3.3.2 Milling equipment

Milling can be carried out in high-energy shaker mills, where small quantities of about 10–20 g are milled at one time, mostly for alloy screening purposes, or in relatively low-energy planetary ball mills where larger quantities of powder (about 200 g) can be milled, or in attritors where even larger quantities can be milled at one time. Industrial mills can process several kilograms of powder at a time. High energy mills such as SPEX mills, medium energy mills such as Fritsch Pulverisette mills, or low-energy mills such as Union Process Szegvari attritors are commercially available. While the shaker mills have a rotation speed of about 1200 rpm, planetary mills rotate at 150–600 rpm and attritors rotate at about 100–200 rpm. Commercial large-sized mills operate at much lower speeds. Thus, it can be clearly seen that large amounts of powder can be processed only at low speeds, while small quantities can be processed at higher speeds. A number of different mills, some of which are variants of the above mills and some with very different designs, have been used for milling purposes. In recent years, cryogenic milling has become a popular method. Facilities for cooling the powders to low temperatures or heating them to high temperatures and monitoring the pressure and temperature during milling are some of the attachments that are currently available, even in some commercial mills.

The milling time decreases with an increase in the energy of the mill. It has been reported that 20 min milling in a SPEX mill is equivalent to 20 h milling in a low-energy mill of the type Invicta BX 920/2 (Yamada and Koch, 1993). As a rule of thumb, it can be estimated that a process that takes only a few minutes in the SPEX mill may take hours in an attritor and a few days in a commercial mill even though the actual details can be different depending on the efficiency of the different mills and the powder characteristics. Similarly, by comparing the powder yield, it was reported that the yield is much higher and that alloying is completed sooner in the Zoz Simoloyer than in the Fritsch P5 planetary ball mill.

3.3.3 Mechanism of alloying

The effects of a single collision on each type of constituent powder particle are shown in Fig. 3.1. The initial impact of the grinding balls causes the ductile metal powders to flatten and work harden. Severe plastic deformation increases the surface-to-volume ratio of the particles and ruptures the surface films of adsorbed contaminants. The brittle intermetallic powder particles are fractured and refined in size. The oxide dispersoid particles are comminuted more severely.

Whenever two grinding balls collide, a small amount of the powder being milled is trapped in between them. Typically, around 1000 particles with

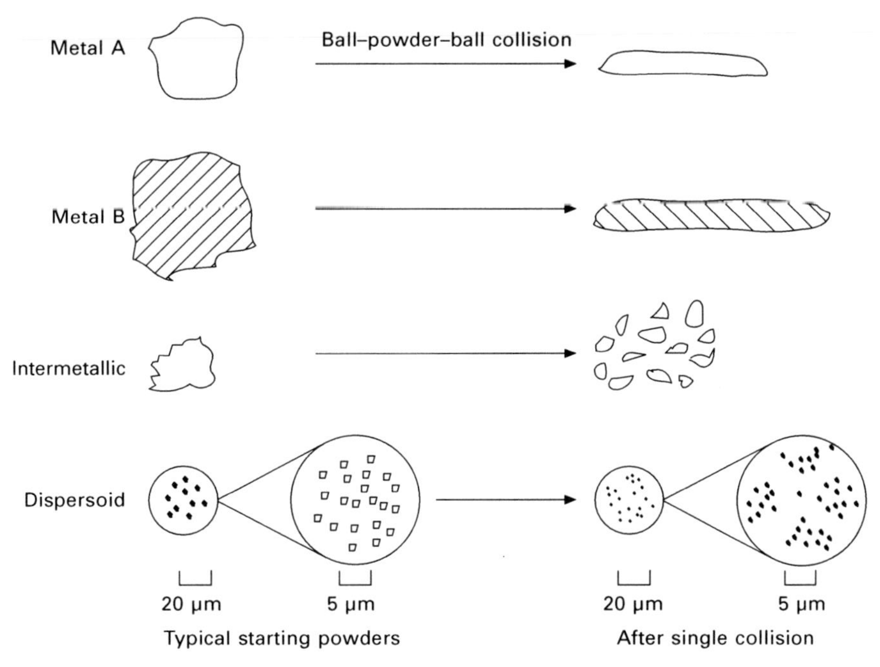

3.1 Deformation characteristics of representative constituents of starting powders in mechanical alloying. Note that the ductile metal powders (metals A and B) are flattened, while the brittle intermetallic and dispersoid particles are fragmented into smaller particles.

an aggregate weight of about 0.2 mg are trapped during each collision (Fig. 3.2). During this process, the powder morphology can be modified in two different ways, depending on whether we are dealing with ductile–ductile, ductile–brittle, or brittle–brittle powder combinations. If the starting powders are soft metal particles, the flattened layers overlap and form cold welds. This leads to formation of layered composite powder particles consisting of various combinations of the starting ingredients. The more brittle constituents tend to become occluded by the ductile constituents and trapped in the composite. The work-hardened elemental or composite powder particles may fracture at the same time. These competing events of cold welding (with plastic deformation and agglomeration) and fracturing (size reduction) continue repeatedly throughout the milling period. Eventually, a refined and homogenized microstructure is obtained and the composition of the powder particles is the same as the proportion of the starting constituent powders.

The presence of a defect structure enhances the diffusivity of solute elements in the matrix. Further, the refined microstructural features decrease the diffusion distances. Additionally, the slight rise in temperature during milling further aids the diffusion behavior and, consequently, true alloying takes place between the constituent elements. While this alloying generally

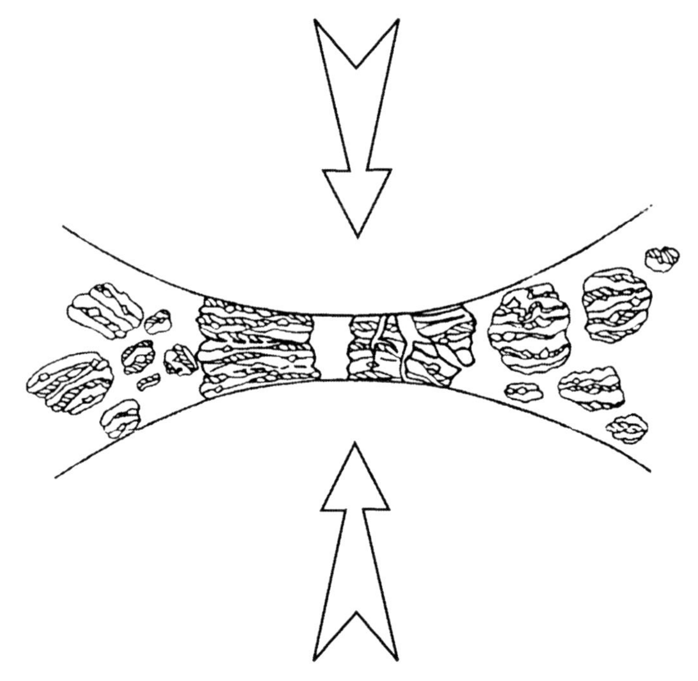

3.2 Ball–powder–ball collision of powder mixture during mechanical alloying.

takes place nominally at room temperature, sometimes it may be necessary to anneal the mechanically alloyed powder at a slightly elevated temperature in order for alloying to be achieved. This is particularly true when formation of intermetallics is desired.

Along with the cold welding event described above, some powder may also coat the grinding medium and/or the inner walls of the container. A thin layer of the coating is beneficial in preventing wear and tear of the grinding medium and also in preventing contamination of the milled powder by the debris. But too thick a layer will result in compositional inhomogeneity of the powder and should be avoided.

3.3.4 Evolution of particle size

As mentioned above, the powder particles are plastically deformed and are cold welded if they are soft. Further, work hardening also takes place, owing to the force of impact of the grinding medium, and eventually the work hardened powder particles fracture. The fresh and atomically clean surfaces created during fracture allow the particles to weld together under cold conditions and this leads to an increase in particle size. Since in the early stages of milling, the particles are soft (if we are using either ductile–ductile or ductile–brittle material combination), their tendency to weld together and form large particles is high. A broad range of particle sizes develops, with some as large as three times bigger than the starting particles (Fig. 3.3). The composite particles at this stage have a characteristic layered structure consisting of various combinations of the starting constituents (Fig. 3.4). Even if single component or prealloyed powder (such as during MM) is being used, the particle size increases owing to cold welding of the smaller particles. With continued deformation, the particles become work hardened and fracture by a fatigue failure mechanism and/or by fragmentation of the fragile flakes. Fragments generated by this mechanism may continue to reduce in size in the absence of strong agglomerating forces. At this stage, the tendency to fracture predominates over cold welding. Owing to the continued impact of grinding balls, the structure of the particles is steadily refined, but the particle size continues to be the same. Consequently, the inter-layer spacing decreases and the number of layers in a particle increases.

After milling for a certain length of time, steady-state equilibrium is attained when a balance is achieved between the rate of welding, which tends to increase the average particle size, and the rate of fracturing, which tends to decrease the average composite particle size. Smaller particles are able to withstand deformation without fracturing and tend to be welded

3.3 Broad distribution of particle sizes in an Al–30 at% Mg powder sample mechanically alloyed for 2 h.

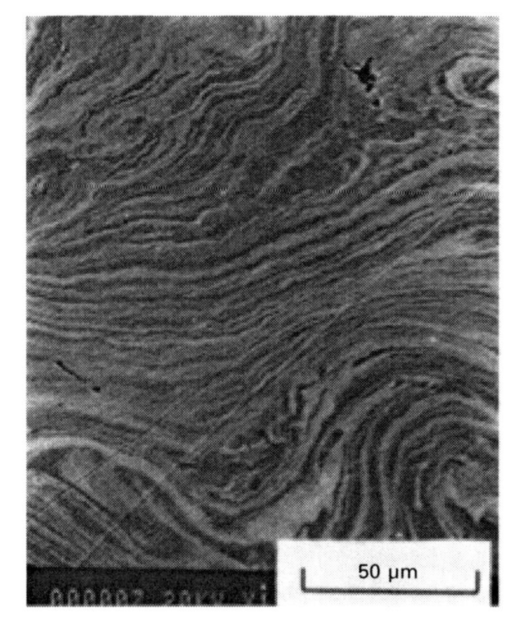

3.4 Scanning electron micrograph depicting the convoluted lamellar structure obtained during milling of a ductile–ductile component system (Ag-Cu).

into larger pieces, with an overall tendency to drive both very fine and very large particles towards an intermediate size. The particle size distribution at this stage is narrow, because particles larger than average are reduced in size at the same rate that fragments smaller than average grow through agglomeration of smaller particles.

3.3.5 Role of defects in alloying

During milling, heavy deformation is introduced, manifested by the presence of a variety of crystal defects such as dislocations, vacancies, stacking faults and an increased number of grain boundaries. It is also noted that lattice strain in the powder increases with milling time. However, it reaches a peak value and then shows a decrease in the last stages of milling. Since the main contribution to lattice strain comes from the dislocations introduced during milling, the strain increases with milling time owing to the increased dislocation density. But, beyond a particular milling time, nanometer-sized grains are produced, further grain refinement ceases owing to the inability of dislocation generation in these small grains and, therefore, the strain remains constant. Since deformation in extremely small grains occurs by grain boundary sliding accompanied by a reduction in overall dislocation density by annihilation at grain boundaries, not by formation and glide of dislocations, the strain could decrease.

It should, however, be remembered that the efficiency of particle size reduction is very low, about 0.1% in a conventional ball mill. The efficiency may be somewhat higher in high-energy ball milling processes, but is still less than 1%. The remaining energy is lost mostly in the form of heat, but a small amount is also utilized in the elastic and plastic deformation of the powder particles.

The specific times required to develop a given structure in any system should be a function of the initial powder particle size and the mechanical characteristics of the ingredients as well as the specific equipment used for milling and the operating parameters of the equipment. But, in most cases, the rate of refinement of the internal structure (particle size, crystallite size, lamellar spacing, etc.) is roughly logarithmic with regard to processing time and therefore the size of the starting particles is relatively unimportant. From a few minutes to an hour, the lamellar spacing usually becomes small and the crystallite (or grain) size is refined to nanometer dimensions (Fig. 3.5). The ease with which nanostructured materials can be synthesized is one reason why mechanochemical processing has been extensively employed to produce nanocrystalline materials. Thus, the mechanochemically processed powders will exhibit increased lattice strain and decreased grain size. As a result, the milled powder is in a highly energetic condition. There have been instances when these ultrafine powders have caught fire on exposure to atmosphere.

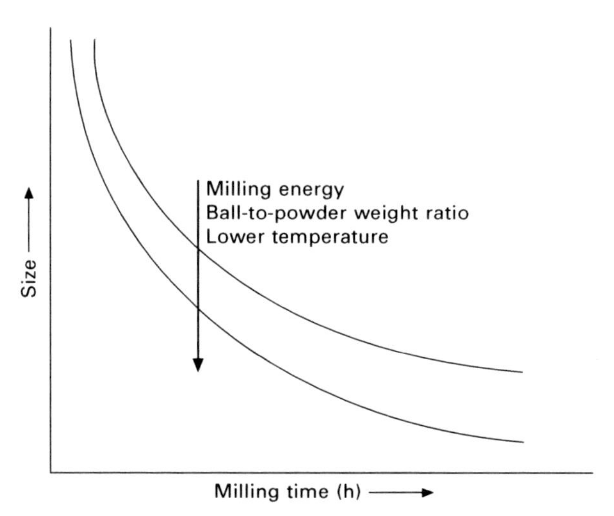

3.5 Refinement of particle/grain/crystallite size with milling time. The rate of refinement increases with increasing milling energy, ball-to-powder weight ratio and lower temperatures.

3.4 Grain size and process variables

3.4.1 Mechanism of grain refinement

Grain sizes with nanometer dimensions have been observed in almost all mechanochemically processed pure metals, intermetallics and alloys (if they continue to be crystalline). The minimum grain size achieved has been reported to be a few nanometers, ranging typically from about 5–50 nm, but depends on the material and processing conditions. Thus, it appears that synthesis of nanostructures by mechanochemical processing is a ubiquitous phenomenon and that nanostructures could be produced in every material. In spite of this, there have not been many detailed investigations to explain why and how nanometer-sized grains are obtained in these materials.

Hellstern *et al.* (1989) have studied the evolution of nanostructure formation in mechanically milled AlRu compound through detailed transmission electron microscopy (TEM) techniques. From high-resolution TEM observations, it was noted that deformation was localized within shear bands in the early stages of milling, owing to the high deformation rates experienced. These shear bands, which contain a high density of dislocations, have a typical width of approximately 0.5–1.0 μm. Small grains, with a diameter of 8–12 nm, were seen within the shear bands and electron diffraction patterns suggested significant preferred orientation. With continued milling, the average atomic level strain increased owing to increasing dislocation density, and at a certain dislocation density within these heavily strained regions, the crystal disintegrated into subgrains that are separated by low-angle grain boundaries. This resulted in a decrease in the lattice strain. The subgrains formed this way had nanometer dimensions and are often between 20 and 30 nm.

On further processing, deformation occurred in shear bands located in previously unstrained parts of the material. The grain size decreased steadily and the shear bands coalesced. The small-angle boundaries were replaced by higher angle grain boundaries, implying grain rotation, as reflected by the absence of texture in the electron diffraction patterns and random orientation of the grains observed from the lattice fringes in the high-resolution electron micrographs. Consequently, dislocation-free nanocrystalline grains were formed. This is the currently accepted mechanism of nanocrystal formation in mechanochemically processed powders.

3.4.2 Minimum grain size

As noted earlier, the grain size of milled materials decreases with milling time and reaches a saturation level when a balance is established between the fracturing and cold-welding events. This minimum grain size, d_{min} is different depending on the material and milling conditions. Some efforts have been made in recent years to rationalize the obtainable d_{min} in different materials in terms of the material properties. The value of d_{min} achievable by milling is determined by the competition between the plastic deformation via dislocation motion, which tends to decrease the grain size, and the recovery and recrystallization behavior of the material, which tends to increase the grain size. This balance gives a lower boundary for the grain size of pure metals and alloys.

The d_{min} value obtained is different for different metals and also is found to vary with the crystal structure. In most of the metals, the minimum grain size attained is in the nanometer dimensions. But metals with a body centered cubic (bcc) crystal structure reach much smaller values in comparison to metals with the other crystal structures. This is probably related to the difficulty of achieving extensive plastic deformation and the consequent enhanced fracturing tendency during milling. Ceramics and compounds are much harder and usually more brittle than the metals on which they are based. Therefore, intuitively, one expects that d_{min} of these compounds is smaller than those of the pure metals, although it does not appear to be always the case.

Trying to relate the d_{min} obtained during milling to the physical and material properties of the material being milled, it is noted that it decreases with an increase in the melting temperature of the metal (Eckert *et al.*, 1992a). This trend is amply clear in the case of metals with close-packed structures. Figure 3.6 shows a plot for metals with a face centered cubic (fcc) structure and a similar trend is observed for metals with a hexagonal close packed (hcp) structure. Another point of interest is that the difference in grain size is much less amongst metals that have high melting temperatures; the minimum grain size is virtually constant. Thus, for the hcp metals Co, Ti, Zr, Hf and Ru, the minimum grain size is almost the same even though the melting temperatures

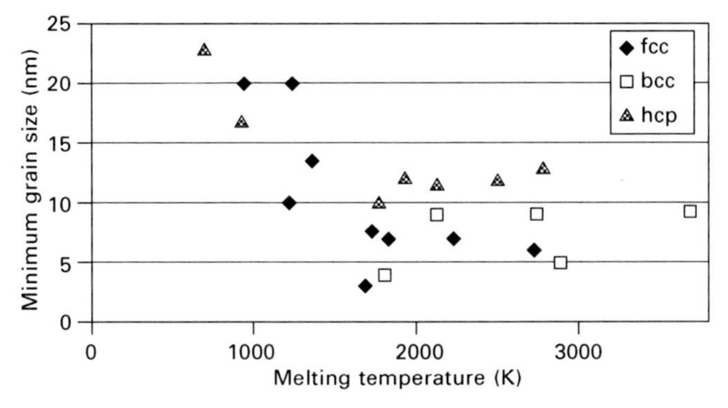

3.6 Minimum grain size obtained by mechanical milling of different pure metals with different crystal structures versus their melting temperature.

vary between 1495°C for Co to 2310°C for Ru. An inverse relation like the one above is less obvious in the case of bcc metals.

The existing data on d_{min} of mechanically milled pure metals was analyzed and correlated with different properties of the metals (Mohamed, 2003; Mohamed and Xun, 2003). It was shown that the normalized minimum grain size, $d_{min}/\boldsymbol{b}$, where $\boldsymbol{b}$ is the Burgers vector of the dislocations in that crystal structure, decreases with increasing melting temperature, activation energy for self diffusion, hardness, stacking fault energy, bulk modulus and the equilibrium distance between two edge dislocations.

3.4.3 Effect of process variables

Mechanochemical processing is a complex process and involves a number of process variables, including the type of mill, milling container, milling energy/speed, milling time, type, size, and size distribution of the grinding medium, ball-to-powder weight ratio, extent of filling the vial, milling atmosphere, process control agent, and temperature of milling. Some information is available in the literature on the minimum grain size achieved under different processing conditions. Amongst these, the effect of milling energy, milling temperature, and alloying effects are discussed below.

Milling energy

Since the d_{min} attained in a metal during milling is expected to depend on its mechanical properties, it is suspected that neither the nature of the mill nor the milling energy will have any effect on the minimum grain size achieved. It was reported (Galdeano *et al.*, 2001) that there was no significant effect of milling intensity on nanostructure formation in a Cu-Fe-Co powder blend. But, in another investigation, it was shown that d_{min} was about 5 nm when the TiNi intermetallic powder was milled in a high-energy SPEX shaker mill, but only about 15 nm in the less-energetic Invicta vibratory ball mill (Yamada and Koch, 1993). Similar trends were noted for variations in the ball-to-powder weight ratios (BPR). The grain size of the niobium metal milled in the Invicta vibratory mill was about 26 ± 2 nm at a BPR of 5:1, but was only 18 ± 1 at a BPR of 10:1 (Koch, 1993). Similar results were also reported for the pure metal Cu: 25 ± 5 nm for a BPR of 5:1 and 20 ± 1 for a BPR of 10:1. Further, the kinetics of achieving this d_{min} value could also depend on the milling energy, although no such studies have been reported so far.

It was also reported that during nanocrystal formation, the average crystal size increased and the internal lattice strain decreased at higher milling intensities owing to the enhanced thermal effects (Kuhrt *et al.*, 1993). In accordance with this argument, the grain size of Si milled at a high energy of 500 kJ g^{-1} was 25 nm, while that milled at a low energy of 20 kJ g^{-1} was only 4 nm (Streleski *et al.*, 2002).

Milling temperature

The grain size of powders milled at low temperatures was smaller than those milled at higher temperatures. For example, the grain size of copper milled at room temperature was 26 ± 3 nm, while that milled at –85°C was only 17 ± 2 nm (Shen and Koch, 1995). Similar results were reported for other metals and the CoZr intermetallic compound. Milling of powders at higher temperatures also resulted in reduced root mean square (rms) strain in addition to larger grain sizes (Hong *et al.*, 1994).

Alloying effects

Alloy composition appears to have a significant effect on d_{min} after milling. Since solid solutions are harder and stronger than the pure metals on which they are based, it is expected that the minimum grain size will be smaller for solid solutions than for pure metals. This can be understood on the basis of increased fragmentation tendency of the harder (and brittle) powders. Accordingly, it was noted that d_{min} in Cu-rich Cu-Fe powders decreased with increasing Fe content. The grain size after milling for 24 h in a SPEX mill decreased from 20 nm for pure Cu to about 9 nm for Cu-55 at% Fe. This was attributed to solid solution strengthening of the Cu matrix by segregation of Fe atoms by their attachment to the stacking faults (Eckert *et al.*, 1992b). For the Fe-rich Fe-Cu powders, d_{min} was reported to decrease from 6 nm for pure Fe to 4.6 nm for Fe-15 at% Cu, again attributed to solid solution hardening.

A similar decrease in grain size with increasing Ni content was also reported for the B2-NiAl phase. The grain size decreased from 10 nm for the Al-rich Al-46 at% Ni to about 5 nm for the Ni-rich Al-60 at% Ni. This observation again confirms that the alloy composition determines the final grain size. Since NiAl has a wide range of homogeneity, this compound can accommodate constitutional vacancies in its lattice. It is known that for the Al-rich compositions, excess vacancies in the Ni sublattice lead to strong vacancy hardening and a large resistance against dislocation motion in coarse-grained polycrystalline material. Further, the large diffusivity enhances dynamic recovery for Al-rich compositions. A combination of these two effects could prevent formation of very small grain sizes on milling the Al-rich Ni-Al alloys. Cross-slip and low diffusivities of Ni-rich compositions hinder recovery. Thus, transgranular fracture behavior coupled with little recovery is reported to be the reason for efficient mechanical attrition and the consequent small grain sizes.

The above observations make it clear that the small grain sizes obtained in solid solution alloys are due to the effects of solid solution strengthening. Therefore, it is logical to expect that the grain size in the milled solid solution powder would be larger if the solid solution is softer than the pure metal and would remain constant if there is no significant change in hardness. Results consistent with this observation were reported in Ni(Co) where the hardness remained the same in the solid solution as in the pure metal (Shen and Koch, 1995) and in Ni(Cu), Fe(Cu) and Cr(Cu) which exhibited solid solution softening (Eckert *et al.*, 1992b).

It was suggested that d_{min} is determined by the minimum grain size that can sustain a dislocation pile-up within a grain and by the rate of recovery (Eckert *et al.*, 1992a). Based on the dislocation pile-up model, the critical equilibrium distance between two edge dislocations in a pile-up, L_c (which could be assumed to be the crystallite or grain size in milled powders), was calculated (Nieh and Wadsworth, 1991) using the equation:

$$L_c = \frac{3G\boldsymbol{b}}{\pi(1-\nu)H}$$

where G is the shear modulus, $\boldsymbol{b}$ is the Burgers vector, ν is Poisson's ratio and H is the hardness of the material. According to the above equation, increased hardness results in smaller values of L_c (grain size) and an approximate linear relationship was observed between L_c and the minimum grain size obtained by milling a number of metals (Eckert *et al.*, 1992b). Other attempts have also been made to predict d_{min} theoretically on the basis of thermodynamic properties of materials (Sun and Lu, 1999).

3.5 Displacement reactions

It has been shown above that nanocrystalline structures have been synthesized in materials subjected to mechanical alloying or milling. Two of the serious limitations of powders produced by the above methods are that they are agglomerated and that they display a wide particle size distribution. Consequently, it is difficult to control their morphology and overall particle size distribution. Another way to produce these nanocrystalline materials is by chemical reactions induced by mechanical activation. Most of the reactions studied have been displacement reactions of the type:

$$MO + R \rightarrow M + RO$$

where a metal oxide (MO) is reduced by a more reactive metal (reductant, R) to the pure metal M. In addition to the oxides, metal chlorides and sulfides have also been reduced to pure metals this way. Similar reactions have also been used to produce alloys and nanocomposites (Suryanarayana, 2001, 2004).

All solid-state reactions involve the formation of a product phase at the interface of the component phases. Thus, in the above example, the metal M forms at the interface between the oxide MO and the reductant R and physically separates the reactants. Further growth of the product phase involves diffusion of atoms of the reactant phases through the product phase, which constitutes a barrier layer preventing further reaction from occurring. In other words, the reaction interface, defined as the nominal boundary surface between the reactants, continuously decreases during the course of the reaction. Consequently, the kinetics of the reaction are slow and elevated temperatures are required to achieve reasonable reaction rates.

Mechanochemical processing can provide the means to increase substantially the reaction kinetics of chemical reactions in general and the reduction reactions, in particular. This is because the repeated cold-welding and fracturing of powder particles increases the area of contact between the reactant powder particles by bringing fresh surfaces into contact repeatedly owing to a reduction in particle size during milling. This allows the reaction to proceed without the necessity for diffusion through the product layer. As a consequence, reactions that normally require high temperatures will occur at lower temperatures, or even without any externally applied heat. In addition, the high defect densities induced by MA accelerate diffusion processes.

Depending on the milling conditions, two entirely different reaction kinetics are possible (Schaffer and McCormick, 1990; Takacs, 2002):

- the reaction may extend to a very small volume during each collision, resulting in a gradual transformation, or
- if the reaction enthalpy is sufficiently high (and if the adiabatic temperature is above 1800 K), a self-propagating high-temperature synthesis (SHS) (also known as combustion reaction) can be initiated. By controlling the milling conditions (e.g., by adding diluents) to avoid combustion, reactions may be made to proceed in a steady-state manner.

The latter type of reaction requires a critical milling time for the combustion reaction to be initiated. If the temperature of the vial is recorded during the milling process, the temperature initially increases slowly with time. After some time, the temperature increases abruptly, suggesting that ignition has occurred and this is followed by a relatively slow decrease in temperature.

In the above types of reaction, the milled powder consists of the pure metal M dispersed in the RO phase, both usually with nanometer dimensions. It has been shown that by an intelligent choice of the reductant, the RO phase can be easily leached out. Thus, by selective removal of the matrix phase by washing with appropriate solvents, it is possible to achieve well-dispersed metal nanoparticles as small as 5 nm in size (Ding *et al.*, 1999). If the crystallinity of these particles is not high, it can be improved by subsequent annealing, without the fear of agglomeration caused by the enclosure of nanoparticles in a solid matrix. Such nanoparticles are structurally and morphologically uniform and have a narrow size distribution. This technique has been used to produce powders of many different metals: Ag, Cd, Co, Cr, Cu, Fe, Gd, Nb, Ni, Ti, V, W, Zn and Zr. Several alloys, intermetallics and nanocomposites have also been synthesized (Suryanarayana, 2004; Suryanarayana and Al-Aqeeli, 2013). Figure 3.7 shows two micrographs of nanocrystalline ceria particles produced by conventional vapor synthesis and mechanochemical processing methods. Even though the grain size is approximately the same

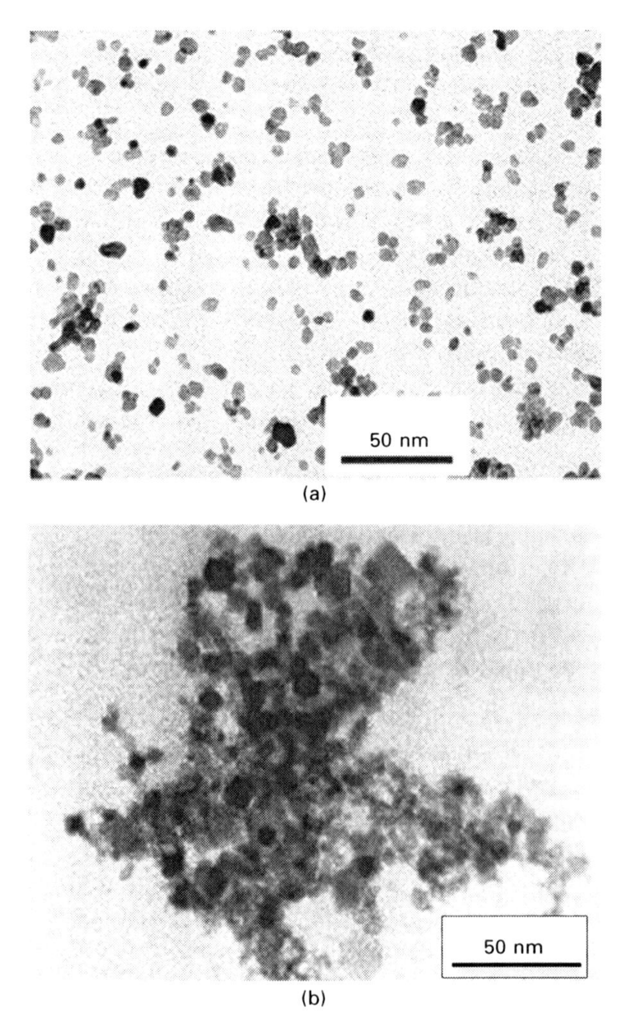

3.7 Transmission electron micrographs of ceria particles produced by (a) mechanochemical processing and (b) vapor synthesis methods. The grain size is about the same in both the methods. Note, however, the agglomeration of particles in the latter case.

in both the cases, it may be noted that the nanoparticles produced by the mechanochemical method are very discrete and display a very narrow particle size distribution, while those produced by vapor deposition are agglomerated and have a wide particle size distribution.

3.6 Consolidation

The products of all the above processes are metal powders. Except in cases of chemical applications, such as catalysis, when the product does not require consolidation, widespread application of mechanically processed powders requires efficient methods of consolidating them into bulk shapes. Successful consolidation of nanocrystalline powders is not a trivial problem since fully dense materials should be produced while simultaneously retaining the nanometer-sized grains without coarsening. Conventional consolidation of powders to full density through processes such as hot extrusion and hot isostatic pressing requires use of high pressures and elevated temperatures for extended periods of time to achieve full densification. Unfortunately, however, this results in significant coarsening of the nanometer-sized grains and consequently the benefits of nanostructure processing are lost. On the other hand, retention of nanostructures requires use of low consolidation temperatures and it is difficult to achieve full interparticle bonding at these low temperatures. Thus, the objectives of consolidation of nanocrystalline powders are (i) to achieve full densification (without any porosity), (ii) to minimize microstructural coarsening (i.e. retention of nanocrystalline state), and/or (iii) to avoid undesirable phase transformations. In fact, early results with excellent mechanical properties, and most specifically superplasticity at room temperature (Karch *et al.*, 1987), attributed to nanocrystalline ceramics, have not been successfully reproduced, because of incomplete consolidation achieved in the material tested. Thus, consolidation to full density assumes even greater importance. Therefore, novel and innovative methods of consolidating nanocrystalline powders are required.

Because of the small size of the powder particles (typically a few micrometers, even though the grain size is only a few nanometers), some special precautions need to be taken to consolidate the metal powders to bulk shapes. However, nanocrystalline powders pose special problems. For example, they possess very high strength and hardness. They also have a high level of interparticle friction. Further, since the nanocrystalline powders have a large surface area, they also exhibit high chemical reactivity.

Successful consolidation of nanocrystalline powders has been achieved by electro-discharge compaction, plasma-activated sintering, shock (explosive) consolidation, hot-isostatic pressing (HIP), hydrostatic extrusion, strained powder rolling and sinter forging. By utilizing the combination of high temperature and pressure, HIP can achieve a particular density at lower pressures than cold isostatic pressing or at lower temperatures than sintering. It should be noted that because of the increased diffusivity in nanocrystalline materials, sintering (densification) takes place at temperatures much lower than in coarse-grained materials. This is likely to reduce the grain growth. Two excellent reviews by Groza (Groza, 1999, 2007) may be consulted for full details of the methods of consolidation and a detailed description of the results.

3.7 Powder contamination

A major concern in metal powders processed by mechanochemical methods has been the nature and amount of impurity that is incorporated into the powder and contaminate it. The small size of the powder particles and consequent availability of large surface area, formation of fresh surfaces during milling, and wear and tear of the milling tools, all contribute to contamination of the powder. These factors will be in addition to the purity of the starting raw materials and the milling conditions employed (especially the milling atmosphere). Thus, it appears as though powder contamination is an inherent drawback of the method and that the milled powder will always be contaminated by impurities unless special precautions are taken to avoid/minimize them. The magnitude of powder contamination appears to depend on the time of milling, intensity of milling, atmosphere in which the powder is milled, nature and size of the milling medium and differences in the strength/hardness of the powder and the milling medium and the container.

3.7.1 Sources of contamination

Contamination of milled metal powders can be traced to (i) chemical purity of the starting powders, (ii) milling atmosphere, (iii) milling equipment (milling container and grinding medium), and (iv) process control agents (PCA) added to the powders during milling. Contamination from source (i) can be either substitutional or interstitial in nature, while contamination from source (ii) is essentially interstitial and that from (iii) is mainly substitutional, even though carbon from the steel milling equipment can be an interstitial impurity. Contamination from the PCA essentially leads to interstitial contamination, since the PCAs used are mostly organic compounds containing carbon, oxygen and nitrogen. Irrespective of whether the impurities are substitutional or interstitial in nature, their amount increases with milling time and with increasing ball-to-powder weight ratio (BPR), and reaches a saturation value. Since the powder particles become finer with milling time, their total surface area increases and consequently the amount of contamination increases. Most commonly, the level of powder contamination is investigated as a function of time. For example, Fig. 3.8 shows a plot of the levels of contamination

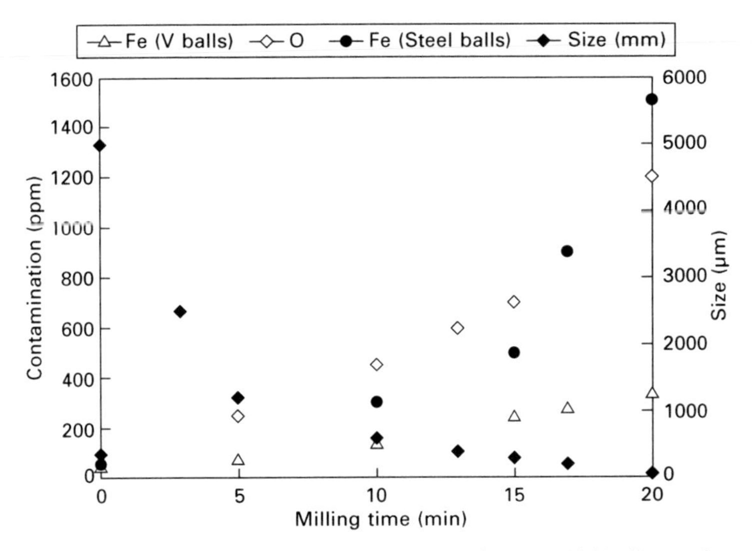

3.8 Levels of powder contamination in a Cr-19 wt% V alloy using steel or vanadium balls. Note that the level of contamination is increasing with increasing milling time and that the Fe contamination is higher when steel balls are used instead of V balls.

of Fe and oxygen in a Cr-19 wt% V alloy powder milled in a water-cooled AGO high-energy ball mill at 1500 rpm using either steel or V grinding medium in steel containers. Powder contamination from both Fe and oxygen is higher with longer milling times. But, it may be noted that Fe contamination is significantly higher when steel balls are used instead of V balls. It may also be noted that the reduction in particle size, and consequent large total surface area at longer milling times, is responsible for the increased powder contamination with milling time. Similarly, with increasing BPR, the wear and tear of the grinding medium increases and consequently contamination is also higher. But when particle refinement reaches the limiting value and a steady state condition is achieved, the level of contamination does not increase any further.

3.7.2 Elimination/minimization of contamination

Several attempts have been made in recent years to minimize powder contamination during milling. An important point to remember is that if the starting powders are highly pure, then the final product is going to be clean and pure, with minimum contamination from other sources, provided that proper precautions are taken during milling. It is also important to keep in mind that all applications do not require ultra clean powders and that the purity desired will depend on the specific application. But it is a good practice to minimize powder contamination at every stage and ensure that additional impurities are not picked up subsequently during handling and/ or consolidation.

One way of minimizing contamination from the grinding medium and the milling container is to use the same material for the container and the grinding medium as the powder being milled. Thus, one could use copper balls and a copper container for milling copper and copper alloy powders (Suryanarayana *et al.*, 1999). If a container of the same material to be milled is not available, then putting a thin adherent coating on the internal surface of the container (and also on the surface of the grinding medium) with the material to be milled will minimize contamination. For example, the extent of iron contamination during milling of blended elemental Ta-Al powders was less in the second run and much less in the third run (Sherif El-Eskandarany *et al.*, 1990). The idea here is to mill the powder once, allowing the powder to be coated onto the grinding medium and the inner walls of the container. The milled loose powder is then discarded and a fresh batch of powder is milled, but with the old grinding media and container. In general, a simple rule that should be followed to minimize contamination from the milling container and the grinding medium is that the container and grinding medium should be harder/stronger than the powder being milled.

The milling atmosphere is another important parameter that needs to be controlled. Even though nominally pure gases are used to purge the glove box before loading and unloading the powders, even small amounts of impurities appear to contaminate the powders, especially if they are reactive. Thus, Goodwin and Ward-Close (Goodwin and Ward-Close, 1993) milled pre-alloyed and blended elemental Ti-24Al-11Nb (at%) powders in two different atmospheres. AT1 refers to research grade argon (99.995% pure) and AT2 refers to higher purity argon (guaranteed 99.998%), which is subsequently passed through moisture and oxygen filter towers to purify it further. The levels of powder contamination in both the atmospheres are listed in Table 3.1. It may be noted that the powders milled in AT2 are much purer than those milled in AT1, owing to the greater would purity of the atmosphere. In fact, the best solution one could think of would be to place the mill inside a

Table 3.1 Oxygen and nitrogen contents of Ti-24Al-11Nb (at%) powders milled for 24 h in a SPEX mill with a BPR of 4:1 using two different types of atmosphere. No PCA was used during milling (after Goodwin and Ward-Close, 1993)

Powder	Oxygen content (wt%)		Nitrogen content (wt%)	
	AT1	AT2	AT1	AT2
Prealloyed Ti-24Al-11Nb (at%)	3.6	0.10	6.8	0.015
Blended elemental Ti-24Al-11Nb (at%)	4.8	0.48	7.6	0.035

chamber that is evacuated and filled with high-purity argon gas. Since the whole chamber is maintained under argon gas atmosphere (continuously purified to keep oxygen and water vapor below 1 ppm each), and the container is inside the mill, which is inside the chamber, contamination from the atmosphere is minimum. Thus, Klassen *et al.* (1997) reported that the oxygen content in their milled $Ti_{51}Al_{49}$ powder was only 0.30 wt% after milling for 100 h in a Fritsch P5 mill, which was enclosed inside a glove box maintained under argon atmosphere. An fcc phase has been reported to form in reactive metals like Ti, Zr and Hf even when milled under nominally high-purity atmospheres (Suryanarayana, 1995b). An experiment as described above has also helped in solving the mystery of why the fcc phase in reactive metals is an impurity phase (Seelam *et al.*, 2009).

Contamination from PCAs is perhaps the most ubiquitous. Since most of the PCAs used are organic compounds, which have low melting and boiling points, they decompose during milling owing to the heat generated. The decomposition products consisting of carbon, oxygen, nitrogen and hydrogen react with the metal atoms and form undesirable carbides, oxides, nitrides, and so on. The consolidated powders may, however, be stronger owing to the dispersion of these compounds in the metal matrix.

3.8 Conclusions

Nanocrystalline materials have been synthesized by a number of different methods in recent years. But mechanochemical processing appears to be a versatile method of producing nanocrystalline powders in a variety of metals and alloys. The process is relatively simple, easy to upgrade to a pilot plant/commercial scale, and a variety of metastable phases can also be produced. The three major issues for powders processed this way are cost, consolidation and contamination. Powder processing is inherently expensive and therefore there is not much that can be done to alleviate this difficulty. But the possibility of producing near-net shape products with tailored properties will certainly offset this concern. The second issue is consolidation. As discussed, it is imperative that the consolidated product is fully dense and should not contain any porosity. This is a challenging problem and unless this is achieved, the integrity of the product is likely to be compromised. But novel and innovative methods are being currently employed to achieve this objective and it is likely that great strides will be made in the near future. If the milled powder can be used without consolidation, for example, for catalytic purposes, mechanochemical processing will be an ideal process for producing such powders. The last concern is about contamination. This is a ubiquitous problem. Even though remedies have been suggested to avoid or minimize contamination in milled powders, these appear to be expensive and, perhaps, not easy to achieve on a commercial scale, without undue expense.

If milled powders could be used in non-critical applications, where some amount of contamination could be tolerated, it is likely that this process will prove highly beneficial and attractive.

3.9 References

Avvakumov E.G. (1986). *Mechanical Methods of Activation of Chemical Processes*. Nauka Publishing House, Novosibirsk, Russia.

Avvakumov E.G., Senna M. and Kosova N.V. (2001). *Soft Mechanochemical Synthesis: A Basis for New Chemical Technologies*. Springer, Boston, MA.

Baláž P. (2008). *Mechanochemistry in Nanoscience and Minerals Engineering*, Springer, Berlin.

Benjamin J.S. (1992). 'Mechanical alloying – History and future potential'. In: *Advances in Powder Metallurgy and Particulate Materials, vol. 7 (Novel Powder Processing)*, Capus J.M. and German R.M. (compilers). Metal Powder Industries Federation, Princeton, NJ, 155–68.

Boldyrev V.V. (2006). 'Mechanochemistry and mechanical activation of solids'. *Russian Chem Rev*, **75**, 177–89.

Carrey-Lea M. (1892). 'On endothermic reactions affected by mechanical force: Part I'. *Amer J Sci*, third series, **43**, 527.

Ding J., Tsuzuki T. and McCormick P.G. (1999). 'Microstructural evolution of Ni-NaCl mixtures during mechanochemical reaction and mechanical milling'. *J Mater Sci*, **34**, 5293–98.

Eckert J., Holzer J.C., Krill, III, C.E. and Johnson W.L. (1992a). 'Structural and thermodynamic properties of nanocrystalline fcc metals prepared by mechanical attrition'. *J Mater Res*, **7**, 1751–61.

Eckert J., Holzer J.C., Krill, III, C.E. and Johnson W.L. (1992b). 'Reversible grain size changes in ball-milled nanocrystalline Fe–Cu alloys', *J Mater Res*, **7**, 1980–83.

Ermakov A.E., Yurchikov E.E. and Barinov V.A. (1981). 'The magnetic properties of amorphous Y-Co alloy powders obtained by mechanical comminution'. *Phys Met Metallogr*, **52**(6), 50–58.

Galdeano S., Chaffron L., Mathon M-H., Vincent E. and De Novion C-H. (2001). 'Characterisation of the ball-milled $Cu_{80}(Fe_{0.3}Co_{0.7})_{20}$ compound and effects of the milling conditions on its nanostructure'. *Mater Sci Forum*, **360–2**, 367–72.

German R.M. (1992). *Powder Metallurgy*. Metal Powder Industries Federation, Princeton, NJ.

Gleiter H. (1989). 'Nanocrystalline materials'. *Prog Mater Sci*, **33**, 223–315.

Goodwin P.S. and Ward-close C.M. (1993). 'Process control in the mechanical alloying of TiAlNb alloys'. In: *Mechanical Alloying for Structural Applications*, deBarbadillo J.J., Froes F.H. and Schwarz R.B. (eds). ASM International, Materials Park, OH, 139–48.

Groza J.R. (1999). 'Powder consolidation'. In: *Non-equilibrium Processing of Materials*, Suryanarayana C. (ed.), Pergamon, Oxford, UK, 347–72.

Groza J.R. (2007). 'Nanocrystalline powder consolidation methods'. In: *Nanostructured Materials: Processing, Properties, and Applications*, Carl C. Koch (ed.). William Andrew Publishing, Norwich, NY, 173–233.

Gutman E.M. (1994). *Mechanochemistry of Solid Surfaces*, World Scientific, Singapore.

Gutman E.M. (1998). *Mechanochemistry of Materials.* Cambridge International Science Publishing, Cambridge, UK.

Heinicke G. (1984). *Tribochemistry*, Hanser Publishers, Munchen.

Hellstern E., Fecht H.J., Garland C. and Johnson W.L. (1989). 'Mechanism of achieving nanocrystalline AlRu by ball milling'. In: *Multicomponent Ultrafine Microstructures*, McCandlish L.E., Polk D.E., Siegel R.W. and Kear B.H. (eds). Material Research-Society, Pittsburgh, PA, vol. 132, 137–42.

Hong L.B., Bansal C. and Fultz B. (1994). 'Steady state grain size and thermal stability of nanophase Ni_3Fe and Fe_3X (X = Si, Zn, Sn) synthesized by ball milling at elevated temperatures'. *NanoStructured Mater*, **4**, 949–56.

Ivanov E. (1993). 'Reactive mechanical alloys in materials synthesis'. *J Mater Synth Process*, **1**, 405–13.

Ivanov E. and Suryanarayana C. (2000). 'Materials and process design through mechanochemical routes'. *J Mater Synth Process*, **8**, 235–44.

Jangg G. (1989). 'Reaction milling of aluminium alloys'. In: *New Materials by Mechanical Alloying Techniques*, Arzt E. and Schultz L. (eds). DGM Informationgesellschaft, Oberursel, Germany, 39–52.

Juhász A.Z. and Opoczky L. (1990). *Mechanical Activation of Minerals by Grinding, Pulverizing, and Morphology of Particles.* Akademia Kiadó, Budapest, Hungary.

Karch J., Birringer R. and Gleiter H. (1987). 'Ceramics ductile at low temperature'. *Nature*, **330**, 556–8.

Klassen T., Oehring M. and Bormann R. (1997). 'Microscopic mechanisms of metastable phase formation during ball milling of intermetallic TiAl phases'. *Acta Mater*, **45**, 3935–48.

Koch C.C. (1993). 'The synthesis and structure of nanocrystalline materials produced by mechanical attrition: A review'. *NanoStructured Mater*, **2**, 109–29.

Koch C.C., Calvin O.B., Mckamey C.G. and Scarbrough C.G. (1983). 'Preparation of "amorphous" $Ni_{60}Nb_{40}$ by mechanical alloying'. *Appl Phys Lett*, **43**, 1017–19.

Kuhrt C., Schröpf H., Schultz L. and Arztz E. (1993). 'Synthesis of nanocrystalline FeAl and NiAl by mechanial alloying'. In: *Mechanical Alloying for Structural Applications*, JJ deBarbadillo J.J., Froes F.H., and Schwarz R.B. (eds), ASM International, Materials Park, OH, 269–73.

Luton M.J., Jayanth C.S., Disko M.M., Matras S. and Vallone J. (1989). 'Cryomilling of nano-phase dispersion strengthened aluminum'. In: *Multicomponent Ultrafine Microstructures*, McCandlsih, L.E., Polk D.E., Siegel R.W. and Kear B.H. (eds). Materials-Research Society, Pittsburgh, PA, vol. 132, 79–86.

McCormick P.G. (1995). 'Application of mechanical alloying to chemical refining'. *Mater Trans Japan Inst Metals* **36**, 161–69.

McCormick P.G. (1997). 'Mechanical alloying and mechanically induced chemical reactions'. In: *Handbook on the Physics and Chemistry of Rare Earths*, Gschneidner, Jr., K.A. and Eyring, L. (eds), Elsevier Sci BV, Amsterdam, Vol. **24**, 47–81.

McCormick P.G., Wharton V.N. and Schaffer G.B. (1989). 'A novel method for the direct reduction of metal powders'. In: *Physical Chemistry of Powder Metals Production and Processing*, Small W.M. (ed.), TMS, Warrendale, PA, 19–34.

Mohamed F.A. (2003). 'A dislocation model for the minimum grain size obtainable by milling'. *Acta Mater*, **51**, 4107–19.

Mohamed F.A. and Xun Y. (2003). On the minimum grain size produced by milling Zn-22% Al'. *Mater Sci Eng*, **A358**, 178–85.

Nieh T.G. and Wadsworth J. (1991). 'Hall-Petch relation in nanocrystalline solids'. *Scripta Metall Mater*, **25**, 955–58.

Ostwald W. (1919). *Handbuch der allgemeine Chemie*, *Leipzig*, **1**, 70.
Schaffer G.B. and McCormick P.G. (1989). 'Reduction of metal oxides by mechanial alloying'. *Appl Phys Lett*, **55**, 45–46.
Schaffer G.B. and McCormick P.G. (1990). 'Displacement reactions during mechanical alloying'. *Metall Trans A*, **21A**, 2789–94.
Seelam U.M.R., Barkhordarian G. and Suryanarayana C. (2009). 'Is there a HCP → FCC allotropic transformation in mechanically milled Group IVB elements?'. *J Mater Res*, **24**, 3454–61.
Shen T.D. and Koch C.C. (1995). 'The influence of dislocation structure on formation of nanocrystals by mechanical attrition'. *Mater Sci Forum*, **179–81**, 17–24.
Sherif El-Eskandarany M., Aoki K. and Suzuki K. (1990). 'Rod milling for solid-state formation of $Al_{30}Ta_{70}$ amorphous alloy powder', *J Less-Common Metals*, **167**, 113–18.
Sopicka-Lizer M. (2010) (ed.). *High Energy Ball Milling*, Woodhead Publishing, Cambridge.
Streleski A.N., Leonov A.V., Beresteskaya I.V., Mudretsova S.N., Majorova A.F. and Butyagin P.Yu. (2002). 'Amorphization and reactivity of silicon induced by mechanical treatment'. *Mater Sci Forum*, **386–388**, 187–92.
Sun N.X. and Lu K. (1999). 'Grain size limit of polycrystalline materials'. *Phys Rev*, **B59**, 5987–89.
Suryanarayana C. (1995a). 'Nanocrystalline materials'. *Int. Mater Rev*, **40**, 41–64.
Suryanarayana C. (1995b). 'Does a disordered γ-TiAl phase exist in mechanically alloyed Ti-Al powders?'. *Intermetallics*, **3**, 153–60.
Suryanarayana C. (1999). *Non-Equilibrium Processing of Materials*, Pergamon, Oxford.
Suryanarayana C. (2001). 'Mechanical alloying and milling'. *Prog Mater Sci*, **46**, 1–184.
Suryanarayana C. (2004). *Mechanical Alloying and Milling*, Marcel Dekker, New York.
Suryanarayana C. (2005). 'Recent developments in nanostructured materials', *Adv Eng Mater*, **7**, 983–92.
Suryanarayana C. and Al-Aqeeli N. (2013). 'Mechanically alloyed nanocomposites', *Prog Mater Sci*, **58**, 383–502.
Suryanarayana C., Ivanov E., Noufi R., Contreras M.A. and Moore J.J. (1999). 'Phase selection in a mechanically alloyed Cu-In-Ga-Se powder mixture'. *J Mater Res*, **14**, 377–83.
Takacs L. (2002). 'Self-sustaining reactions induced by ball milling'. *Prog Mater Sci*, **47**, 355–414.
Tiessen P.A., Meyer K. and Heinicke G. (1966). *Grundlagen der Tribochemie*, Akad. Verlag, Berlin.
Tkacova K. (1989). *Mechanical Activation of Minerals*, Elsevier Science Publishers, Amsterdam.
Witkin D.B. and Lavernia E.J. (2006). 'Synthesis and mechanical behavior of nanostructured materials via cryomilling'. *Prog Mater Sci*, **51**, 1–60.
Yamada K. and Koch C.C. (1993). 'The influence of mill energy and temperature on the structure of the TiNi intermetallic after mechanical attrition'. *J Mater Res*, **8**, 1317–26.

4 Plasma synthesis of metal nanopowders

I. CHANG, University of Birmingham, UK

DOI: 10.1533/9780857098900.1.69

Abstract: This chapter begins by discussing the benefits and applications of metal nanopowders. This is followed by an introduction of plasma technology and a review of various types of plasma methods used in the synthesis of metal nanopowders. The chapter includes a description of the formation of metal nanopowders by a nucleation and growth mechanism from the vapour phase. Finally, it discusses the relationship between metal nanopowder characteristics (e.g. average size and size distribution) and plasma processing parameters.

Key words: arc discharge, metal nanopowders, nucleation and growth, plasma torch, spark discharge.

4.1 Introduction

Conventional powder metallurgy technology is based on the powders with average sizes between a few and hundreds of micrometres. However, there is a rapidly expanding technology known as nanotechnology that is based on materials with a size scale in the nanometres. These nanoscaled materials are produced in many forms (e.g. thin film, powder or bulk) by various processing methods. In the case of powder form, nanopowders are commonly defined as powders with averages sizes below 100 nm. They have attracted enormous attention because of the flexibility of their use as either the final products or as the precursors in the formation of bulk nanostructured materials. One of the most commonly used methods of producing metal nanopowders is known as plasma synthesis. This is a physical method that converts metals from bulk or powdered forms into nanopowders via evaporation and condensation processes. This chapter gives a review of the production of nanopowders using plasma processing technologies, as well as the potential benefits and applications of metal nanopowders.

4.2 Potential benefits and applications of metal nanopowders

Nanopowders exhibit unique and improved optical, thermal, chemical, electrical and physical properties (El-Sayed, 2001; Feldheim, 2002) compared to corresponding bulk materials. These unique properties allow nanopowders to have (i) enhanced chemical reaction, (ii) faster sintering kinetics, (iii) higher electrical resistivity, (iv) superparagnetism (Huber, 2005), (v) microwave absorption (Rittner and Abraham, 1998) and (vi) localized surface plasmon resonances (Harra *et al.*, 2012). They are ideal candidates for catalysts, sintering aids, microwave absorption (Rittner, 2002), magnetic recording media, magnetic fluids, magnetic ink (Liu *et al.*, 2002), rocket propellants (Wood and Scott, 2002; Galfetti *et al.*, 2006), conducting ink/paste (Kim *et al.*, 2007), permeable reactive barriers for soil decontamination (Nurmi *et al.*, 2005), biomarkers (Tartaj *et al.*, 2003) and biosensors applications (Harra *et al.*, 2012). Latest advancement in heat transfer fluids, is the development of nanofluids, which are engineered colloidal mixtures of the base fluids and nano-sized metallic particles (1–100 nm). These nanofluids exhibit increased thermal conductivity and improved heat transfer rate, thereby making them very efficient coolant media (Godson *et al.*, 2010; Chopkar *et al.*, 2006).

4.2.1 What is plasma?

Plasma consists of an equal number of positive and negative charged particles (Dendy, 1990), which are produced from either complete or partial ionization of gas atoms or molecules. It is an electrically neutral but conductive medium and it responds to a magnetic field. Therefore, plasma is considered to be the fourth state of matter because it has properties unlike other states of matter, such as solid, liquid and gas. Plasma can be classified into thermal or non-thermal plasma depending on relative temperatures between the electrons, ions and neutral particles. When a thermal equilibrium is reached between the electrons and heavy particles and the temperatures of electrons and heavy particles are kept the same, this is known as thermal ('hot') plasma. A non-thermal ('cold') plasma occurs when the electron temperature is much hotter than those of ions and neutral particles (e.g. usually at room temperature). In both thermal and non-thermal plasma, the electron temperature can reach several thousand degrees celsius. Hence, conventional material manufacturing processes such as cutting, welding and coating applications have used the intense heat source provided by thermal plasma. The advent of nanotechnology has opened up another use of thermal plasma in the processing of metal nanopowders. This review is focussed on the synthesis of metal nanopowders using a plasma source generated from electrical arc discharge, spark discharge and radio frequency discharge processes.

4.3 Electrical arc discharge synthesis of metal nanopowders

There have been many research activities into the use of electrical arc discharge to produce metal nanopowders. An electrical arc discharge is an electrical breakdown of a gas between two conducted electrodes connected to typical dc welding power supplies, operated at currents up to 120 A. The arc discharge produces a high temperature (~10 000K) thermal plasma region between two electrodes in the presence of a dynamic flow of gas. Two types of electric arc discharge methods have been used in the past to produce metal nanopowders. They are known as transferred arc plasma and non-transferred arc plasma.

4.3.1 Transferred arc plasma in gaseous atmosphere

For the transfer arc plasma, the anode electrode material is made from metal (e.g. Al, Cu, Fe, Co, Ni etc.) and is consumed in the synthesis of metallic nanopowders. The anode material is placed inside a crucible made from either graphite or copper. A copper crucible requires water cooling to avoid excessive heating. The cathode electrode is made from tungsten and is a non-consumable electrode. A gas (e.g. Ar, N_2, He, H_2) is introduced into the process chamber (Fig. 4.1) and the plasma is maintained between electrodes in a small spacing less than 5 mm. Water cooling of the chamber wall is essential to avoid excessive heat transfer via conduction, convection and radiation from the high temperature thermal plasma (Tanaka and Watanabe, 2008).

The plasma heats up the anode material to above its boiling point. This leads to the evaporation of the anode material and subsequent cooling of metal vapour by collision with the background gas species. This reduces the diffusion rate and develops a supersaturated vapour, where the metal atoms diffuse around and collide with each other to form clusters of nuclei. Homogenous nucleation from the supersaturated vapour results in the formation of a nucleation zone and more nuclei are formed, some of which are rapidly cooled and combine to form primary particles by Brownian coagulation and an agglomeration growth mechanism (Lunden and Flagan,

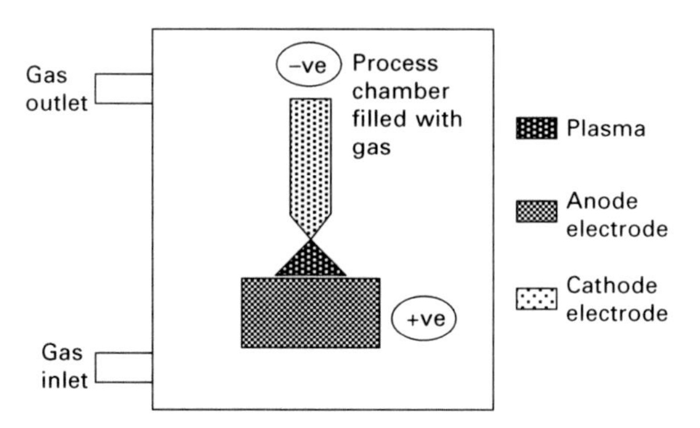

4.1 A transferred arc plasma set up for nanopowder synthesis.

1995; Weber and Friendlander, 1997), thereby developing the growth zone. Figure 4.2 illustrates the evolution of nanopowders from the electric arc discharge process.

Figure 4.3(a) shows typical spherical copper nanopowders produced using the electric arc discharge method operated at an applied current of 10 A, an electrode gap spacing of 5 mm and 10 slpm(standard l min^{-1}) of nitrogen gas at atmospheric pressure. The primary particle size is below 100 nm, as shown in the log-normal cumulative size distribution plot (Fig. 4.3(b)).

The size of nanopowders increases with increasing applied current but decreases with increasing electrode gap spacing (Forster *et al.*, 2012). The increasing current enhances the energy input of the arc to the system and raises the temperature of the anode metal target. This leads to higher vapour pressures and higher vapour concentrations, which in turn results in the growth in the primary particle size, as well as agglomerate size. However, an increase in the electrode gap spacing generates a greater length of the plasma, accompanied by a linear rise of the arc voltage and increasing energy input to the system. The expansion of the plasma causes the 'surface' of plasma to enlarge and radiation heat losses. Therefore, only part of additional energy input is available for heating the molten metal. Furthermore, a large plasma volume dilutes the metal vapour and reduces the metal vapour concentrations, which in turn causes a shrinking mean particle size and less agglomeration of the nanopowder. The size of nanopowder is also influenced by the gas flow rate and gas composition. The larger gas flows enhance quenching of vapour, reduce particle concentration and therefore slow down growth and agglomeration kinetics, leading to smaller primary particle sizes. The change of gas from Ar to N_2 increases both the temperature and voltage of the arc

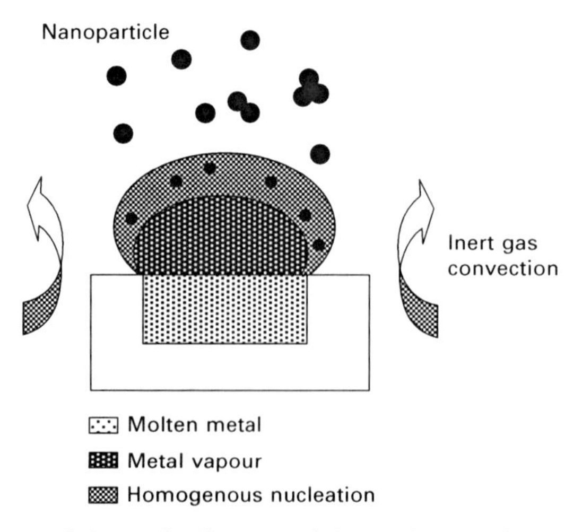

4.2 Schematic diagram of the evolution of metal nanopowders from the vapour phase.

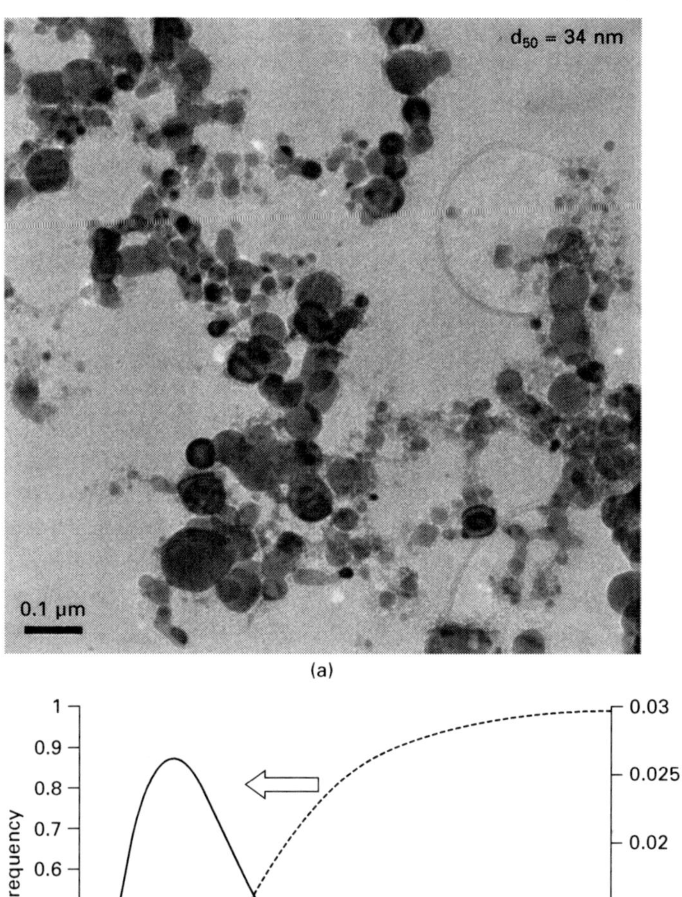

4.3 (a) TEM micrograph with (b) a log-normal cumulative and actual size distribution plots of copper nanopowders produced by an electric arc discharge process in a nitrogen atmosphere with a median particle size of 34 nm (Chang *et al.*, 2012).

plasma, increasing the input energy to the system and enhances the melting and evaporation processes (Savas and Ceyhun, 2011). This results in large particle sizes similar to the particle growth caused by increasing the arc current. Active hydrogen atoms generated in the arc plasma can also enhance the evaporation rate (Lee *et al.*, 2010). This is due to the hydrogen atoms dissolving in the molten metal and forming bubbles. The recombination of atoms releases exothermic heat and raises the temperature of melt. Furthermore, the release of bubbles as they reach the surface enhances the evaporation area, leading to a higher evaporation rate. Consequently, this results in coarse particles size owing to increasing growth and agglomeration. Transferred arc plasma in gaseous atmosphere has been used to produce nanopowders from a range of metallic systems, as summarized in Table 4.1. The nanopowders are usually collected as dry powders using filters.

4.3.2 Transferred arc plasma in liquid medium

The replacement of the gaseous atmosphere with a liquid medium has led to the creation of a submerged arc discharge method for the production of nanopowders. This is done by submerging the transferred arc plasma in a liquid medium (e.g. water, ethylene glycol or cryogenic liquid) to enhance the quenching of metal vapour owing to a higher heat transfer coefficient compared with a gaseous atmosphere. In addition, this method enables the production of a colloidal solution, containing nanopowders dispersed in a liquid medium, in a single step operation. Figure 4.4 shows various types of metal nanopowders produced by submerged arc discharge in cryogenic liquid (Chang and Ren, 2004) using currents of up to 60 A. The primary particles are spherical and have a median size below 100 nm. The primary particle size increases with increasing current owing to enhanced input energy of plasma to the system, which encourages growth and agglomeration.

Ashkarran (2011) has reviewed the production of various metallic (e.g. Ag, Au) nanopowders using submerged electrical arc discharge in water. However, water decomposition (electrolysis) can occur during the production of gold nanopowders by arc discharge in water. Consequently, gaseous hydrogen and oxygen are formed in the water as small bubbles partly dissolved in

Table 4.1 Summary of metal nanopowders produced using the arc discharge method

System	Reference
Cu	Forster (2012); Wei *et al.* (2006); Kassaee *et al.* (2010)
Al-Mn	Lee *et al.* (2010)
Ag	Chen *et al.* (2007)
Co-Cr	Ma *et al.* (2005)
FeCoNiAl	Geng *et al.* (2005)
Fe, Fe-TiN	Sakka *et al.* (2002)
Ni-Cu	Song *et al.* (2010)
Co	Meng *el al.* (2012)

the water. The gold nanopowders are negatively charged by electrons from the cathode during the arc discharge process, which in turn saturate with atomic oxygen and create hydrogen bonds with water particles in a water

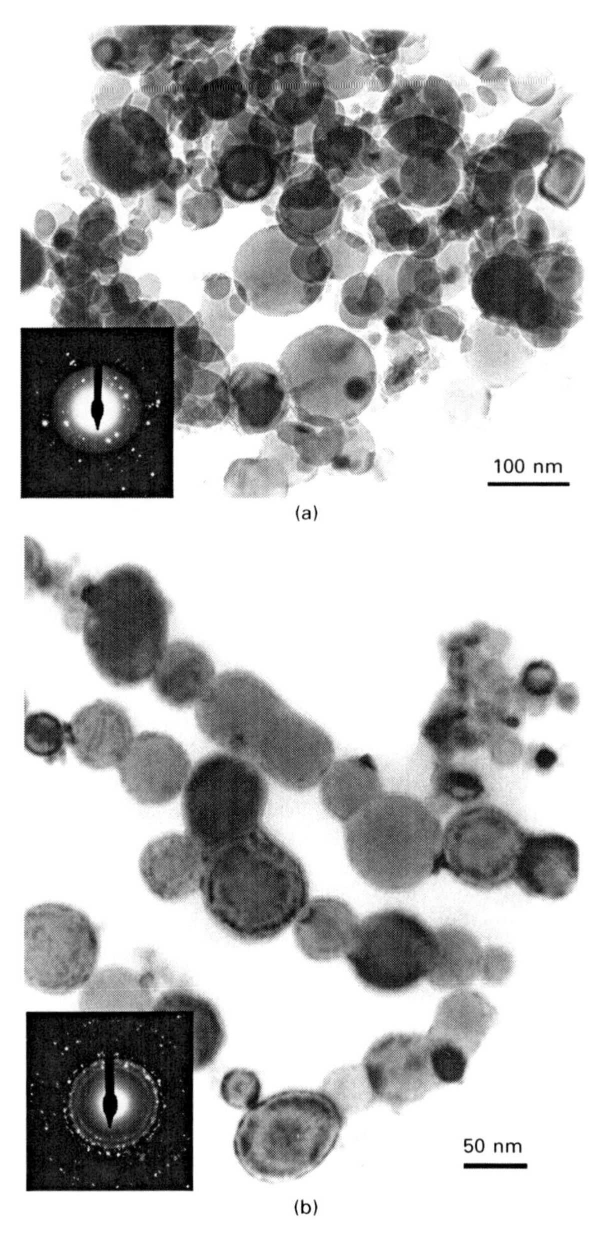

4.4 TEM micrographs of (a) Al (d_{50} = 70 nm), (b) Fe (d_{50} = 44 nm), (c) Cu (d_{50} = 37 nm) and (d) Ni (d_{50} = 44 nm) nanopowders produced by the submerged arc discharge process (Chang and Ren, 2004).

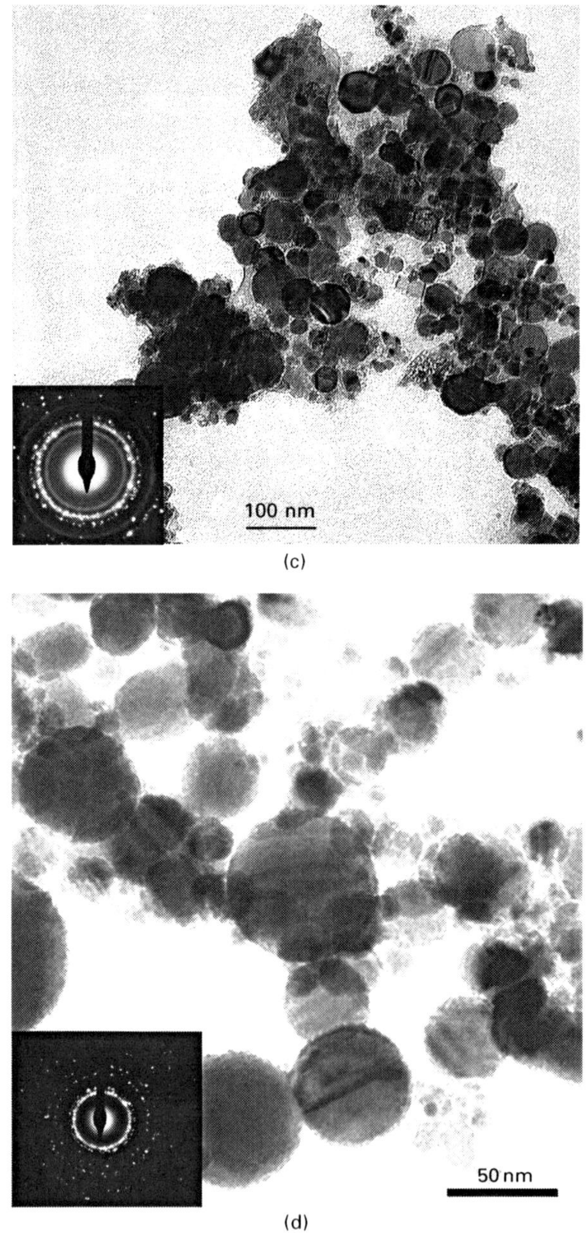

4.4 Continued

environment. As a result, negatively charged 15–35 nm gold nanopowder micelles are formed in the water medium (Lung *et al.*, 2007) owing to hydrogen bonding between the water molecule and the gold nanopowder. Hence a stable suspension is achieved without any stabilizers and surfactants because the negatively charged gold nanoparticles are surrounded by water molecules. Ethylene glycol is another liquid medium that can interact chemically with the aluminium nanopowders formed by the submerged arc process. The ethylene glycol protects the particles' surface against oxidation and self-ignition on exposure to air, as well as serving as a solvent, coating and/or stabilizing agent where the particles are capped by the ethylene glycol molecules, thereby giving monodispersed spherical Al nanopowders with a relatively small size of 26.6 nm (Kassaee and Buazar, 2009).

So far only noble metal nanopowders can be produced using submerged arc discharge in water. However, the use of inert cryogenic liquid can offer the production of a wide range of metal nanopowders.

4.3.3 Spark discharge method

Electrical arc discharge in both gaseous and liquid media yields metal nanopowders with a polydispersed size distribution. A spark discharge method has been developed to produce nanopowders with a narrow size distribution (Vons *et al.*, 2010; Tabrizi *et al.*, 2009). In this method, a high voltage power supply is used to charge a capacitor bank which is connected in parallel to a spark gap between two metal electrodes of the material required to be converted into nanopowders, as shown in Fig. 4.5. Several kilovolts are required to cause an electrical breakdown of the gas between electrodes, over a very short time (<10 ms) interval (i.e. spark discharge), owing to the rapid release of the store energy in the capacitor bank. This results in very high temperatures typically of the order of 20 000K, which is sufficient to evaporate a small amount of the metal electrodes with subsequent condensation of the vapour into metal nanopowders by the surrounding cooling gas. These powders have mean primary particle sizes below 10 nm and are carried away by the inert gas and collected by filters. Several hundreds of discharges per

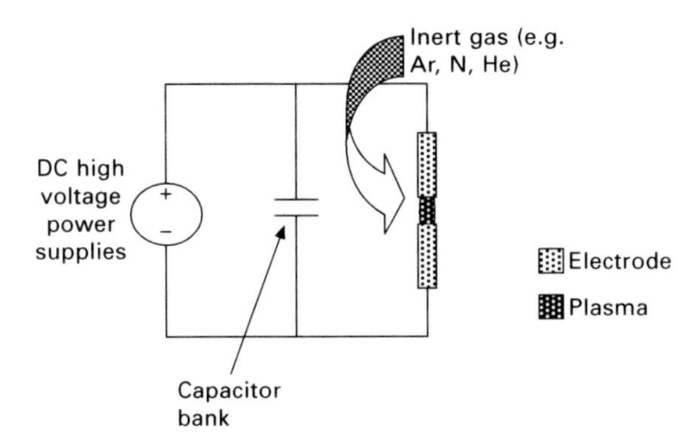

4.5 Schematic diagram of a spark discharge set up for the production of nanopowders.

second (f_s) can be generated depending on the current used to charge the capacitance bank according to:

$$f_s = \frac{I}{CV_b}$$

where I is the applied current (mA), C is the capacitance (nF) and V_b is the breakdown voltage (kV) of gas. If we assume $V_b = 4$ kV, $I = 10$ m and $C = 10$ nF, then an f_s of 250 Hz can be obtained.

The mean primary particle size remains constant as the gas flow rate increased from 5–10 slpm. This is attributed to the fact that particle formation is essentially completed in the high concentration laminar flow region between the electrodes, where dilution is dominated by diffusion and is thus flow independent. Hence the particle diameter that develops in this region is close to the primary particle diameter. However, at very low flow rates below 5 slpm, there is more time for newly formed particles to coagulate in the low-concentration turbulent dilution region outside the electrode gap, resulting in the coarsening of primary particles (Tabrizi *et al.*, 2009a). An increase of the electrode gap from 0.5–3 mm leads to a reduction in evaporation rate due to the increasing volume of the plasma. The mean diameter of primary particle increases with increasing electrode spacing owing to a higher vapour loading (Lehtinen and Zachariah, 2002). The increase in capacitance results in an increase in the spark energy and leads to higher vapour loading in the particle growth region. Consequently, the mean diameter and particle concentration increase with the capacitance. Finally, the spark frequency increases linearly with the nanopowder production rate. However the size distribution widens with increasing repetition frequency as a result of coagulation. At a higher frequency, more coagulation takes places at high concentration before dilution is completed, leading to coarsening of particles. Table 4.2 lists a range of metal nanopowders that have been produced by the spark discharge method.

The transferred arc plasma method offers an ideal research tool for the production of nanopowders from a range of metal and alloy systems using relatively low cost equipment. However, transferred arc plasma methods based on a single pair of electrodes are limited to a small production yield

Table 4.2 Summary of metal nanopowders produced using the spark discharge method

System	Reference
Pd	Vons *et al.* (2010)
Au	Tabrizi *et al.* (2009a)
Cr-Co, Au-Pd, Ag-Pd	Tabrizi *et al.* (2009b)
Ag-Cu, Pt-Au, W-Cu	Tabrizi *et al.* (2010)
Al, Cu	Bau *et al.* (2010)

at a rate of a few milligrams to a few grams per hour depending on the type of material and processing conditions. A recent report using a multiple-electrode configuration in an arc discharge process has dramatically increased the production yield to 329 g h^{-1} (Meng *et al.*, 2012) for the production of cobalt nanopowders. Hence, further development is needed to scale up this method for industrial scale production of nanopowders.

4.3.4 Non-transferred arc plasma in a gaseous atmosphere

In the non-transferred arc method, the anode converts into a nozzle and the plasma can extend beyond one of the electrodes in the form of a plasma jet by gas at a relatively high flow rate. As a result, the electrodes do not participate in the process but have the sole function of plasma generation. Figure 4.6 shows a schematic diagram of a non-transferred arc plasma torch set up for metal nanopowder production. Non-transferred arc plasma torch technology has been developed since the 1970s for deposition of ceramic and hardmetal coatings, chemical waste destruction (Selwyn *et al.*, 1999) and spheriodisation (Kumara *et al.*, 2006) of metal and ceramic powders.

The electrodes used in a non-transferred arc torch are generally water cooled. Non-transferred arc torches are available commercially with levels of power between 1 kW and 6 MW (Heberlein, 1992). The heating efficiency ranges between 50% and 90% and increases with an increase in gaseous flow, which is much higher than in the transferred arc plasma process. This leads to an increase in both equipment and maintenance costs. Gases, liquid reactants or solid feedstock are injected into the plasma to produce gaseous species by either a chemical reaction or an evaporation process. These vaporized species react with either the plasma gas or a quenching gas in the condensation region outside the plasma jet. A high cooling rate of quenching

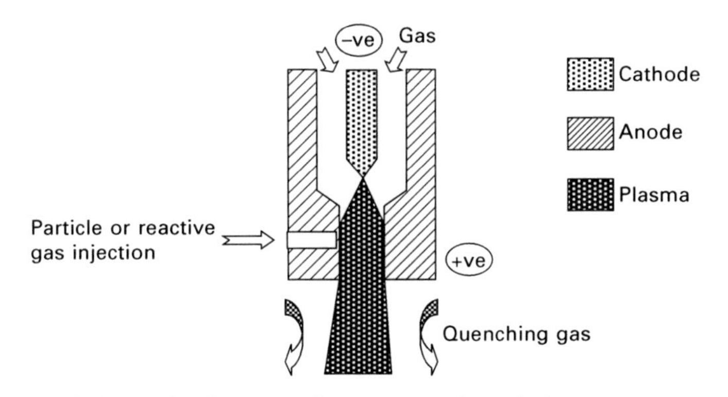

4.6 Schematic diagram of a non-transferred plasma set up.

gas or a fast expansion of the plasma jet encourages homogeneous nucleation. This is followed by coagulation and agglomerate growth processes, producing nanopowders (Fauchais *et al.*, 1997).

A 7–9 kW dc plasma torch is insufficient to cause complete evaporation of a micrometre-sized silver powdered feedstock delivered at a rate in excess of 2 g min^{-1}. This leads to the formation of sub-micrometre silver powders with sizes exceeding 100 nm (Lee *et al.*, 2007). However, the introduction of H_2 gas of up to 0.9 l min^{-1} (lpm) into 15 lpm of Ar plasma gas can reduce the primary particle size of Ag nanopowders to below 100 nm. H_2 gas enhances the heat transfer owing to high thermal conductivity and H_2 plasma gives a higher reactivity than molecules in the gas state (Matsumoto and Kobayashi, 1987; Akatauka *et al.*, 1988). Therefore, the raw silver powder feedstock can be evaporated completely when H_2 gas is mixed into Ar plasma gas. In addition to Ag nanopowders, FeAl nanopowders processed by a dc plasma torch have been reported. 30–70 nm sized FeAl nanopowders are formed after passing the micrometre-sized FeAl powders at a rate of 14 g min^{-1} into a 15.3 kW dc plasma torch using an Ar/N_2 mixture of 221:31 lpm ratio, respectively (Suresh *et al.*, 2008).

However, problems of clogging and cooling of the plasma jet occur when the solid particles are inserted into the plasma jet at the nozzle exit. A central injection is developed to improve the particle feeding into the plasma jet by using a multiple-cathodes configuration system (Marques *et al.*, 2009), as shown in Fig. 4.7. The cathodes require three power supplies and the anode is either a graphite or water-cooled copper ring. Hence, it is important for efficient insertion of powder feedstock and sufficient heating by the plasma jet to enable complete evaporation of the powder feedstock during the synthesis of nanopowders. At the moment, there is a very limited published work on the effects of the processing conditions of a dc plasma torch on the evolution of metal nanopowders. However, an industrial company such as Intrinsiq Nanomaterials Limited (Intrinsiq Nanomaterials Limited.

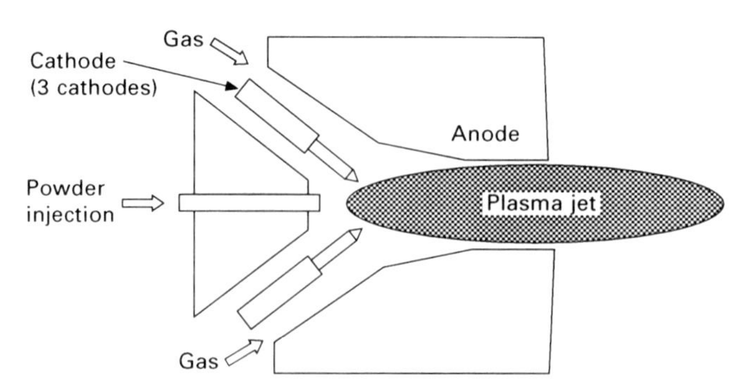

4.7 Schematic diagram of a multiple-cathode configuration plasma torch with central powder injection.

http://www.intrinsiqmaterials.com/Materials.html.) has made use of a twin dc plasma torch technology, known as 'TESIMA' to generate a wide range of metal nanopowders (e.g. Al, Ag, Cu, Ni, W, stainless steel, Co, Ti, Mo) at production rates of kilograms per hour, depending on material and processing conditions.

Metal nanopowders have been produced using radio frequency (RF) plasma torches. These torches are inductively coupled discharges, generated by suitable coils connected to RF power supplies ranging from 30–600 kW operating at working frequencies between 200 kHz and 40 MHz. Figure 4.8 shows a schematic diagram of an RF plasma torch apparatus comprising induction coils, confinement tube, gas distributor (e.g. torch head) and injection probe. The induction coils provide the electromagnetic field to couple the energy to the plasma. The confinement tube is to confine the plasma and is made from an air-cooled quartz (Power <10 kW), a water-cooled quartz (Power <30 kW) or a water-cooled ceramic (30<600 kW). The gas distributor introduces different gas streams into the discharge chamber: sheath gas to protect the walls and a central plasma-forming gas. A water-cooled central injection probe is positioned through the torch head and allows the introduction of gaseous reactants or powders into the central part of the discharge region. This enables either chemical reaction or evaporation of the feedstock to form gaseous species, which are condensed into nanopowders by a quenching gas. The plasma gas does not come into contact with electrodes, thus eliminating possible sources of contamination and allowing the operation of such plasma torches with a wide range of gases including inert, reducing, oxidizing and other corrosive atmospheres. Tekna Plasma Systems Inc. is an equipment

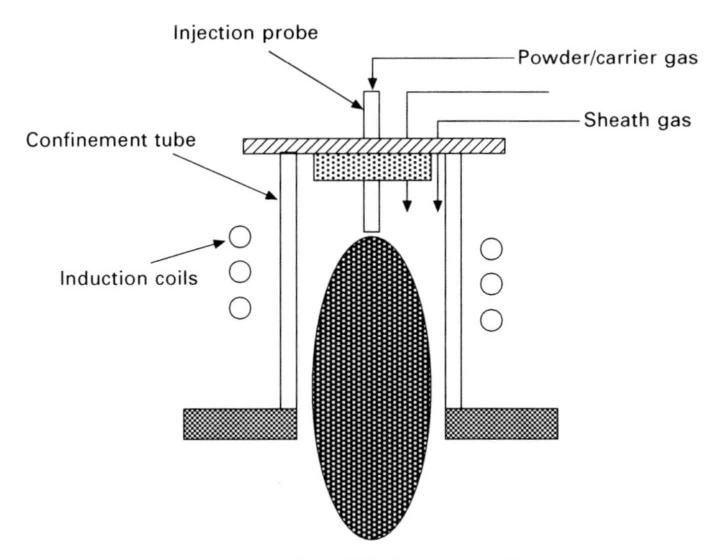

4.8 Schematic diagram of an RF plasma torch set up.

manufacturer in Canada, which supplies turn-key RF plasma torch facilities for the production of nanopowders (Boulos, 2004).

The amount of 40 μm sized copper powder feedstock evaporated is found to increase with increasing feed rates from 0.33–3.5g min^{-1} in a 40 kW RF plasma torch using Ar central gas and Ar+H_2 sheathing gas. Consequently, the average size of plasma processed copper nanopowder increases from 50–280 nm with increasing evaporation rate. This is because of a higher supersaturation of vapour, which leads to greater homogeneous nucleation and heterogeneous condensation, giving coarser nanopowders (Kobayashi *et al.*, 2008). The residence time of copper powder feedstock in the high-temperature region increases with increasing plasma reactor chamber pressures from 39–67 kPa. This leads to an increase in powder evaporation and an increasing amount of copper vapour. A high degree of supersaturation is produced and heterogeneous condensation occurs increasing the size from 139–250 nm at 39 kPa and 67 kPa, respectively. The amount of evaporated copper powder feedstock also increases with increasing H_2 flow rate. This is because of the increase in heat transfer by the introduction of more H_2 gas into the plasma.

Zhang (Zhang *et al.*, 2012) has produced tungsten nanopowders by reacting 20–150 μm sized ammonium paratungstate powder feedstock in a hydrogen plasma provided by a 30 kW RF plasma torch. The reduction of ammonium paratungstate by hydrogen is found to be dependent on the rate of powder feedstock. Only metallic alpha-tungsten nanopowders are produced at a relatively small feed rate of 19 g min^{-1}. As the feed rate increases to 26 g min^{-1} or higher, additional non-equilibrium beta-tungsten and tungsten oxide phases are found to coexist with the alpha-tungsten phase. The reductive process of ammonium paratungstate to metallic tungsten is an endothermic process and when excess ammonium paratungstate is injected into the plasma flame, it has difficulty in absorbing enough energy and reacting completely. This leads to the formation of tungsten oxide. Finally, the average size of tungsten nanopowder can be reduced from 30–20 nm by increasing chamber cooling. This improves the quenching of vapour species, inhibiting the growth and agglomeration process. A similar approach has been used to produce 60–100 nm sized nickel nanopowders by hydrogen reduction of nickel hydroxide or carbonate using hydrogen plasma provided by an RF plasma torch (Bai *et al.*, 2009).

4.4 Conclusions

The thermal plasma technologies offer a flexible and scalable platform for the production of nanopowders from a range of metal and alloy systems, without the need for post-purification or waste handling needed by other chemical synthesis routes. They can produce metal nanopowders that have primary particle sizes below 100 nm with either monodispersed or polydispersed particle size distribution. The as-produced nanopowders can be collected as dry powders or suspensions in liquid media. The technologies can also be applied to other nanomaterials including oxides, carbides, nitrides and composites.

4.5 References

Akatauka, F., Hirose, Y. and Komaki, K. (1988). 'Rapid growth of diamond films by arc discharge plasma CVD'. *Japanese Journal of Applied Physics*, **27**, L1600.

Ashkarran, A. A. (2011). 'Metal and metal oxide nanostructures prepared by electrical arc discharge method in liquid'. *Journal of Cluster Science*, **22**, 233–66.

Bai, L., Fan, J., Hu, P., Yuan, F., Yuan, J. and Tang, Q. (2009). 'RF plasma synthesis of nickel nanopowders via hydrogen reduction of nickel hydroxide/carbonate'. *Journal of Alloys and Compounds*, **481**, 563–7.

Bau, S., Witschger, O., Gensdarmes, F., Thomas, D. and Borra, J. P. (2010). 'Electrical properties of airborne nanoparticles produced by commercial spark discharge generator'. *Journal of Nanoparticle Research*, **12**, 1989–995.

Boulos, M. (2004). 'Plasma power can make better powders'. *Metal Powders Report*, May, 21.

Chang, I. T. and Ren, Z. (2004). 'Simple processing and characterisation of nanosized metal powders'. *Materials Science Engineering A*, **A375–377**, 66–71.

Chang, I. T., Faliceanu, C. L., Su, S. and Kuo, C. H. (2012). 'Fabrication of copper nanoparticles using electrical arc discharge'. *2012 NanosMat Conference*, Prague, Sept. 18–21.

Chen, J., Lu, G., Zhu, L. and Flagan, R. C. (2007). A simple and versatile mini-arc plasma source for nanocrystal synthesis. *Journal of Nanoparticle Research*, **9**, 203–13.

Chopkar, M., Das, P. K. and Manna, I. (2006). Synthesis and characterization of nanofluid for advanced heat transfer applications. *Scripta Materialia*, **55**, 549–52.

Dendy, R. O. (1990). *Plasma Dynamics*. Oxford University Press, Oxford.

El-Sayed, M. A. (2001). Some interesting properties of metals confined in time and nanometer space of different shapes. *Accounts Chemical Research*, **34**, 257–64.

Fauchais, P., Vardelle, A. and Denoirjean, A. (1997). Reactive thermal plasmas: ultrafine particle synthesis and coating deposition. *Surface and Coatings Technology*, 66–78.

Feldheim, D. L. (2002). *Metal Nanoparticles: Synthesis Characterisation and Applications*. Marcel Dekker, New York.

Forster, H., Wolfrum, C. and Peukert, W. (2012). Experimental study of metal nanoparticle synthesis by an arc evaporation/condensation process. *Journal of Nanoparticle Research*, **14**, 926–32.

Galfetti, L., De Luca, L. T., Severini, F., Meda, L., Marra, G., Marchetti, M., Regi, M. and Bellucci, S. (2006). 'Nanoparticles for solid rocket propulsion'. *Journal of Physics: Condensed Matter*, **18**, S1991–S2005.

Geng, D. Y., Park, W. Y., Kim, J. C., Yu, J. H. and Choi, C. J. (2005). Synthesis and characterization of FeCoNiAl nanocapsules by plasma arc discharge. *Journal of Materials Research*, **20**(9), 2534–2543.

Godson, L., Raja, B., Mohan Lal, D. and Wongwises, S. (2010). 'Enhancement of heat transfer using nanofluids–An overview'. *Renewable and Sustainable Energy Reviews*, **14**, 629–41.

Harra, J., Makitalo, J., Siikanen, R., Virkki, M., Genty, G., Kobayashi, T., Kauraner, M. and Mäkelä, J. M. (2012). Size-controlled aerosol synthesis of silver nanoparticles for plasmonic materials. *Journal of Nanopart Research*, (14), 870–880.
Heberlein, J. V. (1992). 'Generation of thermal and pseudo-thermal plasmas', *Pure Applied Chemistry*, **64**(6), 629.
Huber, D. L. (2005). 'Synthesis, properties, and applications of iron nanoparticles'. **1**(5), 482–501.
Kassaee, M. Z. and Buazar, F. (2009). Al nanoparticles: Impact of media and current on the arc fabrication. *Journal of Manufacturing Processes*, **11**, 31–37.
Kassaee, M. Z., Buazar, F. and Motamedi, E. (2010). Effects of current on arc fabrication of Cu nanoparticles. *Journal of Nanomaterials*, 1–5.
Kim, S., Kim, J., Lim, C., Choi, M. and Kang, S. (2007). Nanoimprinting of conductive tracks using metal nanopowders. *Applied Physics Letters*, 143117–20.
Kobayashi, N., Kawakami, Y., Kamada, K., Li, J. G. and Ye, R. (2008). Spherical submicron-size copper powders coagulated from a vapor phase in RF induction thermal plasma. *Thin Solid Films*, **516**, 4402–6.
Kumara, S., Selvarajan, V. and Padmanabhan, P. V. (2006). Spheroidization of metal and ceramic powders in thermal plasma jet: Comparison between experimental results and theoretical estimation. *Journal of Materials Processing Technology*, **176**, 87–94.
Lee, J. G., Choi, C. J. and Dong, X. L. (2010). 'Synthesis of Mn–Al alloy nanoparticles by plasma arc discharge'. *Thin Solid Films*, **519**, 81–5.
Lee, S. H., Oh, S. M. and Park, D. W. (2007). Preparation of silver nanopowder by thermal plasma. *Materials Science and Engineering C*, **C27**, 1286–90.
Lehtinen, K. E. and Zachariah, M. R. (2002). Energy accumulation in nanoparticle collision and coalescence processes. *Journal of Aerosol Science*, **33**, 357–68.
Liu, B. H., Ding, J., Zhong, Z. Y. and Dong, Z. L. (2002). Large-scale preparation of carbon-encapsulated cobalt nanoparticles by the catalytic method. *Chemical Physics Letters*, **358**, 96–102.
Lunden, R. and Flagan, M. (1995). Particle structure control in nanoparticle synthesis from the vapour phase. *Materials Science and Engineering A*, **A204**, 113–24.
Lung, J. K., Huang, J. C., Tien, D. C., Liao, C. Y., Tseng, K. H., Tsung, T. T., Kao, W. S., Tsai, T. H., Jwo C. S., Lin, H. M. and Stobinski, L. (2007). Preparation of gold nanoparticles by arc discharge in water. *Journal of Alloys and Compounds*, **434–435**, 655–8.
Ma, S., Wang, Y. B., Geng, D. Y., Li, J. and Zhang, Z. D. (2005). Structure and magnetic properties of Co–Cr solid-solution nanocapsules prepared by arc discharge. *Journal of Applied Physics*, **98**(9), 1–5.
Marques, J. L., Forster, G. and Schein, J. (2009). 'Multi-electrode plasma torches: motivation for development and current state-of-the-art'. *The Open Plasma Physics Journal*, **2**, 89–98.
Matsumoto, S. and Kobayashi, T. (1987). 'Synthesis of diamond films in a rf induction thermal plasma', *Appied Physics Letters*, **51**, 737.
Meng, H., Zhao, F. and Zhang, Z. (2012). Preparation of cobalt nanoparticles by direct current arc plasma evaporation method. *International Journal of Refractory Metals and Hard Materials*, **31**, 224–9.
Nurmi, J. T., Tratnyek, P. G., Sarathy, V., Baer, D. R., Amonette, J. E., Pecher, K., Wang, C., Linehan, J. C., Matson, D. W., Penn, R. L. and Driessen, M. D. (2005). 'Characterization and properties of metallic iron nanoparticles: spectroscopy, electrochemistry, and kinetics'. *Environmental Science Technology*, **39**, 1221–30.

Rittner, M. N. (2002). Market analysis of nanostructed materials. *American Ceramic Society Bulletin*, **81**(3).

Rittner, M. N. and Abraham, T. (1998). Nanostructured materials: An overview and commercial analysis. *International Journal of Powder Metallurgy*, **34**, 33–6.

Sakka, Y., Okuyama, H., Uchikoshi, T. and Ohno, S. (2002). Synthesis and characterization of Fe and composite Fe–TiN nanoparticles by dc arc plasma. *Journal of Alloys and Compounds*, **346**, 285–91.

Savas, A. and Ceyhun, V. (2011). Finite element analysis of GTAW arc under different shielding gases. *Computational Materials Science*, **51**, 53–71.

Selwyn, G. S., Herrmann, H. W., Park, J. and Henins, I. (1999). Materials processing using an atmospheric-pressure plasma jet. *Physics Division Progress Report 1999–2000*, 189–97.

Song, A. J., Ma, M. Z., Zhang, W. G., Zong, H. T., Liang, S. X., Hao, Q. H., Zhou, R. Z., Jing, Q. and Liu, R. P. (2010). Preparation and growth of Ni–Cu alloy nanoparticles prepared by arc plasma evaporation. *Materials Letters*, **64**, 1229–31.

Suresh, K., Selvarajan, V. and Mohai, I. (2008). Synthesis and characterization of iron aluminide nanoparticles by DC thermal plasma jet. *Vacuum*, **82**, 482–90.

Tabrizi, N. S., Ullmann, M., Vons, V. A., Lafont, U. and Schmidt-Ott, A. (2009a). Generation of nanoparticles by spark discharge. *Journal of Nanoparticle Research*, **11**, 315–32.

Tabrizi, N. S., Xu, Q., van der Pers, N. M., Lafont, U. and Schmidt-Ott, A. (2009b). Synthesis of mixed metallic nanoparticles by spark discharge. *Journal of Nanoparticle Research*, **11**, 1209–18.

Tabrizi, N. S., Xu, Q., van der Pers, N. M. and Schmidt-Ott, A. (2010). 'Generation of mixed metallic nanoparticles from immiscible metals by spark discharge'. *Journal of Nanoparticle Research*, **12**, 247–59.

Tanaka, M. and Watanabe, T. (2008). Vaporization mechanism from Sn-Ag mixture by Ar-H_2 Arc for nanoparticle preparation. *Thin Solid Films*, **516**, 6645–9.

Tartaj, P., del Puerto Morales, M., Veintemillas-Verdaguer, S., Gonzalez-Carreno, T. and Serna, C. J. (2003). The preparation of magnetic nanoparticles for applications in biomedicine. *Journal of Physics D: Applied Physics*, **36**, R182–R197.

Vons, V. A., Leegwater, H., Legerstee, W. J., Eijt, S. W. and Schmidt-Ott, A. (2010). Hydrogen storage properties of spark generated. *International Journal of Hydrogen Energy*, **35**, 5479–5489.

Weber, A. P. and Friendlander, S. K. (1997). In situ determination of the activation energy for restructuring of nanometre aerosol agglomerates. *Journal of Aerosol Science*, **28**, 179–192.

Wei, Z., Xia, T., Feng, W., Dai, J., Wang, Q. and Li, W. (2006). Preparation and particle size characterization of Cu nanoparticles prepared by anodic arc plasma. *Rare Metals*, **25**, 172–6.

Wood, A., and Scott, A. (2002). Nanomaterials: A big market potential. *Chemical Week*, October 16, 17–21.

Zhang, H., Bai, L., Hu, P., Yuan, F. and Li, J. (2012). 'Single-step pathway for the synthesis of tungsten nanosized powders by RF induction thermal plasma'. *International Journal of Refractory Metals and Hard Materials*, **31**, 33–8.

5 Warm compaction of metallic powders

A. SIMCHI and A.A. NOJOOMI, Sharif University of Technology, Iran

DOI: 10.1533/9780857098900.1.86

Abstract: Warm compaction is a cost saving and effective method for obtaining high performance powder metallurgy (PM) parts. This chapter presents the principles of warm compaction and technical aspects of the process. The green and sintered properties of warm compacted parts are discussed and compared with conventionally (cold) produced compacts. The applications of the process for ferrous and non-ferrous PM parts are presented and future trends are outlined.

Key words: densification, ferrous and non-ferrous powders, green strength, powder metallurgy, warm compaction.

5.1 Introduction

For numerous powder metallurgy (PM) applications, high sintered densities, for example over 7.3 g cm^{-3} for ferrous powders, are needed in order to achieve high mechanical strength. Such high densities are difficult to reach using standard compaction and sintering techniques. One development in PM production is warm compaction, which allows the production of higher density ferrous PM parts via a single compaction process. This process utilizes preheated tools and powders during the compaction step. The compaction temperature commonly ranges between 130 and 150°C (260 and 300°F) (Rahman, 2011a), which yields higher green density compared to cold (room temperature) compacted parts. Consequently, the green strength of PM parts is improved significantly. An improvement of 50–100% in the green strength has frequently been reported (Sation *et al.*, 2003; Ngai *et al.*, 2007), which facilitates green machining of PM parts. After sintering, the ultimate density of the product, particularly when a high sintering temperature is afforded, exceeds 94% of the pore free density (Asaka and Ishihara, 2005).

Microstructural studies showed that the pore morphology of sintered parts prepared by warm compaction is different from that of conventionally compacted powders (Chu *et al.*, 2009; Ariffin and Rahman, 2003, Hanejko, 1998). The pores are smaller, rounder and distributed more uniformly throughout the warm-compacted specimens; hence, higher strength and better dimensional tolerance are achievable (Lin and Xiong, 2012). These microstructural features are particularly important for parts under dynamic loads such as fatigue. Research has shown a remarkable improvement in the fatigue endurance limit of various ferrous alloys when warm compaction is employed (Li *et al.*, 2002a). Furthermore, the fatigue endurance limit is even slightly better than compacts prepared by a double-pressing/double-sintering (DP/DS) procedure at the same level of density. Therefore, warm compaction enables cost effective production of high performance PM parts via a two-step compaction/sintering route. Producers have become more and more convinced that warm compaction is one of the most economical and effective procedures for manufacturing high-density PM parts. Estimations show that the overall cost of production is about 25% higher than that of the conventional PM process but about 40% lower than that of forging and about 10% lower than that of DP/DS and Cu infiltration processes (Hanejko, 1998). Case studies on economic competitiveness of PM processing of gear components show that, with a comparable part density and mechanical properties compared with DP/DS, warm compaction has a consistent cost saving over the entire range of production volume (Bocchini, 1999).

In spite of the many advantages of the warm compaction process, a number of issues should be considered and appropriate remedies must be adopted. Since the powder mixture is compacted at relatively high temperatures during the compaction cycle, flowability, die filling, agglomeration and sticking of the particles to the die surface are concerns. Here, the powder lubricant is a key element in the powder mixture for providing good flowability and reducing the ejection force (Simchi, 2003; Feng *et al.*, 2011). Especial attention should also be given to die design as the pressure is exerted at higher temperatures. Thus higher strength materials with more heat resistance to heat (oxidation) should be utilized. It is noteworthy that pressing conditions such as compaction rate or using hydraulic or mechanical pressing could make a large difference in the powder response to the compacting temperature and pressure. These small conditions are of key importance in the optimization of the compaction methods so as to meet the specifications of the target parts. This chapter concisely presents both theoretical and practical aspects of warm compaction and explains future trends in this field.

The first work performed on warm compaction in the late 1980s, revealed the advantages of compaction in the vicinity of 100°C (Musella and D'angelo, 1988). Interestingly, experimental work showed that when bulk lubricated ferrous powders were heated, compressibility was improved compared to the some powders in unheated conditions. However, the first practical application of warm compaction was realized in 1994 by Hoeganaes Inc. with the introduction of Ancordense and Densemix powders (Hanejko and Rutz, 1994).

When a metal powder is heated, the compressive yield strength decreases. This softening could be as high as 30% for iron powder at 150°C. Under these circumstances, a high density close to a pore-free density (PFD) can be obtained with lower compaction pressures. On the other hand, at elevated temperatures, the lubricant can be redistributed from inter-particles to the die-part interface. Consequently, higher density is attained while reducing the ejection force by up to 35%. The amount of required lubricant can thus be decreased owing to better lubricity (Xiaoa *et al.*, 2009). Typically, warm compaction results in 0.1–0.25 g. cm^{-3} increase in the green and sintered density of PM parts while more uniformity in the density distribution, especially in complex-shaped parts, can be achieved. The higher green density is also accompanied by significant improvement in the green strength owing to enhanced particle welding during compaction as a result of greater particle deformation. Additionally, the commercial lubricants like ANCORDENSE (a warm compaction method presented by Hoeganaes Corp) provide better particle bonding (Ngai *et al.*, 2002). An example of practical application of higher green strength of warm compacted parts, which eases green machining, is safety locker parts (Engstrom and Johansson, 1995). Obviously, the green machining ability gives additional flexibility to designers in selecting materials and in part design. Another promising advantage of warm compaction is explored in magnetic parts. The enhanced density achievable by warm compaction increases the saturation induction and permeability with no change in the coercive force (Füzera *et al.*, 2009). A mixture of high strength polymers with magnetic particles can be easily shaped into final parts via warm compaction without the need for sintering. The product can be utilized in alternating current (AC) magnetic applications; since the polymer renders strength and magnetic insulation to the magnetic particles; thus, a variety of high-performance magnetic products can be manufactured (Shokrollahi and Janghorben, 2006; Zhang, 2009). Examples are automotive ignition coils and stators for high-speed electric motors (Guruswamy *et al.*, 1996).

Potential applications of warm compaction in automotive industry are numerous. Densities in excess of 7.25 g cm^{-3} using diffusion alloyed materials and molybdenum pre-alloyed steels can be achieved; hence, parts with mechanical properties comparable to steel forgings and ductile iron castings can be manufactured at reduced cost (Hanejko, 2008). Some examples include turbine hubs for high performance engines, helical gears, lock components and gearing with complex gear forms or spiral gears (Rutz and Hanejko, 1997).

To get a better understanding on how warm compaction can improve densification, it is worth studying the effect of compaction temperature on radial and axial stresses during uniaxial pressing. Since powder is a solid material (not liquid), the pressure exerted by a compacting punch exerts a radial pressure upon the die wall, the magnitude of which is less than the normal pressure. In the elastic region, the relationship between the radial (σ_r) and axial stress (σ_a) in a cylindrical sample can be expressed as (Bocchini, 1998):

$$\sigma_r = \frac{\sigma_a \nu}{1 - \nu} \quad [5.1]$$

where ν is Poisson factor. According to von Misses yielding criteria, plastic yielding occurs when:

$$\sigma_a - \sigma_r \geq \sigma_o \quad [5.2]$$

where σ_o is the yield stress of the metal cylinder. It is noteworthy that the transition from elastic to plastic behaviour occurs gradually in the individual powder particles. In any case, as the deformation temperature increases, the yield stress of the metal powder decreases gradually as a result of facilitated dislocation movement. The effect of temperature on the yield point of iron-based materials is shown in Fig. 5.1. The ease of plastic deformation of the particles causes better pore filling under applied load and thus improved densification. On the other hand, as the resistance of metal powder to plastic flow decreases, the radial pressure increases. Consequently, more homogeneous densification can be achieved.

When the axial pressure is released and the maximum shear stress in the metal cylinder falls below the yielding point, the radial stress can become greater than axial pressure:

$$\sigma_r - \sigma_a \geq \sigma_o \quad [5.3]$$

Therefore, a hysteresis curve for radial stress versus axial stress occurs. The severity of the hysteretic behaviour depends on the resistance of the metal powder to plastic deformation and thus on the compaction pressure. An

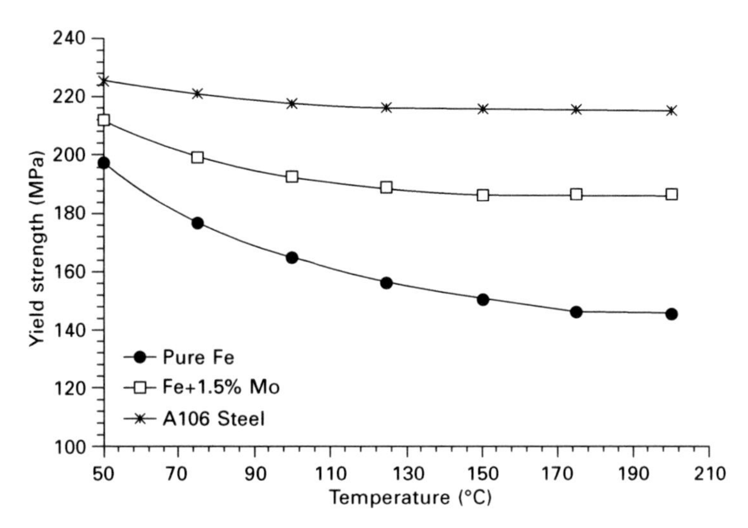

5.1 Yield point of pure iron (Engstrom *et al.*, 1995), Fe+1.5 wt% Mo (Höganäs, 2004) and A106 Steel (Deng and Murakawa, 2006).

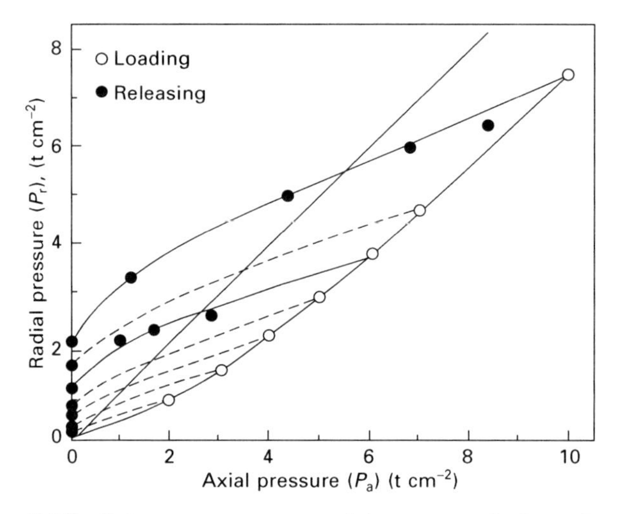

5.2 Radial pressure versus axial pressure during a loading-releasing cycle in a cylindrical die for sponge iron powder (Bockstiegel and Hewing, 1968).

example of this dependency for sponge iron powder is shown in Fig. 5.2. It is of interest to note that after completely releasing the axial pressure, the compact is under compressive stress, magnitude of which can be as high as the yield pressure of the metal cylinder. At elevated temperatures, the residual stress is lower than that of room temperature; thus, the spring back effect is less pronounced in warm compacted specimens. This yields another advantage for achieving better dimensional accuracy and tolerance.

It should be noted that in the aforementioned argument, the effect of die-wall friction (lubrication) was not considered. Generally, the amount of radial pressure and the residual stress after compaction increase at higher frictional forces. The effect of friction on the hysteresis diagram is schematically shown in Fig. 5.3. It should be noted that at elevated temperatures lubrication is usually more effective. Therefore, better particle rearrangement under the applied load, owing to direct influence of the lubricant and/or softening and flattening of asperities of the particle surfaces, provides better densification and more homogeneous density.

5.2 Warm compaction process

Similar to the traditional powder metallurgy compaction process, warm compaction utilizes traditional compaction equipment while the powder and the die assembly are heated to temperatures of about 100–150°C. At higher temperatures, lubricants begin to break down and the oxidation of iron powders occurs more rapidly; hence, the application of the warm

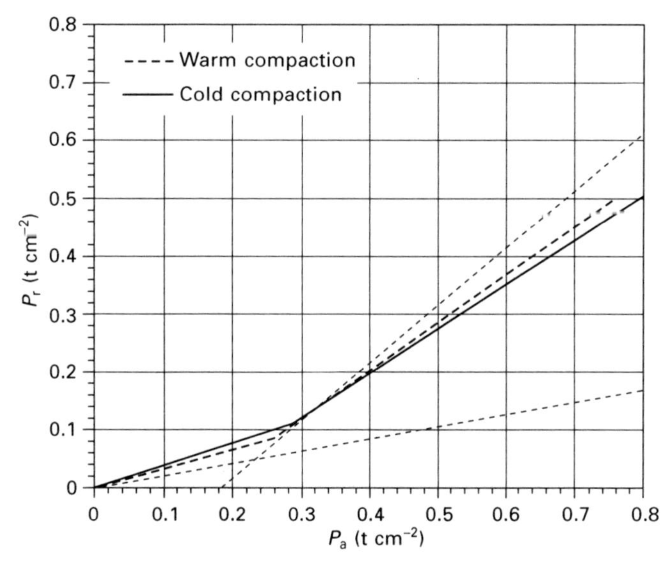

5.3 Scheme showing the effect of die wall friction on the relationship between radial and axial pressure during one compaction cycle of a metal plug inside a rigid die (Bocchini, 1998).

compaction process is technically limited to temperatures <150°C however at too low temperatures (about 100°C) a sufficient compaction effect would not be achieved. Part producers use a specific temperature for each material system that is dependent on the powder compaction technology they utilize. For instance, Höganäs AB supplies Densmix^{TM}, where the powder temperature can be between 125°C and 130°C and the tool temperature between 130°C and 150°C. These temperature ranges were determined to achieve consistent apparent density and flowability, which guarantee close dimensional tolerance and weight scatter of the compacted parts (see Fig. 5.4). Here, it is pertinent to point out that binder treatment of the powder is one of the key components in the success of the warm compaction process. Although a high melting point lubricant ensures good flowability, good compressibility and low ejection forces, much research has been conducted on finding a suitable binder/lubricant system and the lubricant content of the powder mix (Rawlings and Hanejko, 2000).

5.2.1 Tooling and techniques

Since the temperature used for warm compaction affects the quality of the compacted parts, the temperature variations around a chosen value are critical to the consistency of the product. Therefore, a set of heating and cooling systems are required to maintain the temperature in the standard range upon the warm compaction process. These systems include heating the powder and

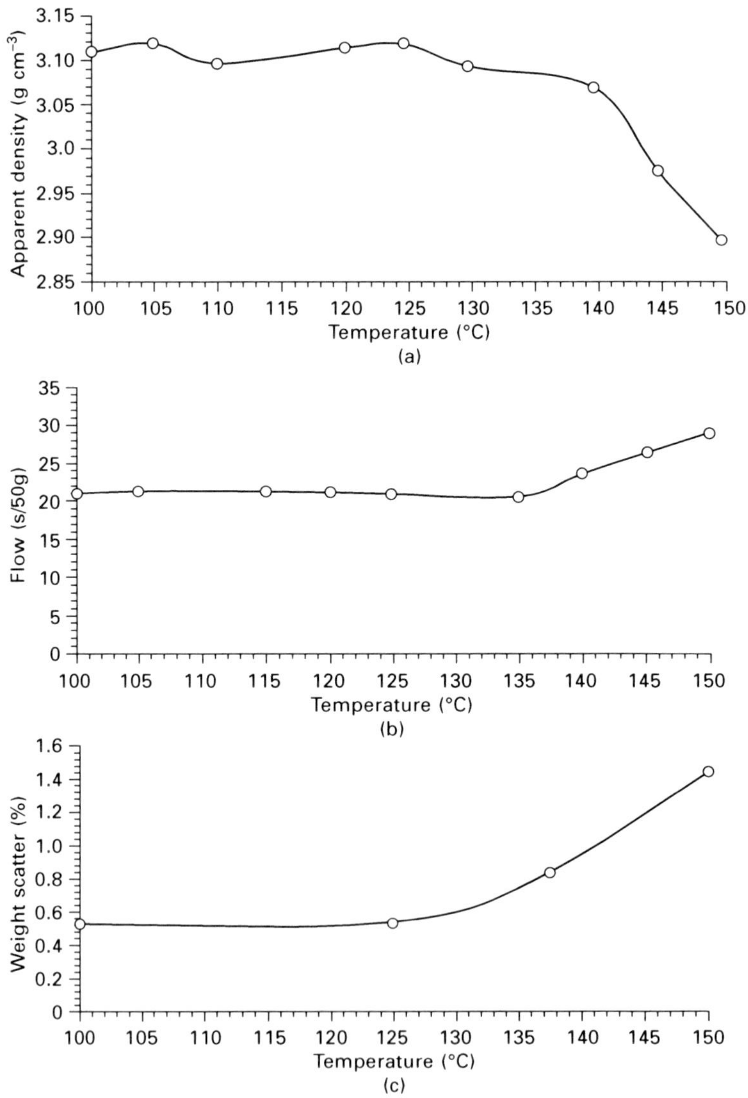

5.4 Effect of temperature on the (a) apparent density (b) flowability and (c) weight scatter of Distaloy AE+0.3%C+0.6% lubricant (Höganäs AB 2004).

the die assembly. Powder particles should be heated to the desired temperature uniformly without excess temperature variation. In practice, three different technologies have been developed and used for powder heating (Hanejko, 1998). Slot heaters utilize heat exchangers where oil is used as the heating medium, having a capacity of 25–30 kg depending on the apparent density of the powder mixture. Resistant heating of the oil is performed in an open tank and the warm liquid is pumped through the heat exchanger. The free flowing powder is then filled in a filling shoe, the temperature of which is controlled via a heating system with a proper insulation to avoid temperature variation. The filling shoe compromises a stationary feeder, which operates like a reservoir for the heated powder, and a moving filling shoe. The stationary feeder fills the moving filling shoe after each compaction cycle. The moving filling shoe moves slightly in the horizontal direction to fill the die cavity and normally contains enough powder for three to four compaction cycles. Cartridge heaters or foil elements are utilized to maintain the desired temperature in the filling shoe. Another method for powder heating is the EL-Temp system. Herein, a heated screw feeder (auger) is used to heat the powder particles to the desired temperature and simultaneously to transport them from the powder feed hopper to the heated shuttle. Obviously, the amount of shear stress exposed to the powder particles (when fed to the filling shoe) is much higher than the slot heater. The auger is resistively heated within a shell while the preheated air, flows through the hollow auger, provides an additional heating path. In the Abbot's heating method, a more sophisticated heating system is utilized. The powder bed is heated by a moving fluid (air), creating a fluidized state. Low-pressure fluidizing air (35 KPa) is heated by resistively heated elements and blown through the powder bed in a sealed reactor. Since the system has no moving parts, maintenance is minimal and the system is portable and can be adapted to any press. The described filling shoe system for the slot heater can be utilized for both the El-Temp system and Abbot's heating method.

In addition to powder heating, the tooling set in the warm compaction process requires adequate heating and cooling systems. The die, adaptor table and upper punch are heated to maintain a steady state (thermal equilibrium) condition. Heat output, locations and the number of the heating devices depend on the size, geometry, die environment and the material, which is used on warm compaction process (Höganäs AB, 2004). Nevertheless, it is imperative to have uniform and symmetrical heating in the die to avoid temperature variations and to maintain the tolerances of the tooling. In fact, a temperature difference in the die assembly will change their dimensions and create tensions and, in the end, perhaps distort the upper punch plate. Considering more stress allowance for the die deflection and using thicker stress rings are also of practical importance. However, the tooling design for warm compaction is essentially the same as regular compaction with typical radial tooling clearance of 0.01–0.02 mm (0.004–0.008 in).

5.2.2 Lubrication

PM lubricants are an indispensable part of powder metallurgical processing. In fact, lubricants have a small but arguably the most significant role in the compaction and ejection cycles. Friction between the die wall and compacts hinders pressure transmission and products density gradients within compacts (Rawlings, 2000). Therefore, the use of a lubricant can decrease density variation by promoting more homogenous pressure transmission. In addition, it influences the die-filling density and flowability of powders, and the consolidation and ejection characteristics during the compaction process as well as the green density and the green strength of PM compacts (Feng *et al.*, 2012; Ngai *et al.*, 2002).

The friction coefficient which is a measure of the frictional interaction between powder particles and the die wall decreases as the compaction temperature increases. This is due to the effect of temperature on the efficiency of lubricant as well as on decreasing resistance of powders to plastic deformation which contributes to the internal friction among particles sliding against each other (Li *et al.*, 2005; Wikman *et al.*, 2000). Different lubricants produce different lubrication effects and their optimal application temperature is different. Most lubricants that are suitable for cold compaction cannot be used for warm compaction, as this would cause increased die wear and produce parts with a low-quality surface finish (Simchi, 2003; Bergkvist, 2002). Conventional lubricants used in cold compaction include metallic stearates, synthetic amide and polyethylene waxes. A lubricant that is resistant at the warm compaction temperature, able to assure a good lubrication, without a decrease in the powder flowing capacity and with a heating system for both powder and tooling is necessary. Several lubricants consisting of a group of metal stearates, paraffins, waxes, natural and synthetic fat derivates, polyamides, and high-density polyethylene have been developed and patented (Thomas, 2000).

The simplest method of lubrication is mixing the lubricant during the powder preparation stage. This method reduces interparticle stresses and extends the working life of costly tooling. However, the presence of lubricant may reduce the green density, in that the lubricant can fill the voids between the particles and prevent pore filling by a plastic deformation mechanism (Rahman, 2011b; Simchi, 2003). Moreover, fewer metal–metal contacts are formed during compaction in the presence of admixed lubricants, which affects the green strength of PM parts. Furthermore, removing of the lubricant in high-density green parts is more challenging; since the gas pressure of evaporated lubricants during sintering may create voids while residual ash may influence densification upon sintering. Therefore, in order to prevent erratic flow, apparent density variability, lower compatibility and burn-off issues associated with the admixed lubrication, the PM industry always tries to reduce the amount of lubricant while maintaining its advantages (Simchi, 2003).

The idea of eliminating the use of admixed lubricants has been a dream of PM practitioners for many years. Die wall lubrication offers the possibility of achieving the required density in single pressing and for parts where a good green strength is required rather than a high density (Simchi, 2003). Although external lubrication of the die walls should significantly increase the green density of PM compacts and reduce the environmental and the other disadvantages of internal lubricants, the difficulty of getting the lubricant onto the die walls remains. Regarding this, pumping oil through channels drilled in the lower punch, forcing lubricant through the pores of PM high-speed steel dies with the densities of 88%, and a spray method based on tribo-static charging of dry lubricant particles in a gun have been developed (Capus 1995, 1998). It has been reported that affording die wall lubrication without using an admixed lubricant or in combination with a lower amount of the admixed lubricant results in improved green density and strength (Li 2002b). For instance, Fig. 5.5 shows the effect of die wall lubrication on the properties of some iron-based PM parts with 0–0.6% lithium stearate at different temperatures and compacting pressures. It can be seen that reduction of admixed lubricant increases the green density and mechanical properties of the sintered parts. It is also clear that die wall lubrication is more effective in increasing green density at higher compaction pressures.

5.2.3 Dewaxing and sintering

During the first stages of the sintering cycle, the compaction lubricant is eliminated, a process which is referred to as dewaxing or delubrication. Warm compacted parts typically contain about 0.6 wt% lubricant, which means a film of the organic material covers the particle surfaces. The organic contaminants and oxides covering the particle surfaces have to be removed so as to form the interparticle necks, responsible for the increased properties of the finished 'sintered' part. The decomposition behaviour of simple lubricants such as metal stearates and ethylene-bis-stearamide (EBS) has been clarified in a number of studies (Takashi, 2011). However, the increased density of the warm compacted parts and the complexity of the chemistry of the high-temperature lubricants have become an issue in the powder metallurgy community. The higher the green density, the lower the number of available dewaxing paths and the higher the gas pressure built up in the powder compact during dewaxing (Gimenez, 2006).

Figure 5.6 shows thermogravimetry analysis (TGA) of Distaloy AE powder in a 96% N_2 and 4% H_2 atmosphere. It seems that the delubrication process takes place in the temperature range 300–550°C with largest weight loss at around 400°C. The dew point and the green density have not been observed to have a significant influence in this temperature range. However, similar to conventional (cold) compaction process, blistering, stains and granular soot during dewaxing should be avoided by adopting suitable processing conditions. A higher temperature gradient (at least 50°C. min^{-1}), a rich hydrogen atmosphere and a counter-current gas-stream are suggested to minimize dewaxing problems.

In order to acquire the strength needed for a particular application, the delubricated green parts are heated to elevated temperatures (below the melting point of the main constituents) to weld the powder particles together

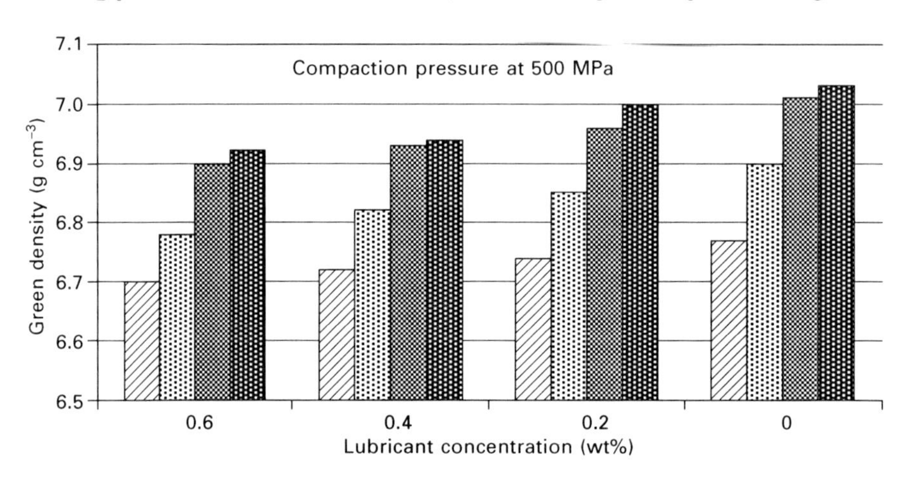

(a)

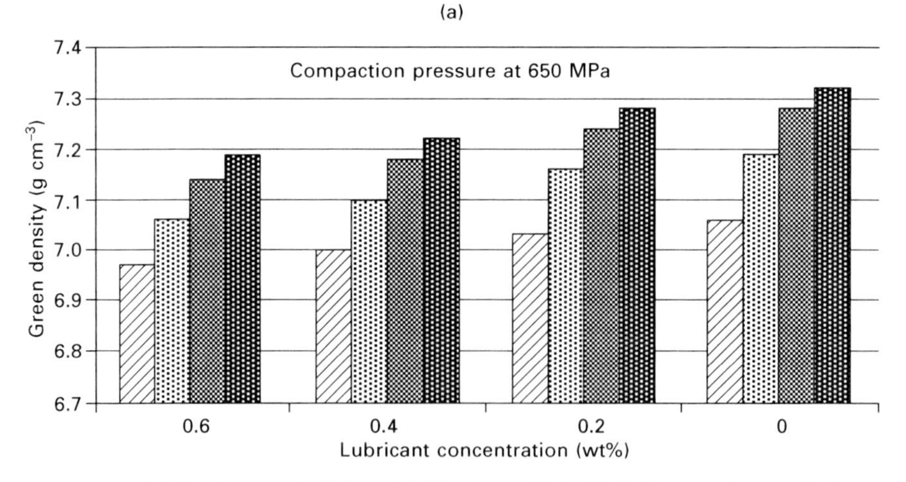

(b)

5.5 Effect of die wall lubrication and the amount of admixed lubricant (lithium stearates) on the properties of some iron-based powders: (a) effect of lubricant content on green density of Astaloy CrM compacted at 500 MPa; (b) effect of lubricant content on green density of Astaloy CrM compacted at 650 MPa; (c) Variation of impact strength of samples with lubricant content dependent on the compaction pressure (Babakhani *et al.*, 2006).

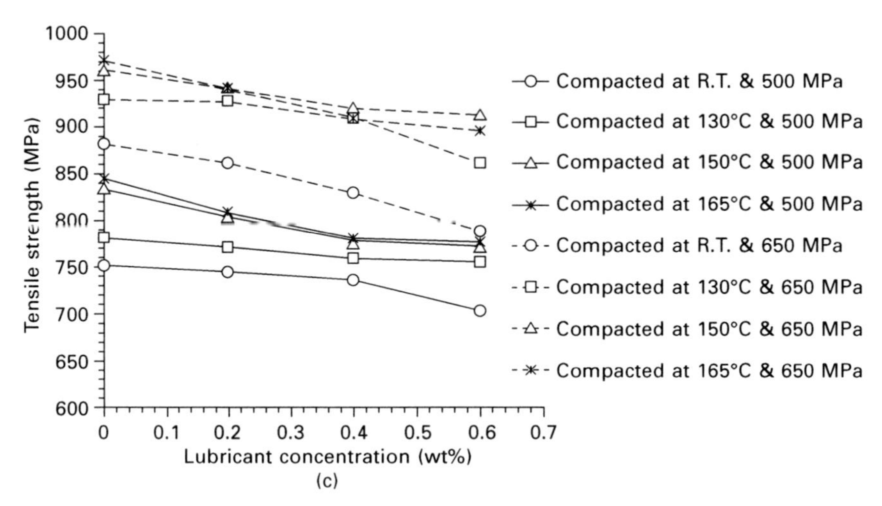

5.5 Continued

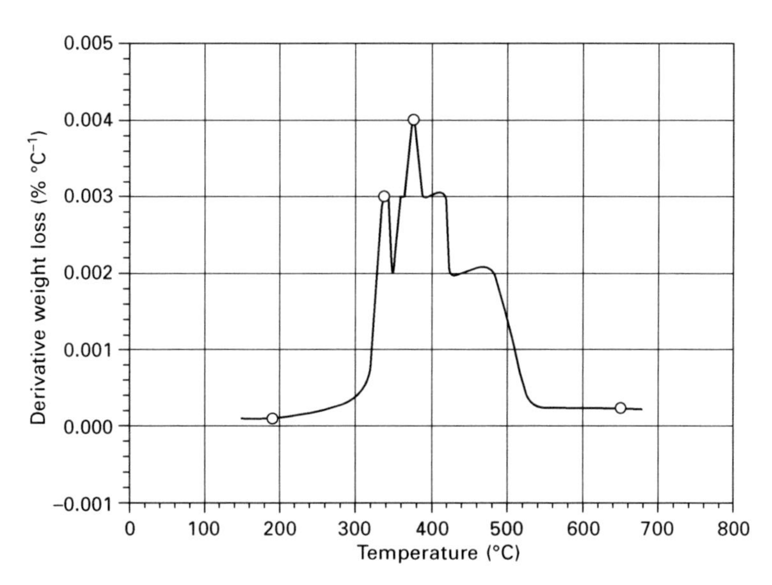

5.6 Thermogravimetric analysis (TGA) curve for Distaloy AE (DensmixTM) at a heating rate of 50°C min^{-1} in a 96% N_2/4% H_2 atmosphere (Höganäs AB, 2004).

(sinter). Warm compaction typically results in a 0.1–0.25 g cm^{-3} increase in the sintered density. The magnitude of the increase in sintered density depends on the material system and the subsequent part processing. For instance, copper-containing premixes exhibit swelling during sintering, eliminating the benefits of warm compaction (Luk *et al.*, 1994). Hence, these premixes are not considered ideal candidates for warm compaction. As will be presented in the next section, the enhanced density of warm compacted parts provides a cost-effective improvement in the mechanical strength making it attractive for many PM applications.

5.3 Properties of warm compacted parts

5.3.1 Green properties

Three important advantages of the warm compaction process on the green properties of PM parts include density, strength and dimensional tolerance. Figure 5.7(a) shows the compressibility curves of some iron-based alloys at

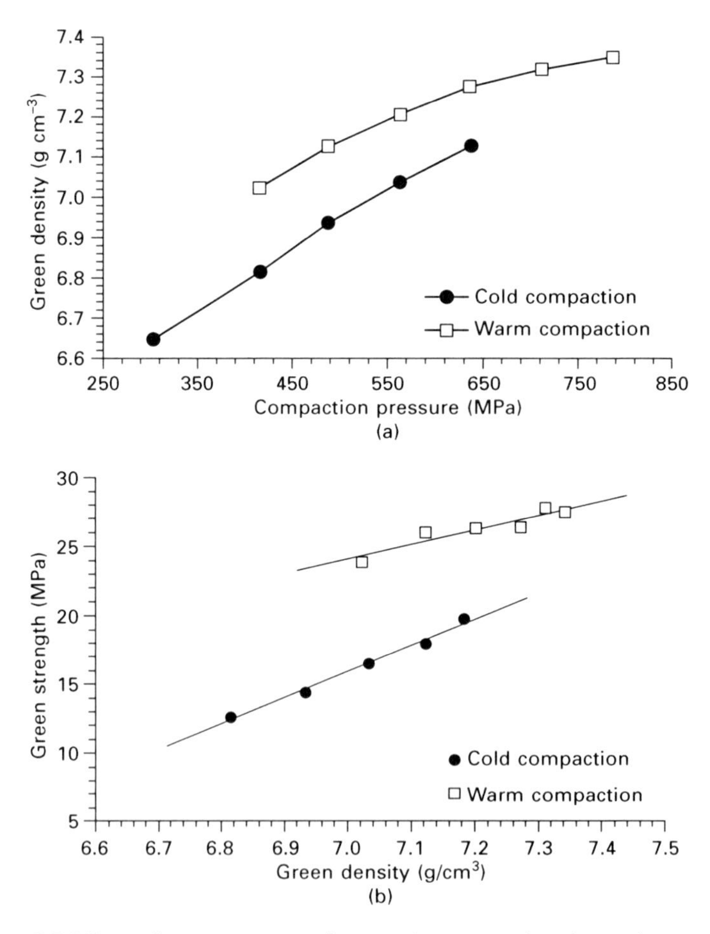

5.7 Effect of warm compaction on the green density and strength of Distaloy 4800A+0.6 w/o graphite+0.6 w/o lubricant: (a) Compressibility curve and (b) green strength as a function of density (Rutz, 1994).

room and elevated temperatures. As shown, depending on the material system, a 0.10 to 0.25 g cm^{-3} increase in the green density is achieved. Owing to better pressure transmission at lower compacting pressures, the beneficial effect of warm compaction is greater than the improvement observed at higher compaction pressures. Interestingly, regardless of the green density, the green strength of the compacts is higher than cold compacted parts (Fig. 5.7(b)). This observation reveals that the origin of higher strength is not only a higher density but also a more uniform density distribution, a finer pore structure and better metal–metal contacts (greater powder deformation and particle welding during compaction). A consequence of the enhanced green strength is the ability to machine the green compact. This feature helps the PM industry to reduce the cost and offers the part designer greater flexibility in part design and material selection. Spring back, which is a measure of the radial green expansion that takes place when a component is ejected from a die, is lower for warm compacted parts than conventional compacted powders. For instance, the radial expansion of Distaloy AE powder compacted at 600 MPa/150°C is about 10% lower than the cold compacted part. It should be noted that the amount of spring back depends on the tooling materials and compaction pressure. The beneficial effect of warm compaction on radial expansion is more pronounced for carbide dies compared to high-speed steels. Although the green expansion increases at higher densities (compaction pressures), the spring back is still lower than when compacting at room temperature.

5.3.2 Sintered properties

Table 5.1 summarizes the mechanical properties of some iron-based powders prepared by conventional (cold) and warm compaction. A 10–15% increase in the tensile strength has frequently been reported for different materials, which corresponds well to the density increase of the compacts. The fatigue endurance limit was also improved by warm compaction; however, no general correlation exists between tensile strength and the fatigue endurance limit (Bergkvist, 2002; Abdoos *et al.*, 2009). Although the same level density can be obtained by the DP/DS procedure, experimental results showed that the endurance limit of warm compacted powders is slightly higher than that of DP/DS parts. This observation reveals that density (the total porosity) is not the only factor determining the dynamic and cyclic properties of PM parts. Microstructural studies revealed the effect of processing conditions on the pore structure of the compacts (Ngai *et al.*, 2007; Rutz *et al.*, 1994).

As Fig. 5.8 shows, smaller pores with more uniform distribution are obtained for warm compaction of Distaloy AE+0.3%C premix at approximately the same density level. Therefore, pore size determines the dynamic properties in absence of other (larger) defects.

Table 5.1 As-sintered tensile properties and fatigue endurance limit of warm compacted iron-based alloys (Babakhani *et al.*, 2006; Hanejko, 1998)

Material	Density (g cm^{-3})	99% fatigue endurance limit (MPa)	Tensile strength (MPa)
Astaloy CrM + 0.2 wt% lithium stearate (compacted at R.T.)	7.10	208	754
Astaloy CrM + 0.2 wt% lithium stearate (compacted at 130°C)	7.21	260	792
Astaloy CrM + 0.2 wt% lithium stearate (compacted at 165°C)	7.32	300	845
Fe – 0.45 wt% P	7.23	185	365
FC-0208[(a)]	7.07	175	576
FD-4805[(a)]	7.19	181	710
A150HP, 2%Ni and 0.6 Gr	7.18	189	641
FLN2-4405[(a)]	7.21	222	632
A41AB	7.16	353	1211
FL-4405[(a)]	7.17	283	1131
FD-0205[(a)]	7.19	315	1192
FN0250[(a)]	7.23	276	1193

[a]The Metal Powder Industries Federation (MPIF) designations, based on MPIF Standard 35, 1997 edition.

5.4 Materials and applications

5.4.1 Ferrous powders

Warm compaction is a single-press and single-sinter process that is ideal for complex multilevel PM parts requiring high mechanical properties. The mechanical properties of low-alloyed PM steels, for example FLN-4505 and FD-0405 at a density of 7.35 g cm^{-3}, produced by warm compaction are almost equivalent to those of wrought alloys and better than cast irons (Feng *et al.*, 2011; Hanejko, 1998). Therefore, components potentially made from these alloys are suitable candidates for the warm compaction if the processing costs were lower. Meanwhile, owing to lower elongation of the PM materials compared to the wrought alloys, proper application of the warm compaction process must consider the reduced elongation and impact energy of the PM part.

Case studies of the economic competitiveness of PM processing of gear components show that, with comparable part density and mechanical properties to DP/DS, warm compaction provides consistent cost savings over the entire range of the production volume. Turbine hubs in automatic transmissions are the most and earliest parts that have been produced by PM processes like Cu-infiltration or forging (which needs considerable machining). However now they can be produced by warm compaction with sintered density of 7.25 g cm^{-3} at a lower cost and with fewer production steps. Warm compaction

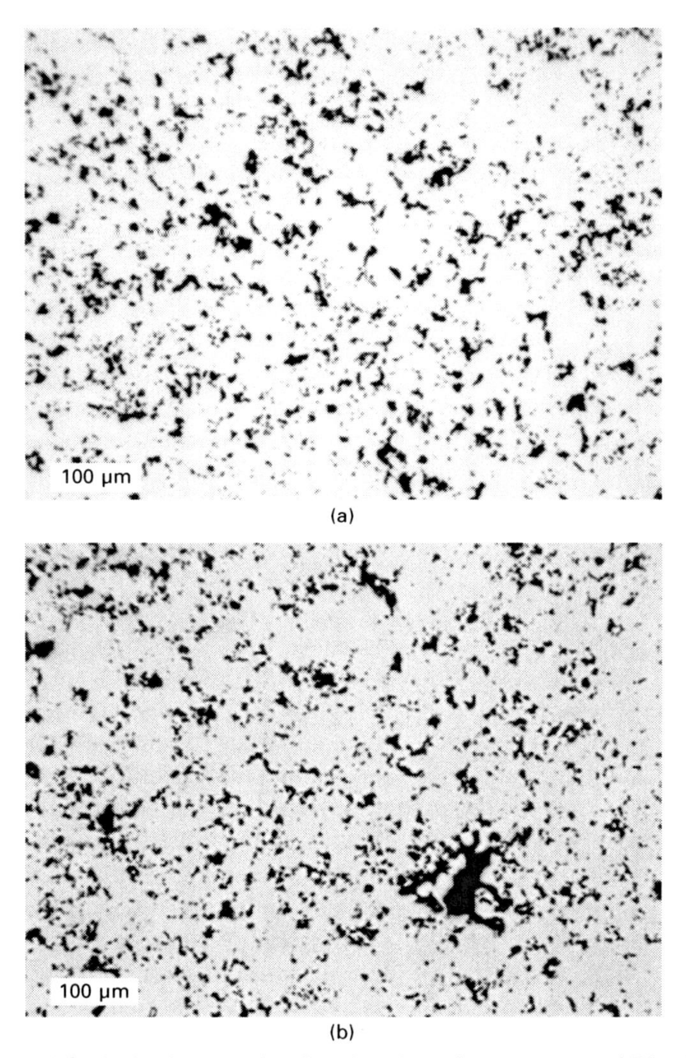

5.8 Optical micrographs showing the microstructure of Distaloy AE+0.3%C powder at 800 MPa with approximate density level of 7.3 g cm^{-3} after sintering at 1120°C for 20 min in a 90%N_2/10% H_2 atmosphere: (a) warm compacted at 600 MPa and 130/150°C powder/tool temperature; (b) cold compacted at 800 MPa (Höganäs AB, 2004).

processing of a high torque automatic transmission turbine hub successfully replaced a forged, machined, heat-treated AISI 1045 steel part (Rutz and Hanejco, 1997).

Recently, warm compaction of 316L stainless steel has been found of interest for obtaining improved tensile strength by 17% compared to conventional pressing (Xia and Shi-ju, 2008). Besides the potential application of the warm compaction process in the fabrication of structural parts, it is of particular importance to explore magnetic materials in order to boost the application of the PM process.

Studies on iron–phosphorous alloys revealed that warm compaction can produce compacts with a density of 7.4 g cm^{-3} and that their magnetic and mechanical properties are the same level as those obtained by AISI 1008 steel forging. Therefore, the PM process is an alternative and suitable replacement for the wrought steel. Warm compaction can also be utilized to introduce a new class of PM materials for alternating current (AC) magnetic applications (Shokrollahi and Janghorban, 2006). Using iron powder and a high-strength polymer as a binder (0.6–0.75 wt%), complex-shaped parts with a density of about 7.2 g cm^{-3}can be produced by warm compaction without the need for sintering (Zhang *et al.*, 2009; St-Laurent and Chagnon, 1999). Manufacturing flexibility and a variety of material options with unique magnetic performance have attracted attention for producing parts for automotive ignition coils and stators for high-speed electric motors.

5.4.2 Non-ferrous powders

In spite of remarkable research work performed on the warm compaction of iron-based powders, the application of this process to non-ferrous alloys has been very limited. Most of the structural non-ferrous PM materials, particularly lightweight aluminium alloys, possess high compressibility when they are utilized as elemental powder mixtures. Therefore, the green density of the powder compacts is not a major concern as high density can be attained via liquid phase sintering. However, with the ongoing trend towards weight saving in automobiles, the application of aluminium alloys is increasing. Some examples of application where the replacement of current ferrous materials is foreseen include in the connecting rod, camshaft sprocket, piston and oil pump gear (Simchi, 2006). However, to meet the high strength required for these applications, a considerable amount of research work has been directed towards employing the advantages of powder metallurgy in establishing the final desired alloy composition in the melt followed by atomization and rapid solidification of the powder. This results in significant improvement on the properties of aluminium alloys owing to microstructural refinements as well as the extended solid solubility of the alloying elements.

Unfortunately, the high strength of pre-alloyed powders also makes them extremely incompressible. Therefore, vacuum degassing of the powder and special consolidation techniques, such as hot pressing or forging, must be used. Clearly, these techniques are not cost-effective for high volume production. The possibility of using warm compaction for non-ferrous powders has also been studied (Simchi, 2006). The benefits of the warm compaction of pre-alloyed and rapid solidified aluminum powders, particularly Al–Si alloys, with regard to in-die compaction behaviour and ejection characteristics have been demonstrated.

Figure 5.9(a) shows the effect of compaction temperature on the green density of some aluminium alloys. The beneficial effect of temperature is more pronounced on the highly alloyed and rapidly solidified powders. The decrease in density at high temperature is attributed to cracking of the compact

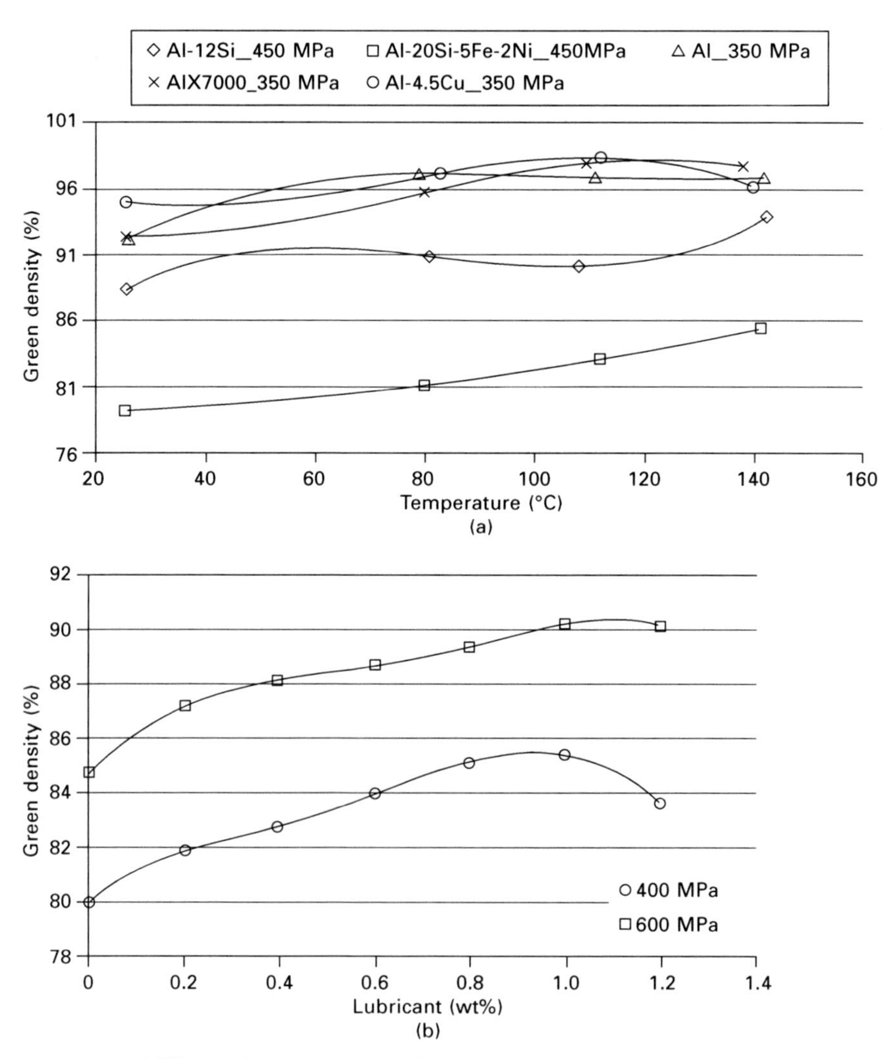

5.9 Effect of warm compaction on the properties of some aluminium alloys: (a) green density with respect to compaction temperature; (b) effect of the admixed lubricant (Microwax C) with respect to green density; (c) sliding coefficient at different compaction temperatures; (d) radial strain as a function of temperature. (Simchi, 2006).

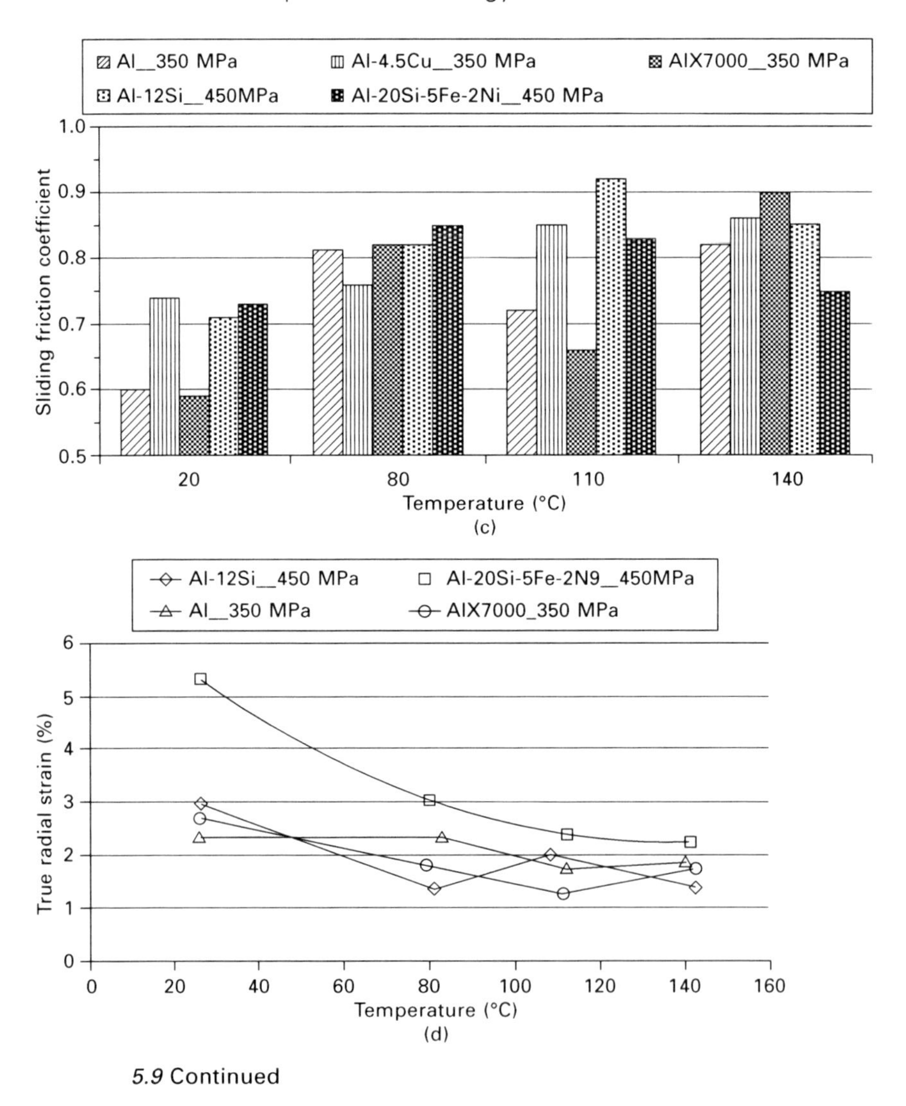

5.9 Continued

pressed at high pressure/temperature. This increase in density was found to be linked to the amount of lubricant (Fig. 5.9(b)). At higher temperatures, lubrication is more efficient, resulting in a higher sliding coefficient (Fig. 5.9(c)). Another advantage is lower radial strain after ejection (Fig. 5.9(d)), which improves dimensional tolerance. Eksi *et al.* (2006) studied the mechanical properties of warm compacted Al123 premix and reported higher tensile and fatigue endurance limit of warm compacted parts. MacAskill *et al.* (2007) investigated the mechanical properties of a hot swaged Alumix 431D and reported gains in densification and tensile properties when Alumix was compacted at the elevated temperature. Therefore, the warm compaction process might be a promising method for manufacturing aluminium alloy parts.

5.5 Future trends and concluding remarks

Powder metallurgy has the distinctive benefit over other processing techniques of enabling near-net shape formation. However, the lower density of the compacted parts relative to corresponding wrought alloys yields low material strength. Warm compaction is a single-press single-sintering route that enables higher densification compared to conventional (cold) pressing with a trade-off of approximately 25% higher cost. Future trends in warm compaction will explore the application of die wall lubrication using a spray method to decrease the amount of the admixed lubricant in order to achieve higher density and strength. The ability to achieve higher green densities at lower compaction pressures and thus minimize the tooling stresses is one of the major trends. On the other hand, the higher green strength of the warm compacted parts makes green machining feasible which is very attractive for cost-effective manufacturing of complex-shaped parts with mechanical properties equivalent to wrought alloys. This benefit is particularly important for incorporating sinter-hardening (one step sintering and hardening) to the production cycle since the machining after sinter-hardening is technically challenging. Of particular importance is the increased density uniformity achieved in the warm compacted parts which enhances their load-carrying capacity under dynamic and fatigue loads with reduced dimensional variations.

5.6 References

Abdoos H., Khorsand H. and Shahani A.R. (2009). 'Fatigue behavior of diffusion bonded powder metallurgy steel with heterogeneous microstructure'. *Materials and Design*, **30**, 1026–31.

Ariffin A.K. and Rahman M.M. (2003). 'Warm metal powder compaction process'. *Advance Material Processes*, **1**, 159–95.

Asaka K. and Ishihara C. (2005). *Technical Trends in Soft Magnetic Parts and Materials*. Hitachi Powdered Metals, Technical Report, 3–9, Hitachi, Japan.

Babakhani A., Haerian A. and Ghambari M. (2006). 'On the combined effect of lubrication and compaction temperature on properties of iron-based P/M parts'. *Materials Science and Engineering A*, **437**, 360–5.

Bergkvist A. (2002). *Warm Compaction of Steel Powders*. Sweden Patent 6,365,095 B1.

Bocchini G.F. (1999). 'Warm compaction of metal powders: why it works, why it requires a sophisticated engineering approach'. *Powder Metallurgy*, **42**, 171–80.

Bockstiegel G., Hewing J. 'Deformation work hardening and side pressure during pressing of metal powders'. *European Symposium on Powder Metallurgy*, 1968.

Capus J.M. 'Replacing internal with external lubricants'. *Metallurgy Powder Report*, **50**, 22–3.

Capus J.M. 'Die wall lubrication aids higher density'. *Metallurgy Powder Report*, **53**, 28.

Chagnon F. and St-Laurent S. (1996). 'Key parameters for warm compaction of high density materials'. *World Congress on Powder Metallurgy & Particulate Materials*. Washington, DC, Parts, Hot Compaction/Warm Pressing.

Chu G., Liu W. and Lang T.-Z. (2009). 'Properties of nano-crystalline copper prepared by vacum warm compaction Method'. *Transactions of Nonferrous Metals Society of China*, **19**, 394–398.

Deng D. and Murakawa H. (2006). 'Prediction of welding residual stress in multipass butt-welded modified 9Cr–1Mo steel pipe considering phase transformation effects'. *Computational Materials Science*, **37**, 209–19.

Eksi A.K., Lipp K., Sonsino C.M.,Veltl G. and Petzoldt F. (2006). 'Static and fatigue properties of the cold and warm compacted sintered aluminium alloy alumix 431'. *Material Science and Engineering Technology*, **37**(5), 374–82.

Engstrom U. and Johansson B. (1995). 'Properties and tolerances of warm compacted PM material'. *European Conference on Advanced PM Materials* (*Euro '95*) Birmingham, UK, October, 23–25.

Engstrom U., Johansson B., Rutz H., Hanejko F. and Luk S. (1995). 'High density materials for future applications'. *Advances in Powder Metallurgy and Particulate Materials*, **3**(11) 106–126.

Feng S.S., Geng H.R. and Guo Zh.Q. (2012). 'Effect of lubricants on warm compaction process of Cu-based composite'. *Composites: Part B, Engineering*, **43**(3) 933–9.

Füzera J., Kollára P., Olekšákováa D. and Rothb S. (2009). 'AC magnetic properties of the bulk Fe–Ni and Fe–Ni–Mo soft magnetic alloys prepared by warm compaction'. *Journal of Alloys and Compounds*, **483**, 557–9.

Gimenez S., Vagnon A., Bouvard D. and Van der Biest O. (2006). 'Influence of the green density on the dewaxing behaviour of uniaxially pressed powder compacts'. *Materials Science and Engineering A*, **430**, 277–84.

Guruswamy S., McCarter M.K., Shield J.E. and Panchanathan V. (1996). 'Explosive compaction of Magnequench Nd–Fe–B magnetic powders'. *Journal of Applied Physics*, **79**(8), 4850–3.

Hanejko F.G. (Höganäs Corporation) (1998). 'Warm compaction'. In *ASM Handbook, Volume 7: Powder Metal Technologies and Applications*, Y. Trudel, R. Iacocca, R.M. German, B.L. Ferguson, W.B. Eisen, K. Moyer, D. Madan, H. Sanderow and P.W. Lee (eds)., ASM International, 376–381.

Hanejko, F.G. (2008). 'Cost efficiency allied to high performance parts the aim for single-run processing'. *Metal Powder Report*, **63**(3), 12–16.

Hanejko F. and Rutz H. (1994). 'High density processing of high performance ferrous material'. *Advances in Powder Metallurgy and Particulate Materials*, **5**, 117–33.

Höganäs, A.B. (2004). *Höganäs Handbook For Warm Compaction*. Höganäs Corporation, Bruksgatan 36, Sweden.

Li Y. Y., Ngai T. L. and Xiao Z-Y (2002a). 'Study on mechanical properties of warm compacted iron-base materials'. *Journal of Central South University of Technology*, **9**(3). 154–8.

Li Y.Y., Ngai T.L., Zhang D.T., Long Y. and Xia W. (2002b). 'Effect of die wall lubrication on warm compaction powder metallurgy'. *Journal of Materials Processing Technology*, **129**, 354–358.

Li Y. Y., Ngai T. L. and Lin, W.S. (2005). 'Effect of lubricant's friction coefficient on warm compaction powder metallurgy'. *Transactions Metallurgy Society*, **19**(1) 14–17.

Lin S. and Xiong W. (2012). 'Microstructure and abrasive behaviors of TiC-316L composites prepared by warm compaction and microwave sintering'. *Advanced Powder Technology*, **23**(3), 419–25.

Luk S., Rutz H. and Lutz M. (1994). 'Properties of High Density Ferrous P/M Materials – A Study of Various Processes'. *Advances in Powder Metallurgy and Particulate Material*, **5**, 135–55.

MacAskill I.A., Heard D.W. and Bishop D.P. (2007). 'Effects of silicon on the metallurgy and sintering response of Al–Ni–Mg PM Alloys'. *Materials Science and Engineering A*, **452**, 688–698.

Musella V. and D'angelo M. (1988). *Process for Pre-heating Metal in Preparation for Compacting Operations*. US Patent 4,955,798.

Ngai T.L., Chen W.P. and Xiao Z. (2002). 'Die wall lubricated warm compaction of iron-based powder metallurgy material'. *Transactions Nonferrous Metallurgy Society of China*, **12**(6), 1095–8.

Ngai T.L., Kuang Y.H. and Li Y.Y. (2007). 'Warm compaction forming of a binder-treated Fe-base material'. *Proceedings of Sino-Swedish Structural Materials Symposium*. 77–81.

Rahman M.M., Ariffin A.K., Nor S.S.M. and Rahman H.Y. (2011a). 'Powder material parameters establishment through warm forming route'. *Materials and Design*, **32**, 264–71.

Rahman M.M., Nor S.S.M. and Rahman H.Y. (2011b). 'Investigation on the effect of lubrication and forming parameters to the green compact generated from iron powder through warm forming route'. *Materials and Design*, **32**, 447–52.

Rawlings A., Luk S. and Hanejko F. (2000). *Engineered Approach to High Density Forming using Internal and External Lubricants*. Hoeganaes Corp, Cinnaminson, NJ. http://www.gkn.com/hoeganaes/media/

Rawlings A. and Hanejko F. (2000). Die wall lubricant utilizing warm compaction methods. *2000 International Conference on Powder Metallurgy and Particulate Materials*, Part 3, Compaction and forming, May 30–June 3, New York.

Rutz H., Hanejko F. and Luk S. (1994). 'Warm compaction offers high density at low cost'. *Metal Powder Report*, **49**(9), 40–7.

Rutz H. and Hanejko F. (1997). 'The application of warm compaction to high density powder metallurgy parts'. *PM2TEC '97 International Conference on Powder Metallurgy and Particulate Materials*, June 29–July 2, Chicago.

Rutz H. and Oliver C.G. (1995a) 'Powder metallurgy in electromagnetic application'. *Advances in Powder Metallurgy and Particulate Materials*, **3**(11), 87–106.

Rutz H., Rawlings A. and Cimino T. (1995b). 'Advanced Properties of High Density Ferrous Powder Metallurgy Materials'. *PM2 TEC '95 World Congress on Powder Metallurgy and Particular Material*. Seattle, WA.

Sation S., Lavakiri M. and Kagaya T. (2003). *Powdered Metals Technical Report*, no. 2, Matsudo-shi, Chiba, Japan, 36–43.

Shokrollahi H. and Janghorban K. (2006). 'The effect of compaction parameters and particle size on magnetic properties of iron-based alloys used in soft magnetic composites'. *Materials Science and Engineering B*, **134**, 41–3.

Simchi A. (2003). 'Effects of lubrication procedure on the consolidation, sintering and microstructural features of powder compacts'. *Materials and Design*, **24**, 585–94.

Simchi A. and Veltl G. (2006). 'Behavior of metal powders during cold and warm compaction'. *Powder Metallurgy*, **49**, 281–7.

St-Laurent S. and Chagnon F. (1999). 'Behaviour of steel powder mixtures processed by warm compaction'. *Metal Powder Report*, **54**(3), 42.

Takashi K., Tomoshige O. and Yukiko O. (2011). 'Analysis of dewaxing behavior of iron powder compacts based on a direct observation of decomposing lubricant during sintering in a furnace. *JFE Technical Report no. 16, Special Issue on Engineering, Steel Section Products and Iron and Steel Powders*, 83–88.

Thomas Y. (2000). *Lubricated Ferrous Powder Composition for Cold and Warm Pressing Applications*. US Patent No. 6,140,278.

Wikman B., Solimannezhad N., Larsson R. and Oldenburg M. (2000). 'Friction coefficient estimation through modeling of powder die pressing experiment'. *Powder Metal*, **43**(2), 132–8.

Xia Y., Shi-ju G. (2008). 'Fe-Mo-B enhanced sintering of P/M 316L stainless steel'. *Journal of Iron and Steel Research*, **15**(1), 10–14.

Xiaoa Z.Y., Kea M.Y., Fanga L., Shaoa M. and Lia Y.Y. (2009). 'Die wall lubricated warm compacting and sintering behaviors of premixed Fe–Ni–Cu–Mo–C powders'. *Journal of Materials Processing Technology*, **209**, 4527–30.

Zhang X, Xiong W., Ye D., Qu J. and Yao Z. (2009). 'Research of warm compaction technology on nylon bonded Nd-Fe-B magnets'. *Acta Metallurgica Sinica (English Letters)*, **22**(3) 174–80.

6 Developments in metal injection moulding (MIM)

I. TODD and A. T. SIDAMBE,
University of Sheffield, UK

DOI: 10.1533/9780857098900.1.109

Abstract: Metal injection moulding (MIM) is a manufacturing process used for small-to-medium-shaped precision components. This chapter provides a detailed overview of MIM and includes descriptions of the principles of the process such as powders, binders, mixing and feedstock analysis, injection moulding, binder removal (debinding), sintering and post-sintering. The chapter concludes with case studies and the design requirements for MIM and its applications.

Key words: debinding, metal injection moulding, non-ferrous alloys, powder metallurgy, sintering, superalloys, titanium.

6.1 Introduction to metal injection moulding

Metal injection moulding (MIM) is a manufacturing process which, in the past 15 years, has been established as an alternative manufacturing process for small-to-medium-shaped precision components that were previously manufactured by other methods at a higher cost. The advantages of MIM have emerged as being able to produce cost-effective, complex shaped parts in both large and small volumes using almost all types of metals and intermetallic compounds.

MIM is a process that was developed from the combination of plastic injection moulding and traditional powder metallurgy and is rightly regarded as a branch of both technologies (German and Bose, 1997; Anwar, 1996). MIM is similar to plastic injection moulding as the material is fed into a heated barrel, mixed and pushed into a mould cavity where it cools and then hardens to the mould die cavity shape. On the other hand, MIM is similar to traditional powder metallurgy in that the procedure is able to compact a lubricated powder mix in a rigid die by uniaxial pressure, eject the compact from the die and sinter it. MIM is also a branch of powder injection moulding (PIM), which is a subject that covers both metallic and non-metallic powder used in the manufacturing of small-to-medium-complex-shaped parts in large numbers. In another branch of PIM, ceramic or carbide materials are used as raw materials in a process called ceramic injection moulding (CIM).

The MIM cycle begins with preparation of a feedstock by mixing fine metallic powder with a binder comprising waxes, polymers, lubricants and surfactants. The resulting feedstock is then granulated. An injection moulding machine is used to heat the feedstock before injecting it into a mould cavity under pressure. The molten feedstock is then allowed to cool and solidify, producing a green part. The binder components are then removed and the green part becomes a highly porous brown part. The brown part is then sintered and shrinks, typically to >95% of the pore-free density (PFD). The flow diagram for the MIM process is shown in Fig. 6.1.

The rheological properties of the feedstock, which consists of the powder and binder mix, are of major importance. The requirement is that the mix flows smoothly into the die cavity without segregation at the moulding temperature and therefore the viscosity should be as constant as possible over a range of temperature.

The as-moulded part, which is also called a green part, contains a high volume percentage of binder and the result is that during sintering a large shrinkage occurs. It is, therefore, a major requirement of the sintering process to ensure that this shrinkage is controlled because this affects the density as well as the mechanical properties. The important parameters in the sintering cycle that affect the final properties of an MIM part are heating and cooling rate, sintering time, sintering temperature and the sintering atmosphere (Nayar and Wasiczko, 1990; Xai and German, 1995; Rawers *et al*., 1996; Khor and Loh, 1994). It is in this regard that MIM has an advantage over traditional powder metallurgy because if the sintering is optimised the shrinkage should also be uniform.

Figure 6.2 shows where MIM is used in comparison with the other technologies (Froes, 2006). The figure shows that MIM is essentially a technology that is used for producing complex shape parts in high quantities. If the shape allows the production of the part by, for example, conventional pressing and sintering, MIM would in most cases be too expensive. A typical competing technology for MIM is investment casting but MIM certainly has advantages compared with investment casting in the case of high part

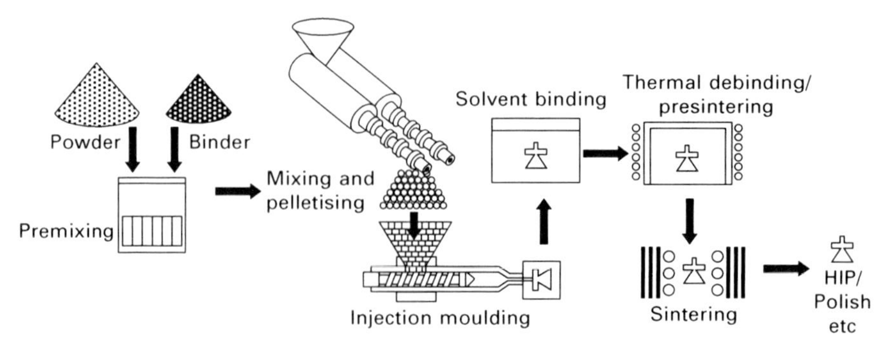

6.1 Metal injection moulding (MIM) flow diagram.

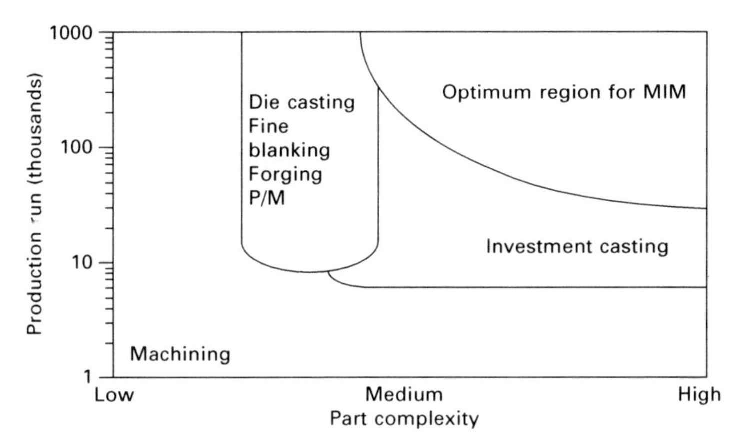

6.2 Use of MIM in comparison with other technologies (Froes, 2006).

numbers of castings and, of course, in non-castable alloys. The effect of the volume production on cost shows that, for example, for the smallest part weighing 4.5 g, the cost per part falls from US$1.4 for an annual production of 250 000 pieces to US$0.2 for 3 million or more (EPMA, 2010).

The successful commercialisation of MIM parts has seen production increase to around 90% of the market for PIM products, which account for an annual growth of between 10–20% worldwide. According to recent data, Asia is the world's largest PIM producing region by sales, followed by Europe and North America (Schlieper, 2009a).

Figure 6.3 shows the regional concentration of MIM applications. Data for Europe and North America relate to 2010 and for Asia to 2008 (PIM, 2010). Figure 6.4 shows MIM sales growth in Europe over the last few years.

The publication of MIM applications has also been increasing proportionally over the years and some of the examples that demonstrate the strengths and potential of this technology are in medical, automotive, electronics, telecommunications, aerospace, consumer products, firearms and defence sectors. Many objects contain MIM parts, such as cars, mobile phones, watches, domestic appliances, cameras and power tools.

6.2 Powders for metal injection moulding

The primary raw materials for MIM are metal powders and a thermoplastic binder. The properties of the powder determine the final properties of the MIM product and therefore the characteristics of the powder used in metal injection moulding are important in the control of the process. The properties that are considered in the powders used in MIM are:

- particle shape: slightly non-spherical with an aspect ratio of 1.2 to 1.5;

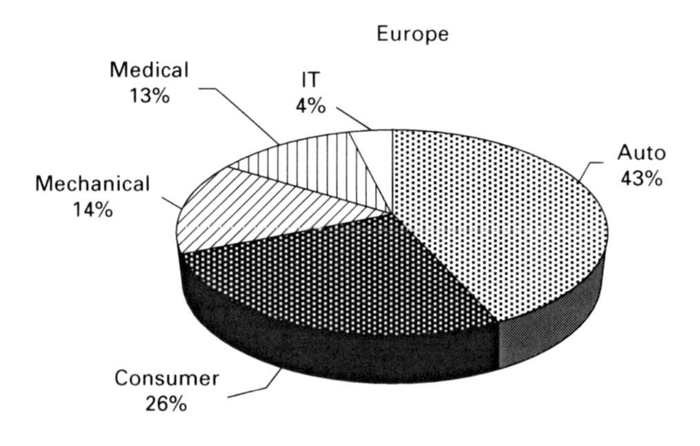

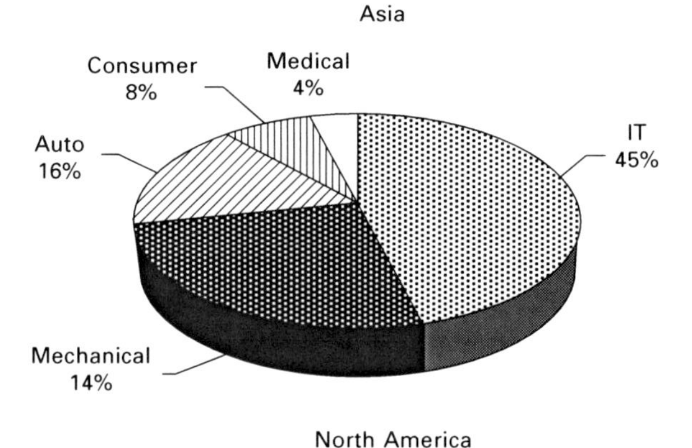

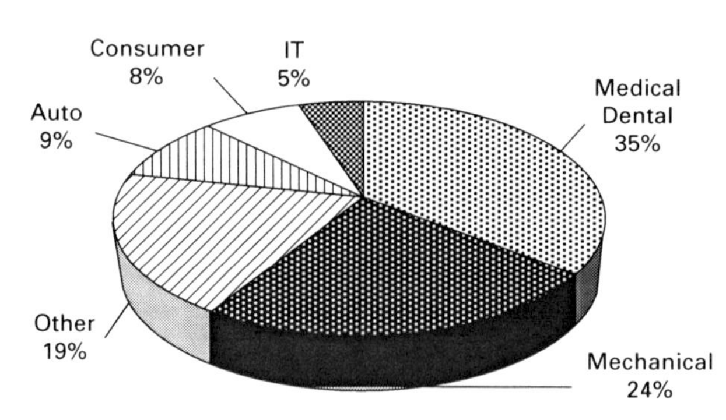

6.3 Regional concentration of MIM applications. Data for Europe and North America relate to 2010 and for Asia to 2008 (PIM, 2010).

- particle size: 0.1–20 μm sizes are recommended;
- mean particle size: 2–8 μm sizes are recommended;
- tap density: the recommended is at least 50% of the theoretical;
- angle of repose: above 45°;

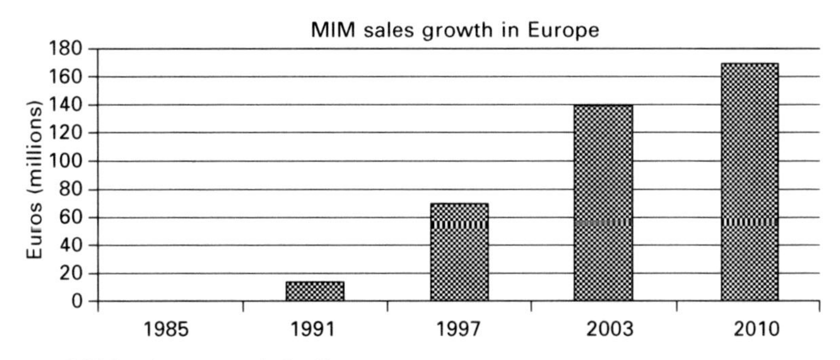

6.4 MIM sales growth in Europe.

- compacted angle of repose: above 55°;
- dense, discrete particle free of voids;
- clean particle surface (Vervoort *et al.*, 1996).

Of these characteristics and properties, the particle size distribution is the most important because it determines the sinterability and surface quality of the final product. The finer powders sinter more readily than coarser powders and it is for this reason that the finer powders are preferred in MIM to coarser powders. There are however a number of limiting factors relating to the different powder production techniques and their relative cost for MIM powders.

The other powder property that is considered to be important is the particle shape of the powder because it is desirable to incorporate as high a proportion of metal as possible (EPMA, 2010). Spherical or near spherical powders fall into this category and, as such, they have a high packing density. However they have poor shape retention capabilities, particularly during debinding because there is no metallurgical bonding between the particles as is the case in a die pressed compact. Other ideal properties include an angle of repose over 55 degrees. The choice of powder is in reality often determined by availability, but the growth in demand has encouraged powder manufacturers to produce powders that meet the requirements of MIM as desired.

Even though almost any metal that can be produced in a suitable powder form can be processed by MIM, there are some metals such as aluminium that are difficult to process via MIM. This is because they have an adherent oxide film that is always present on the surface and this inhibits sintering. Such challenges have been overcome with the help of academic research in the field of MIM. For example, researchers found that mixing the aluminium with small quantities of magnesium overcomes the oxide barrier (Lumley *et al.*, 1999).

In general, the list of metals that are widely used in MIM includes many common and several less common metals and their alloys – plain and low alloy steels, high speed steels, stainless steels, superalloys, intermetallics, magnetic alloys and hard metals (cemented carbides) (EPMA, 2010). The more expensive materials like titanium offer better prospects for economic gain because, unlike alternative processes such as machining, there is practically no waste due to scrap which helps to offset the high cost of producing the powder in the required form. In inexpensive metals it is usual to disregard scrap. Figure 6.5 shows the breakdown of MIM materials in Europe in 2009.

Inevitably, with the growth of MIM has come the growing demand for fine metal powders. In order to supply the special powders for MIM the powder manufacturers have had to improve their powder manufacturing techniques which are mainly centred on different methods of atomisation. There are a number of well established companies supplying atomised powders for MIM, including producers in Europe, North America and Asia (Schlieper, 2009b).

Of all the powders that are shipped by manufacturers, ferrous powders comprise 80% whereas the non-ferrous powders comprise 20% of the shipments. Titanium (Ti) and its alloys are some of the non-ferrous materials that are always considered to have great promise because of their low density, good mechanical properties and corrosion resistance. Commercially, MIM of titanium is gradually making inroads into structural components such as dental and medical tools. This has also been made possible by improvements in methods for production of Ti and Ti alloy powders. The method that has been successfully developed to produce fine spherical Ti powders suitable for MIM is plasma atomisation (Smagorinski and Tsantrizos, 2002).

Figure 6.6 shows scanning electron micrographs (SEMs) of non-ferrous titanium powder produced via the plasma atomisation method (micrographs 1–3). Micrograph 1 shows the morphology of commercially pure (CP-Ti) sub 45 μm, micrograph 2 shows the morphology of the Ti6Al4V (Ti-64) sub 45 μm and micrograph 3 shows the morphology of Ti-64 sub 25 μm. The powders are produced by plasma atomisation by Raymor Industries Inc. (also

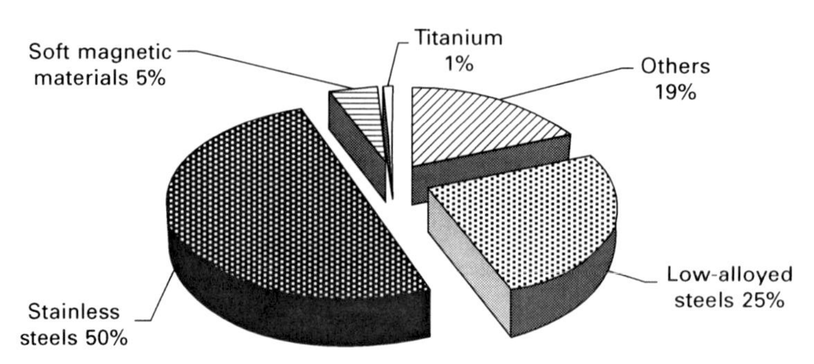

6.5 Breakdown of MIM materials in Europe (Murray, 2009).

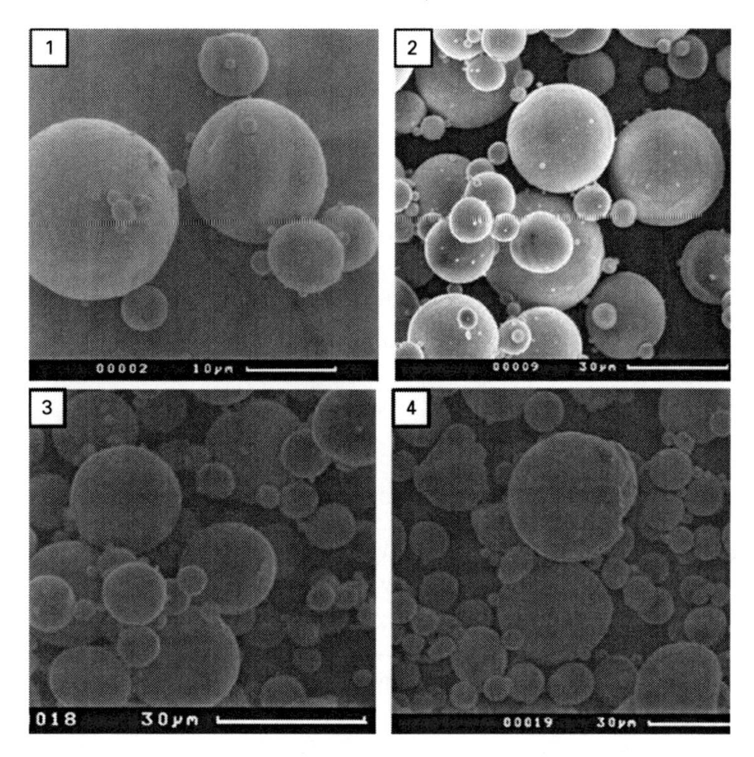

6.6 SEMs of titanium powder. (1) CP-Ti sub 45 μm from AP& C. (2) Ti-64 sub 45 μm from AP& C, (3) Ti-64 sub 25 μm ELI from AP& C and (4) CP-Ti sub 45 μm from TLS.

known as AP& C, Canada). Micrograph 4 shows the morphology of CP-Ti sub 45 μm made by TLS Technik (Germany) produced via the Eiga method (TLS, n.d.). The morphology of all the powders is shown as spherical.

The powder particle size distribution of these powders is shown in Fig. 6.7. The distribution of the powder size is such that more than 90% volume of powder is less than 45 μm and less than 10% volume of powder is over 45 μm for the CP-Ti (TLS), CP-Ti (AP&C) and Ti-64 (AP&C) powders. For Ti-64 ELI (AP&C), the powder size distribution is such that more than 90% volume of powder is less than 25 μm and less that 10% volume of powder is over 25 μm. The powder size distribution conforms to the log normal law, signifying that the particle distributions form straight lines on the log plot (Smith and Jordan, 1964).

6.3 Binders for metal injection moulding

The development of binder compositions has been instrumental in the progress that MIM has made as a technology for manufacturing parts. In MIM, the

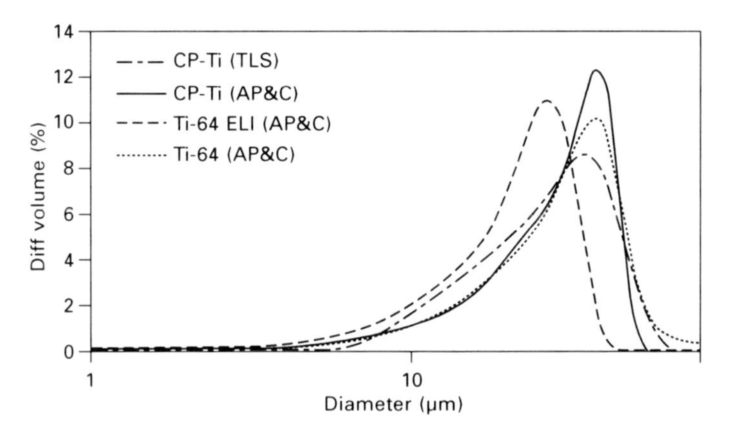

6.7 Powder distribution graphs for titanium powders in Fig. 6.6.

powders are mixed with, in most cases, a thermoplastic binder. The binder material is present in the green part to assist in processing by providing plasticity and it is removed from the products after injection moulding in a process widely known as debinding.

One of the early challenges that presented itself during the early development of MIM was to find suitable compositions of binder. Researchers and developers working with MIM were faced with the challenge of finding binder compositions which fulfil several tasks as listed below:

- To be able to incorporate a high volume of fine metal powders, typically 60% by volume;
- To form a coherent mass that can be plastified and injection moulded at elevated temperature;
- To allow removal of the main binder constituent in a reasonably short, environmentally friendly process;
- To provide enough strength after debinding by means of the 'backbone binder';
- To be supplied in a regular granular form that can easily be fed into an injection moulding machine;
- To have consistent, uniform properties from batch to batch;
- To be able to produce runners and green scrap which are easily recyclable;
- To be cost effective (Schlieper, 2009a).

In general, there are five types of binder used in the MIM process and these are classified according to the following categories:

- thermoplastic compounds
- thermosetting compounds

- water-based systems
- gellation systems, and
- inorganics (German, 1990).

Although there have equally been improvements in starting powders and sintering furnace technologies to support the growth of MIM, some of the widely used polymer and wax binders known for their ability readily to thermally unzip to their starting monomers (e.g. polymethylmethacrylate (PMMA), polypropylene carbonate, poly-α-methylstyrene) still tend to introduce impurities into the sintered MIM bodies because their depolymerisation occurs close to those temperatures where impurity uptake initiates. These binders on the other hand are preferred because they are relatively easy to mould.

Other binder systems based on catalytic decomposition of polyacetals have required expensive capital equipment to handle the acid vapour catalyst as well as a suitable means of eliminating the formaldehyde oligomers that form as polymer decomposition by-products (Froes, 2005). The use of catalytic debinding of MIM feedstocks on the other hand ensures shorter moulding cycle times, a high green strength, no deformation during debinding and a quick debinding treatment.

Some of the solutions to the improvement of debinding systems have been targeted at the decomposition mechanisms and chemical composition of binder which play a major role in levels of residual contamination. Therefore in order to reduce contamination of some metals, the amounts of decomposable substances in the binder have had to be reduced. Alternative binders, such as the solid polymer solution (SPS) have filled this role and thus proven to have an advantage over the widely used waxed-based binders because they can be rapidly removed and the binder system is also believed to cost less as well as being environmentally friendly. The SPS binder system uses water-soluble polyethylene glycol (PEG) as the major component (70 wt%) with PMMA used as a backbone binder (30 wt%) and was first reported by Cao *et al.* in 1992 (Cao *et al.*, 1992) (Fig. 6.8).

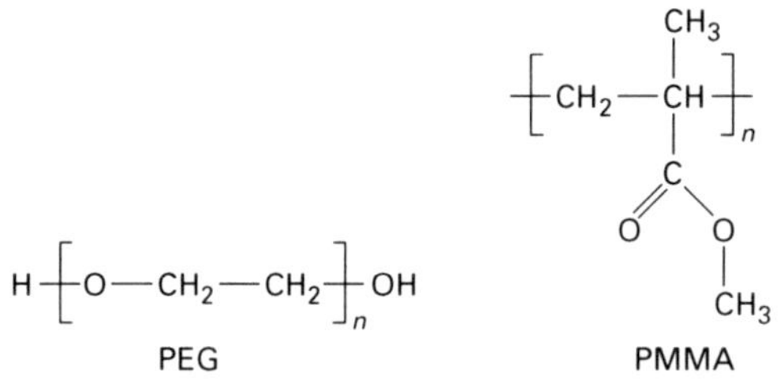

$$H{-}\left[O{-}CH_2{-}CH_2\right]_n{-}OH$$

PEG

$$\left[CH_2{-}C(CH_3)(COOCH_3)\right]_n$$

PMMA

6.8 PEG and PMMA chemical compositions.

6.4 Mixing and feedstock analysis

As mentioned in the previous section, the primary raw materials for MIM are metal powders and binder materials which are mixed to produce the feedstock. The feedstock is produced by mixing it with plasticised binder at elevated temperature using a mixing device such as an extruder, kneader or shear roll extruder. The feedstock is then granulated with granule sizes of several millimetres, similar to the method used in the plastic injection moulding industry.

Feedstock preparation plays a central role in the MIM process. The availability of various mixing devices has allowed different methods of mixing to be used to produce MIM feedstock. These range from sigma blade mixers, twin screw extruders, and kneaders to planetary mills. Regardless of the methods of mixing used, the feedstock mixture should be homogeneous and possess a pseudo-plastic behaviour.

A few of the technologies used to blend and compound powders and binder constituents to produce injection moulding feedstock are described below:

- Twin screw extruder: This has two co-rotating screws which give better mixing at lower melt temperatures. The screws and barrels are made up of smaller segments (mixing, conveying, venting and additive feeding) and the design can be changed to meet the production and product needs.
- Z-blade mixer: This mixer has a Z-shaped blade and is ideal for mixing, kneading and dispersion of high viscosity feedstock.
- Planetary mixer: This mixer has two offset mixing blades that rotate around individual shafts and then further rotate around the centre axis. The net effect is intermixing and stirring, as well as shear (Sidambe *et al.*, 2010).
- Twin cam mixer: This has counter-clockwise rotating cams in a small chamber.
- Non-contact mixer: An example is the centrifugal mixer. It mixes in minutes at a low temperature, removing air bubbles while homogenising. This cannot be used for binder systems with high melting temperatures.

The mix produced is usually granulated and converted into solid pellets which can be stored and fed into the moulding machine as required.

Commercially produced MIM feedstock is now becoming widely available, giving the MIM parts manufacturers the choice of whether to produce their own feedstock or to purchase ready-made feedstock. There are various ways in which MIM feedstock can be characterised and these include capillary rheometry for rheological behaviour, differential scanning calorimetry (DSC) for heat flow, thermogravimetric analysis (TGA) to study depolymerisation and other optical techniques which include SEM.

Figure 6.9 shows Ti64 commercial feedstock pellets that are ready for metal injection. The feedstock is available under the product name polyMIM®. Figure 6.10 is an SEM showing the surface of CP-Ti feedstock that has been produced in-house at the University of Sheffield (UK) using a non-contact centrifugal Speedmixer™. The micrograph shows the CP-Ti powder coated in binder after mixing with PEG, PMMA and stearic acid at speeds of 800, 1200, 1400 and 1600 rpm within a space of 10 min and without need to apply heat energy.

Figure 6.11 shows phase transitions within the same CP-Ti feedstock. The DSC graph shows the heat flow of CP-Ti feedstock illustrating the point at which the melting of the PEG occurs. This information is useful in establishing the parameters that will be used to extract the binder material from the feedstock.

Studies of the way in which the PMMA unzips in the feedstock are usually carried out using TGA. Testing is performed on feedstock samples to determine changes in weight in relation to change in temperature. Figure 6.12 is a graph of the TGA trace showing weight loss of CP-Ti feedstock representing the decomposition of the PMMA in the Sheffield University feedstock.

6.9 Ti64 commercial feedstock pellets that are ready for metal injection. The feedstock is available under the product name polyMIM®.

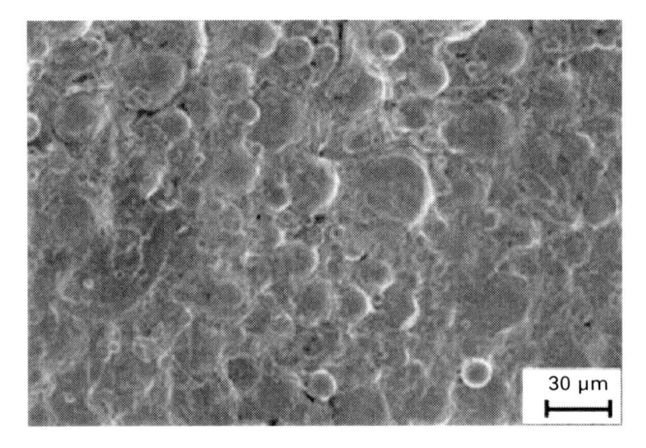

6.10 SEM showing CP-Ti feedstock after speed mixing.

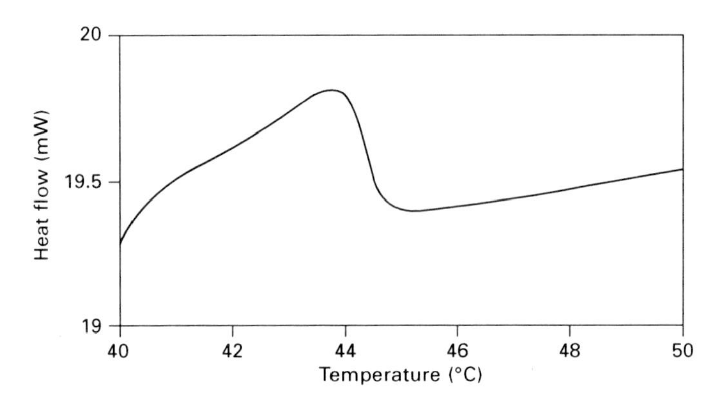

6.11 DSC graph showing the heat flow of CP-Ti feedstock.

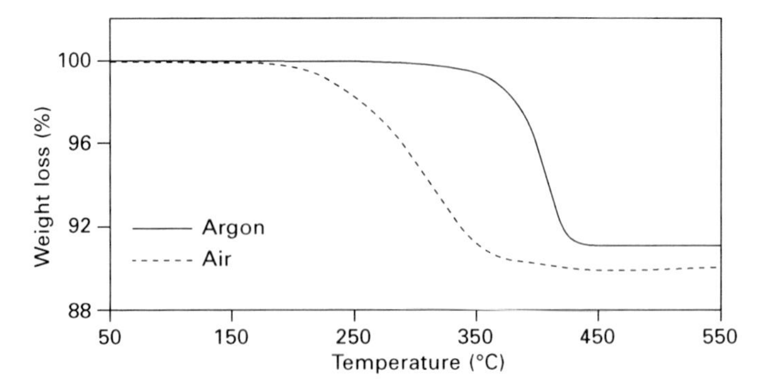

6.12 TGA trace showing weight loss of CP-Ti feedstock.

The flow of the feedstock is usually characterised by a capillary rheometer at different temperatures and shear rates. Figure 6.13 shows the evolution of the viscosity with temperature of the CP-Ti powder mixed with PEG, PMMA and SA. The shear viscosity vs shear rate plot indicates that the flow is pseudoplastic with the viscosity decreasing with increasing shear rate, the viscosity being lower at the higher temperature of 120°C and increasing at lower temperature. It is desirable that the viscosity of the feedstock should decrease quickly with increasing shear rate during injection with no dilatant behaviour (Sidambe *et al.*, 2010). The viscosity data of the feedstock is also used to set the injection moulding temperature, pressure and speeds for MIM feedstock by analysing the mouldability index.

Another important property that describes the behaviour of the MIM feedstock is the temperature dependence of viscosity. This relation can be expressed by an Arrhenius equation as follows (German, 1990):

$$\eta(T) = \eta_0\left[\left(\frac{E}{R}\right)\left(\frac{1}{T} - \frac{1}{T_0}\right)\right] \qquad [6.1]$$

where E is the flow activation energy, R is the gas constant, T is the temperature and η_0 is the reference viscosity at the reference temperature T_0. The flow activation energy, E, determines the sensitivity of the viscosity to temperature. Large values indicate high sensitivity to temperature changes (Agote *et al.*, 2001).

Figure 6.14 is a plot of the natural logarithm of the viscosities against inverse temperature at three different shear rates of Ni-Ti powder mixed with PEG, PMMA and SA. All curves were fitted to straight lines with a high correlation coefficient. It is clearly shown that the E value is very sensitive to shear rate. It is presumed that the feedstock is very sensitive especially at a high shear rate and defects such as powder binder separation will probably occur especially at a narrow path in the cavity. Therefore optimising the injection moulding parameters is needed in order to keep the value of E as

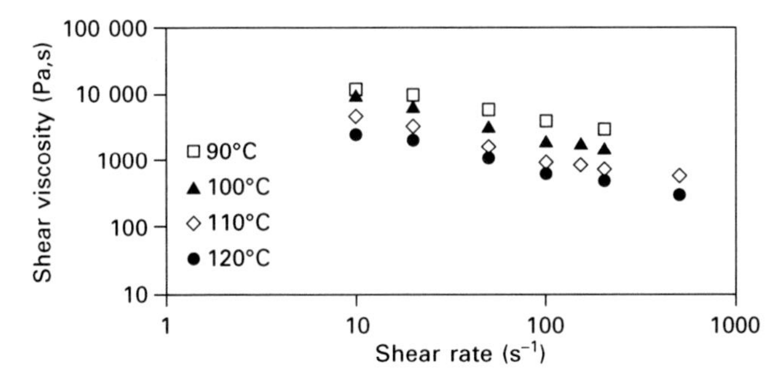

6.13 Apparent viscosity vs apparent shear rate of CP-Ti feedstock at different temperatures.

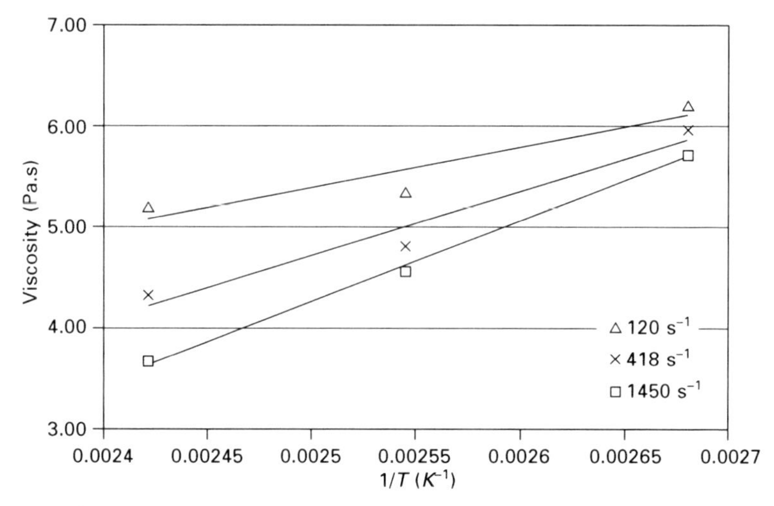

6.14 Correlation of viscosity and temperature at different shear rates.

low as possible. Owing to the complexity of the MIM die cavity, the feedstock should exhibit as low an E value as possible, to minimise the effects of any changes of temperature and shear rate on the flow. In addition, defective moulded parts can also be controlled (Ismail *et al.*, 2010).

6.5 Injection moulding

The MIM industry can therefore rely on the vast experience and technical background of the plastics injection moulding industry. The idea to plastify powdered raw materials with the help of thermoplastic additives and subsequently use injection moulding to form complex components was first developed for ceramic components. The machines used for this part of the MIM process are substantially the same as those in use in the plastics industry with small modifications to reduce wear (hardened screw and barrel) and assure homogeneity (modification of screw design) during the plastification. Therefore, standard plastic injection moulding machines undergo modifications to optimise them for MIM with injection cylinders being adapted for the particularly abrasive properties of the materials used. The screw geometry is also modified. Compared to a standard thermoplastic screw, a lower compression rate, typically 1.4:1 to 1.6:1 is used in MIM, along with screw material. Another modification is an extended compression zone. These modifications allow lower shear heating and therefore reduce wear.

The MIM industry can also rely on the availability of many experienced third-party mould makers, as well as mould design software, parts handling devices, and so on. Therefore, the 'green' MIM parts are formed in an injection moulding process equivalent to the forming of plastic parts. The varieties of part geometries that can be produced by this process are similar to the great variety of plastic components.

The metal injection moulding process is a cyclic process of forming material into a desired shape by heating the feedstock to its molten state. The hopper is filled with the material to be injection moulded. Heat energy is supplied by the barrel heaters and by shear heating. Shear heating occurs between the rotating screw and barrel and the shear forces acting on the material as it is conveyed along by the rotating screw. Figure 6.15 shows a schematic of the injection moulding process.

The molten material is conveyed along the screw to the shot zone, which is the region between the nozzle and the screw tip. As the melt is delivered to the shot zone, the screw moves backwards to create the necessary volume. The volume of the shot is dependent upon the piston travel. Pressure is applied to force the molten feedstock material into a mould cavity. The powder and binder feedstock material is allowed to solidify to the desired shape, which is dictated by the mould cavity geometry. The injection moulding cycle can be broken into various stages as shown in Fig. 6.16.

The injection moulding process is therefore subject to control when the parameters such as injection pressure, injection speed, holding or packing pressure and screw decompression are optimised such that the mould is progressively filled by the material. Figure 6.17 is a plot of the variation of the pressure on the feedstock material in the mould and shows some typical and constantly recurring characteristic points within that mould cavity. The effects of varying pressure at each point or section in time are explained below.

1. Start of injection of melt into cavity. There is a rise in the machine hydraulic pressure as the screw begins to move forward but no rise in cavity pressure.
2. Melt flows with the mould and the cavity pressure gradually begins

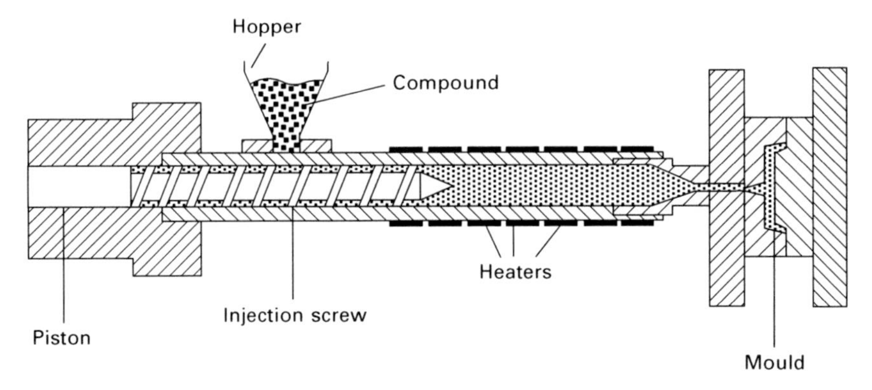

6.15 Schematic of a basic injection moulding machine.

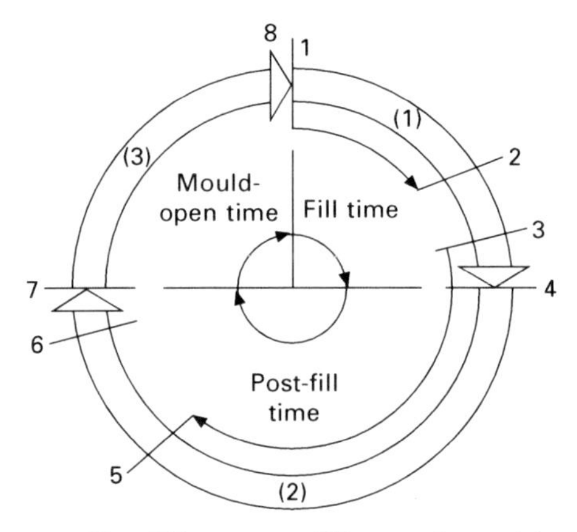

6.16 The filling, post-filling and mould opening stages of the injection moulding cycle.

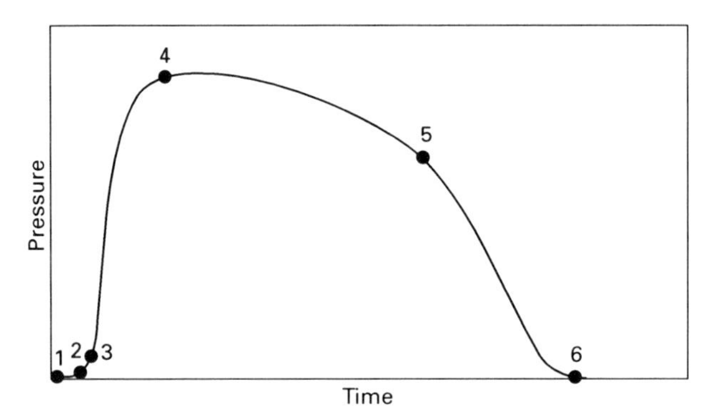

6.17 A plot of the variation of the pressure on the feedstock material in the mould.

to rise. As the cavity fills, the flow resistance determines the filling pressure. If the cavity pressure rise is too flat, this is reflected by slow filling which leads to low internal stress with no pressure peak. If the pressure rise is too steep (fast filling), there is material degradation and high internal stress in moulding.

3. Start of melt compression. The pressure rise from this point should not be low because this would result in low compression, no pressure peak, smooth transition with low internal stress and voids in moulding. Whereas if the rise is too steep (over-injection), this leads to high compression on melt, pressure peaks and high internal stress in moulding.
4. Maximum pressure is reached and this depends on material characteristics, injection speed and packing/holding pressure set points.
5. Sealing point is reached and melt in the gate is frozen.

6. Start of process of cooling and pressure returns to atmospheric pressure. Fluctuation normally indicates changing dimensions and a residual pressure reflects a delayed switchover to packing. The part is thereafter ejected.

Figure 6.18 summarises the effect of processing conditions on the resultant pressure profiles for a typical injection moulding process involving metallic powder.

Defects that can occur during the injection moulding process can be managed in many various ways. Computer aided design (CAD) of the injection moulding process is one of the methods that is used. This allows every detail of mould filling to be studied in 3D. Design engineers use simulation to analyse the filling dynamics in thick and thin areas on screen, study the effects of gate position, thermal gradients in the mould and phenomena such as jetting and binder segregation. This is done by studying the effects of common injection moulding parameters such as injection pressure, speed and fill time and so on.

CAD is also widely used to optimise the kinetics of mould filling and to determine shrinkage and distortion of the green compact relative to the mould dimensions. The CAD analysis and optimisation of the mould cavities can reduce design time and cost while significantly improving yield and quality (Schlieper, 2009a).

Figure 6.19 is a scanning electron micrograph of a moulded surface showing the distribution of the Ni-Ti powder within the PEG/PMMA/SA

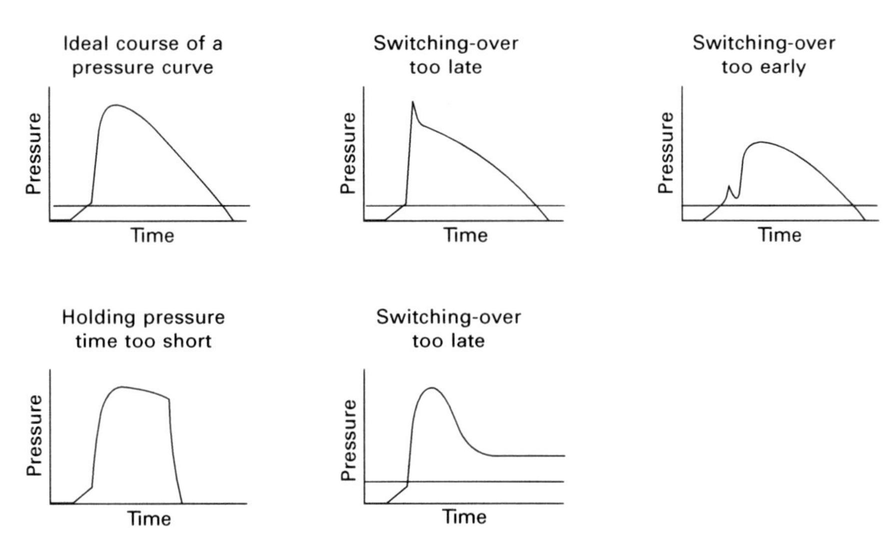

6.18 Effect of processing conditions on the resultant pressure profiles (Sidambe, 2004).

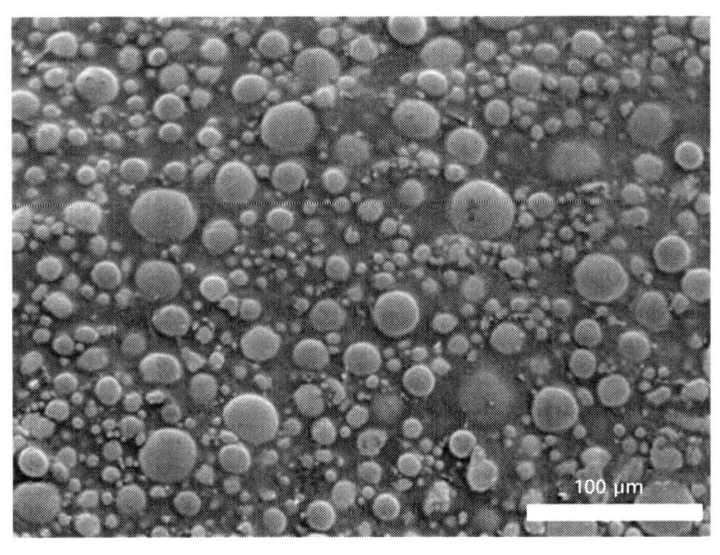

6.19 Scanning electron micrograph showing the morphology for an as-moulded Ni-Ti component showing uniform distribution of powder and mixed binder.

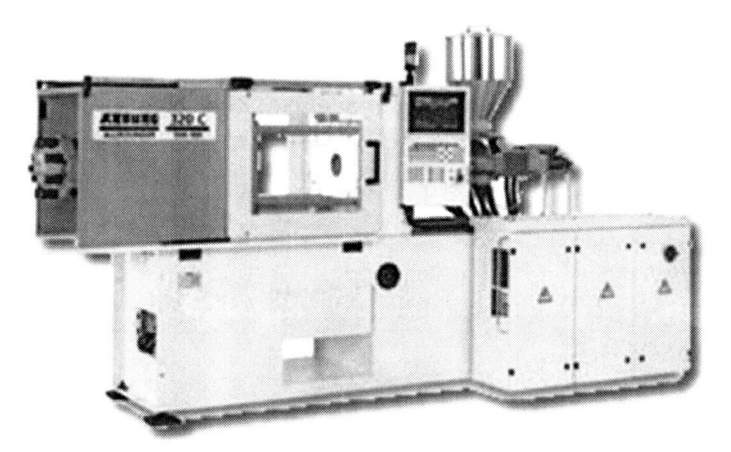

6.20 Injection moulding machine manufactured by Arburg.

binder system. The binder, as can be seen, is evenly distributed throughout the sample body.

Figure 6.20 shows an example of the injection moulding machine manufactured by Arburg.

6.6 Binder removal (debinding)

As has been mentioned in Section 6.3, the binder material in MIM green components is only an intermediate processing aid and it is always removed from the products after injection moulding. Removal of the binder from the green part is also considered a key stage of the process and one that requires most careful control. The stage at which the binder is removed is known as debinding. The manner in which the binder is extracted consists of the heating of the green compact in order to melt, decompose, and/or evaporate the binder. This binder extraction has to be optimised so that there is no disruption of the as-moulded part. The process normally takes several hours, depending on the thickness of the component.

It has been the challenge for MIM developers to reduce and optimise the times for debinding. There are different methods which serve to obtain parts with the required interconnected pore network without destroying the shape of the components in the shortest possible time. Different commonly used debinding methods applied in the MIM industry are further explained below.

6.6.1 Thermal debinding

Binders that usually lend themselves to this process are polymers such as polyethylene or polypropylene, a synthetic or natural wax and stearic acid (Schlieper, 2009a). The MIM feedstocks based on these type of binders are easy to mould, but the removal of the binder requires very careful and slow heating in a thermal pyrolysis process. The debinding time lasts 24 or more hours and is therefore considered costly. In order to overcome the long and costly debinding times associated with thermal debinding, other methods have been adopted for use in conjunction with the process such that the MIM components are debound in multi-stages. Thermal debinding is now widely used as a second stage of debinding to remove organic binder material prior to sintering.

6.6.2 Solvent debinding

Thermal debinding is now often used as a second stage of debinding in systems where the first stage is solvent debinding. Solvent debinding involves immersing the MIM compact in liquid that dissolves the binder material. The binder composition includes a constituent that can be dissolved in the liquid at low temperature. Acetone or heptane is sometimes used as the solvent although water-soluble binder compositions are preferred since it is easier to handle aqueous solvents than organic solvents. The times for debinding during solvent extraction are considered to be intermediate, that is they are shorter than thermal debinding times but take longer than catalytic binder removal, which is discussed below. The investment and operating costs are lower so that total processing costs are competitive (Schlieper, 2009a).

6.6.3 Catalytic debinding

Catalytic debinding of the binder is a process where most of the binder is attacked by a catalytic acid vapour (Krug *et al.*, 2002) such as highly concentrated nitric or oxalic acid. Binder removal is done using a vapour catalyst at relatively low temperatures of approximately 120°C, that is below the softening temperature of the binder and has the advantage of reducing thermal defects. The acid acts as a catalyst in the decomposition of the polymer binder. Reaction products are burnt in a natural gas flame at temperatures above 600°C. The binder material that is mainly used with this process is known as polyoxymethylene (POM) and it belongs to a grade of polymers known as polyacetals. These MIM feedstocks based on this binder are also easy to mould and possess excellent shape retention but there are hazards associated with acid catalysts and additional material costs.

6.6.4 Supercritial debinding

The most recent innovation in binder removal techniques has been the use of supercritical carbon dioxide. Binder extraction is carried out using supercritical fluids which have an extremely low viscosity allowing the molecules to penetrate into the fine pore channels that are created during debinding. Substances, such as carbon dioxide become supercritical fluids above but near their critical temperature and above their critical pressure. For carbon dioxide the critical temperature is 31°C and the critical pressure is 7.8 MPa (da Silva Jorge, 2008). Under these conditions the density of CO_2 is approximately 0.5 g cm^{-3}, that is less than in the liquid state but much higher than in the gaseous form. The processing time for debinding MIM parts is claimed to be about three hours with minimum defect formation. The process however requires precise temperature and pressure control in high cost equipment and is therefore expensive (Chartier *et al.*, 1995).

Other less commonly used binding processes use gelation, for example with mixtures of cellulose and gums, and freezing of an aqueous slurry also containing organic ingredients. During debinding the strength of the compact decreases markedly and great care is necessary in handling the 'brown' parts (EPMA, 2010).

In order to optimise the debinding process, such as the solvent debinding method, it is possible to carry out related studies and then illustrate the binder extraction using different types of analysis. Figure 6.21 shows the schematics of solvent PEG binder distributions at the (1) as-molded and (2) initial solvent debinding based on water extraction (3) intermediate and (4) final stages of solvent debinding. The PEG starts dissolving when the water concentration in the PEG and water mix is larger than the equilibrium water content (EWC) of PEG (Yang *et al.*, 2003).

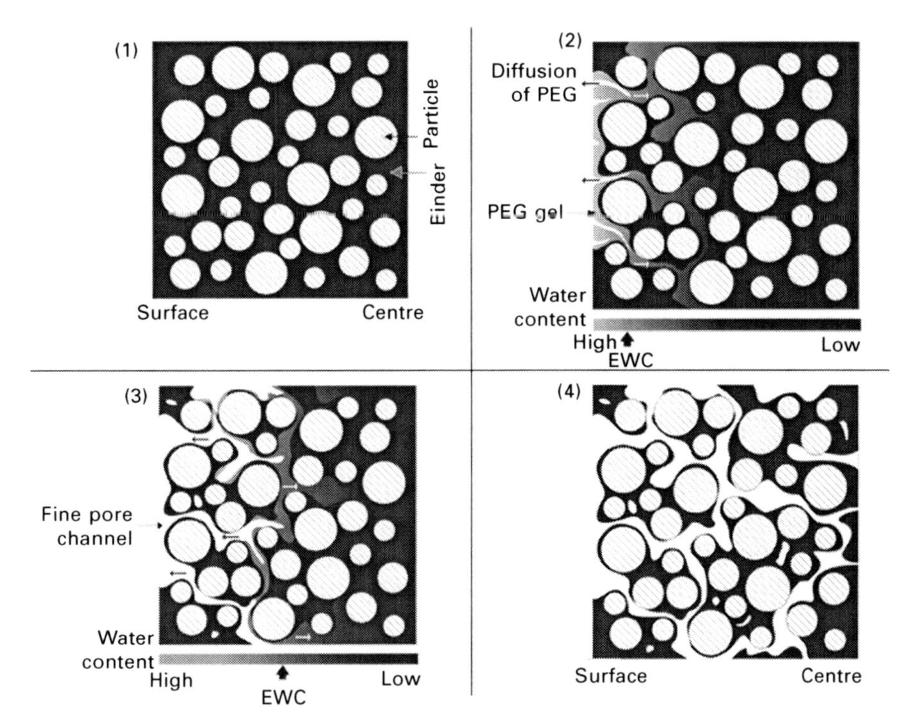

6.21 Schematics of binder distributions at the (1) as-molded and (2) initial solvent debinding based on water extraction (3) intermediate and (4) final stages of solvent debinding based on water extraction.

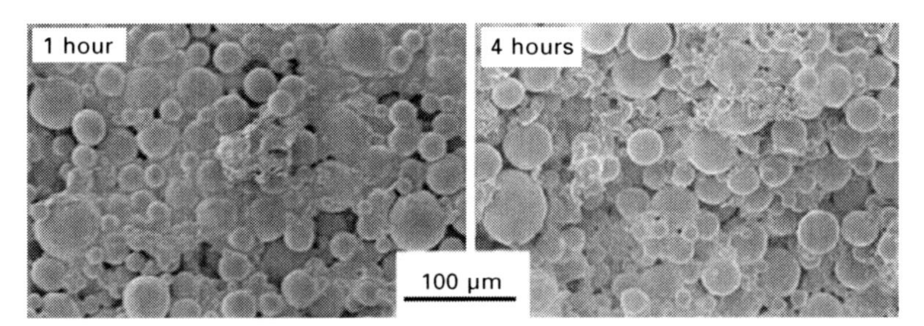

6.22 SEMs showing the development of pores during the removal of PEG at 60°C.

Solvent debinding kinetics of titanium feedstock made using a major fraction of PEG and a minor fraction of PMMA as backbone is shown in Fig. 6.22. The two SEMs illustrate the development of pores during the removal of PEG in heated distilled water at 60°C. After an hour of solvent debinding, it can be seen that there is still some residual PEG within the Ti/PMMA matrix, whereas after four hours of the first stage of debinding, the SEM shows virtually no PEG, with the remaining strands of binder consisting of PMMA.

This form of solvent debinding is a two-stage process consisting of dissolution and diffusion. Initially, solvent dissolves the polymer phase, thus forming a porous surface. The solvent then gets into the pores by capillary action. This is followed by diffusion of dissolved polymeric substances out of the green body. The process can be formulated using Fick's diffusion-based model, reported by Kim *et al.* (2007):

$$\ln\left(\frac{1}{F}\right) = \frac{D_e t \pi^2}{(2L)^2} + K \qquad [6.2]$$

where F is the fraction of the remaining soluble polymer, D_e is the inter-diffusion coefficient of polymer and solvent, t is time, $2L$ is the thickness of the specimen and K stands for the change in the mechanism controlling the debinding behaviour.

Equation [6.2] can be utilised to explain solvent debinding behaviour of PEG for the injection moulded compact. In order to demonstrate that solvent debinding is a two-stage process, results have been included in Fig. 6.23 for the debinding temperatures of 40°C and 75°C where $\ln(1/F)$ is plotted against leaching time. At 40°C it is clear that the solvent debinding is a two-stage process. The dissolution of PEG is the rate limiting step in the beginning of debinding up to leaching time of three hours. As the process proceeds, a longer diffusion distance through porous channels formed after initial debinding slows down the process and diffusion becomes the rate-determining step.

Figure 6.24 shows the amount of PEG that was removed by water leaching from mouldings plotted against the leaching time at the different temperatures on a Ti MIM component.

The backbone PMMA component of the binder remains after water leaching is removed via thermal debinding, that is by heating the parts in an atmosphere such as flowing argon. Table 6.1 summarises the thermal pyrolysis debinding stages.

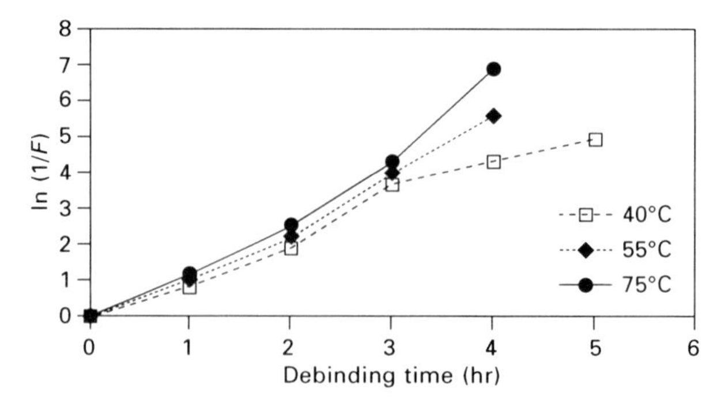

6.23 ln(1/*F*) with leaching time at 40, 55 and 75°C. *F* is the remaining fraction of PEG.

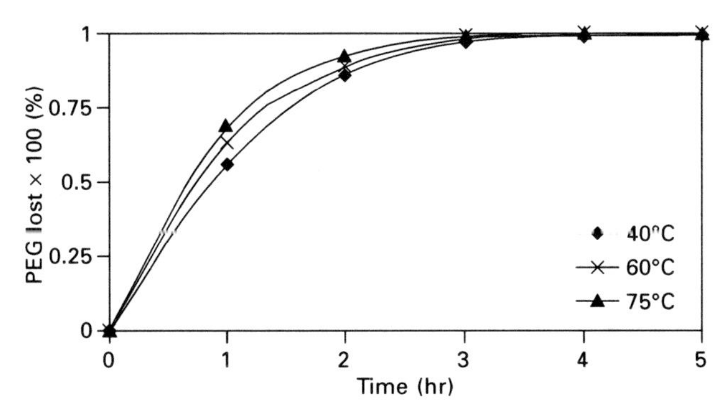

6.24 Graph showing the PEG lost as a function of temperature and time.

Table 6.1 Thermal debinding stages

Step	Process
1	Pore formation and growth
2	Diffusion of low molecular weight components to interface between pore and binder
3	Evaporation of low molecular weight components
4	Diffusion of generated gases through the formed pores to the surface
5	Flushing by a stream of inert gas

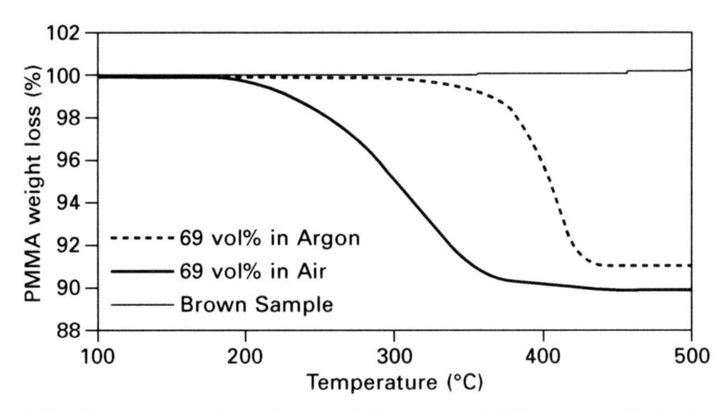

6.25 TGA trace showing weight loss of Ti/binder mix before and after thermal pyrolysis at 69 vol%.

Figure 6.25 is a thermogravimetric analysis trace showing weight loss of Ti/binder mix before and after thermal pyrolysis. It can be seen that there is no recorded weight loss from the brown part after thermal pyrolysis which confirms the complete removal of the PMMA binder component.

The backbone PMMA is removed by decomposition. The decomposition was explained by Jellinek (1978) who stated that the PMMA depolymerises into monomers with the production of gaseous product and leaves minimal levels of residue within the component. There is 95–100% conversion to monomer (MMA) in the process in vacuum or inert atmospheres. The residual oxygen and carbon levels obtained in the sintered samples suggest that the titanium surfaces do not interact significantly with PMMA or its degradation products. Several mass transport processes occur, often simultaneously, during thermal binder removal but in the case of the Ti/PMMA body it is the pore structure of the partially debound body which influences the resistance to mass transfer. The pre-existing porosity allows for fast vapour transport, making the body relatively easy to debind.

In addition to the pore structure, carefully controlling the heating rate limits defect formation. Thermal pyrolysis is carried out in a gas atmosphere and this creates an even flow and no shadow effects, much as if the parts were submerged in a liquid (Joens, 2005).

The transport of gaseous species through empty pores occurs by either Knudsen slip, or viscous (Poiseuille) flow (German, 1979). The flow mechanism was also used by Tsai (1991) when analysing pressure build up and internal stresses during binder burnout and in modelling the effect of gas transport during thermal debinding. The Knudsen flow, which was originally formulated by Knudsen, occurs when the average pore radius (r) is much smaller than the mean free path (λ_g), that is $r/\lambda_g < 0.1$ (Knudsen, 1909). Slip flow occurs when gas transport is intermediate between viscous flow and Knudsen flow, that is r/λ_g is in the range 0.1–10 and, finally, viscous flow occurs when $r/\lambda_g > 10$ (Lewis, 1997).

The process of gas transport through the pores can be estimated using an expression developed by Wakao *et al.* (1965) to estimate K, the transport coefficient for a single gas through a capillary of radius r:

$$K = \frac{1}{RT}\left[\frac{\frac{2r}{3}\sqrt{\frac{8RT}{\pi M_w}}}{1 + 2r/\lambda} + \frac{1}{1 + \lambda(2r)}\left(\frac{\pi r}{6}\sqrt{\frac{8RT}{\pi M_w}} + \frac{r^2 P}{8\eta}\right)\right] \quad [6.3]$$

where R is the gas constant, T is temperature, M_w is molecular weight of the gaseous species, P is pressure and η is the viscosity of the gas given by:

$$\eta = \frac{M_w \bar{v}}{3\sqrt{2} N_0 \pi \sigma^2} \quad [6.4]$$

where $\bar{v}$ is the mean molecular speed, N_0 is the Avogadro constant and σ is the effective molecular collision diameter. This term K can then be adjusted to predict the effective transport coefficient (K_{eff}) of a single gas through a porous body. Using this approach should show that the presence of a porous, PMMA free outer layer gives rise to a moving boundary with a variable concentration of diffusant that depends upon the surface flux, gas transport coefficient and thickness of the porous layer (Lewis, 1997).

6.7 Sintering

Sintering is the heating process in which the brown part which consists of separate powder particles is consolidated to provide the necessary strength in the finished product. The process is carried out at elevated temperatures below the melting temperature of the metal. Figure 6.26 shows the representation of neck formation caused by sintering. Metals are susceptible to oxidation at elevated temperatures and it is for this reason that sintering is carried out in controlled atmosphere and vacuum furnaces. Some of the factors for control during sintering are sintering temperature, heating rate, sintering time, sintering atmosphere and cooling rate (Sidambe *et al.*, 2012).

In order to avoid oxidation of the metal, reducing atmospheres and selected temperatures are generally used. Reducing atmospheres also have the further advantage of reducing any oxide existing on the surfaces of the powder particles which tend to be greater with the finer powder particles. The choice of the sintering parameters is determined by the metal that is being sintered.

The fact that the powders used are very much finer in MIM than those used in PM means that sintering takes place more readily because of the higher surface energy of the particles (Ji *et al.*, 2001). MIM parts also endure a very large shrinkage compared to traditional PM parts because the brown part has more porosity. Linear shrinkage is usually as high as 15–20%. The sintering temperature must therefore also be very carefully controlled in order to avoid slumping and loss of shape and to attain the required densification.

Sintered MIM are expected to have some residual porosity and the final parts usually have densities ranging from 95–99% of the theoretical density. Therefore the mechanical properties of the MIM components are degraded in comparison to those of the wrought materials of the same composition, but not significantly. These mechanical properties can be improved by further heat treatment, consolidation methods such as hot isostatic pressing (HIP) and surface treatments. Figure 6.27 shows a cause-and-effect diagram illustrating

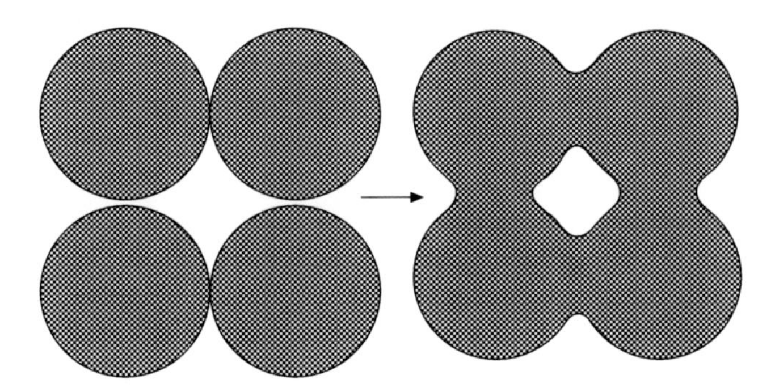

6.26 Representation of neck formation caused by sintering.

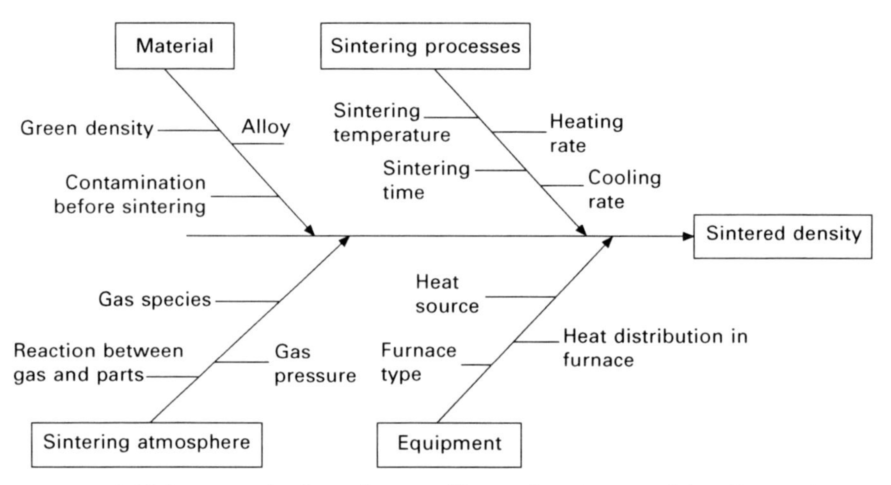

6.27 Cause-and-effect diagram illustrating the possible effects of sintering factors on the sintered density.

the possible effects of the sintering factors on the sintered density (Ferri *et al.*, 2010). These factors and parameters must be coordinated in order to achieve the target properties of the MIM component being sintered.

Sintering can be optimised by selecting the appropriate sintering parameters listed in Fig. 6.27. For example, if the particles in the powder are decreased, it is expected that there will be faster sintering but this will increase the operational cost and also increase the level of impurities in the final part. Increasing the sintering time also increases the cost, leads to grain growth and reduces production output. If the decision is made to increase the sintering temperature, the final part should have more shrinkage and therefore will be denser, with pore coarsening. However, this will also lead to grain growth on parts and higher operational costs. The cost implications associated with the sintering process depend on the material being sintered and the type of sintering atmosphere. For example, titanium MIM parts are usually sintered in furnaces which are equipped with diffusion pumps with minimal use of gases, whereas some steels are sintered in hydrogen gas atmospheres. Whilst hydrogen gas may not be costly, the safety measures required for hydrogen use can bring the cost up. Therefore, the economics of sintering depends on the cost of gas as well as the running cost of the sintering furnace.

Figure 6.28 is a photograph showing the shrinkage undergone by an IN713C MIM tensile test bar before and after sintering. Figures 6.29 and 6.30 show examples of the microstructure in sintered MIM components. The pore structure of a CP-Ti MIM component is shown in Fig. 6.29 indicating a low void ratio of round unconnected pores. Figure 6.30 is an example of a sintered MIM component showing the pore structure of an IN718 MIM component.

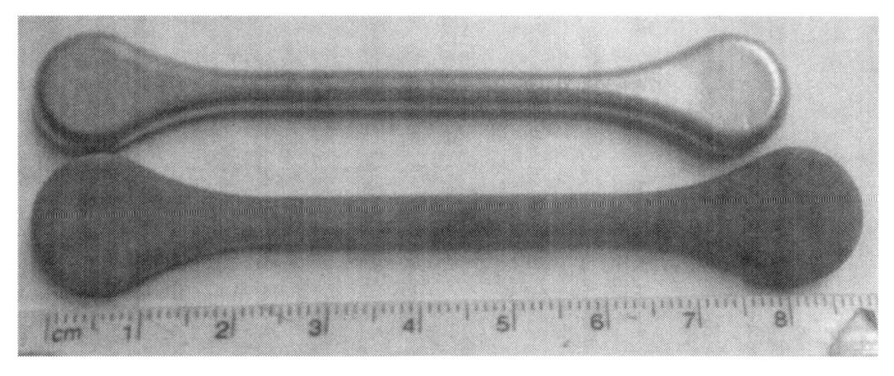

6.28 Photograph showing the shrinkage undergone by an IN713C MIM tensile test bar before and after sintering.

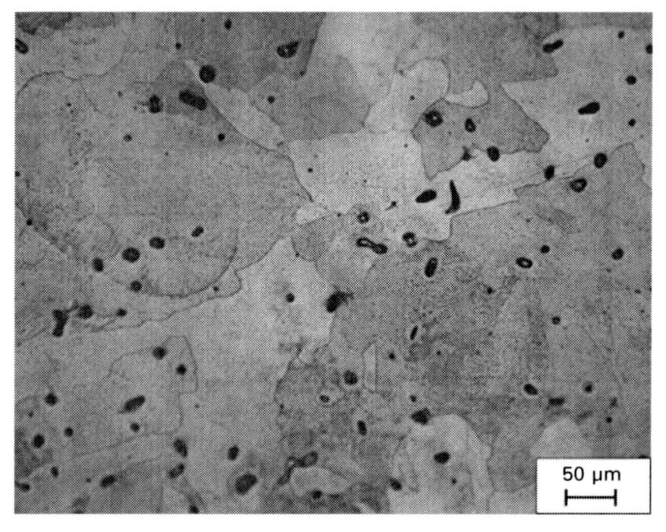

6.29 Pore structures and microstructure of the CP-Ti MIM component.

Therefore it is considered to be the final stage of the MIM process which is also irreversible. It is not possible to correct defects that have come about as a result of poor mixing, injection moulding or debinding. Many defects become more visible after sintering.

6.8 Post-sintering

After sintering, it is possible to improve the properties of MIM components using various methods which include heat treatment. The treatment applied to the sintered components will depend on the application. Post-sintering processes include assembling of parts, heat treatment, densification and

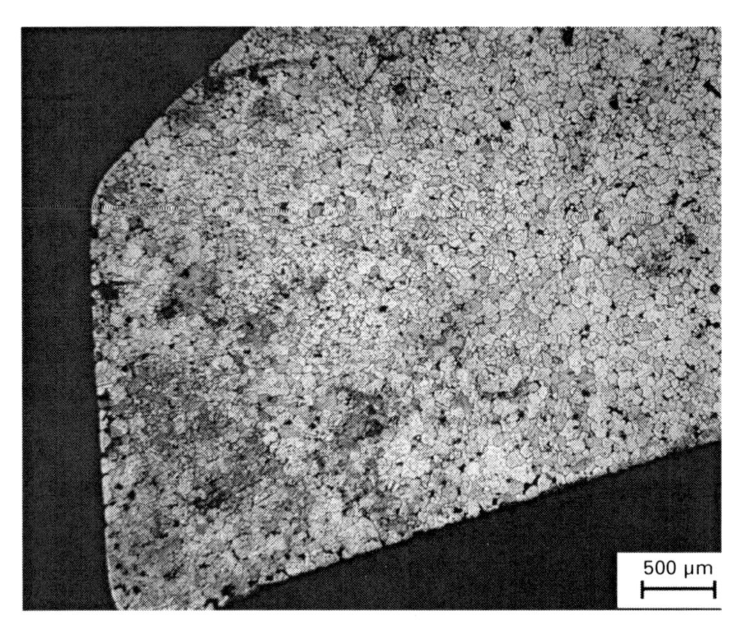

6.30 Pore structure of an IN718 MIM component.

finishing. As MIM parts are typically small, this post-sintering processing can be relatively cost effective for critical components where, for example, densification methods such as HIP are used.

As has been mentioned in the previous section, MIM parts that have been sintered under solid state have densities ranging between 95 and 99% and therefore some residual porosity. There are some applications which require fully dense parts in order to meet the performance criteria. HIP is the most common method used for eliminating or reducing the residual porosity, thereby improving the mechanical properties. The HIP process subjects a component to both elevated temperature and isostatic gas pressure in a high pressure containment vessel. Other densification methods include forging, impregnation and infiltration of the pores but this requires an open network of pores. Figure 6.31 shows the effect of HIP on a Ti-64 MIM component. The HIP in this instance was carried out at a temperature of 920°C and a pressure of 100 MPa for 120 min and it can be seen that after the HIP process, residual porosity is eliminated.

Inevitably, the presence of pores in MIM samples contributes to the degradation of the surface quality. This is determined by the powder particle size or the contents of the binder during the MIM process. Finishing and polishing post-sintering processes are applied to eliminate these surface problems. The processes commonly used include shot peening, anodising, painting, bead blasting and polishing. The mechanical properties of MIM parts are improved by these methods of processing because they eliminate crack

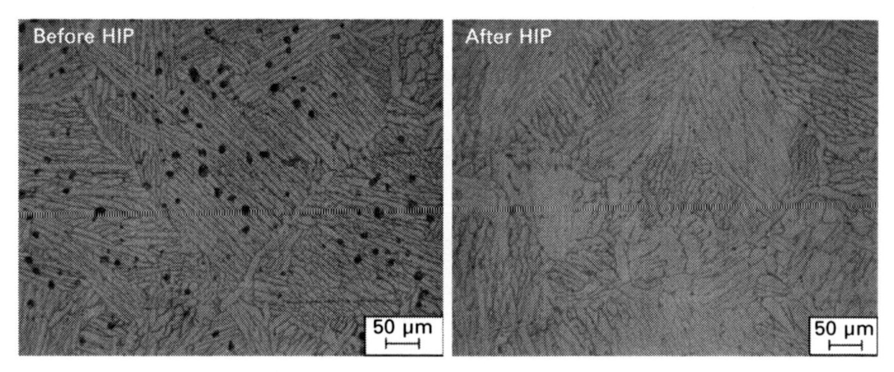

6.31 Effect of HIP on a Ti-64 MIM component.

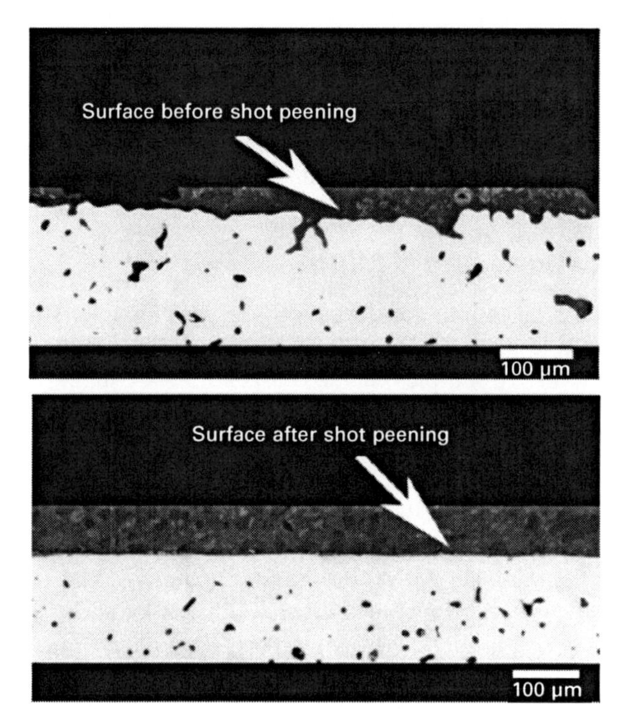

6.32 Surface of an MIM Ti64 component before and after shot peening (Ferri *et al.*, 2010).

initiation areas. Figure 6.32 shows the surface of an MIM Ti64 component before and after shot peening which is intended to improve the fatigue properties. Figure 6.33 is a photograph showing the effect of polishing on an MIM sample showing the aesthetic effect.

Heat treatment of MIM parts is usually used with the purpose of modifying the microstructure. The methods for this include, but are not limited to,

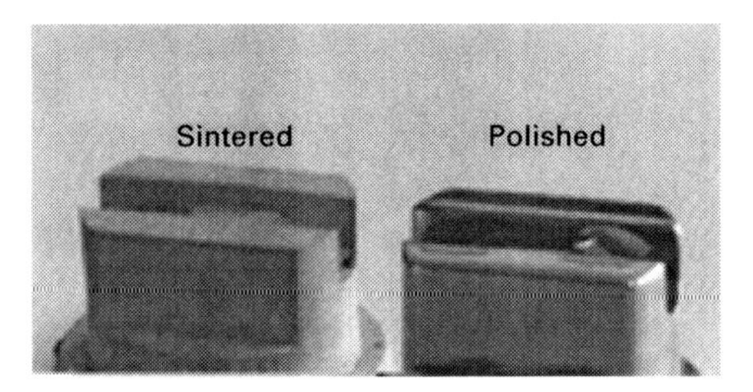

6.33 Effect of polishing on an MIM sample showing the aesthetic effect.

precipitation hardening and thermochemical treatments. The grain size of MIM parts made from titanium, for example, can be reduced ten-fold. Assembling of parts is usually required because MIM components are manufactured for complex systems and machines where they are functional in conjunction with other components. Secondary operations used during assembling also include straightening of parts, coining, grinding or polishing in order to achieve precise dimensions.

6.9 Applications and design

Design of MIM parts takes into consideration various characteristics which include size, shape, tolerances and cost. It has already been mentioned before that MIM is suitable where complex-shaped parts need to be manufactured in very large quantites. Therefore MIM parts generally have to satisfy several criteria which include part size, weight and section thickness, as well as the required properties (German, 1990). Figure 6.34 (German, 2009) shows the recommended design and possible geometries for MIM parts.

As can be seen from Fig. 6.34 there are some limitations to the shapes that can be manufactured via MIM. Furthermore, dimensional tolerances have to be attained with parting lines, ejector pin marks and gates taken into consideration. Costing models of MIM parts also have to include and be justified against the cost of tooling.

Design in MIM is also very much dependent on the material and material properties. It has been shown in Section 6.2 (see Fig. 6.5) that stainless steels dominate the MIM applications in Europe, mainly because of their properties, which include high strength and corrosion resistance. Table 6.2 shows room temperature mechanical properties of selected MIM metals and alloys. Included in the table are the densities, yield strength, tensile strength, elongation and hardness which serve as indicators of the suitability of the metals for different applications.

The applications of MIM in terms of sector are summarised in Fig. 6.35, which shows the breakdown of applications in Europe for 2010, in Asia for

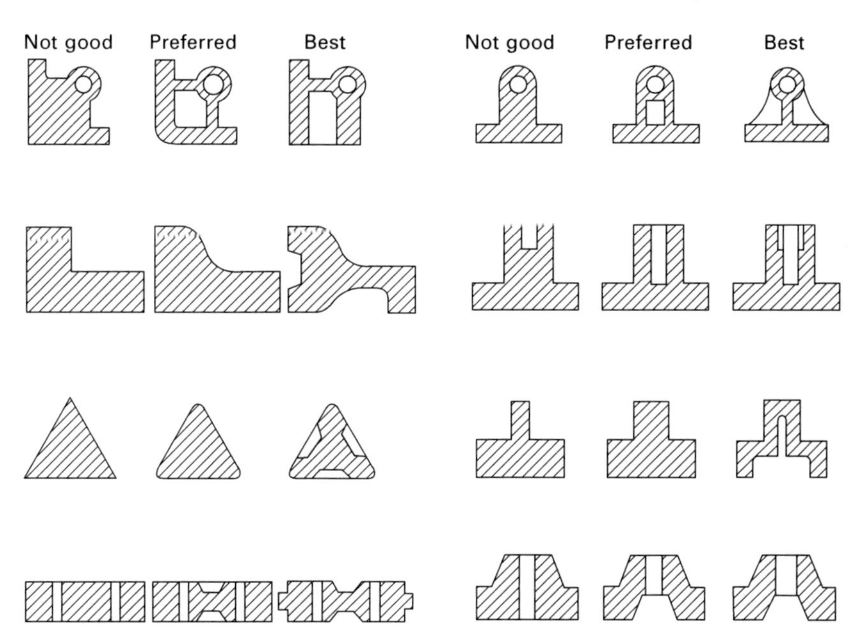

6.34 Design recommendations for MIM components (PIM, 2009).

Table 6.2 Room temperature mechanical properties of selected MIM metals and alloys (HT = heat treated, HIP = hot isostatically pressed, density is % alloy theoretical density)

Material (designation)	Density (%)	Yield strength (MPa)	Tensile strength (MPa)	Elongation (%)	Hardness (Vhn)
Aluminium	93	–	200	10	25
Cobalt–chromium	99	550	880	4	265
Cobalt–chromium (HIP)	100	520	1000	40	260
Hastelloy X (HT)	98	303	675	74	300
Inconel 718 (HIP, HT)	100	1130	1330	14	390
Stainless 17-4 ph	96	750	900	10	265
Stainless 17-4PH (HT)	96	1090	1185	6	340
Stainless 17-4PH (HIP)	100	1103	1137	13	370
Stainless 316L	96	220	510	45	135
CP-Ti	98	500	580	12	215
CP-Ti (HIP)	100	–	800	25	–
Ti6Al4V	98	800	880	12	350
M2 Tool steel (HT)	99	1000	1100	1	750

2008 and USA for 2010. Figure 6.35 shows that whilst automotive remains the single biggest segment in Europe, more than half the market is now in non-automotive applications. In 2008, MIM applications in Asia were dominated by information technology but there were expectations for strong

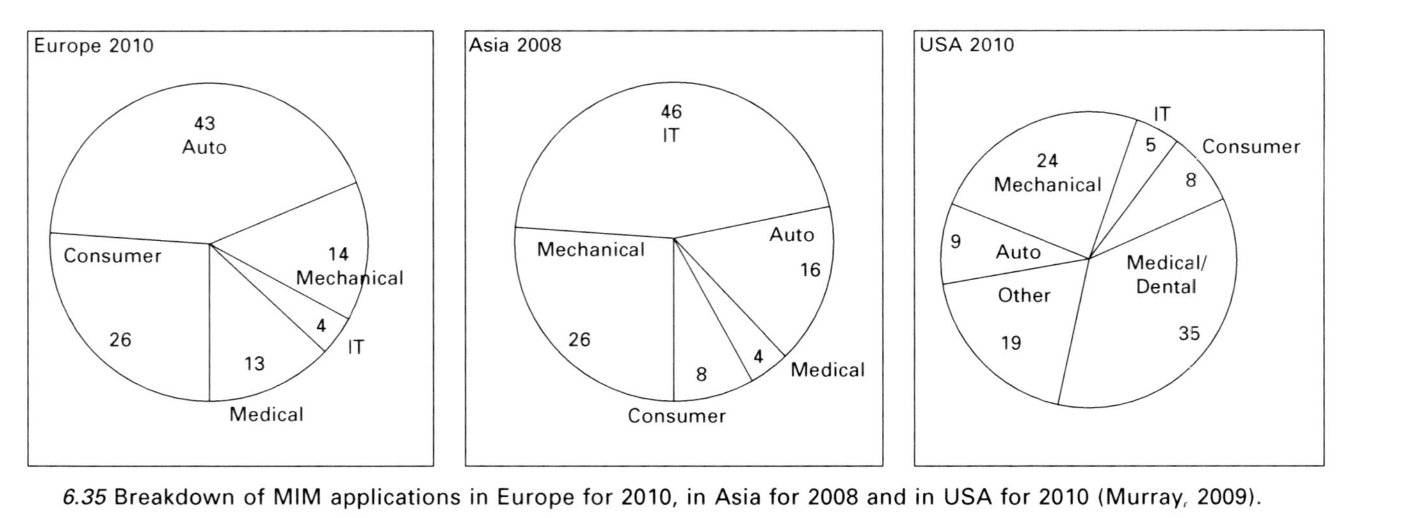

6.35 Breakdown of MIM applications in Europe for 2010, in Asia for 2008 and in USA for 2010 (Murray, 2009).

growth in the automotive market, driven by local markets in China and India. In the USA, the importance of medical and dental markets remain strong.

In the medical and dental markets, the typical components are within the areas orthodontic components, surgical instruments and implants. Orthodontic brackets were one of the first major MIM products and today they are still a major product in the industry.

Figure 6.36 shows some examples of MIM parts that are used in medical or surgical applications. The lower jaw is a critical component in suturing that places uniform stitches on a torn rotator cuff, allowing independent grasping and suturing with one hand. Secondary operations which include reaming three holes, coining and sizings are performed on the part. The tungsten radiation shield is an MIM part used as a shield in medical applications of radioactive isotopes. It is made from a two-phase tungsten heavy alloy with 3 wt% Fe and 4 wt% Ni. The part covers the syringe and screens radiation from injection of a radioactive medicine. The third MIM component is a 17-4PH stainless steel distal channel retainer. The complex, multi-level part is the main distal-side component of an articulation joint in an articulating mechanical stapler/cutter used in endoscopic surgery.

Applications for MIM in the automotive industry include parts for engines such as a rocker arm manufactured using low alloy steel. Around 4.5 million parts are produced annually using a hardenable 50NiCrMo2.2 steel powder alloy (Schlieper, 2009a). Other examples of MIM components that are used in automotive applications are shown in Fig. 6.37. The axle holder is an MIM part made from AISI steel 17-4 PH (17Cr-4Ni with 4% Cu) designed to hold a steady pressure. The part was originally made by a three-stage process involving machining, wire EDM and coating. The MIM version results in a cost saving of 70%. Also shown in Fig. 6.37 are automotive transmission MIM parts made from iron–2% nickel powder which are designed to help in synchronising the reverse gear. The third MIM component is used in a

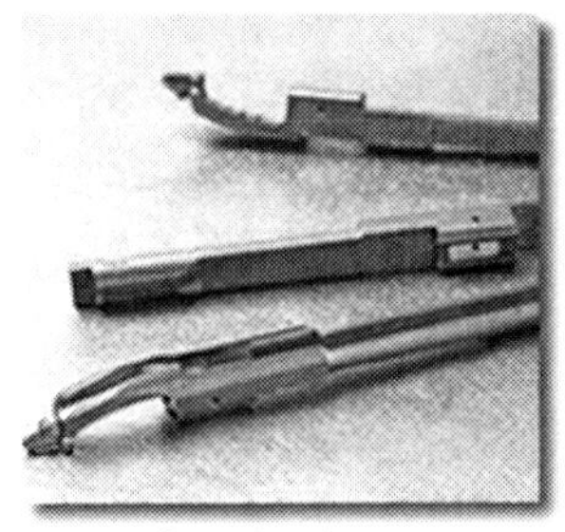

Suturing jaws (Kinetics Inc.) Courtesy of MPIF.

Tungsten radiation shield (IFAM, GmBH) Courtesy of EPMA.

Distal channel retainer (Kinetics Inc.) Courtesy of MPIF.

6.36 Examples of MIM components used within the medical sector.

V8-cylinder engine for piston cooling. It is assembled in the engine block in order to direct a defined oil jet stream onto the diameter of the piston bottom side. The tube is brazed into the appropriate bore after sintering. The small tolerances and required braze slot could only be economically realised by MIM (Whittaker, 2007).

In IT, electronics and telecoms there is a demand for small, precision, high volume components and this has allowed MIM to thrive in the sector. Thin-walled (<1 mm or 0.04 in.) intricate fibre optic transceiver housings are made from 17-4 PH stainless steel via MIM. Cold plates and heat sinks take full advantage of the MIM process. They have complex shapes that give the designer the maximum flexibility in the design concept such as intricate oval shaped pin geometry which provides optimal thermal performance (Schlieper, 2009a). These are shown in Fig. 6.38. Also shown in Fig. 6.38 is the highly complex shape 24-pin print head used in a high speed printer made from 3% silicon steel in order to achieve the desired soft magnetic properties.

Applications for MIM also exist in the consumer products sector in parts such as watch cases and related components such as eyewear components,

Axle holder
(Iscar Ltd, Israel)

Automotive transmission
(GKN sintermetals GmbH)
All courtesy of EPMA.

Piston cooling nozzle
(GKN sintermetals GmbH)

6.37 Examples of MIM components used in the automotive sector.

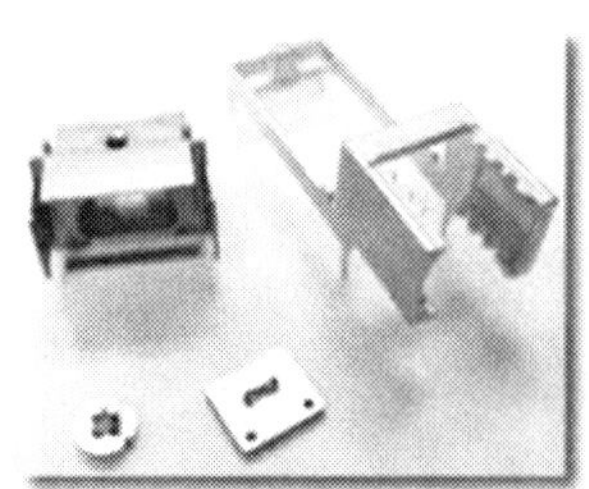

Fibre optic transceiver
housing
(AMT, Singapore)
Courtesy of PIM Int.

MIM copper cold plate
(Acelent technologies,
(Singapore)
Courtesy of PIM Int.

Dot matrix head printer
(IMPAC Technologies,
France)
Courtesy of EPMA.

6.38 Examples of MIM components used in the IT sector.

camera tripods and casings for an automatic tuner. Figure 6.39 shows three examples of MIM components used in consumer applications over the past 20 years. The photographs shown are for the first watch made by the MIM process using titanium and a set of MIM gears used in the rotating bristles of an electric toothbrush made from austenitic stainless steel 316L. MIM has allowed these corrosion resistant parts to be made to a consistent high quality at a relatively low cost. The third picture is of four parts that function in the flip assembly of a Motorola cell phone which are made from 17-4PH and 316L stainless steel powder. As can be seen from the picture, post-sintering polishing has been performed on the parts.

Firearms and defence applications make use of MIM parts such as triggers and rotors. Figure 6.40 shows some of the MIM components that are used in weaponry and firearms applications.

The rear sight for rifles shown is used on sporting and military rifles such as the AR15, M4 and M16 models and is made from nickel and steel. The seven MIM parts for .22- and .38-caliber revolvers include the thumb piece,

Sports watchcase (Hitachi) Courtesy of MPIF.

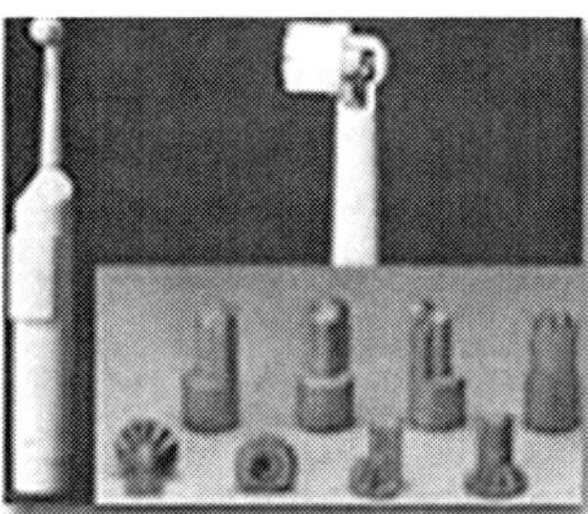

Electric toothbrush gear (Schunk, GmBH) Courtesy of EPMA.

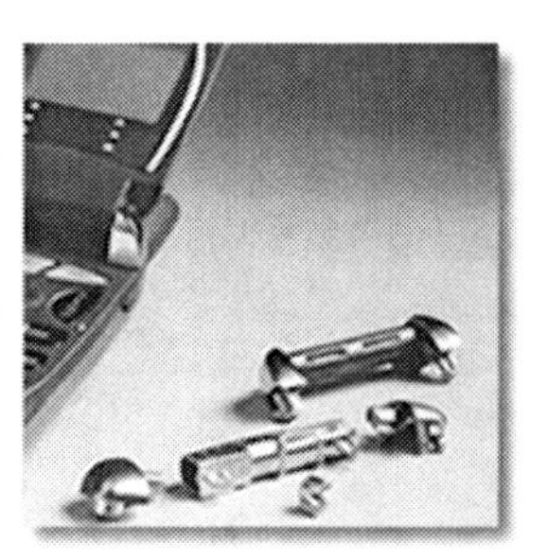

Cellphone hinge (Motorola) Courtesy of MPIF.

6.39 Examples of MIM components used in consumer applications.

Rifle rear sight (Megamet solid metals, Inc.) Courtesy of MPIF.

Revolver parts (Taurus International) Courtesy of MPIF.

Titanium trigger (Mimest Spa, Italy) Courtesy of PIM Int.

6.40 Examples of MIM components used in weaponry and firearms applications.

Stator vane (Rolls Royce) | Seat belt component (MIM Ecrisa, Spain) All courtesy of PIM International. | A380 Valve holder (Metal Injection Mouldings Ltd, UK)

6.41 Examples of MIM components used in aerospace applications.

rear-sight base, rear-sight blade, bolt, hand, barrel and the frame, which alone weighs 240 g, also made from nickel-steel as well as 4140. Also shown is a photograph of a titanium gun trigger MIM part.

Following the early successes for MIM in the late 1970s, MIM has found a number of applications in the aerospace sector, including seatbelt components, latches and fittings, spray nozzles and vane adjustment levers (Schlieper, 2009a). Figure 6.41 shows some examples of MIM components used, or with a potential to be used, in aerospace applications. The MIM stator vane shown is an IN718 component that was recently announced by Rolls Royce (PIM, 2011) who have investigated the feasibility of replacing the current seven-stage forging process by MIM. The stator vanes are used in the compressors of Rolls Royce aero engines. Seatbelt components are made via MIM using Fe7Ni0.6C steel. A small MIM component for the giant A380 Airbus is the valve holder. The parts were previously made from pressed metal, small machined pillars, washers, nuts, and so on, using 22 items, which have been replaced by just two MIM parts.

6.10 Conclusion

The applications examples given above show that MIM is being used to manufacture parts that are encountered in everyday life. They illustrate that there is an economic advantage where the parts are small, complex and can be manufactured in large volumes. The materials choice range is equally broad for different applications. The examples also show that dimensional accuracy can be achieved in MIM where close tolerances are required.

6.11 References

Agote, I., Odriozola, A., Gutierrez, N., Santamaria, A., Quintanilla, J., Coupelle, P. and Soares, J. (2001). 'Rheological study of waste porcelain feedstocks for injection moulding'. *J. Eur. Ceram. Soc.*, **21**(16), 2843–53.

Anwar, M. Y. (1996). *Injection Moulding of Various Particulate Materials Using Novel Emulsion-based Binder System.* PhD Thesis. University of Sheffield, UK.

Cao, M. Y., O'Connor, J. W. and Chung, C. I. (1992). A New Water Soluble Solid Polymer Solution Binder for Powder Injection Molding. Metal Powder Industries Federation (MPIF) San Francisco, CA.

Chartier, T., Ferrato, M. and Baumard, J. F. (1995). 'Supercritical debinding of injection molded ceramics'. *J. Am. Ceram. Soc.*, **78**(7), 1787–92.

da Silva Jorge, H. R. C. (2008). *Compounding and Processing of a Water Soluble Binder for Powder Injection Moulding*, PhD Thesis, Universidade do Minho.

EPMA (2010). *Metal Injection Moulding.* http://www.epma.com/New_non_members/MIM.htm, Shrewsbury.

Ferri, O. M., Ebel, T. and Bormann, R. (2010). Influence of surface quality and porosity on fatigue behaviour of Ti–6Al–4V components processed by MIM. *Mater. Sci. Eng., A*, **527**(7–8), 1800–5.

Froes, F. H. (2006). Getting better: big boost for titanium MIM prospects. *Met. Powder Rep.*, **61**(11), 20–3.

Froes, F. H. (2007). 'Advances in titanium metal injection molding'. *Powder Metall. Metal. Ceram.*, **46**, 303–10.

German, R. M. (1979). Gas flow physics in porous metals. *Int. J. Powder Metall.*, **15**(1), pp. 23–30.

German, R. M. (1990). Powder Injection Molding. Metal Powders Industries Federation, New Jersey.

German, R. (2009). Optimising your design for MIM production [Online]. Shrewsbury, UK. Available: http://www.pim-international.com/designing_for_PIM/design [Accessed 01 March 2013].

German, R. M. and Bose, A. (1997). Injection Molding of Metals and Ceramics. Metal Powders Industries Federation (MPIF), New Jersey.

Ismail, M. H., Sidambe, A. T., Figueroa, I. A., Davies, H. A. and Todd, I. (2010). Effect of Powder Loading on Rheology and Dimensional Variability of Porous, Pseudo-elastic NiTi Alloy Produced by Metal Injection Moulding (MIM) Using a Partly Water Soluble Binder System. EPMA, Florence, 347–54.

Jellinek, H. H. G. (1978). Degradation and depolymerization kinetics. In: Aspects of Degradation and Stabilization of Polymers. Elsevier, Lausanne/New York. 1–37.

Ji, C. H., Loh, N. H. and Tor, S. B. (2001). Sintering study of 316L stainless steel metal injection molding parts using Taguchi method: final density. *Mat. Sci. Eng., A*, **311**, 311(1–2), 78–82.

Joens, C. J. (2005). Laminar gas flows ensure 'clean sweep' in sintering. *Met. Powder Rep.*, March, **60**(3), 52–5.

Khor, K. A. and Loh, N. H. (1994). Dilatometry Studies on Water Atomized Stainless Steel 316L Powders. EPMA, France.

Kim, D. H., Lee, Y. W., Yoo, K. P. and Lim, J. S. (2007). Separation of paraffin wax from ceramic injection molded part. *Theories Applications Chem. Eng.*, **13**(1), 235–8.

Knudsen, M. (1909). 'Kinetic theory of gases'. *Ann. Phys.*, **28**, 75.

Krug, S., Evans, J. R. G. and ter Maat, J. H. H. (2002). Reaction and transport kinetics for depolymerization within a porous body. *AlCheE Journal*, **48**(7), 1533–41.

Lewis, J. A. (1997). Binder removal from ceramics. *Annu. Rev. Mater. Sci.*, **27**, 147–73.

Lumley, R., Sercombe, T. B. and Schaffer, G. B. (1999). Surface oxide and the role of magnesium during the sintering of aluminum. *Metall. Mater. Trans. A*, **30**(2), 457–463.

Murray, K. (2009). Overview of the European MIM Market, Euro PM Congress and Exhibition, EPMA, Copenhagen.

Nayar, H. S. and Wasiczko, B. (1990). Nitrogen absorption control during sintering of stainless steel parts. *Met. Powder Rep.*, September, **45**(9), 611–14.

PIM. (2010). 'PM2010 showcases the latest innovations in powder metallurgy and powder injection moulding'. PM 2010 World Congress Review, PIM International, December, **4**(4), 43–7.

PIM (2011). Rolls Royce investigates MIM super alloy stator vanes. *PIM International*, September, **5**(3), 24.

Rawers, J., Croydon, F., Krabbe, R. and Duttlinger, N. (1996). Tensile characteristics of nitrogen enhanced powder injection moulded 316L stainless steel. *Powder Metall.*, **39**(2), 125–9.

Schlieper, G. (2009a). Metal Injection Moulding (MIM), Ceramic Injection Molding (CIM): An introduction [Online]. Shrewsbury, UK: Innovar Communications Ltd. Available: http://www.pim-international.com/what_is_metal_injection_moulding/introduction_to_metal_injection_molding_MIM_PIM [Accessed 01 March 2013].

Schlieper, G. (2009b). Powders for Metal Injection Moulding [Online]. Shrewsbury, UK. Available: http://www.pim-international.com/aboutpim/sintering.

Sidambe, A. T. (2004). Surface Engineered Polymer Bonded Magnets, PhD Thesis. Cranfield University, Bedford.

Sidambe, A. T., Figueroa, I. A., Hamilton, H. G. C. and Todd, I. (2010). Metal injection moulding of Ti-64 components using a water soluble binder. *PIM Int.*, **4**(4), 56–62.

Sidambe, A. T., Figueroa, I. A., Hamilton, H. G. C. and Todd, I. (2012). Metal injection moulding of CP-Ti components for biomedical applications. *J. Mater. Proc. Tech.*, **212**(7), 1591–7.

Sidambe, A. T., Choong, W. L., Hamilton, H. G. C. and Todd, I. (2013). 'Correlation of metal injection moulded Ti6Al4V yield strength with resonance frequency (PCRT) measurements'. *Mater Sci Eng A-Struct*, **568**, 220–7.

Smagorinski, M. E. and Tsantrizos, P. G. (2002). Production of Spherical Titanium Powder by Plasma Atomization. Metal Powders Industries Federation, Orlando, 248–60.

Smith, J. E. and Jordan, M. L. (1964). 'Mathematical and graphical interpretation of the log-normal law for particle size distribution analysis'. *J. Colloid Sci.*, **19**, 549–59.

TLS, TLS Technik GmbH and Co. *The EIGA Process* [Online] Available at: http://www.tls-technik.de/e_2.html. [Accessed February 2012].

Tsai, D. S. (1991). 'Pressure build up and internal stresses during binder burnout: numerical analysis'. *AI ChE J*, **37**, 547–54.

Vervoort, P. J., Vetter, R. and Duszczyk, J. (1996). Overview of powder injection molding. *Adv. Perf. Mater.*, **3**(2), pp. 121–51.

Wakao, N., Otani, S. and Smith, J. (1965). Significance of pressure gradients in porous materials: Part I, Diffusion and flow in fine capillaries. AI ChE J, **11**(3) 435–9.

Whittaker, D. (2007). Powder injection moulding looks to automotive applications for growth and stability. *PIM Int.*, **1**(2), 14–22.

Xai, L. X. and German, R. M. (1995). Powder injection molding using water-atomized 316L. stainless steel. *Int. J. Powder Metall.*, **31**(3), 257–264.

Yang, W. W., Yang, K.-Y., Wang, M.-C. and Hon, M.-H. (2003). 'Solvent debinding mechanism for alumina injection molded compacts with water-soluble binders'. *Ceram. Int.*, **29**, 745–56.

Part II

Materials and properties

7
Advanced powder metallurgy steel alloys

H. DANNINGER and C. GIERL-MAYER,
Vienna University of Technology, Austria

DOI: 10.1533/9780857098900.2.149

Abstract: Ferrous powder metallurgy (PM) makes up the majority of powder metallurgy products with regard to tonnage. Improving performance is the main trend for pressed and sintered parts, in particular the introduction of cost-effective alloy elements such as Cr and Mn. Furthermore, much can be gained in ferrous PM by elaborate secondary operations. In metal injection moulding (MIM) products, there is a clear trend towards increasingly complex shapes and microsized parts. PM tool steels offer a much finer and fully isotropic microstructure compared to their wrought counterparts and the carbide content may be much higher, resulting in excellent application properties.

Key words: ferrous powder metallurgy parts, metal injection moulding (MIM), micro-MIM, powder metallurgy tool steels, sintered alloy steels.

7.1 Introduction

Powder metallurgy (PM) is the technology – in some cases even the art – of producing metallic materials and components from powders rather than through the classical ingot metallurgy route. PM products can be largely divided into two major groups. One group comprises 'powder metallurgy materials', that is, in this case the PM route is selected to obtain materials that are not accessible by other routes or at least not with the specific properties required; a typical example is the WC-Co hardmetal with its peculiar microstructure (see Fig. 7.1(a)) which cannot be produced by ingot metallurgy. The other group includes 'powder metallurgy precision parts'. Here, PM offers economical manufacturing of complex-shaped parts in large numbers; a classical example is the automotive camshaft belt pulley (Fig. 7.1(b)). Thus, in the first case there are technical reasons that favour PM while the second case predominantly favours a more economical production route. Of course, in the end all PM products have to meet economical criteria in order to remain competitive; in many cases, however, the PM product itself may be more expensive than a competing one, but its performance is so much higher that the cost savings attained more than offset the higher purchasing cost.

Strictly speaking there should be a third product group that is a synthesis of the other two: this is net-shape or at least near-net shape manufacturing

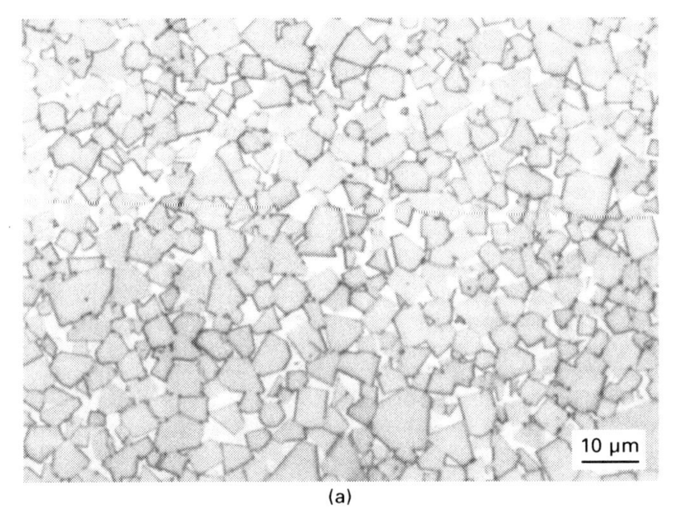

(a)

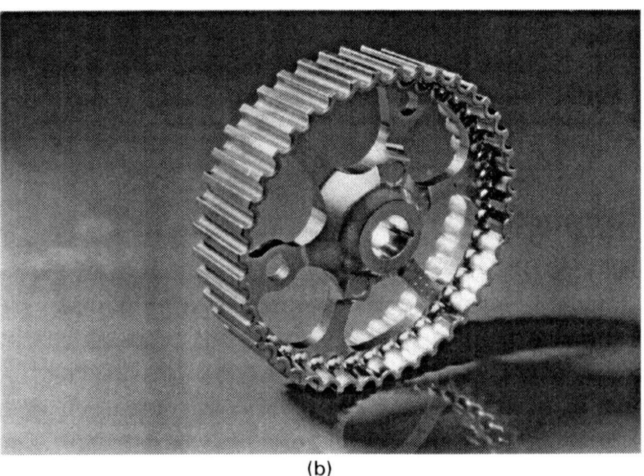

(b)

7.1 Typical powder metallurgy products: (a) microstructure of WC-Co hard metal (b) camshaft belt pulley (photo: MIBA).

of parts from specific materials. Indexable hard metal inserts for metal cutting can be mentioned here as well as self-lubricating sintered bearings, PM manufacturing being well suited to obtain products with high material utilization and energy efficiency, which is particularly useful for expensive materials.

This structure of PM products in general also holds for ferrous powder metallurgy products. The group of sintered steel parts produced by pressing and sintering form the single largest group of PM products with regard to tonnage. Here the complexity of the shape and geometrical precision required combined with large production volumes are the main reasons for selecting PM and not more classical metalworking techniques such as casting, machining or hot forging. The complexity of PM parts can be further extended through shaping by metal injection moulding (MIM); in this case, three-dimensional geometries can be attained that are inaccessible by standard uniaxial die compaction; also for MIM parts, however, large production volumes are a precondition for success on the market.

The other, more material related product group is represented mainly by powder metallurgy tool steels. These high alloy steels are fully dense, in contrast to pressed and sintered parts and to most MIM products. They are available mainly as semi-finished products, that is in this case net shape manufacturing is not a topic except for special cases. In principle these tool steels are similar to the respective classes of ingot metallurgy tool steels – cold work tool steels, high speed steels and so on – but they are typically at the upper end of alloy element content and performance and also of cost. Accordingly they are used mainly for these applications when the material cost is low compared to the 'system cost', such as tool manufacturing or machine downtime.

Although powder metallurgy is a very modern technology, ferrous PM strictly speaking dates back more than 4000 years, since steel has been manufactured by a related technology – very similar to modern powder forging – right into the late middle ages, it being impossible to melt iron up until then.[1] When the more productive blast furnace technology emerged, ferrous PM went into oblivion and only in the 1930s was manufacturing of precision-shaped ferrous components introduced. From then, PM parts production has increased dramatically and today the total consumption of iron and steel powders for parts production can be estimated to be in the range of 600–700 kt annually.[2,3] This is still a very small fraction of total steel production, but the added value in PM parts is considerable and the high energy and raw material efficiency of PM manufacturing must be regarded as a definite asset, in particular when considering the increasing cost of raw materials and energy in latter years, which must be expected to continue further.

7.2 Composition of advanced pressed and sintered steel components

In principle, powder metallurgy steels follow the same pattern as wrought steels, and most of the information about wrought steels available in the literature can also be used for sintered steels. There are, however, some pronounced differences that originate mostly from the powder manufacturing route, that is the very large specific surface of the starting materials and also the fact that in PM it is possible to start from chemical inequilibrium.

In powder metallurgy, plain iron plays a considerable role, for example for soft magnetic parts. Manufacturing plain iron is easier in PM than in ingot metallurgy, owing to the fact that plain iron still forms the majority of starting powders in ferrous PM, in other words while in ingot metallurgy, the alloy elements present from the pig iron have to be removed to obtain plain Fe, in PM plain Fe is available at the beginning and alloy elements are added, in a defined way.

As with wrought steels, carbon is the most important alloy element and carbon steels form a considerable proportion of PM ferrous parts. Nevertheless, the large majority of PM steel parts are alloyed and traditionally a major difference between wrought steels and sintered steels has been the choice of alloy elements. While in wrought structural steels, Cr, Mn, Si and V are the main alloy elements, in powder metallurgy Cu, Ni and Mo are commonly employed; iron–copper–carbon parts make up about half of all PM ferrous parts. The reason for this different choice of alloy elements is, as stated above, the large specific surface and resulting high reactivity of the PM specimens. Metal powders and parts pressed from such powders are sensitive to reactions with the atmosphere, in particular oxidation, and every thermal treatment – in particular sintering – has to be carried out in protective atmosphere. Furthermore, all metal powders that have ever been exposed to air, and this holds for virtually all commercial grades, are covered by oxide layers and these have to be removed in the early stages of sintering since they inhibit formation of stable interparticle bridges. If alloy elements with high oxygen affinity are used, such as Cr or Mn, removal of the oxide layers is difficult, requiring in part high temperatures,[4] and the risk of oxygen pickup is fairly high and at the very least, the requirements for purity of the protective atmosphere are much higher than for less oxygen sensitive alloy elements like Cu or Ni. This can be seen to advantage from the Richardson–Ellingham diagram (Fig. 7.2), which depicts the Gibbs free energy of formation of metal oxides as a function of the temperature. The oxides of the elements Cu, Ni and Mo have a similar or even lower free energy of formation than those of iron, that is in any atmosphere in which iron can be sintered without oxidation, these alloy elements are also not oxidized. The elements Cr, Mn, Si and V, in contrast, show a much more negative ΔG, that is they are preferentially oxidized compared to iron and, if oxidized, are more difficult to reduce, and the purity of the atmosphere has to be adjusted to these elements, not to iron.

In the last decade, progress in furnace and atmosphere technology has enabled successful sintering of Cr alloyed steels[6] and this is one of the major directions of research in ferrous PM. Nevertheless, sintering Cr alloyed systems requires considerable skill and experience and the choice of the sintering atmospheres is restricted, for example the frequently used endogas atmosphere is not feasible since it is oxidizing for Cr. The main drivers towards development of Cr, and more recently Mn as well, alloyed steel are

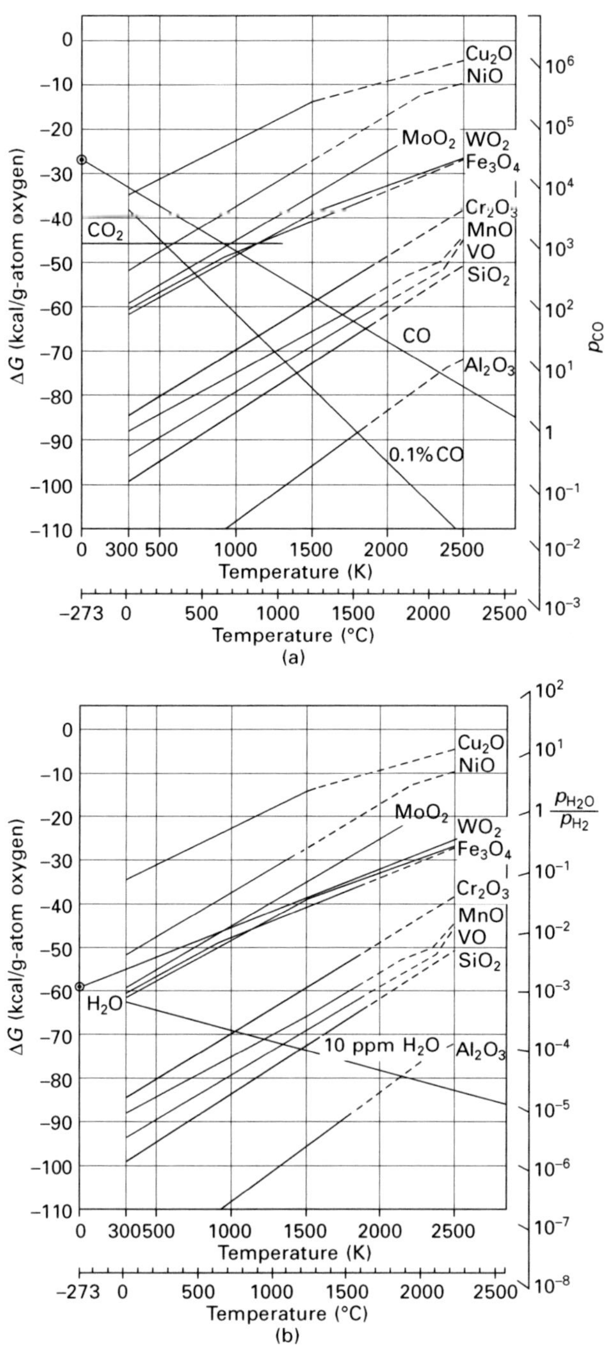

7.2 Richardson–Ellingham diagrams for the alloy elements in wrought and PM steels, with (a) C/CO and (b) H_2 as reducing agents (after Glassner[5]).

the dramatic and unpredictable cost increases for Cu, Ni and Mo (Fig. 7.3) that have affected the material cost of PM steels markedly more than those of wrought steels. Furthermore, health hazards linked to Ni powder as well as recycling problems with Cu in steels promote development of other alloy systems.

In addition to the chemical aspects, the influence of the alloy elements on sintering also has to be considered. Admixed Cu tends to cause expansion of the pressed parts during sintering ('copper swelling')[8] while admixed Ni promotes shrinkage.[9,10] These effects can be used to attain clearly defined dimensional changes during sintering, thus minimizing the risk of distortion and ensuring close dimensional tolerances. If Cu and Ni are to be replaced by other alloy elements, these aspects have to be considered.

Generally it must be recognized that in powder metallurgy the way alloy elements are introduced has a decisive influence on processing and properties, since there are more options than in ingot metallurgy (Fig. 7.4). In ingot

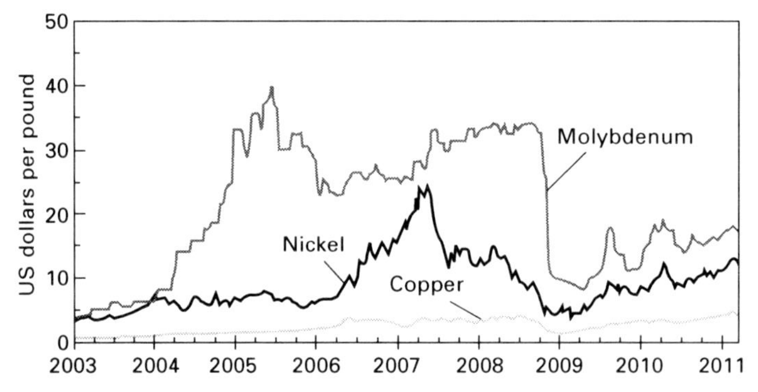

7.3 Cost of the alloy elements Mo, Ni and Cu (after Donaldson *et al.*[7]) (1 pound = 454 g).

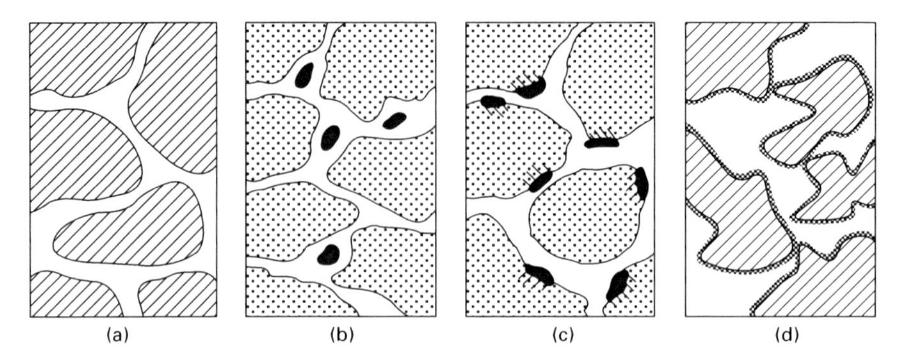

7.4 Alloying variants for powder metallurgy steels (after Engström[10] and Zapf and Dalal,[86] modified): (a) prealloyed, (b) mixed, (c) diffusion bonded, (d) coated.

metallurgy, the initial state – that of the melt – is in chemical equilibrium, and this can also be attained in PM, by atomizing a melt through water or gas jets, which results in 'prealloyed' powders. In PM, however, other states are possible, a blend of elemental powders, which might be regarded as the 'state of maximum inequilibrium'. From this state, by suitable tuning of the sintering parameters, the process towards equilibrium may be stopped arbitrarily and thus microstructures with defined heterogeneity can be attained that are not accessible by ingot metallurgy.

Both variants, 'prealloyed' and 'blended elemental' have their pros and cons. While prealloyed powders are usually harder, owing to solid solution strengthening, and thus less compressible, elemental powder mixes show a tendency to segregate. This problem can be eliminated while retaining the compressibility by 'diffusion bonding', which is slight sintering of the fine alloy element particles to the coarser base iron powders. Which alloying variant is used depends on the alloy element to be introduced. Cr and Mo have only a slightly adverse effect on the compressibility, so these elements are frequently added through the prealloying route. Cu and Ni are usually admixed or diffusion bonded, both regarding compressibility and their positive effects when admixed, as stated above. The only element that is almost exclusively admixed is carbon, which is added as fine, mostly natural, graphite, since prealloying would result in quite hard and poorly compressible powders. Combining different alloying routes in the same powder grade results in 'hybrid' types; such grades are based, for example, on an Fe-Mo prealloy powder and contain Cu and/or Ni diffusion bonded. Thus, the best features of the different alloying routes are combined.

7.3 Manufacturing routes for sintered steel components

7.3.1 Powder grades available for base ferrous powders and additives

Powders used for manufacturing of iron and steel parts are manufactured either by water atomizing, which is dispersion of a melt by high pressure water jets,[11] or by chemical reduction, coke being used as reducing agent. Prealloyed steel powders are always produced by atomization. Both powder types are then screened to form fractions which are stored in large containers, and from the powder fractions a 'synthetic' powder with a defined particle size distribution is produced. Thus, slight variations between the different production runs can be eliminated and a consistent quality of the powder can be afforded as required by the parts producer. These synthetic blends are finally reduced in continuous furnaces to lower the oxygen content and enhance the compressibility. The particle size range is typically 40–140 μm, with d_{50} = 80 μm (see Fig. 7.5); this combines good flowability with high compressibility and sintering activity. In particular with regard to these properties, PM powder grades are high purity products, with very low, highly reproducible levels of C, S and O as well as of Si, Mn and Al. Atomized powders consist of irregularly shaped, fully dense particles, while chemically reduced powders ('sponge iron') contain internal porosity. Some characteristics of typical powders are shown in Table 7.1.

Alloy elemental powders are usually finer than the base powders (see Fig. 7.6), typically <45 μm. For Cu, both atomized and electrolytic grades are used, while the carbonyl grades, manufactured by decomposition of $Ni(CO)_4$, are common for Ni. Mo powder is obtained by hydrogen reduction of MoO_3, but ferroalloy grades may also be used, as for Cr and Mn.

7.5 Scanning electron microscopy (SEM) images of some commercial ferrous powder grades for parts manufacturing. (a) Atomized iron powder, (b) reduced (sponge) iron powder, (c) atomized Fe-Cr-Mo (prealloyed), (d) diffusion bonded Fe-Ni-Cu-Mo.

(c)

(d)

7.5 Continued

Diffusion bonded steel powder grades contain mostly 1.8–4% Ni, typically 1.5% Cu and 0.5% Mo; hybrid types with 1.5% Mo prealloyed, plus 2% Cu or Ni or 4% Ni–2% Cu are also available. As prealloyed grades, Fe-Mo types with 0.3–1.5% Mo are on the market, as well as Cr and Cr-Mo types with up to 3% Cr and 0.2–0.5% Mo. Recently, Cr-Mn[7] and Cr-Mo-Si-containing variants[13] were presented, Mn-Si alloy steels having already been investigated in the 1980s;[14] carefully balanced multiple alloyed steel powders have also shown promising results.[15] Generally, a thorough understanding of the effects of the various alloy elements is now aimed for, compared to the more or less indiscriminate addition of more and more alloy elements done in previous decades, which is impractical today because of rising alloy element costs.

As stated above, carbon is added as fine graphite. Mostly natural graphite grades with ash contents below 4% are used; recently very good synthetic graphite types have also become available.[16,17] For uniaxial die compaction, pressing lubricants are required that prevent cold welding of the pressed

Table 7.1 Properties of some typical ferrous powders[12]

	Apparent density ($g\ cm^{-3}$)	Flow (s 50g)	Compressibility ($g\ cm^{-3*a}$)	C-content (mass%)	O-content (mass%)
ASC 100.29 (Fe, water atomized)	2.98	24	6.79 (400 MPa) 7.19 (600 MPa)	<0.01	0.08
SC 100.26 (Fe, sponge iron powder)	2.65	29	6.91 (500 MPa) 7.16 (700 MPa)	<0.01	0.20
Astaloy Mo (Fe-1.5% Mo, water atomized)	3.10	25	6.70 (400 MPa) 7.13 (600 MPa)	<0.01	0.08
Astaloy CrM (Fe-3%Cr-0.5%Mo, water atomized)	2.95	25	6.47 (400 MPa) 6.96 (600 MPa)	<0.01	0.20
Distaloy AE (Fe-1.5%Cu-4%Ni-0.5Mo, diffusion alloyed)	3.05	24	6.75 (400 MPa) 7.15 (600 MPa)	< 0.01	0.10
Distaloy HP (Fe-2%Cu-1.47% Mo, hybrid powder)	3.15	23	6.69 (400 MPa) 7.13 (600 MPa)	< 0.01	0.08

[a]Lubricant: 0.6% Kenolube P11.

compact to the tool components, in particular the die; here different grades are on the market, based mainly on the organic compounds Zn stearate and/or ethylene bisstearamide. These lubricants are admixed to the metal powder(s) as fine powder, typically <10 μm, at contents of 0.5–0.8 mass%.

7.3.2 Powder blending and conditioning

The powders are weighed and then blended in special mixers, mostly double cone mixers. In order to avoid segregation and in particular agglomeration, heating the bulk powder must be avoided. Therefore the mixing times are in the range of minutes and the alloy elements and other additives are added stepwise since, in particular, graphite and lubricant together tend to form agglomerates, which tendency can be further aggravated by fine metal powders.

The fine, flaky and specifically light graphite (density 2.26 g cm^{-3}) shows a tendency towards 'dusting', in particular at higher contents, an unwelcome phenomenon that results in wide variation of the carbon content in the sintered parts; this can be alleviated by bonding the graphite to the base iron or steel powder with a well defined amount of organic additive. Such 'bonded' powder mixes are available from the big powder manufacturers (e.g. brand names 'Starmix' or 'Ancorbond').

7.3.3 Compacting techniques

The standard shaping technique in ferrous PM is uniaxial die compaction (see also Chapter 5 and Chapter 14, Section 14.3). The powder mixes are compacted in rigid tools at pressures of 300–800 MPa, resulting in absolute green density levels of 6.5–7.2 g cm^{-3}, corresponding to a residual porosity of 16–8%. Higher density levels are difficult to attain because of increasing work hardening of the powders; furthermore, the space taken by graphite and in particular by the lubricant has to be taken into account since it

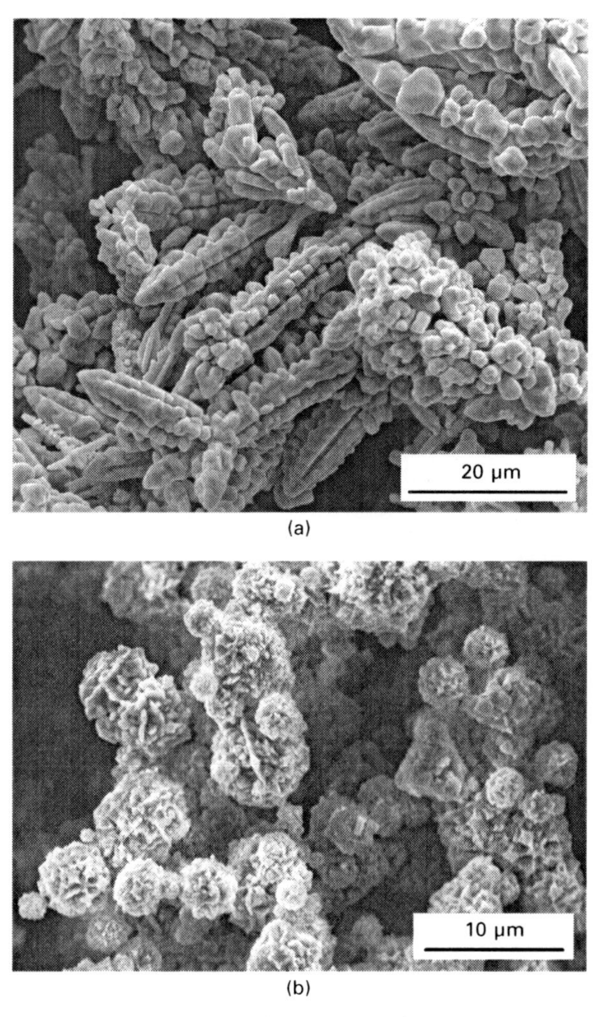

7.6 Alloy element powders used in ferrous powder metallurgy. (a) Electrolytic Cu powder, (b) carbonyl Ni powder, (c) hydrogen reduced Mo powder, (d) natural graphite.

(c)

(d)

7.6 Continued

lowers the theoretical (= pore-free) density of the powder mix. While plain iron has a theoretical density of 7.86 g cm^{-3}, addition of 0.8% lubricant (a typical content) lowers this value to 7.45 g cm^{-3}, with a resulting adverse effect on the compressibility and thus the final density since the space taken by the lubricant is turned to porosity during sintering when the lubricant is evaporated. Generally it has to be taken into account here that in PM parts production, densification during sintering is usually not aimed for, for tolerance reasons, in other words the density, or porosity, attained during pressing is more or less that in the final part, and since the pores tend to lower the mechanical properties, higher density is the most effective way to produce high strength PM components. Therefore, lowering the amount of lubricant added has long been a target of research; the final aim is to avoid completely the admixing of lubricants and to lubricate only the pressing tool ('die wall lubrication'). This works well in the laboratory, but in large scale industrial production the die wall lubrication systems available have only lowered the amount of admixed lubricant to about 0.4%; nevertheless, this is also helpful for attaining a higher green density.

In particular at lower lubricant content, high pressure compaction is effective in attaining higher density levels. Today, pressures up to 1000 MPa are applied in series production, although tool life is a critical factor here. The classical approach for high density has been double pressing; the second pressing operation applied for improving the density being termed 'repressing'. Since the limiting factor for green density is work hardening of the powder particles, sintering proper is not required between the first and second pressing operation but rather a soft annealing treatment is applied at moderate temperature to remove work hardening without dissolving admixed carbon too much, which would result in loss of compressibility for the second pressing.[18] Typically, repressing enables density levels of 7.3 to be attained, in some cases 7.4 g cm^{-3} compared to 7.1 g cm^{-3} maximum for single pressing; it is a competing technology with warm compacting (see below). The inherent drawback of double pressing is the markedly higher cost. In contrast to competing technologies for attaining high density, here the cost penalty compared to single pressing is not due to special powder and equipment but simply the additional processing steps of annealing and second pressing.

Since the yield strength of iron markedly decreases with higher temperatures, compacting at elevated temperature has been studied for a long time. High temperature pressing involves the major problems of lubrication, organic additives no longer being applicable, and of powder flow, since highly pure powders tend to stick, resulting in unacceptable flow behaviour. Furthermore, the risk of oxidation has to be considered. However it was found that slightly above room temperature, in the temperature range 135–150°C, some gain in green density can be obtained while oxidation and flow problems do not yet exist and organic lubricants can be used. This 'warm compaction' came to the market in the mid-1990s and has slowly been introduced for high density parts.[19] In addition to the higher density, the very much improved green strength has shown to be an asset that enables 'green machining', that is cutting of pressed parts, for example to introduce features that cannot be die pressed, such as threads or undercuts.

Still higher density levels can be obtained by high velocity compaction (HVC), applying the load not slowly as in standard pressing, but as a shock wave, through a hydraulic hammer.[20] HVC can be applied to the powder itself or, in the fashion of double pressing, to further density a compact; density levels of up to 7.6 g cm^{-3} can be attained. So far, however, HVC has been used only for single-level parts such as gears; for complex multi-level parts tool loading still seems to be a major problem.

This limitation of HVC refers to a general feature of die compaction:[21] for each level of the part in the axial direction a separate lower punch is required, since the ratio of the height of bulk powder column to the height of pressed powder must be constant for all levels, otherwise widely varying density, and most probably distortion during sintering, and probably also cracking,[22] might occur. This is shown schematically in Fig. 7.7. Multi-level parts thus require complex, multi-component pressing tools whose various parts have

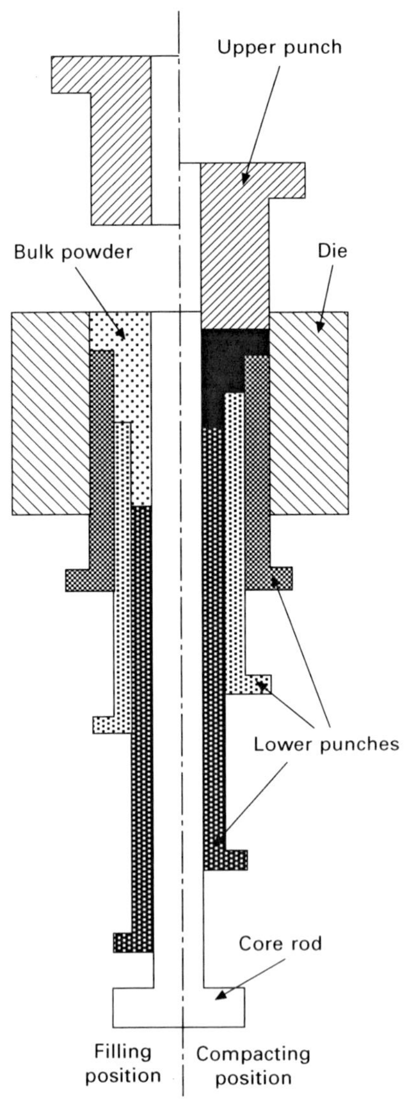

7.7 Pressing tool for a multi-level part.

micrometre clearance between each other. This is a major cost factor in PM; the high tooling cost is also the reason why PM parts manufacturing is cost effective only in large production lots, typically >100,000 parts.

7.3.4 Sintering processes

Sintering is a heat treatment below the melting point of the main constituent; for ferrous products sintering temperatures are typically between 1120 and 1280°C. During sintering the fragile interparticle contacts formed during pressing are transformed into sound metallic bridges that can bear substantial mechanical loads. As stated above, in ferrous parts production dimensionally stable sintering is desirable, with a maximum shrinkage of 1%; in contrast to hardmetals, considerable porosity is retained in the final parts.

In addition to the formation of stable sintering contacts, chemical homogenization also occurs,[23] that is admixed alloy elements that are soluble in the iron matrix are gradually dissolved, depending on their diffusivity in iron. Carbon could be dissolved above 730°C from the thermodynamical view, but in reality temperatures >900°C are required to ensure fast and complete dissolution. For metallic alloy elements, much higher temperatures are required, owing to the slower diffusion, and admixed Ni does not fully homogenize even at the highest temperatures, owing to the very slow solid state diffusion in Fe.

Homogenization can be accelerated, for example by formation of a transient liquid phase, as in the case of Cu: when attaining the melting point of Cu, all admixed Cu particles melt and the liquid phase formed fills capillaries in the iron matrix, leaving more or less spherical pores ('secondary pores') behind (Fig. 7.8). The large contact area between the Cu melt and the iron

7.8 Secondary pores in sintered Fe-3% Cu.

matrix results in rapid solidification of the melt through formation of a solid solution; complete Cu homogenization within the Fe particles requires solid state diffusion and is usually not attained during industrial sintering practice (which is however not necessary, nor even desirable, as will be shown below). Penetration of the pressing contacts by the Cu melt results in expansion ('copper swelling', see, for example, Dautzenberg and Dorweiler[8]) (Fig. 7.9(a)) which can be used to compensate for the natural shrinkage of the iron body. Such transient liquid phases and the related effects of secondary pores

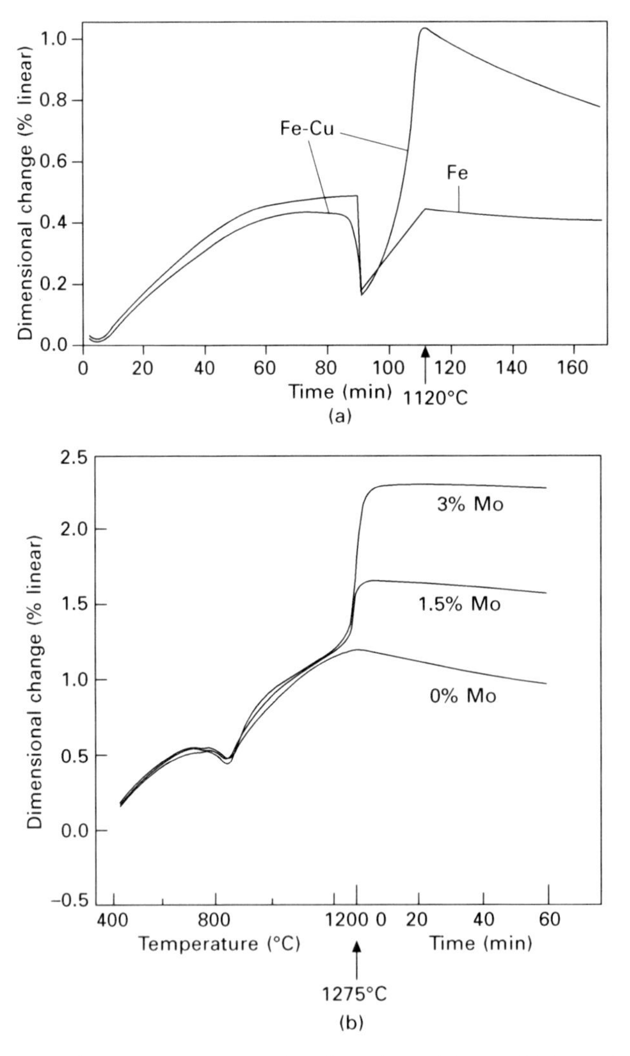

7.9 Dilatometric graphs for sintering different PM steels. (a) Fe and Fe-Cu, (b) Fe-*x*% Mo-0.7% C, (c) Fe-2 %Cr-0.7 %C, different Cr fractions, (d) Fe-*x*% Mn-0.3% C.

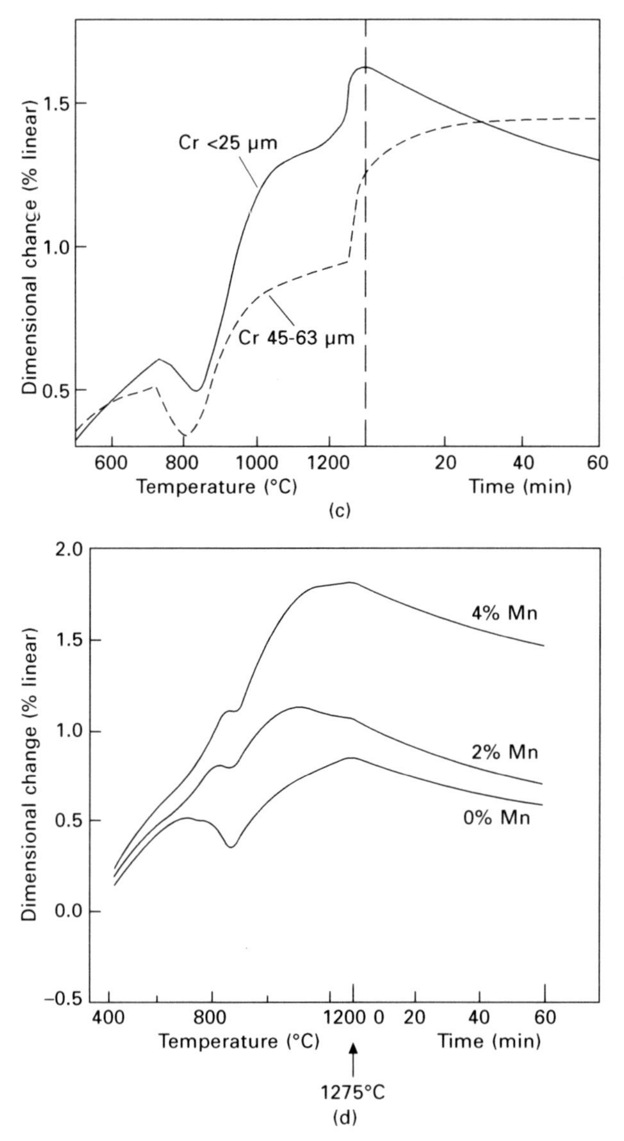

7.9 Continued

and swelling, appear not only in Fe-Cu and Fe-Cu-C but also in systems with admixed carbide-forming elements such as Mo, Cr, W and V (Fig. 7.9(b) and (c)), but only above a given temperature threshold that depends on the carbon content.[24,25]

A special case of homogenization occurs in the case of Mn. This element has a markedly higher vapour pressure than other metals, and admixed Mn or Fe-Mn starts to evaporate below the ferrite–austenite transformation temperature.[26,27] The Mn vapour condenses at the iron surfaces, forming solid solution there, and also at the sintering necks, which results in swelling, but gradually, not within a very narrow temperature interval as in the case of a transient liquid phase (Fig. 7.9(d)). Also here, complete homogenization requires solid state diffusion within the individual iron particles, which is not faster than for Ni.

In addition to the metallurgical aspects, the chemical processes during sintering are of crucial importance. The first process is dewaxing or lubricant burnout, which means evaporation of the admixed lubricant. This is done in the first stage of sintering, typically at temperatures up to 800°C. Basically it is a physical process, but reactions between the lubricant and the powder compacts may occur, resulting in unwelcome carbon pickup. Slightly decarburizing atmospheres may be useful in preventing this effect ('rapid burnoff'), but care must be taken to keep such atmospheres away from the high temperature sintering zone.

After further heating, the oxides covering the powder particles are reduced (the adsorbed humidity having already been desorbed during dewaxing); these processes can be followed by thermoanalytical techniques combined with mass spectrometry (Fig. 7.10). In reducing atmospheres, for example N_2-H_2, the iron oxides on the surfaces can be removed at $T<500°C$;[28] even with Cr prealloy steels, some reduction is observed, since the oxide layers on these powders are in part iron oxides and in part stable Cr- or Cr-Mn oxides. In neutral atmospheres, N_2 or vacuum, reduction requires the presence of carbon, and this carbothermic reduction starts at about 700°C for Fe-C and Fe-Mo-C, but at least 1000°C for Cr prealloy steels. The internal oxides, within the powder particles and in the sintering contacts, need higher temperatures – 900–1100°C for Fe-C and Fe-Mo-C and typically 1250°C for the Cr prealloy variants. For chemically heterogeneous powder mixes, for example steels prepared from base iron powder with masteralloy addition, an approach that has been followed frequently,[29–31] the different affinities for oxygen have to be considered, reduction of the iron oxides on the base powder resulting in oxygen transfer to the masteralloy particles ('internal getter' effect), and the oxides formed there are much more difficult to reduce.[32]

Impure sintering atmospheres, in particular O_2 and H_2O, result in oxidation of the compacts at low to moderate temperatures and in surface decarburization at high temperatures; both phenomena are highly unwelcome. Carbon removal through reaction with hydrogen is less critical; there are some alloy elements such as Mn and Cr that seem to catalyze this reaction,[33] but in general decarburization observed after sintering in H_2-containing atmospheres is caused by humidity in the atmosphere, following the reaction:

$$C + H_2O = CO + H_2$$

Admixed Mn has a special effect in contact with the atmosphere: the Mn

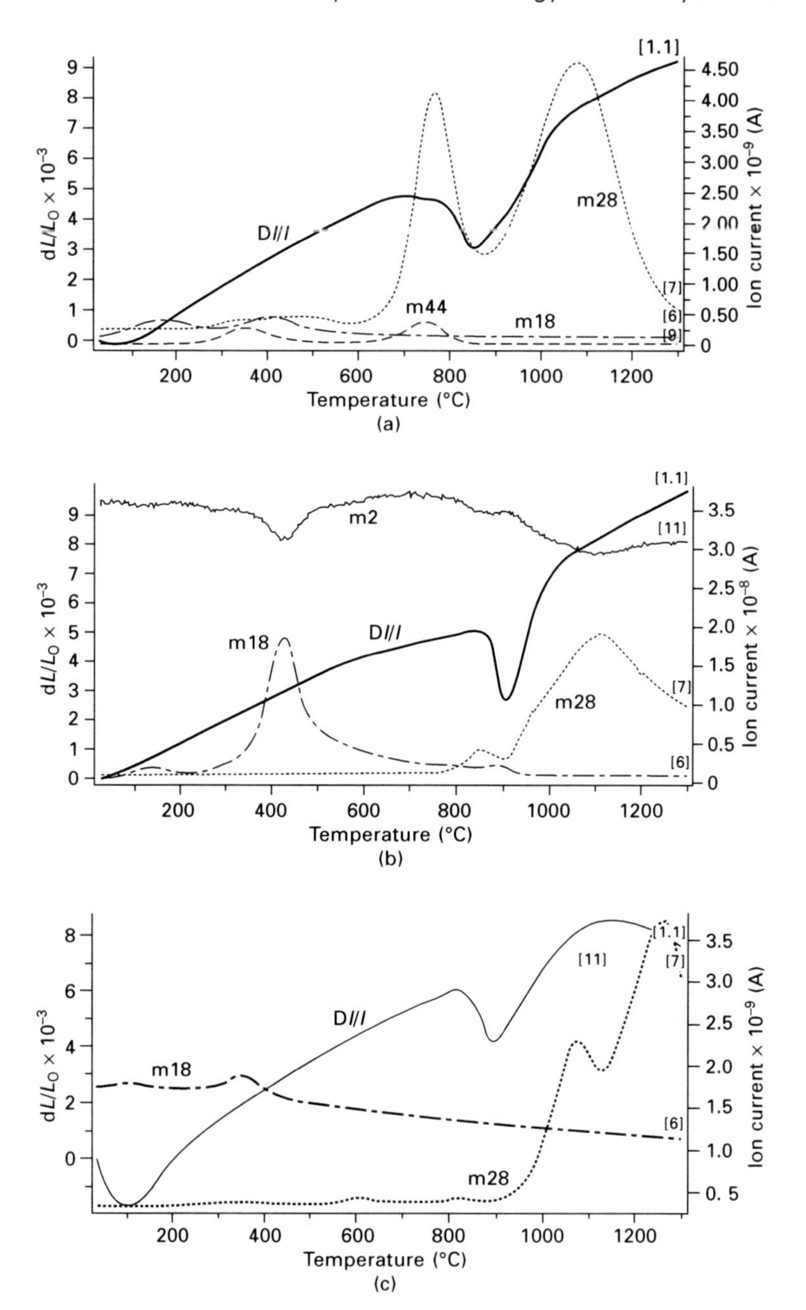

7.10 Dilatometry + MS graphs for sintering Fe-C and Fe-Cr-Mo-C in inert compared with reducing atmospheres.[27] (a) Fe-C, argon, (b) Fe-C, hydrogen, (c) Fe-Cr-Mo-C, argon, (d) Fe-Cr-Mo-C, hydrogen. *Dl/l* is the dimensional change relative to the green dimensions; the *m* numbers indicate the mass signals detected by mass spectrometry.

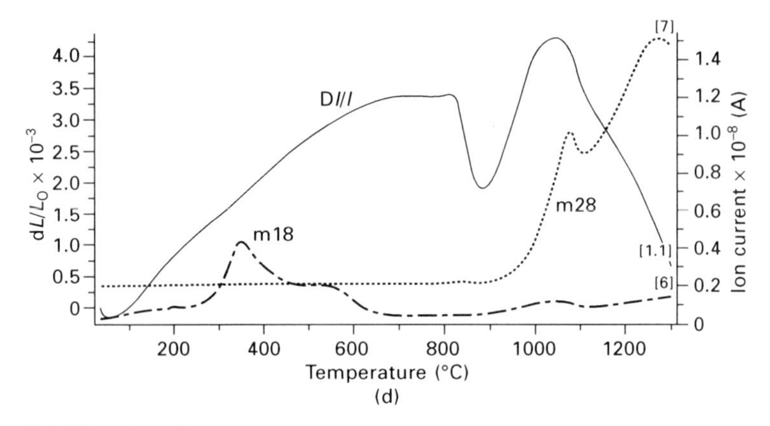

7.10 Continued

vapour formed at higher temperatures forms a 'shell' around the compact that acts as a 'getter', preventing penetration of the O_2 and H_2O molecules into the surface of the specimen.[26] This causes Mn loss at the surface, which is mostly unwelcome since in many parts the strengthening effect of Mn would be needed most at the surface. It has been reported that Mn evaporation is more pronounced in the laboratory than in industrial production,[24] however, the effect of Mn vapour on furnace linings and heating elements is still a matter of concern.[34]

The sintering furnaces used in practice are continuous types,[35] either mesh belt furnaces, which are cheaper to buy and operate, but are usually limited to 1150°C maximum and, being open, are more prone to O_2 and H_2O penetration into the furnace atmosphere. For higher temperatures, walking beam furnaces are used that can operate up to 1400°C in high purity atmospheres, being equipped with locks at the entrance and exit. The sintering atmosphere is commonly N_2-H_2 with 5–25% H_2; occasionally plain N_2 is also used. The gases are of high purity, at least at the furnace entrance; contamination of the sintering atmosphere usually originates from leaks in the furnace or from the sintered parts, not from the introduced gases.

7.3.5 Post-sintering treatments

In some cases, PM parts can be used immediately after sintering; mostly however one or more secondary operations are required, in part to improve the geometrical precision, but more often to modify the microstructure and thus the properties. Many of these operations can be related to those common for wrought steels, but have to be adapted for powder metallurgy products. There are, however, also secondary operations that are applicable only for PM parts, mostly because they rely on the presence of porosity.

Sintering, as a high temperature operation, usually results in some dimensional change in the parts which, mainly owing to anisotropy induced during die compacting, is not fully isotropic and homogeneous, resulting in some distortion, that is loss of precision. This can be rectified by a further pressing operation that corrects these distortion effects without markedly affecting the overall density. This operation is called 'sizing' and is done either in the pressing tool (the die usually being turned 180° to expose that side of the cavity that exhibits a slight taper) or in a separate tool. Commonly the die dimensions are slightly smaller than those of the sintered part and the part is therefore pressed into the cavity (taper of the die being required here) resulting in some local surface densification, so the porosity of the part is necessary to enable local deformation. If the porosity is to be retained at the surface, another sizing variant can be used, the die in this case being slightly larger than the part and the operation resulting in cross flow of the material. This variant of sizing is sometimes referred to as 'coining'. Successful sizing requires parts with low to moderate as-sintered hardness, usually 180–200 HV are defined as the threshold. Parts with higher hardness levels can also be sized in principle but die wear tends to become excessive, with resulting cost penalties. The precision of sintered parts can be improved considerably by sizing, in particular in the dimensions defined by the die (i.e. perpendicular to the pressing dimensions); typical precision values are shown below in Section 7.4.1.

The most successful recent innovation for ferrous PM parts is surely selective surface densification.[36,37] This technique relies on the fact that with many PM parts the maximum mechanical strength is required only locally, mostly at or near the surface. Through densification of these surfaces on sintered parts, mostly combined with thermochemical treatment, the application properties of these parts can be improved to the level of fully dense wrought steel parts but with weight savings – the remaining part being porous – and vibration damping capability. Typical products are gears that are pressed, sintered and then selectively densified by cross rolling the gears between master gears made of tool steel (Fig. 7.11(a)). Thus the tooth flanks and roots are brought to full density to a depth of typically 0.7–1 mm (Fig. 7.11(b)). After carburizing or carbonitriding, this results in high pitting resistance at the flanks and higher fatigue strength at the tooth roots. Such gears are already in service in automotive engines, for example for driving double overhead camshafts,[36] but even tests of PM gears in truck transmissions have been done successfully.[37]

Heat treatments of sintered steels can be done in a similar way as for wrought steels. However, the sometimes different composition has to be taken into account, not only the use of Ni, Cu and Mo as alloy elements but also the lower Mn and Si levels compared to wrought steels which results in lower hardenability. The main feature of PM parts is the open porosity

(a)

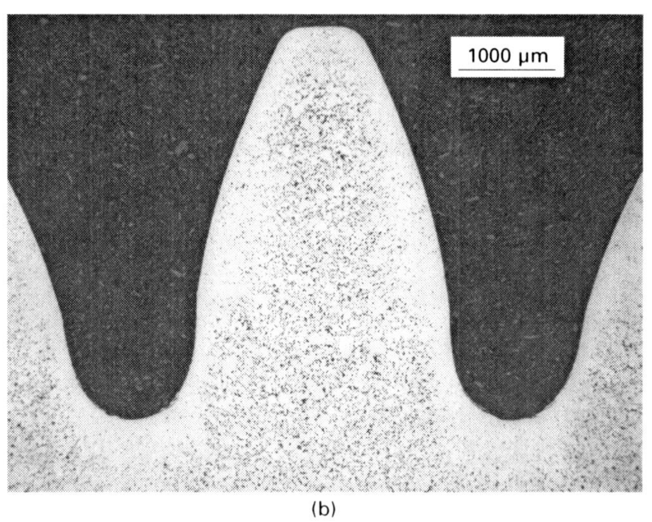

(b)

7.11 Surface densification of PM steels. (a) Surface densification by cross rolling of sintered gear between master gears (photo: MIBA). (b) Surface densified tooth of a gear (photo: MIBA).

that is present in virtually all parts (except those with density >7.4 g cm^{-3}, which are however still rare). The effect of the porosity on the thermal conductivity, and thus on the hardenability, is usually overestimated (see also Section 7.4.1), but open porosity results in intrusion of the quenching media into the pores, which at higher porosity levels has been claimed to improve quench rates[38] but in any case contaminates the parts especially in case of oil or salt bath quenching; removal of quenching oils from the pores is extremely difficult even in sophisticated washing operations.

Therefore, gas quenching has been used intensively in the last few years, usually as sinter hardening, quenching the parts directly after leaving the high temperature zone of the heating furnace. For this purpose, the sintering furnaces are equipped with quenching zones in which cold N_2 gas is blown at the parts. The cooling rates thus attained are lower than for oil quenching, typically 2–3 K s^{-1}, and this means that the parts have to be higher alloyed to obtain fully martensitic structures. Originally, steels containing mainly Ni, Cu and Mo were used, but today grades containing Cr, in part Mn and Si, are employed, mainly for cost reasons. In the beginning, sinter hardening could be done only in belt furnaces, the sintering temperatures being limited to a maximum of 1150°C, but today walking beam furnaces with gas quench facilities are available[39] and combining high temperature sintering with sinter hardening has been shown to be a very promising approach, in particular for Cr alloyed steel parts. Regarding the dimensional precision, it has to be considered that sinter hardened parts are almost impossible to size and particular care must be taken to avoid distortion during sintering which would necessitate expensive grinding operations.

Thermochemical treatments such as carburizing, carbonitriding or nitriding also are strongly affected by the presence of open pores which act as entrance sites for the reactive gas molecules, resulting in a tendency to overcarburizing or even through carburizing; the same holds for nitriding. By adapting the process parameters, for example lower temperatures and/or shorter times, these problems can be alleviated but only for parts with homogeneous porosity. When treating selectively densified parts, for example carburizing of gears, it is not possible to attain proper carburizing of the densified surfaces without overcarburizing the non-densified ones. Here, plasma-assisted treatments have been shown to be successful, in particular plasma nitriding; compared to gas nitriding the nitrogen is introduced as N_2 and not as NH_3, and the reactive atomic nitrogen is not generated by thermal dissociation of NH_3 but by dissociation of N_2 in the plasma above the surface of the parts. In the pore channels, any atomic N quickly recombines to inert N_2, thus precluding internal nitriding. Plasma nitriding thus has become a standard procedure in ferrous PM despite the cost penalties of equipment and of the necessity of electrically contacting the parts to the furnace.[40]

For carburizing or carbonitriding, plasma treatments have been less successful; here, low pressure carburizing (also known as vacuum carburizing) has been used, combined with gas quenching.[41,42] Here, the much larger mean free path of the active molecules means that upon entering a pore channel they rapidly contact the pore walls, reacting there, and penetration into the part core is highly improbable, resulting in a desired well-defined surface which is also the case with fairly high porosity for which standard gas carburizing techniques would result in through carburizing (Fig. 7.12). This porosity-independent carburizing effect is particularly attractive for

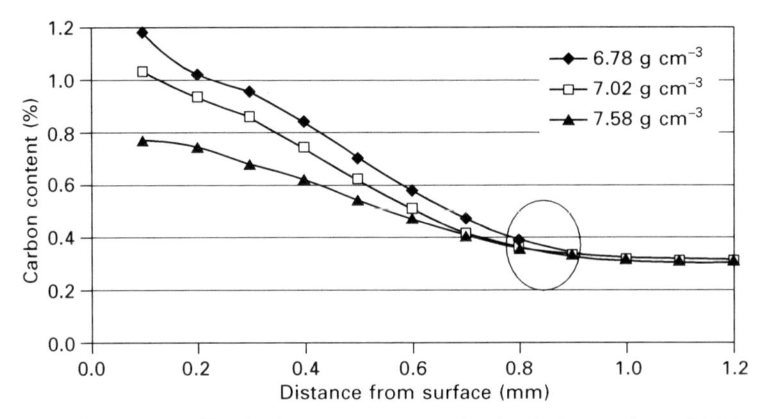

7.12 Carbon profiles in low pressure carburized sintered steel billets with varying density.

parts with graded porosity such as the surface densified gears mentioned above.[43] Control of the carburizing process is more tricky than for standard gas carburizing since the reactive gases are mostly acetylene or propane which carburize through a decomposition reaction and not in equilibrium, as CO does in gas carburizing; however, today processing is sufficiently advanced to enable successful low pressure carburizing of Cr alloyed sintered steels that are sensitive to formation of grain boundary carbides. Gas quenching usually combined with LP carburizing requires the use of steels with sufficient hardenability; on the other hand the hardened parts are clean, bright and shiny, thus avoiding the extensive washing operations necessary for oil quenched parts.

Since not all geometrical features are accessible through uniaxial die compaction, threads being the most frequent case, machining of sintered parts is common in PM production. It is well known that machining ferrous PM parts is difficult.[44] In part this is due to the chemically heterogeneous microstructure with areas of widely differing hardness, diffusion bonded Ni-Cu-Mo steel being notorious for poor machinability. The main feature is, however, the porosity. It has been supposed that interrupted cutting occurs, from matrix to pores and so on,[45] but more recent studies[46] have shown that the porous sintered steel is densified ahead of the cutting edge, with resulting pronounced work hardening, and the edge of the tool thus always meets work hardened material. This is corroborated by the fact that impregnation of the part with resin (i.e. with a material with much lower hardness than the steel matrix) nevertheless results in improved machinability since resin-filled pores of course do not collapse in front of the tool edge, thus avoiding work hardening. Furthermore, addition of machining aids[47] such as MnS or CaF_2 are used to improve the machining characteristics, in particular to act as chip breakers[48] and to eliminate built-up edges.[49] Adapting the cutting parameters and using proper tool geometries is beneficial for improving the machining of PM parts.

Another way to increase the range of PM part geometries is to manufacture separate parts with simpler geometry and then combine them by suitable joining techniques. This also enables production of parts consisting of different materials, such as ferromagnetic-diamagnetic materials. Many joining techniques known from standard metalworking are also applicable for PM; arc welding is, however, avoided since it tends to cause large cavities in the weld, due to pore coalescence. Rather, fusion welding techniques with very concentrated energy and resulting narrow weld seams, such as laser welding, are preferred,[50] while resistance (projection) welding is the most widely used welding technique for PM steel components. Brazing requires special fillers that solidify in contact with the steel; otherwise, the liquid filler would infiltrate the pores, leaving an empty joint. Such reactive fillers are commercially available today, mostly based on Ni-Cu-Mn-Fe-B,[51] but cast-iron type fillers have also been successfully tested.[52] A technique specific for PM is 'sinter bonding', applicable for parts with concentric features: here, the inner part is prepared from a material that swells during sintering, such as Fe-Cu, while the outer one shrinks, for example Fe-C, and during sintering the components form a solid metallic joint. For details please see Chapter 11.

7.4 Properties, microstructures and typical products

7.4.1 Attainable properties

Geometrical precision

The powder metallurgy press-and-sinter route is one of the most effective methods for producing complex-shaped precision parts. Typically precision in the dimensions perpendicular to the pressing axis, that is those dimensions that are defined by the die ('mouldable dimensions'), is better than parallel to the axis as shown by Assinter.[53] In the former case the clearance between the tool components – die, punches, core rods – has to be considered, which limits the precision of concentric features. The precision attainable for PM compared to other standard metal shaping techniques is shown in Fig. 7.13 (from Rübenach[54]), defined as international tolerance (IT) grades. As stated above, secondary operations such as sizing and double pressing, improve the dimensional precision, in particular in the mouldable dimensions, from a standard IT 10 as-sintered to IT 7 as-sized, and for double pressed and double sintered steel parts from IT 9 to IT 6 after sizing.[53]

The precision and in particular the effect of sizing also depends on the material, in particular the hardness, as indicated in Table 7.2;[55] the closer

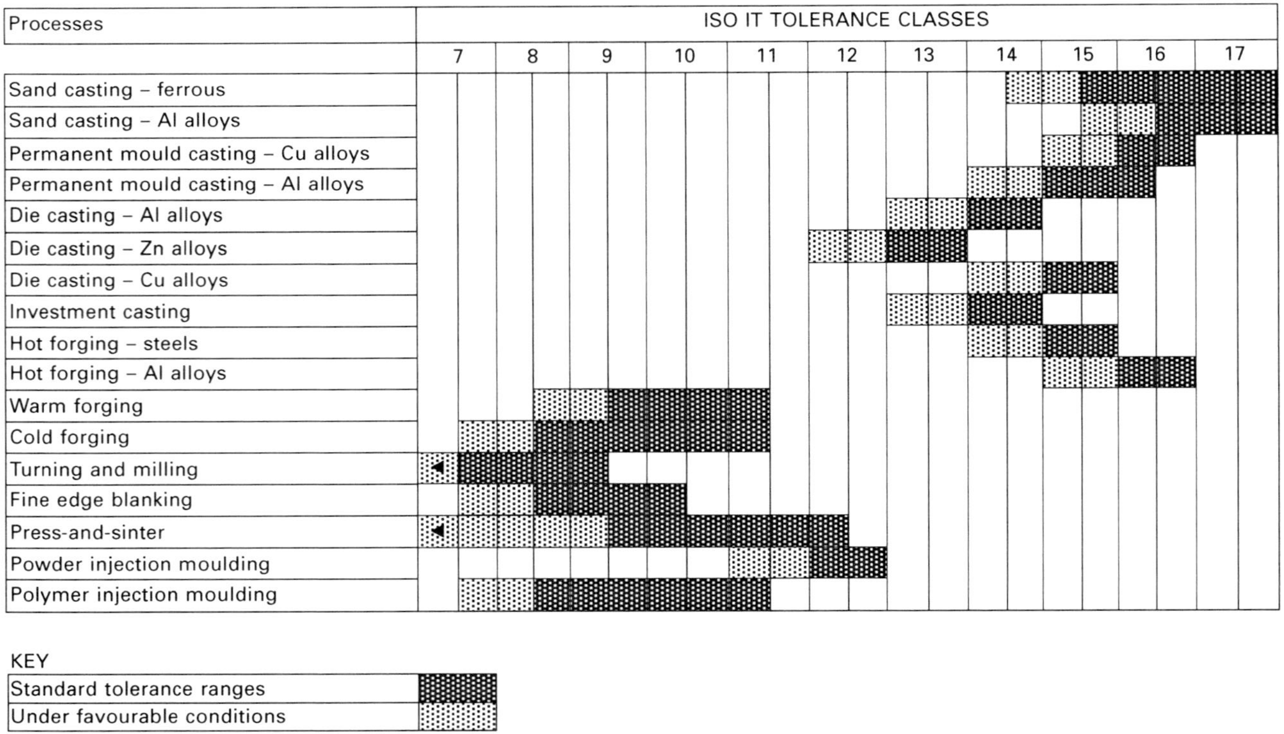

Processes

ISO IT TOLERANCE CLASSES

7 8 9 10 11 12 13 14 15 16 17

Sand casting – ferrous
Sand casting – Al alloys
Permanent mould casting – Cu alloys
Permanent mould casting – Al alloys
Die casting – Al alloys
Die casting – Zn alloys
Die casting – Cu alloys
Investment casting
Hot forging – steels
Hot forging – Al alloys
Warm forging
Cold forging
Turning and milling
Fine edge blanking
Press-and-sinter
Powder injection moulding
Polymer injection moulding

KEY
Standard tolerance ranges
Under favourable conditions

7.13 Geometrical precision attainable by press-and-sinter routes compared to other metalworking techniques (after Rübenach[54]).

Table 7.2 Tolerances attainable as-sintered and as-sized, respectively, in different sintered steel grades[55]

Designation DIN 30910	C (mass%)	Cu (mass%)	Ni (mass%)	Cr (mass%)	Mo (mass%)	Fe (mass%)	Density (g cm^{-3})	Sintered (°C)	Hardness (HB)	IT sintered	IT sized
C 00	< 0.3	< 1.0				balance	6.4 ... 6.8	1120	> 35	9	5 ... 6
C 10	< 0.3	1 ... 3				balance	6.4 ... 6.8	1120	> 40	9	6
D 10	< 0.3	1 ... 3				balance	6.8 ... 7.2	1120	> 50	9	6
E 10	< 0.3	1 ... 3				balance	> 7.2	1120	> 80	9 ... 10	7
C 11	0.4 ... 1.5	1 ... 3				balance	6.4 ... 6.8	1120	> 85	9	7
D 11	0.4 ... 1.5	1 ... 3				balance	6.8 ... 7.2	1120	> 100	9	7 ... 8
D30	< 0.3	1 ... 3	1 ... 2		0.3 ... 0.8	balance	6.8 ... 7.2	1120	> 80	9	7
D30	< 0.3	1 ... 3	1 ... 2		0.4 ... 0.8	balance	6.8 ... 7.2	1250	> 80	10	7
D39	0.3 ... 0.6	1 ... 3	1 ... 2		0.4 ... 0.8	balance	6.8 ... 7.2	1120	> 140	9	8
D39	0.3 ... 0.6	1 ... 3	1 ... 2		0.4 ... 0.8	balance	6.8 ... 7.2	1250	> 160	10	
D39	0.3 ... 0.6	1 ... 3	2 ... 5		0.4 ... 0.8	balance	6.8 ... 7.2	1250	> 160	10	
C40	< 0.08		10 ... 14	16 ... 19	2 ... 4	balance	6.4 ... 6.8	1140	> 100	10 ... 11	7
D40	< 0.08		10 ... 14	16 ... 19	2 ... 4	balance	6.8 ... 7.2	1140	> 130	10 ... 11	8

tolerances attainable with softer materials are clearly evident. Generally it must be stated that the IT data given here should always be regarded as a general rule, the tolerances attainable for a specific sintered part depending on composition, density, sintering parameters and, of course, the geometry. Therefore the data available about precision are relatively scarce, parts manufacturers being reluctant to offer detailed information to the public.

The criteria defined by the automotive industry for their supply chain regarding the stability and reliability of the PM processes, are standard nowadays and six-sigma capability as well as ISO 900x are essential conditions.

Mechanical properties

The mechanical properties of sintered steels differ from those of wrought steels insofar as there are many more parameters that affect the properties. For wrought steels, the composition and heat treatment state are relevant; for PM steels, the total porosity, the pore morphology (mostly defined by the sintering parameters) and the chemical homogeneity additionally have a strong effect on the properties (Fig. 7.14). Therefore, properties given for sintered steels should always be regarded as indicative, but even for the same composition and nominal density, variations in the properties from one manufacturer to another may occur. In the respective standards for sintered steels, minimum and typical properties are given, but frequently the parts producer and the customer agree upon additional definitions for composition and properties.

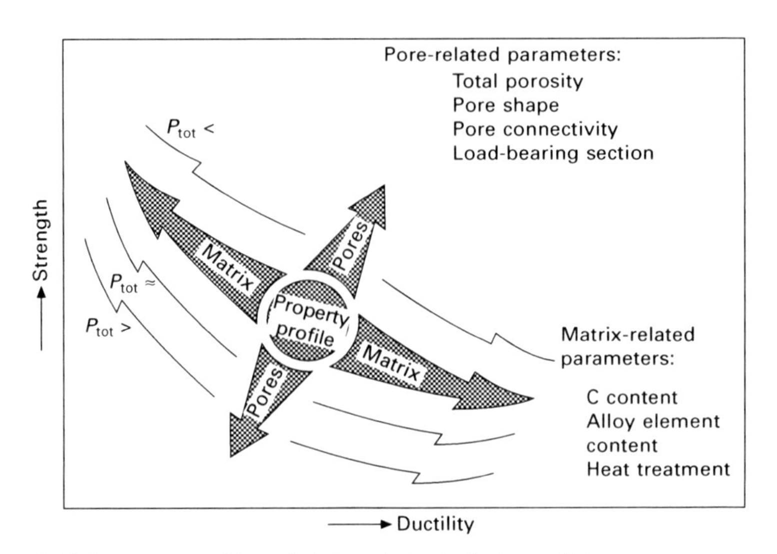

7.14 Property profiles of sintered steels (schematic).

In particular for the benefit of designers, an extensive database of PM materials has been established by the leading powder metallurgy associations of North America (MPIF), Europe (EPMA) and Japan (JPMA). This database, which is accessible free of charge,[56] was made public in 2004 and has been consistently revised and updated.

The main parameter that governs the mechanical properties of sintered steels is the relative density or, looking from another perspective, the total porosity.[57] There are, however, different relationships between the density and other properties, depending on the relative effect of the particle cores and the sintering contacts: while for tensile strength, hardness and the dynamic Young's modulus an almost linear relationship is described, the graphs of property versus density show an increasingly convex, almost exponential, shape in the sequence 'fatigue endurance strength – elongation – impact energy', indicating the progressive dominance of the sintering contacts (Fig. 7.15). It can further be stated that this sequence also holds for the effect of the sintering conditions. While there is an only marginal effect of the sintering temperature, for example on Young's modulus,[58] the impact energy is strongly influenced not only by the density but also by the sintering temperature and

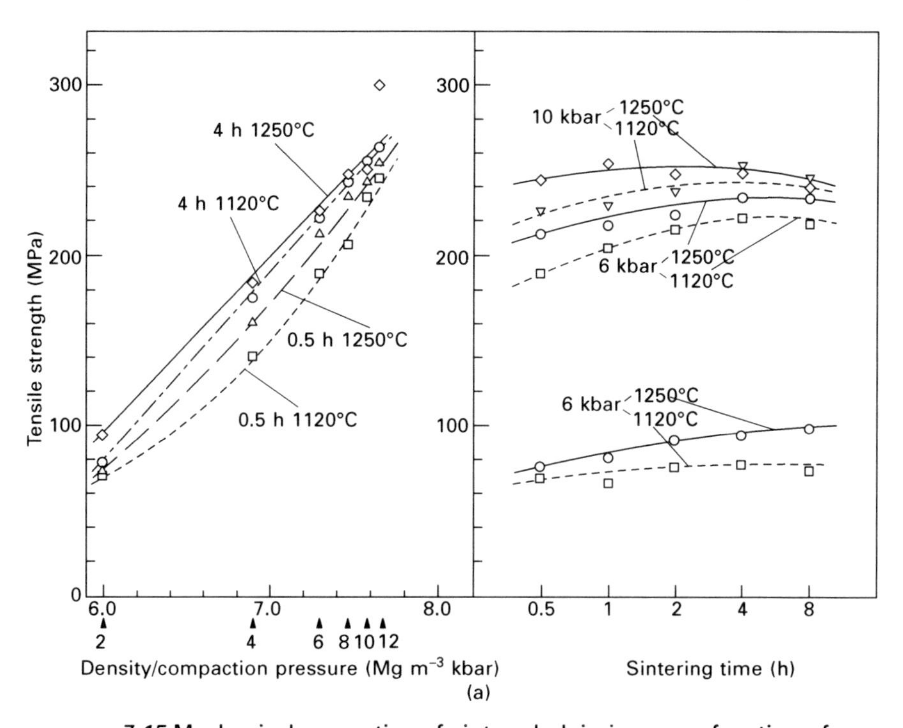

7.15 Mechanical properties of sintered plain iron as a function of compacting pressure and sintering parameters: Density in Mg m^{-3} and compacting pressure in kbar. (a) tensile strength, (b) impact energy.

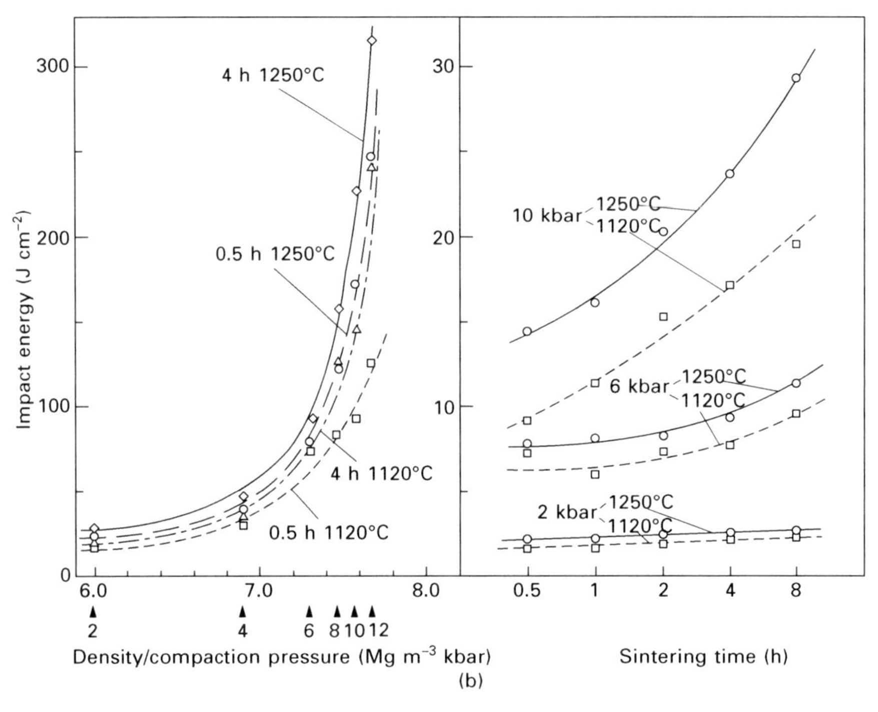

7.15 Continued

time; however, this effect is noticeable mainly at high density/low porosity levels (Fig. 7.15(b): plain Fe). Generally the best mechanical properties are attained when combining high density, accessible for example by warm pressing, with high temperature sintering.

It has been stated[59] that up to a certain density level there is a positive relationship of the fracture toughness with the yield strength, that is for low to moderate strength levels, higher yield strength also means higher fracture toughness, both properties being dominated by the quality of interparticle bonding. At high density, and strength, levels, however, the fracture toughness remains roughly constant even if the strength increases (Fig. 7.16); this change of the behaviour is linked to a transition from ductile rupture to transgranular cleavage fracture.

The fatigue endurance strength of sintered steels has been studied under widely varying loading conditions. The relationship to the density as well as the effects of the mean stress have been described, enabling the prediction of the fatigue strength from one single well documented *S–N* curve.[60–62] Zafari and Beiss[62] note that the effectively loaded volume is a crucial parameter; the smaller it is the higher is the endurance strength. The effect of various surface treatments has also been studied. Since many PM parts are used in automotive applications, for which service loading cycle numbers easily

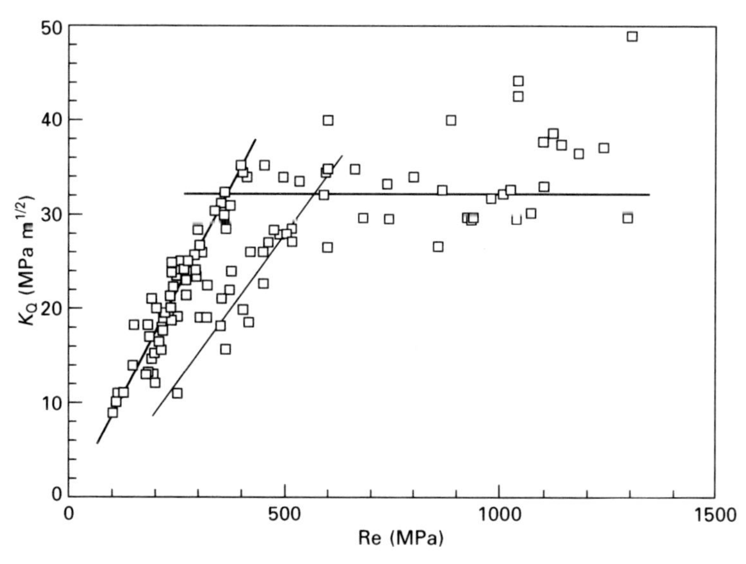

7.16 Relationship between fracture toughness and yield strength of sintered steels.[59]

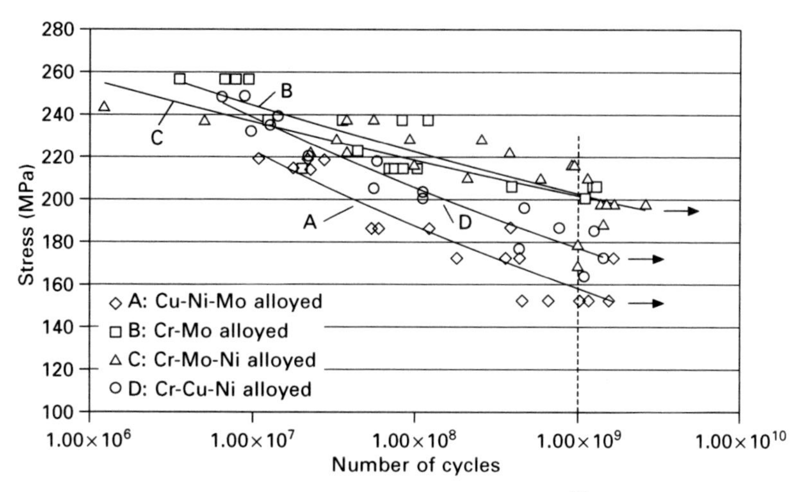

7.17 S–N curves of various sinter hardened steels.[64]

exceed 10^8, gigacycle fatigue tests have been systematically performed.[63, 64] These showed that even when testing up to 10^9 cycles and more, there is no fatigue 'limit', that is the *S–N* curves drop consistently with higher *N* (Fig. 7.17). This behaviour is, however, not restricted to sintered steels but has also been documented for wrought high strength steels,[65] for example for bearing or spring steels. A typical feature of gigacycle testing, both in sintered and wrought steels, is its sensitivity to singular defects such as inclusions, large pores or pore clusters;[66] if these defects are present in the loaded volume, fatigue cracking invariably starts from the largest defect (Fig. 7.18). For sintered steels this means that care must be taken during handling the powders to avoid introduction of impurities or formation of lubricant agglomerates that would result in crack-initiating defects; for wrought steels inclusions act in the same unwelcome way (e.g. Furuya[67]).

Cyclic fracture mechanical studies have shown that crack growth in a stable (Paris–Erdogan) regime proceeds faster for porous sintered steels than for wrought ones, while the threshold stress intensity factor is virtually the same.[68] This indicates that for low to medium cycle fatigue loading, wrought steels are superior while for the high to ultra high cycle regime, the sintered steels are competitive. A further aspect is alloy element distribution:[69] in particular for Ni containing diffusion bonded variants with heterogeneous microstructures, slower crack growth is reported compared to homogeneous grades.[70]

Thermal and thermophysical properties

The database for thermal properties of sintered steels is rather limited and in part contradictory, in particular with respect to the influence of porosity. It has been claimed by several authors that the coefficient of thermal expansion

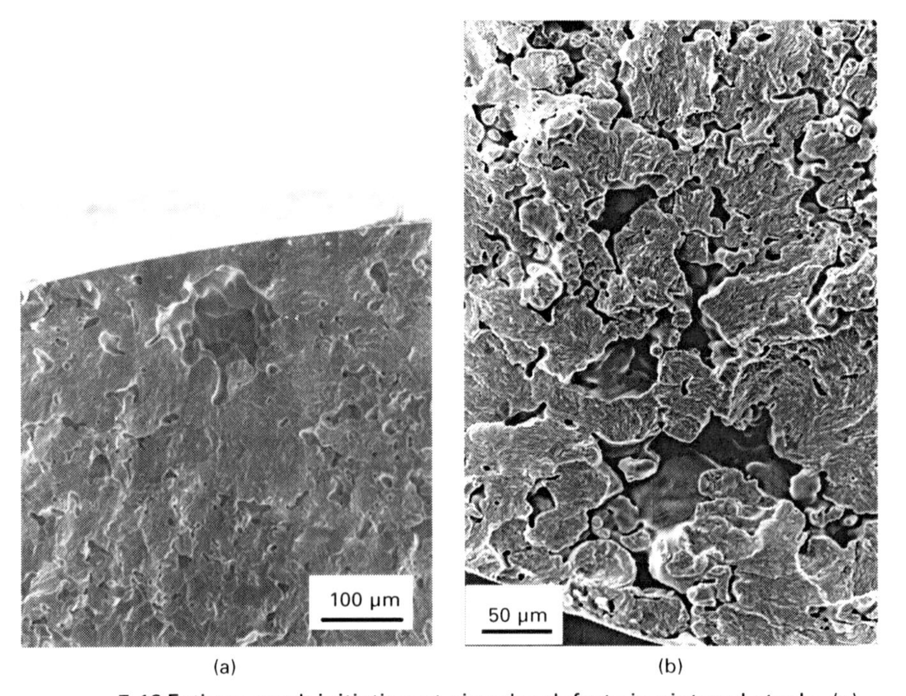

7.18 Fatigue crack initiation at singular defects in sintered steels. (a) secondary pore, (b) pore cluster.

(CTE) of sintered ferrous materials decreases with increasing porosity. Recent investigations[71] have, however, shown that these findings have probably been artifacts, caused by slight sintering shrinkage during measurement. CTE studies of various types of sintered steels did not show any dependence on the porosity (Fig. 7.19), which is not surprising considering that the thermal expansion of a steel grid is the same as a massive steel body, so data from wrought steels can be transferred to sintered steels, although, of course, the sometimes slightly different composition has to be taken into consideration.

The thermal conductivity has also been a matter of discussion, mainly with regard to quenching during heat treatment. It has been assumed that the pores lower the hardenability significantly. However, it has been shown recently[71] that the effect of the porosity on the thermal conductivity should not be overestimated. At least at room temperature the effect of alloy elements tends to be more pronounced than that of the porosity (Fig. 7.20 and 7.21); at higher temperatures the former effect tends to decrease while that of the pores remains, but in any case it is relatively moderate if the steels have been properly sintered. Only for green or presintered specimens is the thermal conductivity pronouncedly lower,[72] which is of relevance for designing the sintering process.

Electrical and magnetic properties

The electrical conductivity of sintered steels is related to the thermal conductivity by Wiedemann–Franz' law:

$$\lambda/\kappa = kT$$

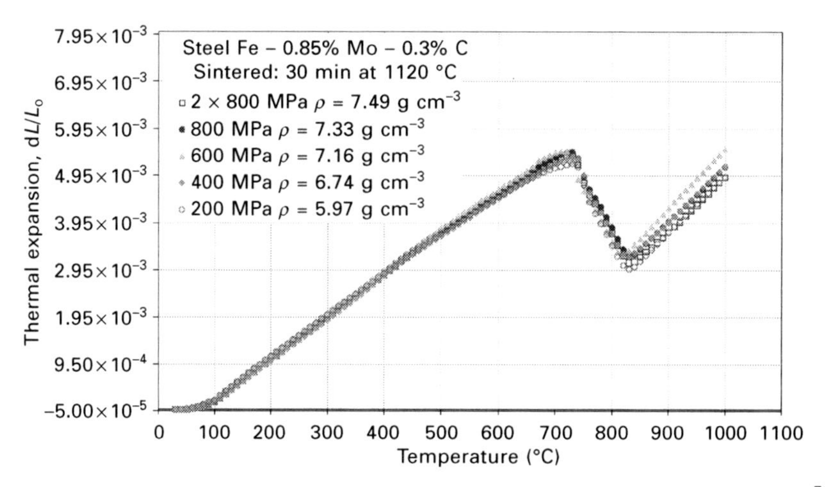

7.19 Dilatometric graphs of sintered steel with varying porosity.[71] (2 × 800 MPa indicates double pressed specimens.)

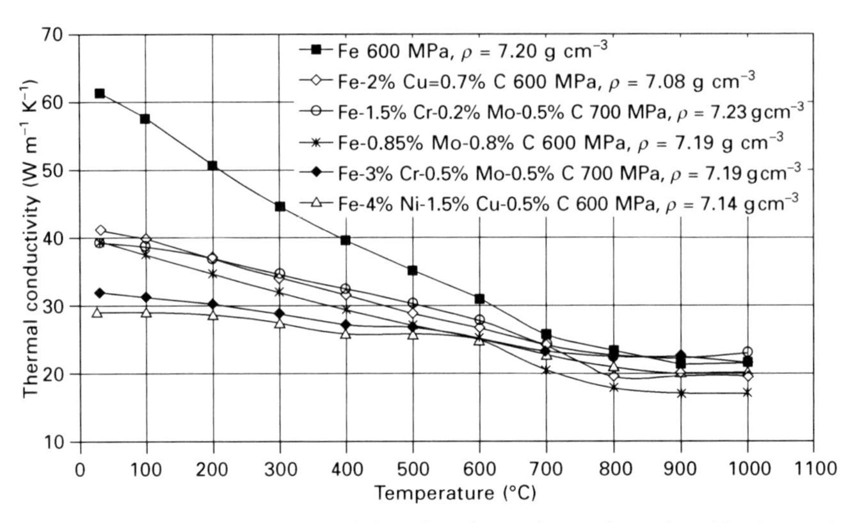

7.20 Thermal conductivity of various sintered steels with sintered density 7.1 ... 7.2 g cm^{-3}.[71]

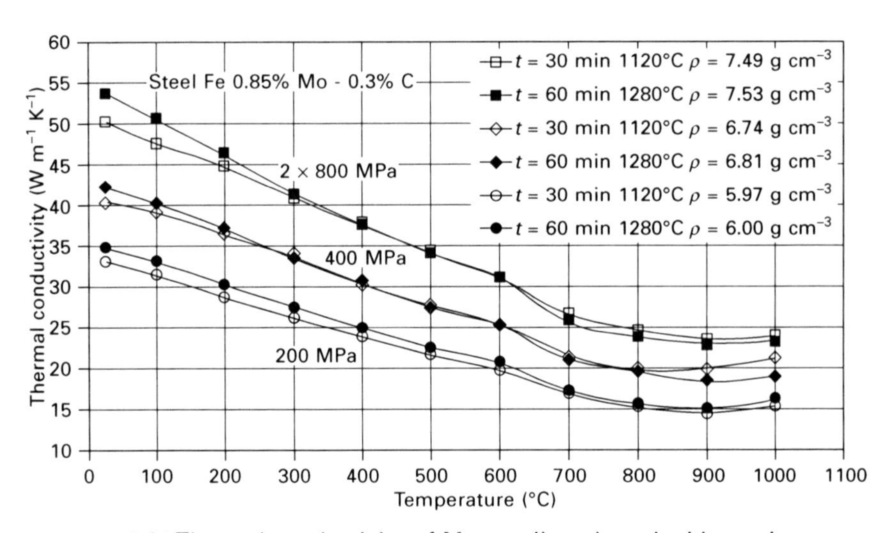

7.21 Thermal conductivity of Mo prealloyed steel with varying densities.[71]

where λ is the thermal conductivity and κ is the electrical conductivity.

Therefore the statements given above also hold for the electrical conductivity. It has been shown[73] that the conductivity is low for green compacts, in particular if pressing lubricant is used which has an insulating effect; presintering tends to increase the conductivity rapidly and for well-sintered materials the effect of the pores is rather insignificant. This can be related to the fact that in contrast to several mechanical properties, the electrical conductivity is determined not only by the sintering contacts but also by the cores of the original powder particles[74], so the effect of the load-bearing cross section is 'dampened', as has been found for Young's modulus.[75]

Sintered ferrous materials are used exclusively for soft magnetic applications, typical compositions being plain Fe, Fe-P, Fe-Si or Fe-Si-P. Therefore, high saturation maximum magnetization B_{max}, magnetic remanence B_r, and magnetic permeability μ, and low coercive force, H_c, are required. B_{max}, B_r and μ depend on the volume fraction of the ferromagnetic phase, that is, on the relative density of the material, while for the coercive force H_c the pore geometry is also of importance, well rounded pores, as obtained by high temperature sintering or by addition of ferrite stabilizers as P, being preferred to angular ones.[76,77] For reproducibly low H_c, the absence of fine carbides and nitrides is also essential and therefore sintering in high purity H_2 is frequently done which is otherwise avoided owing to cost reasons. PM is also well suited for high cost materials such as Ni and Co alloy irons, owing to its excellent material usage.[78]

For AC applications, eddy currents also play a major role, and lowering eddy current losses means increasing the resistivity of the material. This is done by adding Si, as is also common for wrought transformer sheets, and in PM higher Si levels are possible since the material need not be deformed after sintering. For high frequency applications, so-called soft magnetic composites (SMCs) have been introduced[79,80] which are parts consisting of powder particles that are insulated against one another by inorganic or organic layers. These SMCs have lower B_{max} and μ values than wrought materials but result in very low eddy current losses in an isotropic structure, and they enable special designs for electrical engines. These SMCs are surely one of the most promising fields of ferrous PM.

7.4.2 Microstructures

The microstructures of sintered steels differ from those of their wrought counterparts by the presence of pores, which in contrast to cavities in cast parts are fine and regularly distributed and therefore their effect on the properties can be reliably assessed. A second difference, at least for sintered steels prepared from mixed or diffusion bonded powders, is the chemical heterogeneity that may result in the presence of numerous microstructural constituents in a small area; ferrite, pearlite, martensite and retained austenite may be present within a few micrometres. Some typical microstructures are shown in Fig. 7.22; more detailed information about metallographic preparation techniques and microstructures can be found in Chapter 11.

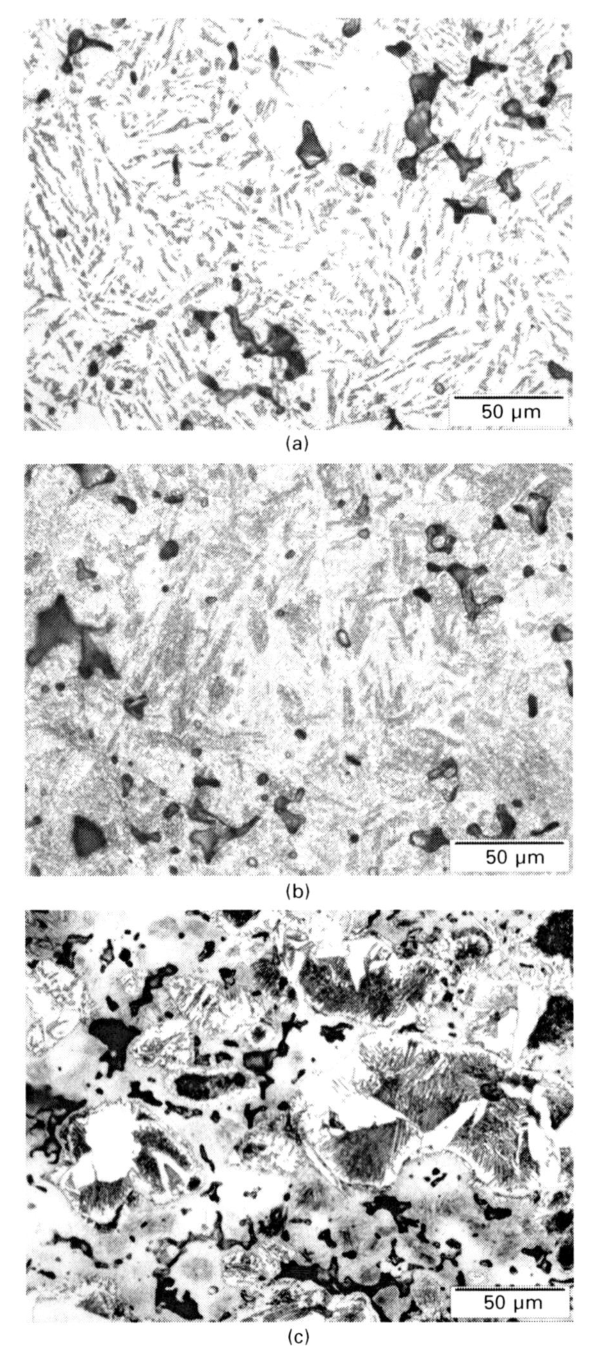

7.22 Microstructures of various sintered alloy steels; as-sintered: (a) Fe-0.85% Mo-0.3% C, prealloyed, (b) Fe-3% Cr-0.5% Mo-0.5% C, prealloyed, (c) Fe-4% Ni-1.5% Cu-0.5% Mo-0.5% C, diffusion bonded.

7.4.3 Typical products from sintered alloy steels

Sintered alloy steels are employed mainly for highly stressed parts as used for example in automotive engines and transmissions. Sprockets and gears for the camshaft drive and main bearing caps as well as synchronizer rings, hubs and sleeves are typical products and are predominantly manufactured by PM, at least for passenger cars. Some parts are shown in Fig. 7.23; more detailed information and further examples are given in Chapter 17.

(a)

(b)

(c)

7.23 Some typical PM steel parts produced by pressing and sintering: (a) PM helical gear (PMG), (b) PM sleeve (PMG), (c) PM main bearing cap (MIBA).

7.5 Powder injection moulded steel components

7.5.1 Powder injection moulding process

Powder injection moulding (PIM) is a PM manufacturing technique that combines the shaping capabilities of polymer injection moulding with the material flexibility of powder metallurgy. It is particularly suited for producing small, extremely complex-shaped metallic parts in large volumes. In short, the PIM process (Fig. 7.24) includes intense mixing of metal powder with a polymeric binder, consisting of at least two components, to a so-called 'feedstock', injection moulding it to a 'green part', removing

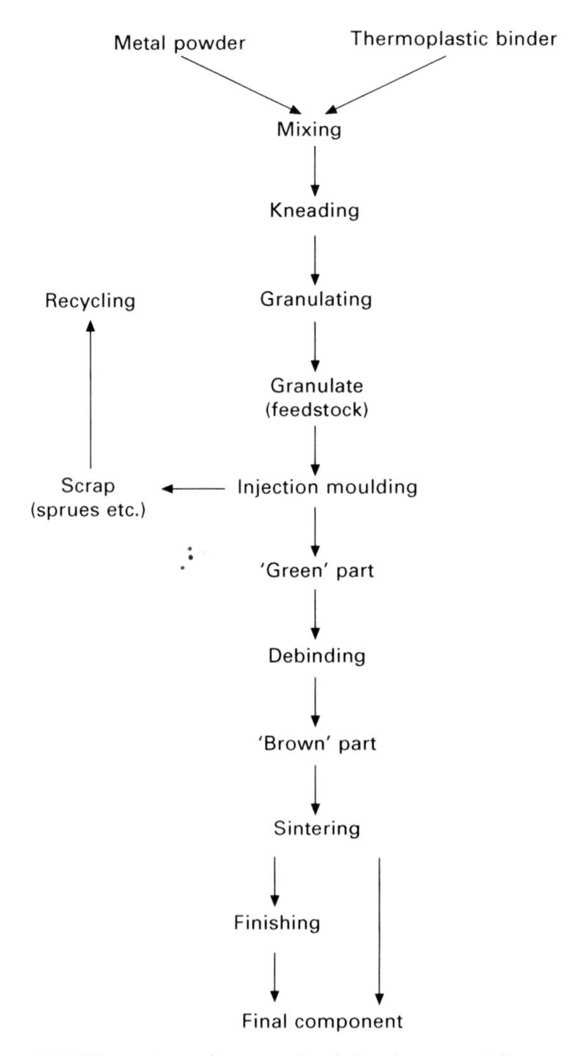

7.24 Flow sheet for powder injection moulding.

the major fraction of the polymer binder (mostly by solvent extraction or catalytical decomposition) to obtain a 'brown' part, thermal removing of the remaining binder, the 'backbone', and finally sintering the part to high density, typically >95%. The last stages, backbone removal and sintering, are usually done within a single run in the sintering furnace. Since the 'brown' compact has a large porosity, 40–50%, the volume originally taken by the polymeric binder, sintering it to <5% porosity means huge shrinkage and controlling this shrinkage without distortion is one of the major challenges of the process.

7.5.2 Powder systems used, alloying techniques

The high densification to be attained during sintering necessitates high sintering activity of the powders used and therefore fine powders are employed, with d_{50} typically <25 μm. Spherical powders are preferred that give good flow behaviour of the feedstock during injection moulding even at fairly high 'powder loading', that is the volume fraction of the metal powder in the feedstock (Fig. 7.25). PIM alloy steels are manufactured predominantly from prealloyed powders, manufactured by atomizing techniques. Since only the lower fractions of the atomizing runs can be used, the rest being recycled, PIM powder grades are fairly expensive, which is one of the major drawbacks of the technology. When using mixed powders, homogenization during sintering is not a problem owing to the short diffusion distances in these fine powders, but the risk of segregation during processing, in particular during injection moulding, should not be ignored. In industrial practice, stainless steel 316L makes up the bulk of PIM steel parts, but low to medium alloy steels are also used.[81]

The binder systems usually consist of a major component that has to provide flowability during injection moulding. Waxes or polyoxymethylene are frequently used here. The second component is the 'backbone' which has to provide some strength to the 'brown' part at least up to a temperature at which metallic contacts have already been formed. Here, polyethylene or polypropylene are commonly used. Finally, a wetting agent, for example stearic acid, is added to ensure complete covering of each metal particle with binder, which is an essential requirement for successful injection. The debinding procedure, the removal of the major constituent, depends on the compound chosen, as given below.

Homogeneity of the feedstock is a must for successful production and therefore intense mixing and kneading are done to ensure homogeneous products. Z-blade mixers and twin-screw extruders are usually employed. While in earlier days, PIM companies produced their own feedstock, with specific formulations, today it is very common to use commercial feedstocks and the suppliers of the feedstocks usually also supply know-how for processing.

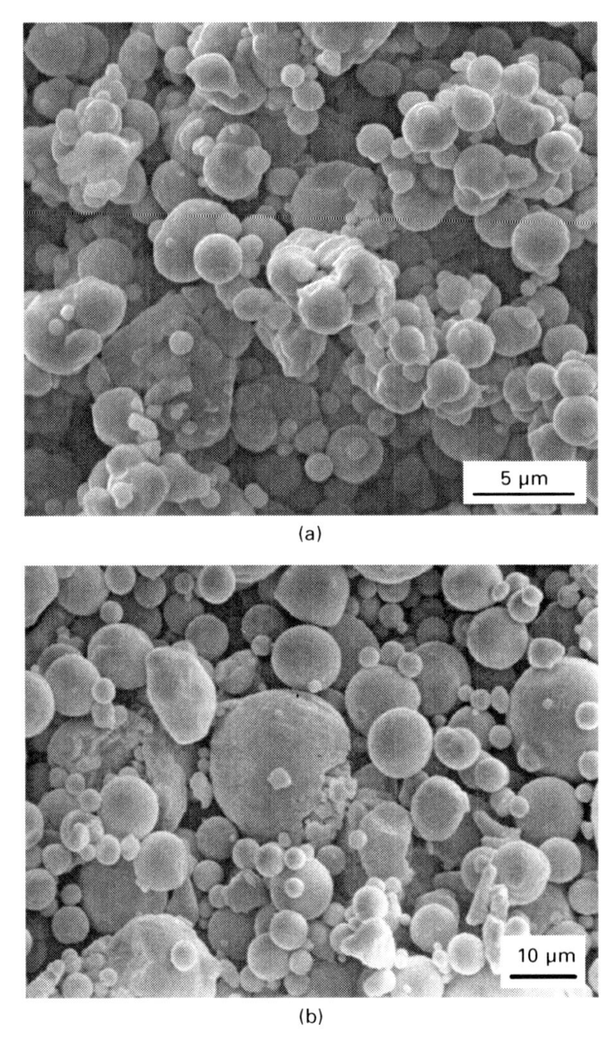

7.25 Spherical ferrous powders used for powder injection moulding: (a) carbonyl iron powder, (b) stainless steel 316L (photo: IFAM).

7.5.3 Moulding, debinding

Injection moulding of a feedstock is done on standard injection moulding machines which have to be modified for the specific behaviour of the PIM feedstock. In particular the much higher thermal conductivity has to be considered, which involves the risk of 'freezing' in narrow cross sections. Therefore such sections should be avoided and the gates of PIM tools are generally larger than in polymer shaping. A further peculiarity is the much higher density of a PIM feedstock compared to a polymer, which enhances inertia effects such as 'jetting'. In general, injection moulding defects are very difficult to recognize in the green parts and frequently they become visible only after debinding or even after sintering.

As stated above, the debinding process used is defined by the chemistry of the binder, in particular of the major constituent. For waxes, solvent debinding, for example extraction in acetone, is done, while for polyoxymethylene, catalytic debinding by HNO_3 or $(COOH)_2$ at temperatures slightly above 100°C is effective.

7.5.4 Sintering processes

As stated above, sintering of PIM parts includes as a first stage the removal of the 'backbone' binder component by thermal decomposition and then densifying the highly porous part to near full density. Thermal debinding means controlled heating and soaking procedures up to about 600°C at which temperature the backbone is commonly removed while first metallic contacts are formed, preventing disintegration of the now binderless part. In particular for stainless steel parts, the temperature–time profile has to be adjusted in such a way that carbon pickup is safely avoided; carbon residues would result in chromium carbide formation at the grain boundaries and thus in a tendency to intergranular corrosion.

Sintering itself is done at the upper end of the common temperature 'window' to ensure as effective densification as possible. For stainless steel parts, temperatures of 1350°C are quite usual. Both batch-type and continuous furnaces are used, the latter gaining ground with increasing volumes of PIM parts production. The atmospheres used depend on the material; for stainless steel, plain H_2 is frequently used resulting in very low carbon and nitrogen levels. Vacuum is also feasible but involves the risk of Cr evaporation, the vapour pressure of Cr at the high temperatures used here being quite considerable. N_2-H_2 mixes are not favourable for stainless steels, resulting in Cr nitride formation, but may be employed for low alloy structural steels. Argon is absolutely detrimental, as Ar trapped in the pores inhibits densification.

Secondary operations for PIM steel parts are more or less the same as for similar wrought steels; the problems encountered with die pressed parts that originate from open porosity do not occur here since the pores are fully closed.

7.5.5 Attainable properties

With regard to dimensional precision, PIM materials are slightly inferior to pressed and sintered parts; considering however the much higher shrinkage involved, the precision attained today is outstanding. Furthermore it must be considered that PIM parts are isotropic with regards to shrinkage, while for pressed parts the dimensional change is always different perpendicular and parallel to the pressing axis.

The mechanical, electrical and magnetic properties of PIM products are much more similar to those of wrought ones than those of pressed and sintered ones. This is mainly due to the lower porosity which is fully isolated; the very small, well rounded pores in PIM parts affect the properties markedly less than the interconnected pore networks in standard uniaxially compacted parts. This also holds for the chemical properties: the corrosion resistance of PIM parts is the same as that of wrought parts, occasionally even better owing to the finer and more homogeneous microstructure.

7.5.6 Microstructures

As stated above, the microstructures of PIM parts contain only fine, well rounded and fully closed pores. Heterogeneous alloy element distribution is rare since the short diffusion distances caused by the fine starting powders combined with the usually high sintering temperatures result in complete homogenization also in the case of mixed powders. Grain size depends on the porosity: usually the grain size is larger the lower the porosity (and the higher the sintering temperature). For low alloy steels, common heat treatments are applicable; here, the problem of quenching fluids entering the pores does not occur. Microstructures for PIM products are shown in Chapter 11.

7.5.7 Typical MIM ferrous products

As stated above, PIM is a process for small, complex shaped parts that are required in large quantities. Applications range from automotive to household or medical. Further information can be found in Chapter 6.

7.6 Powder metallurgy tool steels

7.6.1 Typical steel grades and compositions

Powder metallurgy tool steels make up only a tiny proportion of the tool steels manufactured worldwide, but they are the top grades available with regard to performance (and also cost). There are two main arguments for the PM production route. On one hand, PM enables manufacturing of steels with very fine and fully isotropic microstructures, thus avoiding the coarse carbide stringers common in wrought tool steels (originating from the cast eutectic carbide network) that give the steel the clearly defined texture microstructure and properties. With PM steels the carbides are much finer and more rounded and there is no orientation of the microstructure. This results in generally better properties and the designer need not take care of the carbide orientation in the starting workpiece.

The second advantage is that by the PM route, compositions are possible (see Table 7.3) that cannot be obtained by ingot metallurgy since the cast ingots would not be workable but would disintegrate during hot working. This holds both for the volume fraction of carbide and the type. PM steels may contain much higher carbide levels than IM ones and VC rich steels can be produced with up to 15 mass% vanadium. Such steels are totally inaccessible by ingot metallurgy.

PM tool steels grades include cold work tool steels, for example VC rich grades (CPM9V, CPM15V; K390) as well as high speed steels. The standard M2 grade is available also in a PM variant, but the attractive grades are high W/Mo/V types such as S290 and S390. Furthermore, injection moulding steel grades are available that combine high wear and corrosion resistance.

7.6.2 Manufacturing routes

The usual production route for PM tool steels, resulting in semi-finished products, mainly bars, was developed in Sweden in the 1960s, the so-called ASEA-STORA process. A melt with a defined composition is gas atomized to spherical powder (Fig. 7.26(a)), N_2 being commonly used. This spherical powder is then canned, by filling the powder into a mild steel container, usually without exposure to air, and the container is sealed by welding, evacuated through a pipe at one end and welded gas tight. The container is then placed into a hot isostatic press and subjected to a pressure of at least 1000 bar argon and a temperature of 1000–1100°C. By this treatment the bulk powder in the container is fully densified to a massive steel block without significant carbide growth. The steel may be used in this state without further deformation, but in practice the cylindrical block is usually swaged and then hot rolled to the dimensions required in the market. The remains of the mild steel container usually peel off during hot working.

The second production route for PM tool steels, which is employed mainly for shaped parts, involves water atomization with a subsequent reducing anneal of the very hard and oxidized powder. This irregular-shaped powder (Fig.

Table 7.3 Composition of various commercial PM tool steel grades

Grade	Supplier	Application	C	Cr	W	Mo	V	Co
CPM9V	Crucible	Cold work	1.80	5.25	–	1.30	9.00	–
CPM15V	Crucible	Cold work	3.40	5.25	–	1.30	14.50	–
K390	Böhler	Cold work	2.47	4.20	1.00	3.80	9.00	2.00
S290	Böhler	HSS	2.00	3.80	14.30	2.50	5.10	11.00
S390	Böhler	HSS	1.64	4.80	10.40	2.00	4.80	8.00
ASP 2030	Erasteel	HSS	1.28	4.20	6.40	5.00	3.10	8.50
M390	Böhler	Moulding	1.90	20.00	0.60	1.00	4.00	–

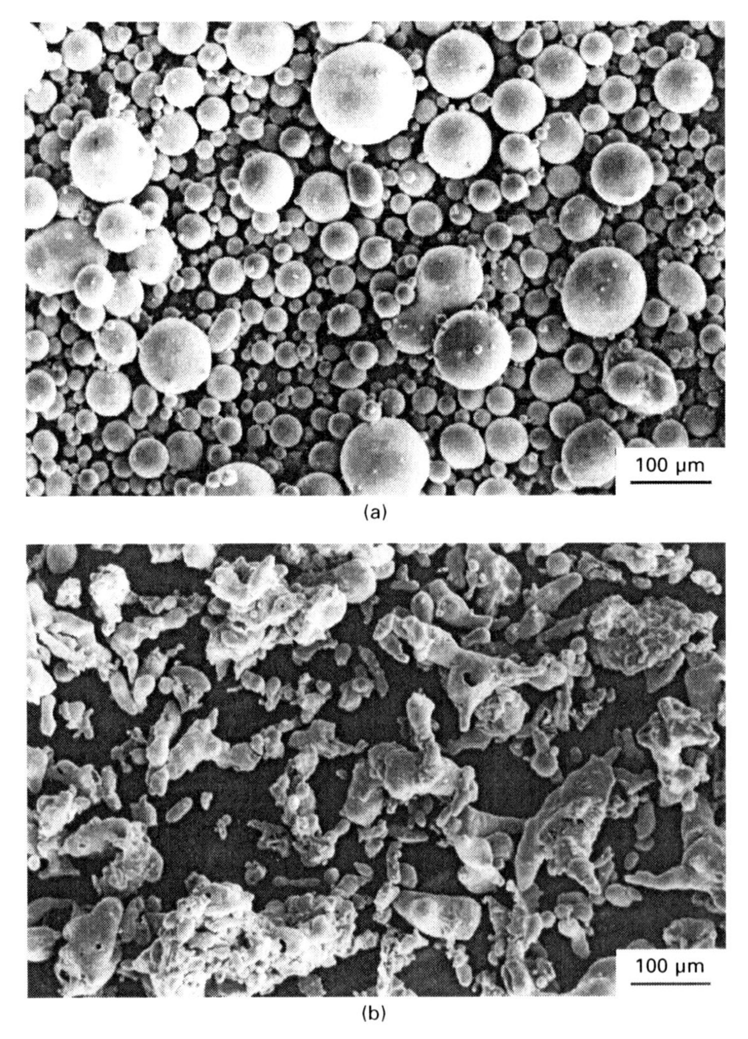

7.26 Atomized tool steel powders: (a) gas atomized. (b) water atomized.

7.26(b)) can then be uniaxially compacted in the standard way, although the compactibility remains moderate and the green density is accordingly low. The green compacts are then vacuum sintered to full density with a very small amount of spontaneously generated liquid phase, this 'supersolidus' sintering being an extremely crucial process that has to be done within a very narrow temperature window, for M2 typically ±3K, which is a tough requirement for a typical 500 kg batch. A slightly too low sintering temperature results in retained porosity and unacceptable strength while too high sintering causes carbide growth and may even result in eutectic melting, with disastrous effects on the toughness. The process is further complicated by the fact that the optimum temperature for the 'supersolidus' sintering strongly depends on the carbon content of the steel. Slight carburizing or decarburizing during heating may result in incorrect sintering and in the loss of an entire sintering batch. Therefore, precise control of temperature and furnace chemistry are essential requirements for successful production.

7.6.3 Microstructures of PM tool steels compared to conventional grades

As stated above, the main difference between ingot metallurgy (IM) and powder metallurgy (PM) tool steels is the pronounced microstructural texture in the former grades, with pronounced differences between longitudinal and cross sections, while the latter are isotropic. This is clearly visible from Fig. 7.27 in which a conventional wrought cold work tool steel is shown compared to a PM grade. Notice that the composition of both steels is different, the IM grade being Cr alloyed while the PM grade is VC rich; however, it is typical in the use of PM steels that simply mimicking IM compositions is not the best way but also that the much wider variety of compositions possible with PM is exploited.

The second difference is the much lower carbide size, as evident from Fig. 7.27 (please note the widely different magnification), which also has a strong impact on the mechanical behaviour, as will be shown in the next Section 7.6.4. This originates from the fact that solidification of tool steels as ingots is very slow, resulting in a coarse eutectic microstructure, with accordingly large carbides, while in PM the droplets solidify very fast, forming 'micro-ingots' with a microstructure that is eutectic but extremely fine. During hot isostatic pressing this cellular eutectic structure is dissolved into isolated carbides but the fine carbide size is retained. The particulate route also offers the chance to increase the volume fraction of carbides by admixing either carbides or carbide-forming additives (e.g. see Wewers[82]), with a resulting increase in particular of the wear resistance.

7.6.4 Properties of PM tool steels versus ingot metallurgy grades

Heat treatment of PM tool steels is similar to that of wrought grades, although the frequently higher content of VC has to be considered, which may necessitate fairly high peak austenitizing temperatures to obtain maximum secondary hardening. The hardness levels are similar to those of the wrought grades, but the toughness is far superior and, most important, is isotropic. The higher toughness originates from the finer microstructure, less coarse carbides being present as crack initiation sites. This is particularly evident

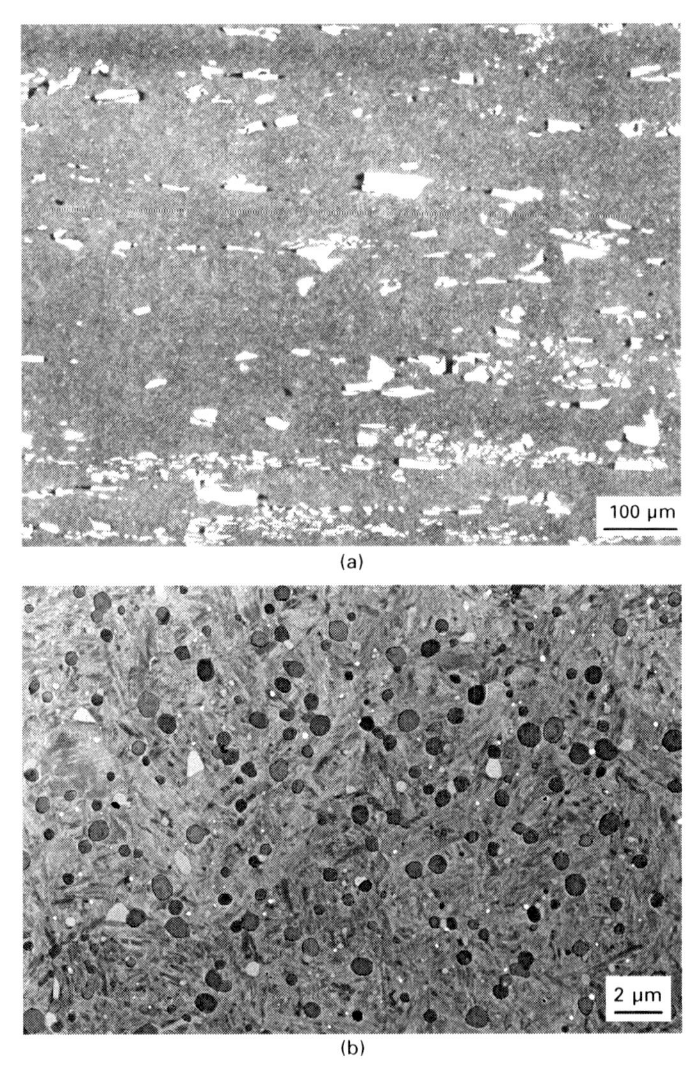

7.27 Microstructures of wrought versus PM cold work tool steels (photos: Böhler Edelstahl): (a) wrought Cr alloy tool steel, (b) PM V alloy tool steel.

for the cyclic properties, as has been shown in a recent study on gigacycle fatigue behaviour of tool steels (Fig. 7.28). However, it also means that PM tool steels are more sensitive to singular defects such as inclusions and it should be remembered that the first series of PM tool steels, which came to the market in the early 1970s, did not meet expectations owing to presence of slag inclusions that caused unpredictable failure. Progress in particular with regard to the atomizing techniques, especially avoiding recharging of the tundish during atomization, has dramatically lowered the presence of

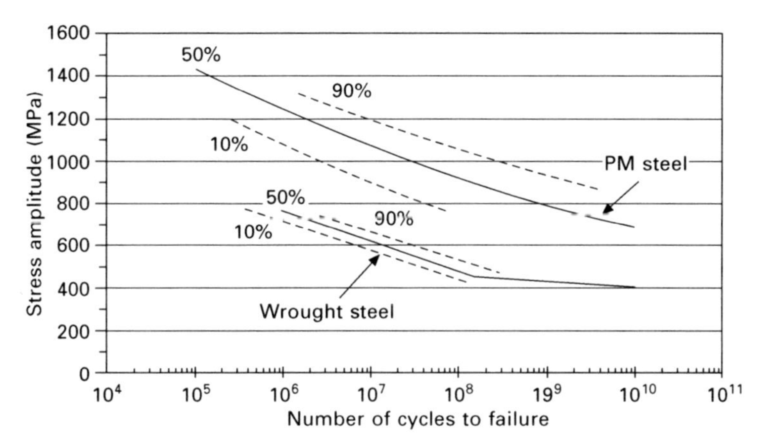

7.28 S–N curves for wrought versus PM cold work tool steels.[84]

larger inclusions;[83] nevertheless a recent study[84] showed that while IM tool steels failed through fracture of large carbides or carbide clusters, in PM the last remaining non-metallic inclusions were the sites of crack initiation, however at a markedly higher stress level. Here it should be recognized that for the *S–N* (Wöhler) curves the level does not depend on the composition or hardness of the steel, but only on the manufacturing route: cold work tool steels with 62 HRC resulted in virtually the same *S–N* curves as high speed steels with 67 HRC. This means that the fatigue here is actually defect controlled and not so much affected by the matrix properties. Today the high cleanliness of PM tool steels results in the fact that frequently the limiting factor for the mechanical properties is the surface finish and not so much the material itself.[85,86]

7.7 Trends in ferrous powder metallurgy

When observing the technical progress in ferrous PM it can be stated that the trend towards higher density has persisted, in part through processes such as warm compaction or high velocity compaction, in part simply by increasing the compacting pressures applied routinely. This latter has been made possible by improved pressing control, improved hardware and software as well as instrumentation of the tool components and also by the use of powder metallurgy tool steels for tool components; leading PM parts manufacturers are not using IM tool steels any more but only PM grades, owing to their higher strength, toughness and reliability. The real success story is definitely surface densification which has enabled PM the inroad to highly stressed engine and transmission components such as gears and sleeves, even gears in truck transmission having been successfully tested.

With the trend to alternative drives, there is definitely a risk that many PM parts will disappear since, for example, an electrical car needs fewer moving parts and much simpler transmissions. However, on the other hand PM functional materials such as SMCs can be expected to enter these markets and if the manufacturers of electrical engines recognize the potential of such PM materials for advanced engine concepts, further progress of PM can be anticipated, although not so much mechanical as electrical and magnetic properties will be required.

For MIM ferrous components, further growth can be predicted, since for MIM the strong dependence on automotive applications does not hold and there are many other fields such as medical and household where small complex shapes parts are required. Further increased complexity can be expected, in part as a consequence of progress in simulation of the PIM process, in particular moulding, which enables much more focused design of MIM tools. A strongly growing field is also 'micro-MIM', manufacturing of very tiny parts that are used for example in micro electromechanical systems (MEMS); here, suitably fine powders have to be available at an acceptable cost. Furthermore, two-component MIM has been successfully introduced, for example combining ferromagnetic and paramagnetic stainless steels within one single component by two-stage injection moulding and then co-debinding and -sintering.

For PM tool steels, finally, still higher hard phase content and finer microstructure as well as further reduced frequency of inclusions can be expected, resulting in higher strength and abrasion resistance but in particular in higher toughness and fatigue endurance strength. PM tool steels will never make up the bulk of tool steel applications, they are simply too expensive, but if superior properties are needed, PM steels are definitely the first choice.

7.8 Acknowledgements

The authors wish to thank for supplying information and illustrations: M.Dlapka, J.Seyrkammer, MIBA Sinter Austria GmbH, Vorchdorf, Austria; P.Beiss, RWTH Aachen, Germany; E.Dudrova, IMR-SAS Kosice, Slovakia; P.Delarbre, PMG, Füssen, Germany; B.Lindsley, Hoeganaes Corp., Cinnaminson NJ, USA; F.Petzoldt, IFAM Bremen, Germany; I.Siller, Böhler Edelstahl, Kapfenberg, Austria.

7.9 Further reading

Books

ASM Handbook Vol.7: Powder Metal Technologies and Applications, ASM, Materials Park OH, 1998.

R.M. German, *Powder Metallurgy and Particulate Materials Processing*, MPIF, Princeton NJ, 2005.
A. Upadhyaya and G.S. Upadhyaya, *Powder Metallurgy – Science, Technology and Materials*, Universities Press (India), Hyderabad – CRC Press, Boca Raton FL, 2011.
W. Schatt, K.P. Wieters and B. Kieback, *Pulvermetallurgie*, Springer-Verlag, 2007 (in German).
L.F. Pease III and W.G. West, *Fundamentals of Powder Metallurgy*, MPIF, Princeton NJ, 2002.
A. Salak, *Ferrous Powder Metallurgy*, Cambridge International Science Publishers, Cambridge, UK, 1995.
A. Salak and M. Selecka, *Manganese in Powder Metallurgy Steels*, Cambridge International Science Publishers, Cambridge UK /Springer, 2012.
R.M. German, *Powder Metallurgy of Iron and Steels*, MPIF, Princeton, NJ, 1996.
R.M. German and A. Bose, *Injection Molding of Metals and Ceramics*, MPIF, Princeton, NJ, 1997.
R.M. German, *Powder Injection Molding – Design and Applications*, MPIF, Princeton, NJ, 2003.
G. Roberts, G. Krauss and R. Kennedy, *Tool Steels*. 5th edition, ASM, Materials Park, OH, 1998.
A. Bose and W.B. Eisen, *Hot Consolidation of Powders and Particulates*, MPIF, Princeton, NJ, 2003.

Journals

Powder Metallurgy, edited by Maney Publications on behalf of the Institute of Materials, London, five issues annually.
International Journal of Powder Metallurgy, edited by MPIF, Princeton NJ, six issues annually.
Metal Powder Report, issued by Elsevier, 11 issues annually.
Powder Metallurgy Progress, issued by the Slovak Academy of Sciences, Kosice; four issues annually, open access through: www.imr.saske.sk/pmp/index.htm
Science of Sintering, issued by the Serbian Academy of Sciences, Belgrade, three issues annually, open access through www.iiss.sanu.ac.rs/journal.htm
Powder Injection Moulding International, edited by Inovar Communications Ltd., Shrewsbury UK, four issues annually.

Websites

Metal Powder Industries Federation (MPIF), Princeton NJ: www.mpif.org

European Powder Metallurgy Association (EPMA), Shrewsbury, UK: www.epma.com
Japan Powder Metallurgy Association (JPMA), Tokyo; www.jpma.gr.jp
Asian Powder Metallurgy Association, Tokyo, www.apma.asia
Global Powder Metallurgy Property Database: www.pmdatabase.com

7.10 References

1. R. Kieffer and W. Hotop, *Sintereisen und Sinterstahl*, Springer, Wien, 1948.
2. J. Capus, *Powder Metall.*, 2011, **54**(5), 553–556.
3. M.E. Lutheran, *Int. J. Powder Metall.*, 2011, **47**(4), 45–48.
4. H. Danninger, C. Gierl, S. Kremel, G. Leitner, K. Jaenicke-Roessler and Y. Yu, *Powder Metall. Prog.*, 2002, **2**(3), 125–140.
5. A.R. Glassner, *The Thermochemical Properties of the Oxides, Chlorides, and Fluorides to 2500°K*. US Atomic Energy Commission Reports ANL-5750, Washington DC, 1957.
6. B. Lindqvist, *Proc EuroPM2001 Nice*, European Powder Metallurgy Association (EPMA), Shrewsbury, 2001, Volume 1, 13–21.
7. I. Donaldson, M. Marucci and B. Lindsley, *Advances in Powder Metallurgy and Particulate Materials* 2011, (Proceedings Powder Met 2011, San Francisco), compiled by I.E. Anderson and T.W. Pelletiers, MPIF, Princeton, NJ, 2011, Part 7, 54–63.
8. N. Dautzenberg and H.J. Dorweiler, *Powder Metall. Int.*, 1985, **17**, 279–83.
9. F.V. Lenel, *Powder Metallurgy – Principles & Applications*. MPIF, Princeton NJ 1980.
10. U. Engström, in: Powder Metallurgy-State of the Art, W.J. Huppmann, W.A. Kaysser, G. Petzow (eds), Verlag Schmid, Freiburg 1986, 41–70.
11. A. Lawley, *Atomizing*. MPIF, Princeton NJ 1992.
12. *Hoganas Handbook for Sintered Components*, Volume 1, *Material and Powder Properties*, Hoganas AB, Hoganas, Sweden, 1998.
13. C. Schade, T. Murphy, A. Lawley and R. Doherty, *Advances in Powder Metallurgy and Particulate Materials*, (Proceedings Powder Met 2010, Ft. Lauderdale FL), compiled by M. Bulger and B. Stebick, MPIF, Princeton, NJ, 2010, Part 7, 50–63.
14. A.N. Klein, R. Oberacker and F. Thümmler, *Powder Metall. Int.*, 1985, **17**, 13–16, 71–4.
15. S. Saccarola, M. Zanon. A. Karuppannagounder and F. Castro, *Advances in Powder Metallurgy and Particulate Materials*, (Proceedings Powder Met 2011, San Francisco), compiled by I.E. Anderson, T.W. Pelletiers, MPIF, Princeton NJ 2011, Part 7, 45–53.
16. H. Danninger, G. Frauendienst, K.-D. Streb and R. Ratzi, *Mat. Chem. Phys.*, 2001, **67**, 72–77.
17. L. Alzati, R. Gilardi, G. Pozzi, and S. Fontana, *Advances in Powder Metallurgy and Particulate Materials*, (Proceedings Powder Met 2011, Ft. Lauderdale FL), compiled by I.E. Anderson, T.W. Pelletiers, MPIF, Princeton NJ 2011, Part 7, 11–18.
18. M. Azadbeh, H. Danninger and C. Gierl, *Powder Metall. Prog*, 2007, **7**(3), 128–38.
19. 'Warm Compaction', *Hoganas Handbook for Sintered Components*, Volume 4, Hoganas AB, Hoganas, Sweden, 1998.

20. P. Skoglund: *Powder Metall.*, 2001, **44**(3), 199–201.
21. *Powder Metallurgy Design Manual*, 3rd edition, MPIF, Princeton, NJ, 1998.
22. D.C. Zenger and H. Cai, Handbook of the Common Cracks in PM Compacts. MPIF, Princeton NJ 1997.
23. H. Danninger and C. Gierl, *Mater. Chem. Phys.*, 2001, **67**, 49–55.
24. H. Danninger, R. Pöttschacher, S. Bradac, A. Salak and J. Seyrkammer, *Powder Metall.*, 2005, **59**, 23–32.
25. H. Danninger, in, *Pulvermetallurgie in Wissenschaft und Praxis Bd. 22*, H. Kolaska (ed), Fachverband Pulvermetallurgie, Hagen 2006, 21–47.
26. A. Salak, *Powder Metall. Int.*, 1986, **18**(4), 266–70.
27. A. Salak, M. Selecka and R. Bures, *Powder Metall. Prog*, 2001, **1**(1), 41–58.
28. H. Danninger and C. Gierl, *Sci. Sintering*, 2008, **40**, 33–46.
29. S. Banerjee, G. Schlieper, F. Thümmler and G. Zapf, *Modern Dev. Powder Metall.*, 1981, **13**, 143–157.
30. P. Beiss and R. Wassenberg, *Proceedings EuroPM2005 Prague*, EPMA, Shrewsbury, 2005, Volume 1, 143–50.
31. R. De Oro Calderon, M. Campos, C. Gierl, H. Danninger and J.M. Torralba, *Proceedings PM2010 Powder Metallurgy World Congress & Exhibition, Florence*, EPMA, Shrewsbury, UK, 2010, Volume 2, 275–82.
32. H. Danninger, M. Jaliliziyaeian, C. Gierl, E. Hryha and S. Bengtsson, *Proceedings PM2010 Powder Metallurgy World Congress & Exhibition, Florence*, EPMA, Shrewsbury, 2010, Volume 3, 3–10.
33. G. Gierl, H. Danninger and R. De Oro Calderon, in, *Pulvermetallurgie in Wissenschaft und Praxis Bd.27*, H. Kolaska (ed), Fachverband Pulvermetallurgie, Hagen, 2011, 215–36.
34. J.A. Sicre Artalejo, M. Campos Gomez, J.M. Torralba, J. Zbiral, H. Danninger and P. Pena Castro, *Boletin de la Sociedad Espanola de Ceramica y Vidrio*, 2008, **47**(5), 305–10.
35. P. Beiss, I. Cremer, D. Geldner and F. Sarnes, in, *Pulvermetallurgie in Wissenschaft und Praxis Bd.27*, H. Kolaska (ed), Fachverband Pulvermetallurgie, Hagen, 2011, 179–214.
36. C. Sandner, J. Dickinger, H. Rößler and P. Orth, *Proceedings PM2004 Powder Metallurgy World Congress*, H. Danninger and R. Ratzi (eds), EPMA, Shrewsbury, 2004, Volume 5, 657–62.
37. L. Forden, S. Bengtsson and M. Bergström, *Proceedings PM2004 Powder Metallurgy World Congress*, H. Danninger and R. Ratzi (eds), EPMA, Shrewsbury, 2004, Volume 5, 641–7.
38. S. Saritas, R.D. Doherty and A. Lawley, *Proceedings EuroPM2001 Nice*, EPMA, Shrewsbury, 2001, Volume 1, 257–65.
39. www.cremer-ofenbau.de.
40. E. Santuliana, C. Menapace, S. Libardi, G. Lorenzi and A. Molinari, *Int. J. Powder Metall.*, 2011, **47**(6), 38–45.
41. S. Kremel, H. Danninger, H. Altena and Y. Yu, *Powder Metall. Prog*, 2004, **4**(4), 119–131.
42. S. Bengtsson, T. Marcu and A. Klekovkin, *Proceedings PM2008 Powder Metallurgy World Congress & Exhibition, Washington DC*, MPIF, Princeton, NJ, 2008, on CD-ROM.
43. M. Dlapka, C. Gierl, H. Danninger, H. Altena, P. Orth and G. Stetina, *Berg- und Hüttenmaenn. Mh.*, 2009, **154**, 200–04.

44. A. Salak, M. Selecka and H. Danninger, Machinability of Powder Metallurgy Steels. Cambridge International Science Publishers, Cambridge UK (2005).
45. R. Koos and G. Bockstiegel, *Prog. Powder Metall.*, 1981, **37**, 145–164.
46. A. Salak, M. Selecka, K. Vasilko and H. Danninger, *Int. J. Powder Metall.*, 2008, **44**(2), 49–61.
47. R.J. Causton, *Advances in Powder Metallurgy and Particulate Materials*, (Proceedings PowderMet2011, San Francisco), compiled by M. Phillips and J. Porter, MPIF, Princeton, NJ, 1995, Volume 2, Part 8, 149–170.
48. U. Engström, *Powder Metall.*, 1983, **16**, 137–144.
49. M. Gagne, M. Gaumont and G. Olschewski, *Proceedings EuroPM2000 Conference on Material and Processing Trends for PM Components in Transportation, Munich.* EPMA, Shrewsbury, 2000, 25–32.
50. R. Ratzi, in *Pulvermetallurgie in Wissenschaft und Praxis*, Volume 18, H. Kolaska (ed), Fachverband Pulvermetallurgie, Hagen, 2002, 187–97.
51. W.V. Knopp, *Adv. Powder Metall. & Partic. Mater.* (Proceedings Powder Metallurgy World Congress 1996, Washington DC), T.M. Cadle and K.S. Narasimhan (eds), MPIF, Princeton, NJ, 1996, Part 11, 167–70.
52. H. Danninger, J.M. Garmendia Gutierrez, R. Ratzi and J. Seyrkammer, *Powder Metall. Prog.*, 2010, **10**(3), 121–132.
53. Assinter Information Brochure, *Competitivita' dei Componenti Sinterizzati. Guida alle alternative tecnologiche*, Assinter, Milano, 1995.
54. F. Rübenach, *Powder Metall. Int.*, 1972, **4**, 85–88.
55. E. Ernst and R. Schmitt, in: *Pulvermetallurgie in Wissenschaft und Praxis Bd.12*, R. Ruthardt (ed), DGM, Frankfurt, 1996, 55–94.
56. www.pmdatabase.com.
57. G.F. Bocchini, *Int. J. Powder Metall.*, 1986, **22**, 185–202.
58. M. Azadbeh, H. Danninger and C. Gierl, *Powder Metall. Prog*, 2006, **6**(1), 1–10.
59. E. Dudrova, R. Bures, M. Kabatova, H. Danninger and M. Selecká, *Proceedings EuroPM1997 Munich*, EPMA, Shrewsbury, 1997, 373–380.
60. P. Beiss, in *Pulvermetallurgie in Wissenschaft und Praxis*, Volume 19, H. Kolaska (ed), Fachverband Pulvermetallurgie, Hagen, 2003, 3–24.
61. A. Zafari and P. Beiss, *Powder Metall. Prog.*, 2008, **8**(3), 200–209.
62. P. Beiss, A. Zafari, K. Lipp and J. Baumgartner, *Int. J. Powder Metall.*, 2012, **48**(1), 19–34.
63. H. Danninger and B. Weiss, *Powder Metall. Prog.*, 2001, **1**(1), 19–40.
64. M. Dlapka, H. Danninger, C. Gierl, B. Weiss, G. Khatibi and A. Betzwar-Kotas, *Powder Metall. Prog.*, 2011, **11**(1–2), 69–77.
65. C. M. Sonsino, Konstruktion 2005, **5**, 87–92.
66. H. Danninger, D. Spoljaric and B. Weiss, *Int. J. Powder Metall.*, 1997, **33**(4), 43–53.
67. Y. Furuya, S. Matsuoka and T. Abe, *Metall. Mater. Trans.*, 2003, **34A**(11), 2517–26.
68. C.M. Sonsino, *Fatigue Design Concepts*. MPIF, Princeton NJ 2003.
69. N. Chawla and X. Deng, *Mater Sci. Eng.*, A 2005, **390**, 98–112.
70. P. Engdahl, B. Lindqvist and J. Tengzelius, *Proceedings PM'90 London*, Volume 2, 1990, 144–54.
71. H. Danninger, C. Gierl, G. Mühlbauer, M. Silva Gonzalez, J. Schmidt and E. Specht, *Int. J. Powder Metall.*, 2011, **47**(3), 31–42.
72. H. Danninger and G. Leitner, *Advances in Powder Metallurgy and Particulate*

Materials, (*Proceedings PM²TEC2002*, Powder Metallurgy World Congress, Orlando), Arnhold, C.-L.Chu, W.F. Jandeska Jr., H.I:Sanderow (eds), MPIF, Princeton, NJ, 2002, Part 13, 157–67.

73. A. Simchi and H. Danninger, *Powder Metall.*, 2000, **43**(3), 209–218.
74. A. Cytermann, *Powder Metall. Int.*, 1987, **19**(4), 27–30.
75. H. Danninger, G. Jangg, B. Weiss and R. Stickler, *Powder Metall. Int.*, 1993, **25**(4), 170–3 and **25**(5), 219–23.
76. G. Jangg, M. Drozda, H. Danninger and R.E. Nad, *Powder Metall. Int.*, 1983, **15**, 173–177.
77. G. Jangg, M. Drozda, G. Eder and H. Danninger, *Powder Metall. Int.*, 1984, **16**, 61–64.
78. M. Dougan, *Proceedings PM2010 Powder Metallurgy World Congress & Exhibition, Florence*, EPMA, Shrewsbury, UK, 2010, Volume 5, 229–36.
79. K. Narasimhan, S. Clisby and F. Hanejko, *Proceedings PM2010 Powder Metallurgy World Congress & Exhibition, Florence*, EPMA, Shrewsbury, UK 2010, Volume 5, 301–4.
80. L. Hultman: *Proceedings PM2004 Powder Metallurgy World Congress*, H. Danninger and R. Ratzi (eds), EPMA, Shrewsbury, 2004, Volume **4**, 591–8.
81. D. Whittaker, *Powder Injection Moulding Int*, 2011, **5**(3) 55–60.
82. B. Wewers and H. Berns, *Mat.-wiss. u. Werkstofftech.* 2003, **34**, 453–463.
83. C. Tornberg and A. Fölzer, *Metal Powder Rep.*, 2005, **60**(6), 36–40.
84. C. Sohar, *Lifetime Controlling Defects in Tool Steels*, Springer, Heidelberg-Dordrecht-London-New York, 2011.
85. S. Sundin, *Metal Powder Rep.*, 2007, **62**(5), 8–9.
86. G. Zapf and K. Dalal, 'The manufacture and properties of metal powders', *Powder Metallurgy Lecture Series*, Lecture 2, EPMA, Shrewsbury, 1992.

8
Powder metallurgy of titanium alloys

F.H. (SAM) FROES, Consultant, USA

DOI: 10.1533/9780857098900.2.202

Abstract: The major reason that there is not more widespread use of titanium and its alloys is the high cost. In this paper, developments in one cost effective approach to fabrication of titanium components – powder metallurgy – is discussed with respect to various aspects of this technology. These aspects are the blended elemental approach, prealloyed techniques, additive layer manufacturing, metal injection molding, spray deposition, far from equilibrium processing (rapid solidification mechanical alloying and vapor deposition) and porous materials. Use of titanium powder for sputtering targets, coating, as a grain refiner in aluminium alloys and fireworks are not addressed.

Key words: additive layer manufacturing, blended elemental, compaction techniques, far from equilibrium processing, mechanical properties and shape production, metal injection molding, prealloyed, porous materials, powder metallurgy, spray deposition, titanium.

8.1 Introduction

Titanium alloys are amongst the most important of the advanced materials which are key to improved performance in aerospace and terrestrial systems.[1–5] This is because of the excellent combinations of specific mechanical properties (properties normalized by density) and outstanding corrosion behavior[6–11] exhibited by titanium alloys. However, negating widespread use is the high cost of titanium alloys compared to competing materials. This has led to numerous investigations of various potentially lower cost processes,[1–3] including powder metallurgy (PM) techniques.[1–2,6–10,12,13] In this paper the titanium PM scenario will be reviewed, dividing the various technologies into the categories shown in Table 8.1. Basically the power metallurgy techniques to be discussed are the blended elemental (BE) approach, prealloyed (PA) methods, additive layer manufacturing (ALM), metal injection molding (MIM), spray deposition (SD) and far from equilibrium processing (rapid solidification mechanical alloying and vapor deposition). Powders to be attached to the surface of body implants (to improve the bonding between artificial devices and the human body) will only be discussed briefly, and for fireworks, sputtering targets, coatings and as a grain refiner in aluminium alloys will not be addressed. An indication of where PM fits into the broad scenario of fabrication techniques is shown in Fig. 8.1.

Table 8.1 Categories of titanium powder metallurgy

Category	Features	Status
Additive manufacturing	Powder feed melted with a laser or other heat source	Pilot production
Powder injection molding	Use of a binder to produce complex small parts	Production
Spraying	Solid or potentially liquid	Research base
Near net shapes	Prealloyed and blended elemental	Commercial
Far from equilibrium processes	Rapid solidification, mechanical alloying and vapor deposition	Research base

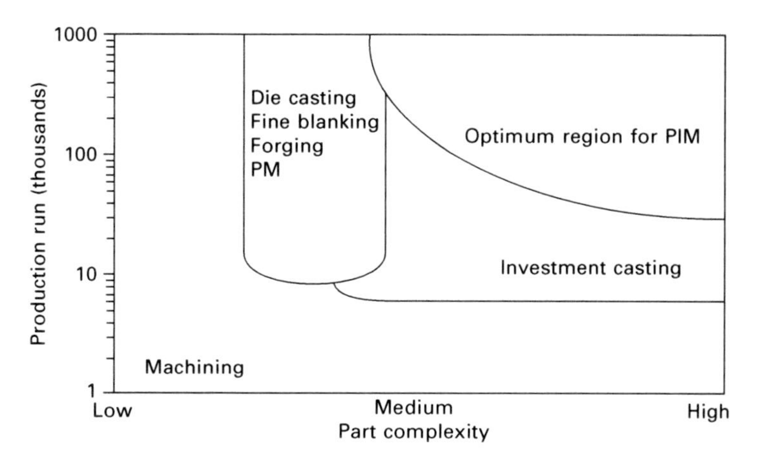

8.1 Diagram showing where powder metallurgy (PM) in general and powder injection molding (PIM) in particular, fit in with other fabrication processes (courtesy of Krebsöge, Radevormwald).

In publications over the past few years[1–3] the cost of fabricating various titanium precursors and mill products has been discussed and it has been pointed out that the cost of extraction is a small fraction of the total cost of a component fabricated by the cast and wrought (ingot metallurgy) approach (Figure 8.2). To reach a final component the mill products shown in the figure must be machined, often with very high buy-to-fly ratios (which can reach as high as 40:1). The generally accepted cost of machining a component is that it doubles the cost of the component (with the buy-to-fly ratio being another multiplier in cost per pound), Fig. 8.3. This means that anything that can be done to produce a component which is closer to the final configuration will result in a cost reduction and hence the attraction of near net shape powder metallurgy components.

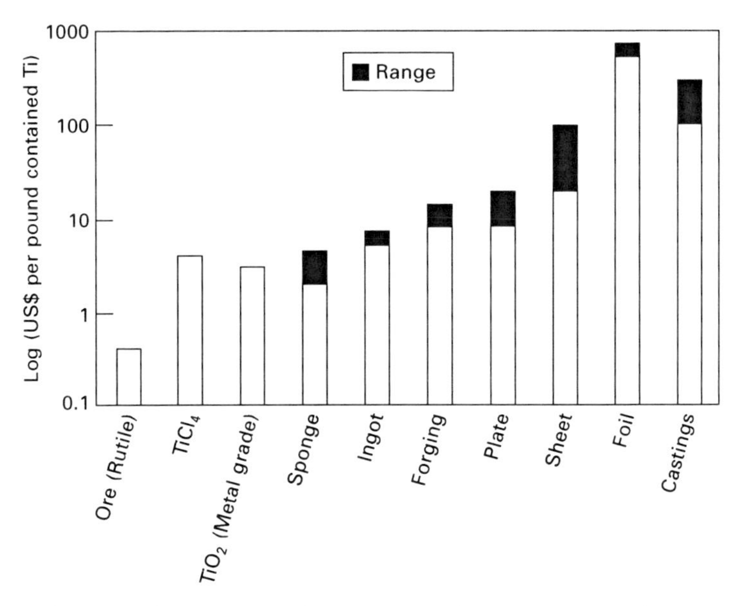

8.2 Cost of titanium at various stages of a component fabrication.

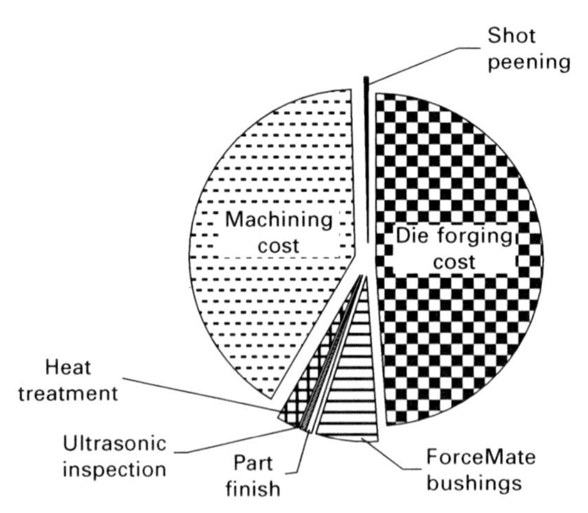

8.3 Manufacturing cost breakdown for Boeing 787 side-of-body chord. (courtesy Boeing).

8.2 Powders

Table 8.2 shows the characteristics of the different types of titanium powders that are either available or under development today. This table is based in part on a recent review of powder production methods co-authored by McCracken.[14] The oxygen level of the hydride–dehydride (HDH) powder can be reduced by deoxidizing with calcium.[14] It is also possible to convert

Table 8.2 Characteristics of different types of titanium powders (modified from Abkowitz *et al.*[15])

Type/process	Elemental or prealloyed	Advantages	Status/ disadvantages
Hunter process (pure sodium)	Elemental	Low cost, excellent for cold press and sinter	Limited availability High chlorido
HDH[a] Kroll process (pure magnesium)	Elemental	Lower cost Good compactibility Readily available Low chloride	
HDH powder produced from alloys	Prealloyed	Readily available	High cost Fair compactibility
Atomized	Prealloyed	High purity available	High cost Not cold compactable
REP/PREP[b]	Prealloyed	High purity	High cost Not cold compactable
Armstrong/ International titanium (ITP) powder	Both	Compactable Moderate cost	Processibility/quality Production Scale-up
Fray	Both	TBD	Developmental
MER[c]	Both	TBD	Developmental
CSIRO TiRO[d]	Both	TBD	Developmental

[a]Hydride-dehydride. [b]Rotating electrode powder/plasma rotating electrode powder. [c]MER Corp., Tucson, AZ. [d]CSIRO Melbourne, Australia.

the angular HDH to a spherical morphology using the Tekna process (see later).

The development of new titanium production methods such as the ITP/ Armstrong, Fray, CSIRO and MER processes (see later) shown in Table 8.2 is aimed at lowering the cost of PM titanium powder. However, these powders will not be available for some time and their relative cost and processing characteristics are yet to be established.

There are a number of processes which produce pre-alloyed spherical titanium powder including the following:

- ATI Powder Metals (formerly Crucible Research Center): spherical gas atomized alloy powder, 100 pounds (45 kg) capacity melting furnace, 50 pounds (23 kg) of –100/+325 (–150/+45 μm) Ti-6Al-4V at US$130.00 per pound (US$260 per kg).
- Advanced Specialty Metals: spherical plasma rotating electrode process (PREP) –100/+325 (–150/+45 μm) Ti-6Al-4V for US$189.00 per pound (US$416 per kg).

- Raymor (now includes Pyrogenesis): spherical plasma atomized Ti-6Al-4V powder, –450 to +60 mesh (–30/+250 μm) powder available. Ti-6Al-4V US$118 per pound (US$260 per kg), oxygen 0.09 wt%.
- Baoji Orchid Titanium: spherical PREP, Ti-6Al-4V –70/+325 (–210/+45 μm), 0.13 oxygen max. US$84 per pound (US$185 per kg).
- ALD Vacuum Technologies: spherical gas atomized Ti-6Al-4V electrode induction melting gas atomization.
- Sumitomo Sitex: gas atomized Ti-6Al-4V, oxygen 0.08–0.13 wt%.
- TLS Technik: gas atomized Ti-6Al-4V with 0.13 oxygen. Ti-6Al-4V 100-270 mesh (53–150 μm), O_2 0.13 wt%, US$73 per pound (US$ 161 per kg).
- Affinity International GA and PREP: but they seem to have gone out of business.
- Iowa State University/Ames Lab: experimental gas atomization, cost effective very fine spherical powder produced using a close-coupled high pressure supersonic gas (less than 325 mesh, 45 μm). Plans are to commercialize the process under a company called Iowa Powder Atomization Technologies.
- Tekna induction plasma spheriodization process converts irregular shaped titanium powders to a spherical morphology. Typically an irregular powder with –100/+400 mesh (–150/+37 μm) is converted to a spherical powder in the same size range (but with a significant improvement in tap density and flow rate).
- Quad Cities Manufacturing Laboratory: to establish capabilities for PREP, GA, HDH and the Tekna induction plasma spheriodization process (to convert HDH powders).

The atomized powders are generally prealloyed and spherical (Figure 8.4(a)), sponge fines (a by-product of sponge production) are 'sponge-like' in nature and contain remnant salt (which prevents achievement of full density and adversely affects weldability) and are angular (Fig. 8.4(b)). The HDH powders, which are generally also prealloyed, are angular in nature (Fig. 8.4(c)).[16] Conversion to a spherical morphology using the Tekna process is shown in Fig. 8.4(d).

Four non-melt processes appear to have the greatest potential for scale-up, with an additional process being developed by Advance Materials (ADMA) Products also of potential commercial interest. These four processes are the FFC Cambridge approach, the MER technique, the Commonwealth Scientific and Industrial Research Organization (CSIRO) methods and the ITP/Armstrong process.

In the FFC Cambridge approach, titanium metal is produced at the cathode in an electrolyte (generally $CaCl_2$) by the removal of oxygen from the cathode. This technique allows the direct production of alloys such as Ti-6Al-4V at a

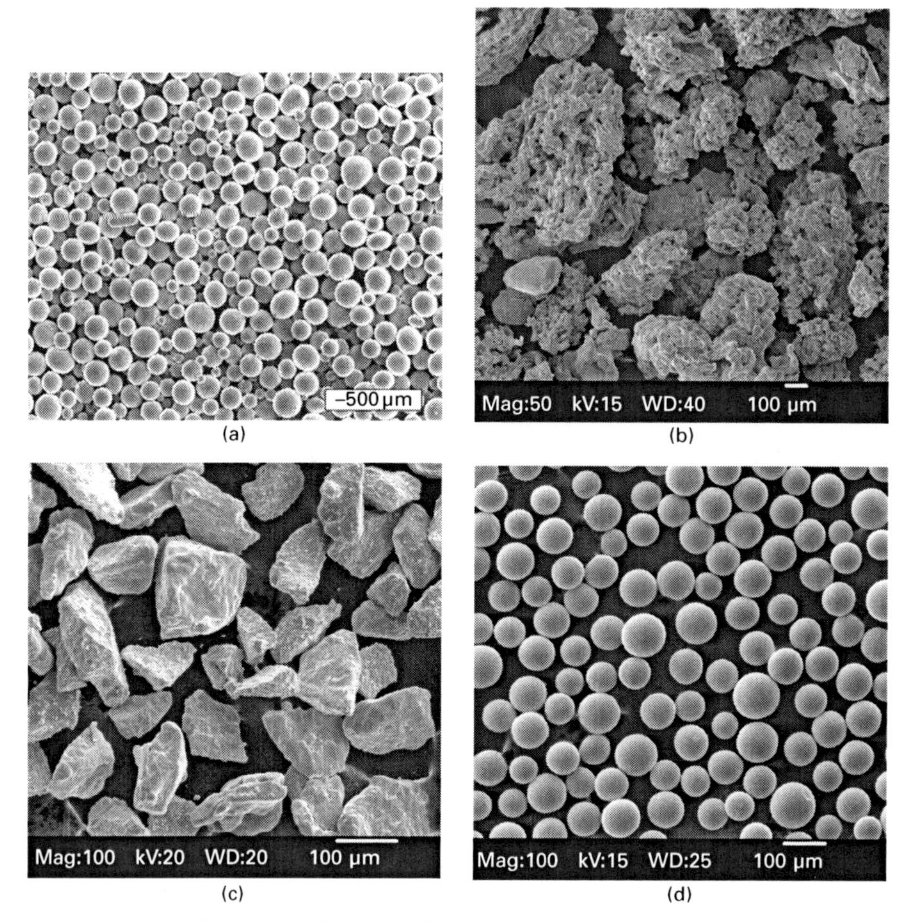

8.4 (a) Scanning electron microscopy (SEM) photomicrograph of a gas atomized prealloyed spherical Ti-6Al-4V (courtesy of Affinity International). (b) SEM photomicrograph of sponge fines produced by the Kroll process (courtesy Ametek). (c) SEM photomicrograph of angular HDH titanium powder. (d) SEM photomicrograph of spherical powder produced by processing angular HDH titanium to a spherical morphology using the Tekna technique.

cost which could be less than the product of the conventional Kroll process.[17] The process is being developed by Metalysis in South Yorkshire, UK.

The MER approach is an electrolytic method, which uses a composite anode of TiO_2, a reducing agent and an electrolyte, mixed with fused halides. Projections are for titanium production at a significantly lower cost than the conventional Kroll process.[18]

The CSIRO technique[19] builds upon the fact that Australia has some of the largest mineral and sand deposits in the world. In this approach, cost-effective commercially pure titanium is produced in a continuous fluidized bed in which titanium tetrachloride is reacted with molten magnesium (the TiRO™ process). They also have a proprietary process for producing alloys, although details are unavailable at the present time. Continuous production of a wide range of alloys including aluminides and Ti-6Al-4V has been demonstrated on a large laboratory scale. The commercially pure titanium powder produced has been used to fabricate extrusions, thin sheet (Fig. 8.5) by continuous roll consolidation, and cold-sprayed complex shapes including ball valves and seamless tubing (see later). Commercialization of the process is now in the planning stage with a decision to proceed to the pilot plant stage likely to be taken in 2013.

The ITP/Armstrong method[1–3] is continuous and uses molten sodium to reduce titanium tetrachloride, which is injected as a vapor. The resultant powder does not need further purification and can be used directly in the conventional ingot approach.The powder is most efficiently utilized in the powder metallurgy technique. A range of alloys can be produced (including the Ti-6Al-4V alloy) as a high quality homogeneous product suitable for many applications. ITP currently operates an R&D facility in Lockport, Illinois, USA and has broken ground on a four million pound (1.8 million kg) per year expansion in Ottawa, Illinois which is expected to ramp-up production

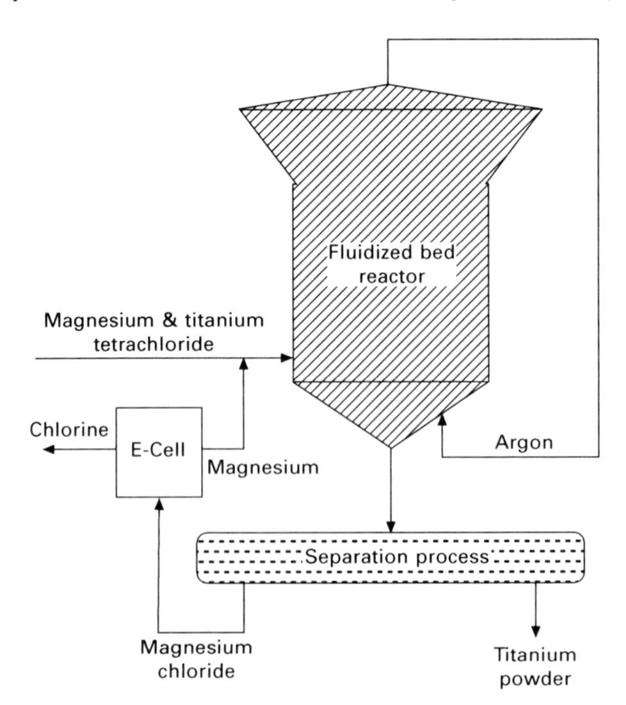

8.5 Schematic of the CSIRO process for producing commercially pure titanium powder.

throughout 2013 and will produce both commercially pure titanium and Ti-6Al-4V alloy powder.

In the ADMA Products approach[20] sponge titanium is cooled in a hydrogen atmosphere rather than the conventional inert gas. The hydrogenated sponge is then easily crushed and in the hydrogenated condition can be compacted to a higher density than conventional low hydrogen sponge, with subsequent hydrogen removal easily accomplished with a simple vacuum anneal. The remnant chloride content of the hydrogenated sponge is reported to be at low levels (helping to avoid porosity and enhancing weldability). There are 14 patents covering this approach.

Estimates of the powder shipments (in all cases per year) that have been made as HDH are: worldwide 1000–2500 MT, USA 200–400 MT and as spherical are: worldwide 150–350 MT, USA 20–50 MT.

8.3 Near net shapes

The techniques generally available for production of near net shapes (NNS) are amenable for use with various types of titanium powders; these include conventional press-and-sinter, elastomeric bag cold isostatic pressing (CIP), and ceramic mold or metal can hot isostatic pressing (HIP). For convenience, NNS will be divided into those produced using blended elemental (BE) powders and those produced from prealloyed (PA) powders.

8.3.1 Blended elemental

The blended elemental (BE) approach is potentially the lowest cost titanium PM process, especially if any secondary compaction step (e.g. HIP) can be avoided.[15,21] In the BE approach, angular titanium sponge fines (or titanium hydride powder) and master alloy composition (generally the 60:40 Al:V variety to produce the Ti-6Al-4V composition) are blended together, cold pressed and sintered to near full density. Use of titanium hydride allows densities very close to 100% to be obtained in components such as an auto connecting rod (Fig. 8.6) with mechanical properties at ingot metallurgy levels.

The blended element PM technology using hydride–dehydride (HDH) titanium powder produced by a Kroll sponge process is the key to the commercial success of Dynamet's PM process.[15] This process is producing a wide range of affordable PM mill product forms and components. Dynamet has developed critical specifications for its titanium and master alloy powders that control morphology, particle size, particle distribution and chemistry. The properties of the PM materials can be adjusted by modifications in these process parameters. The new powders that are under development may provide an opportunity to reduce the costs of PM product further if they can

8.6 Auto connecting rod fabricated via the blended elemental approach using hydrogenated titanium powder (courtesy Orest Ivasishin, Ukrainian Academy of Sciences).

8.7 Toyota Altezza, 1998 Japanese car of the year, the first family automobile in the world to feature titanium valves. Ti-6Al-4V intake valve (left) and TiB/Ti-Al-Zr-Sn-Nb-Mo- Si exhaust valve (right) (courtesy Toyota Central R & D Labs, Inc).

be processed to the necessary density levels with properties equivalent or superior to baseline PM and wrought titanium. Finally, the cost of producing components from those powders must be competitive.

Examples of how the BE approach has been used to produce valves for production models, the Toyota Altezza family automobile, golf club heads and softball bats are shown in Fig. 8.7, 8.8 and 8.9,[1–3] respectively.

Currently, ADMA Products hydrogenated titanium powder manufacturing capacities are 50 000–60 000 lbs/year (22680–2722 kg/year), but they are installing a pilot scale unit which will triple output.[22] Meanwhile, the major aircraft companies and the US Departments of Energy and Defense (DOE

8.8 Titanium metal matrix composite golf club head (reinforced with TiB) (courtesy Toyota Central R & D Labs, Inc).

8.9 Susan Abkowitz of Dynamet Technology, Inc. holds a softball bat with a powder metallurgy titanium alloy outer shell.

and DOD) agencies have tested this material and reported that the properties of the PM Ti alloys meet Aerospace Materials Specification (AMS) and meet or exceed those of titanium wrought alloys made by conventional ingot metallurgy approaches.

8.3.2 The CHIP process

The CIP-Sinter or CHIP (CIP-Sinter-HIP) process,[15] Fig. 8.10, is used by Dynamet Technologies to produce near net shape parts for finish machining

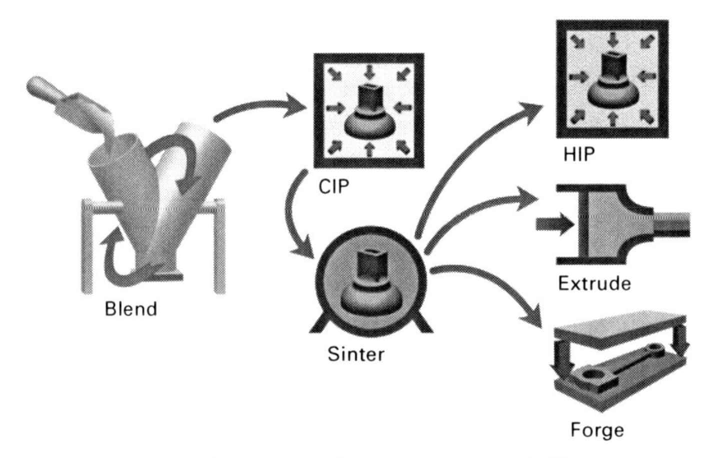

8.10 Schematic of Dynamet Technologies Inc CHIP process.

to high tolerance configurations. These processes can also be used to make forging preforms or mill product shapes for subsequent processing such as billets for casting, extrusion or hot rolling. In the case of as-sintered material, full density is achieved during subsequent processing.

The CHIP process is a green manufacturing technology[15] that has proven to be an acceptable process for producing military, industrial and medical components. This advanced PM process uses titanium powder, typically Kroll process HDH powder, blended with master alloy powder such as an aluminum–vanadium master alloy powder. The blended powder is compacted to shape by cold isostatic pressing (CIP) in elastomeric tooling. With proper selection of powders, well designed CIP tooling and appropriate pressing conditions, a shaped powder compact can be produced and readily extracted from the PM tooling with sufficient green strength for handling. It must also have sufficient uniformity and intimate contact of the powder particles for densification and homogenous alloying in the subsequent sintering process.

A wide range of shapes has been produced with size only limited by the capacity of the equipment. The size of the CIP is usually the limiting factor since vacuum furnaces and HIP units are available in larger sizes than are high pressure CIP units. Size capability also depends on the powder fill characteristics, product configuration and by tooling parameters. Successful products can range from a few grams to hundreds of kilos.

The major cost benefits of this clean, energy efficient manufacturing process are that it uses relatively low cost raw materials, avoids costly melt processes and results in relatively little material lost during processing. The capability to produce to a near net shape conserves raw material and also reduces costs for machining to finished parts. The cold pressed compacts are sintered in vacuum to high or nearly full density. Alloying of the titanium with the desired other elements is accomplished by solid state diffusion during the sintering process. By selecting the proper powders and sintering parameters, a homogeneous alloyed material with sufficiently high density, free of interconnected porosity, is achieved.

The sintering process was historically established to reach a minimum density level at which the material had no interconnected porosity. At this density threshold the material could be hot isostatically pressed (HIP) without the processing expense of HIP encapsulation, making the HIP process economically viable. Through recent developments the capability to reach greater than 98% sintered density has been achieved. This results in as-sintered tensile properties (Table 8.3)[15] that are equivalent to wrought properties and superior to castings. This reduces the need for the HIP operation and further strengthens the economic advantage of this PM CIP-Sinter manufacturing technology.

8.3.3 ADMA Products hydrogenated titanium process

The use of titanium hydride powder instead of titanium sponge fines has led to the achievement of essentialy100% density, using a simple cost-effective press-and-sinter technique, in complex parts.[20,21] In this work, hydrogenated non-Kroll powder (by cooling the sponge produced in a Kroll process with hydrogen rather than the conventional inert gas, a lower cost titanium hydride powder has been produced by ADMA Products) was utilized along with 60:40 Al:V master alloy to produce components made from the Ti-6Al-4V alloy. The press-and-sinter densities achieved using this novel fabrication technique are shown in Fig. 8.11. The associated microstructure and typical mechanical properties are shown in Fig. 8.12 and Table 8.4 (after cold pressing, sintering, forging and annealing), respectively. The mechanical properties compare well with those exhibited by cast-and-wrought product. The low cost of this process in combination with the attractive mechanical properties make this approach well suited to the cost-obsessed automobile industry. The parts shown in Fig. 8.13 have already been fabricated and a cost estimate of less than US$3.00 for an 0.32 kg (0.705 lb) connection link has been made.[23]

Table 8.3 Ti-6Al-4V alloy: ASTM E-8 tensile properties

	Theoretical density (%)	Ultimate tensile strength (MPa) (ksi)	Yield strength (MPa) (psi)	Elongation (%)
AMS 4928 (min)		896 (130)	827 (120)	10
Typical wrought		965 (140)	896 (130)	14
Typical PM CIP-Sinter	98%	951 (138)	841 (122)	15
Typical PM CHIP	100%	965 (140)	854 (124)	16

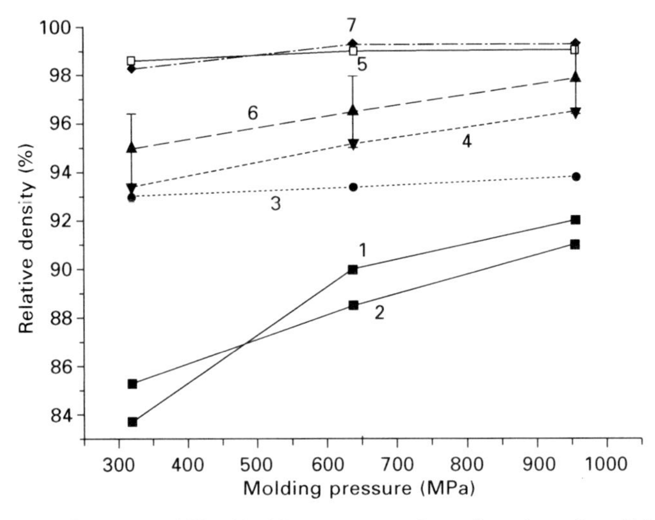

8.11 Density of Ti-6Al-4V compacts after sintering. Conditions 5 and 7 used hydrided powder and show by far the highest and most uniform densities.

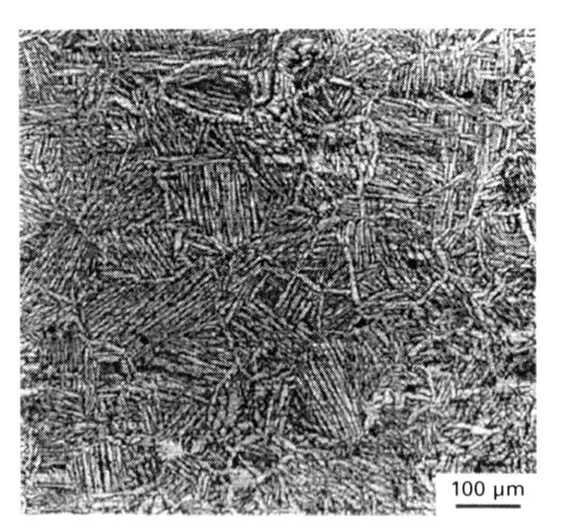

8.12 Microstructure of sintered Ti-6Al-4V material.

Table 8.4 Room temperature tensile properties of a hydrogenated titanium compact (after dehydrogenation)

PM Ti-6Al-4V	Ultimate tensile strength (MPa) (ksi)	Yield strength (MPA) (ksi)	Elongation (%)	Reduction of area (%)
3.5 cm (1.376") thick	994–1028 (144–149)	911–938 (132–136)	14.0–15.5	34–38
ASTM	897 (130)	828 (120)	10	25

8.13 Ti-6Al-4V parts produced using a press-and-sinter approach and titanium hydride: (1) connecting rod with big end cap, (2) saddles of inlet and exhaust valves, (3) plate of valve spring, (4) driving pulley of distributing shaft, (5) roller of strap tension gear, (6) screw nut, (7) embedding filter, fuel pump, and (8) embedding filter (courtesy Ukrainian Academy of Sciences).

In Kroll's process, the removal of the Ti sponge from the retort and its subsequent crushing is time and energy intensive. In comparison, ADMA's process produces TiH_2 which, unlike Ti sponge, is very friable (see Fig. 8.14) and easily removed from the retort with no need for an expensive sizing operation. ADMA's vacuum distillation processing time is also at least 80% less than in Kroll's process since phase transformations/lattice parameter changes of the hydride sponge, in the presence of hydrogen, accelerate the distillation removal of $MgCl_2$. Finally, atomic hydrogen is released during sintering–dehydriding of the TiH_2 powder and acts as a

8.14 TiH_2 powder (courtesy of ADMA Products).

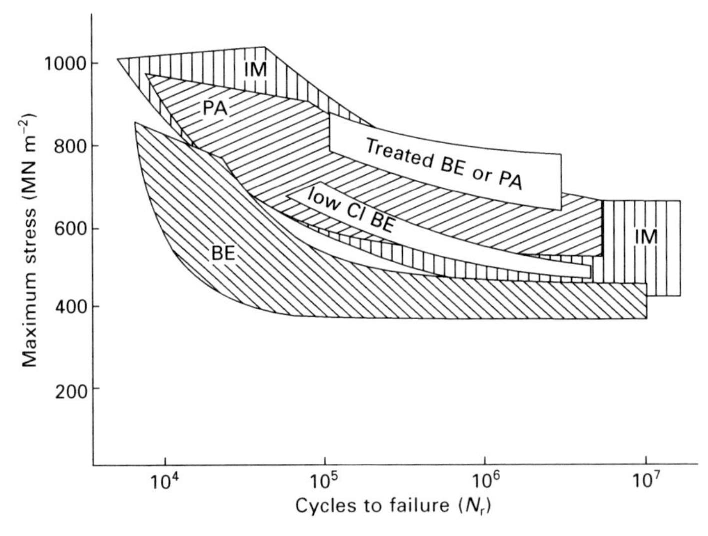

8.15 Fatigue data scatterbands of conventional BE, low chloride BE, treated low chloride BE, and PA, compared with wrought annealed material.

scavenger for impurities (e.g. oxygen, chlorine, magnesium etc) resulting in titanium alloys with low interstitials that at least meet the properties of ingot metallurgy alloys.

A comparison of the *S–N* fatigue behavior of BE and prealloyed material with cast-and-wrought product is shown in Fig. 8.15.[12]

Powders can be subsequently fabricated to other product forms, such as titanium sheet, (Fig. 8.16). Alloy sheet can be fabricated in a similar manner by adjusting the feedstock to a mixture of titanium powder and alloying additions.

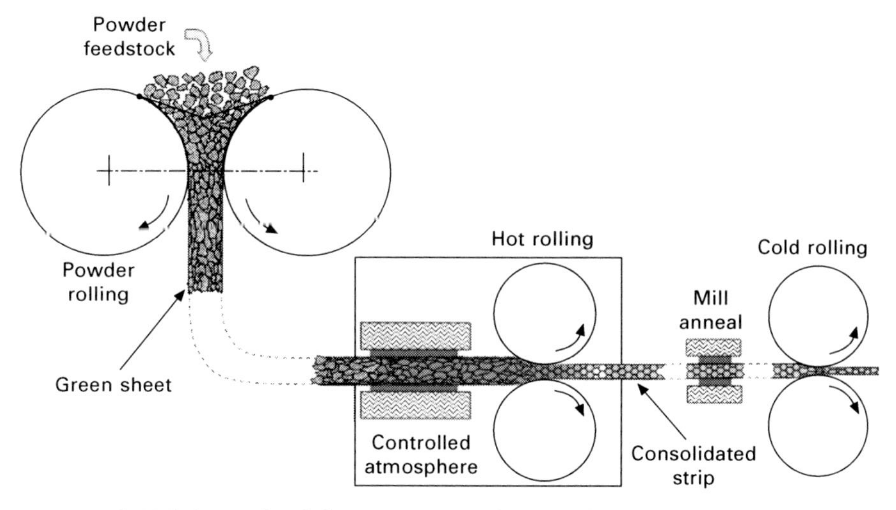

8.16 Schematic of the process used to produce commercially pure titanium sheet at CSIRO.[19]

8.17 Finished titanium MMC ring for spin pit testing (courtesy IMT-Bodycote).

8.3.4 Metal matrix composites (MMC)

Both continuously and discontinuously reinforced titanium components have been produced using PM approaches. Figure 8.17 shows a finished titanium MMC ring for spin pit testing fabricated from HIP densified plasma sprayed tapes.

The blended elemental technique has also been used by Dynamet Technologies for the fabrication of MMCs utilizing particulate and a combined cold and hot isostatic pressing (CHIP) combination or forging, extrusion and rolling of the CHIP preform.[15] The CermeTi family of titanium alloy matrix composites incorporates particulate ceramic (TiC or TiB2) (Fig. 8.18) or intermetallic (TiAl) as a reinforcement with minimal particle/matrix interaction. The mechanical properties of CermeTi material are shown in comparison with PM Ti-6Al-4V in Table 8.5.[15] A seven layer armor and a dual hardness gear have been made by Dynamet Technologies from CermeTi material.

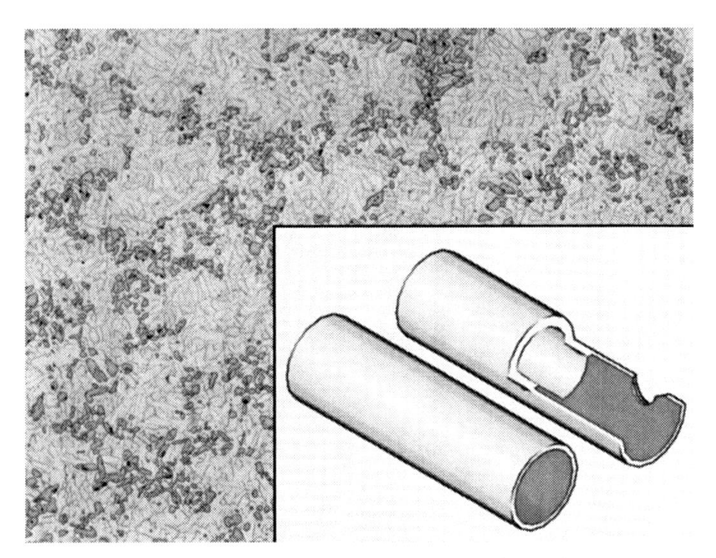

8.18 Microstructure of CermeTi material, TiC reinforcement (top) and parts fabricated from this material (bottom) (courtesy Dynamet Technology Inc.).

Table 8.5 Typical properties of CermeTi® versus Ti-6Al-4V

	Ultimate tensile strength (MPa) (ksi)	Yield strength (MPa)	Elongation (%)	Hardness (Rc)
Ti-6Al-4V PM	965 (140)	896 (130)	14	36
CermeTi®-C MMC (Ti-64+TiC)	1034 (150)	965 (140)	3	42

8.3.5 Prealloy

The prealloyed approach (PA) involves use of prealloyed powder, generally spherical in shape, which has been produced by melting, either by a technique such as the plasma rotating electrode processing (PREP) or gas atomization (GA), followed by hot consolidation (generally by hot isostatic pressing).[12] The mechanical properties are superior to ingot material (because of the refined microstructure and lack of directionality, see later). Powders are poured either into a metal can (with metal inserts) or into a ceramic mold (sealed in a can filled with a secondary pressing media) and compacted in a hot isostatic processing (HIP) unit.

The mechanical properties of PA compacts, fabricated using the ceramic mold process, for example the Ti-6Al-4V alloy, are at least at cast and wrought (ingot metallurgy) levels, including fatigue behavior, Fig. 8.15.[12] The PA approach has been used to produce large and complex parts such as the nacelle frame, impeller and engine mount support shown in Fig. 8.19.

More recently, the advantage to be gained by the near net shape PM approach for difficult to process alloys such as the intermetallic Ti-Al-type compositions has been recognized.[24,25] Components produced from this type of alloy include the shapes shown in Fig. 8.20. The shapes are demonstration parts, the links, which demonstrate the amenability of the process to production of composite concepts, are in production. Another item produced from gamma-material is sheet produced from powder via a pack-rolling approach [26] which has been produced for systems such as the X series of NASA vehicles (Fig. 8.21).

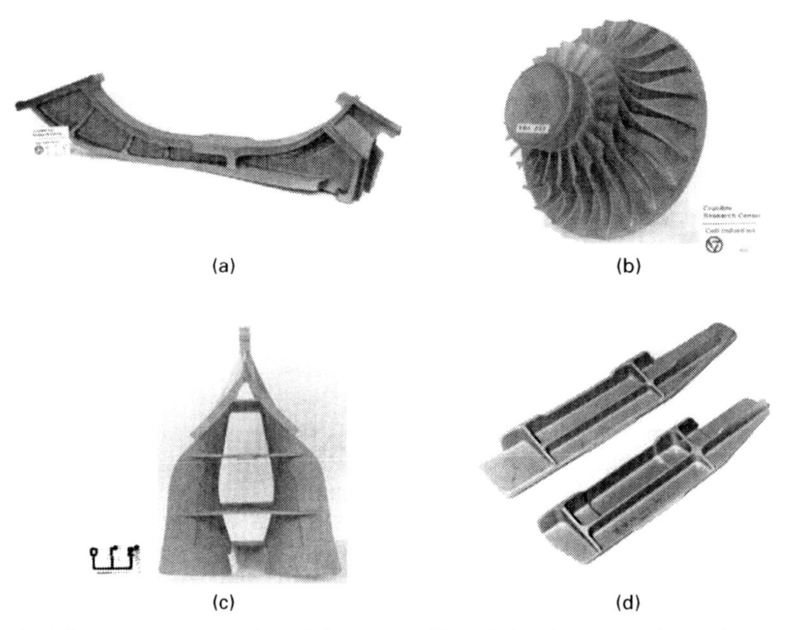

8.19 Components produced from prealloyed titanium powder, using HIP and the ceramic mold process; (a) a nacelle frame for F14A, Ti-6Al-6V-2Sn, (b) radial impeller for F107 cruise missile engine, Ti-6Al-4V, (c) a complex airframe component for the stealth bomber, Ti-6Al-4V and (d) engine mount support, Ti-6Al-4V (courtesy of Crucible Materials Corporation).

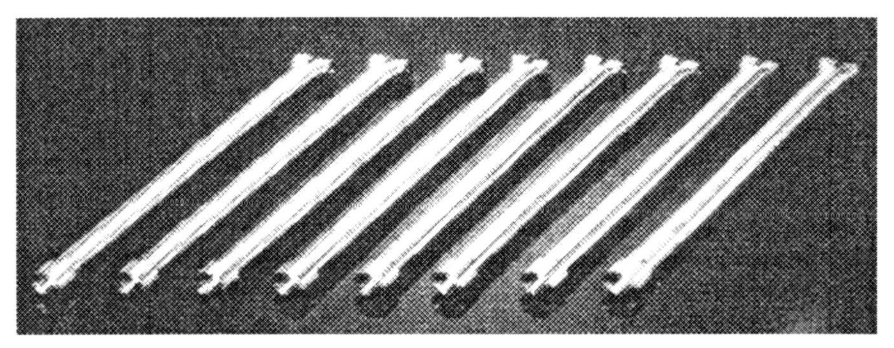

(a) (b) (c)

8.20 (Top) Gamma titanium aluminide shapes made using prealloyed gas atomized powder followed by HIP (left to right), (a) billet for subsequent forging, (b) forging or to be machined, (c) a near net sonic shape for an engine application; and (bottom) exhaust nozzle compression links for the F110 engine (power system for the F-16 Falcon), consisting of continuous SiC fibers in a Ti-6Al-2Sn-4Zr-2Mo matrix.[25]

8.21 PM titanium aluminide sheet (courtesy Plansee Aktiengesellschaft).

Partly because of concerns that ceramic particles could get into the titanium parts fabricated using the ceramic mold process, no parts are currently in production using this approach. However parts produced using a shaped metal can and removable mild steel inserts (removed by chemical dissolution), Figs 8.22 and 8.23, are production ready.[27]

Despite the 30–35% volume shrinkage (typical for HIP of PA powders), advanced process modeling allows 'net surfaces' to be achieved and minimal machining stock on the 'near net surfaces'. Also these near net shape titanium parts can be made up to the size of existing HIP furnaces, that is up to 2 m, which is considerably larger than the capabilities of the other technologies discussed in this paper.

These parts exhibit mechanical properties superior to conventional cast and wrought (ingot metallurgy) components (Figs 8.24 and 8.25 and Table 8.6). Figure 8.24 shows actual tensile strength levels obtained in cast and

8.22 Near net shape Ti-6Al-4V engine component fabricated using the prealloyed metal mold method (courtesy Synertech PM).

8.23 Selectively net shape ELI Ti-6Al-4V impeller for a turbo-pump of a rocket engine. Fabricated using the prealloyed metal method (courtesy Synertech PM/P&W Rocketdyne).

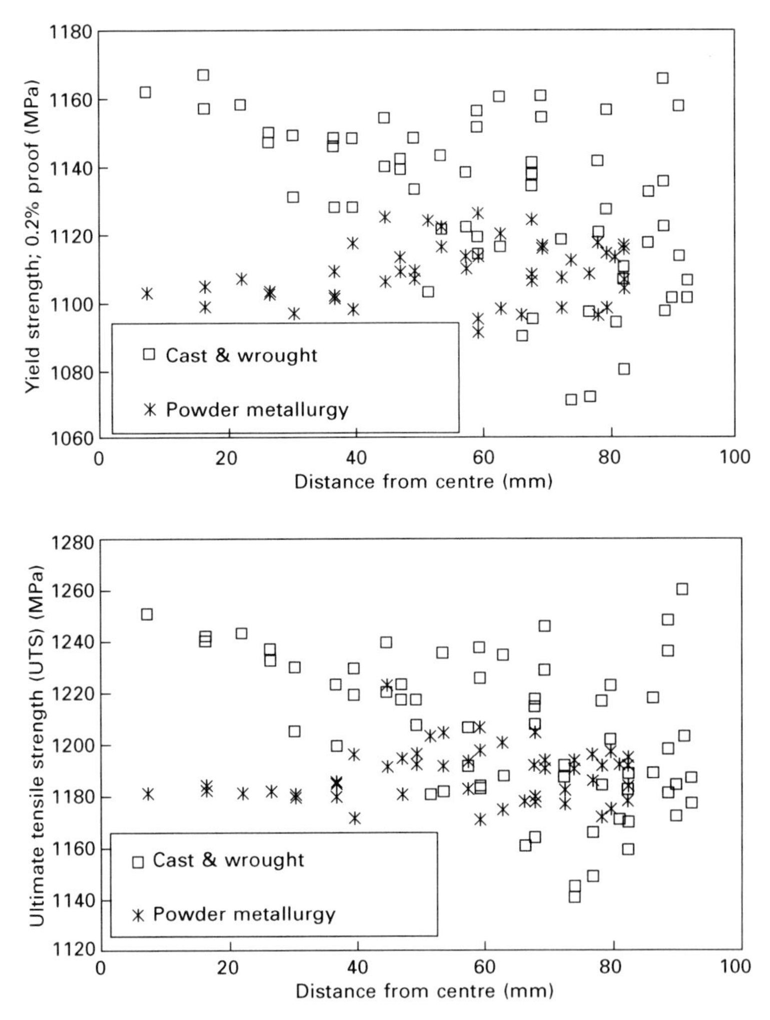

8.24 Comparison of ingot and powder metallurgy tensile properties (courtesy Prof. Igor Polkin, VILS, Russia).

wrought product compared to data from PM product. However the minimum values (which are used in design) for the PM material is above that for the conventionally fabricated material (Fig. 8.25). Fracture toughness of the PM product is superior to cast and wrought material (Table 8.6).

8.4 Additive layer manufacturing and powder injection molding

In the additive layer manufacturing approach, powder is laid down in successive layers and melted under the control of a computer to produce

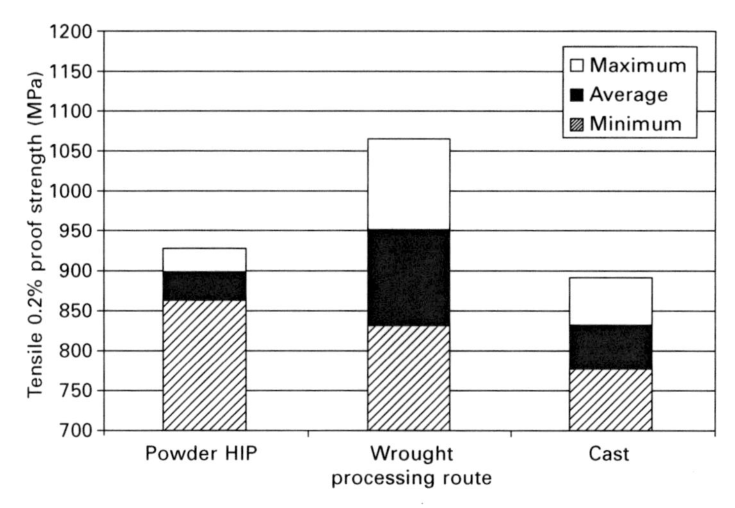

8.25 Comparison of Ti-6Al-4V powder HIP and wrought tensile properties (courtesy of Wayne Voice, Rolls-Royce).

Table 8.6 Fracture toughness of PM material and conventionally forged material (courtesy Dr. Wayne Voice, Rolls-Royce, UK)

Specimen	K_{IC} (MN $m^{-3/2}$)
1	94.0
2	96.5
3	92.5

Forged Ti-6Al-4V K_{IC} = 55 $MNm^{-3/2}$

virtually any shape; the mechanical properties are still being optimized. AeroMet Corporation's Lasform technique (Figs 8.26–8.27) features shorter lead times and greatly reduced buy-to-fly ratios (minimal machining required). This PM fabrication approach has been studied by Boeing to produce large 10 foot x 3 foot substrates, which would then be built-up.[28] However the Aeromet equipment is no longer in operation.

The strength of ALM Ti-6Al-4V is 160 ksi (1104 MPa) with 5–6% elongation 'as-formed' and after a HIP step is 140 ksi (965 MPa) with 15% elongation, equivalent to cast and wrought levels (data courtesy of B. Dutta, the POM Group).

The *S–N* fatigue performance of ALM is at or even a little above conventional material levels. However the main issue influencing growth in deployment of ALM for titanium alloys relates to raw material supply. First, with material cost being typically 40–50% of total manufacturing cost for ALM titanium, material cost is a major issue. The material supply chain is an issue for both powder and wire; sustainable sources are not always

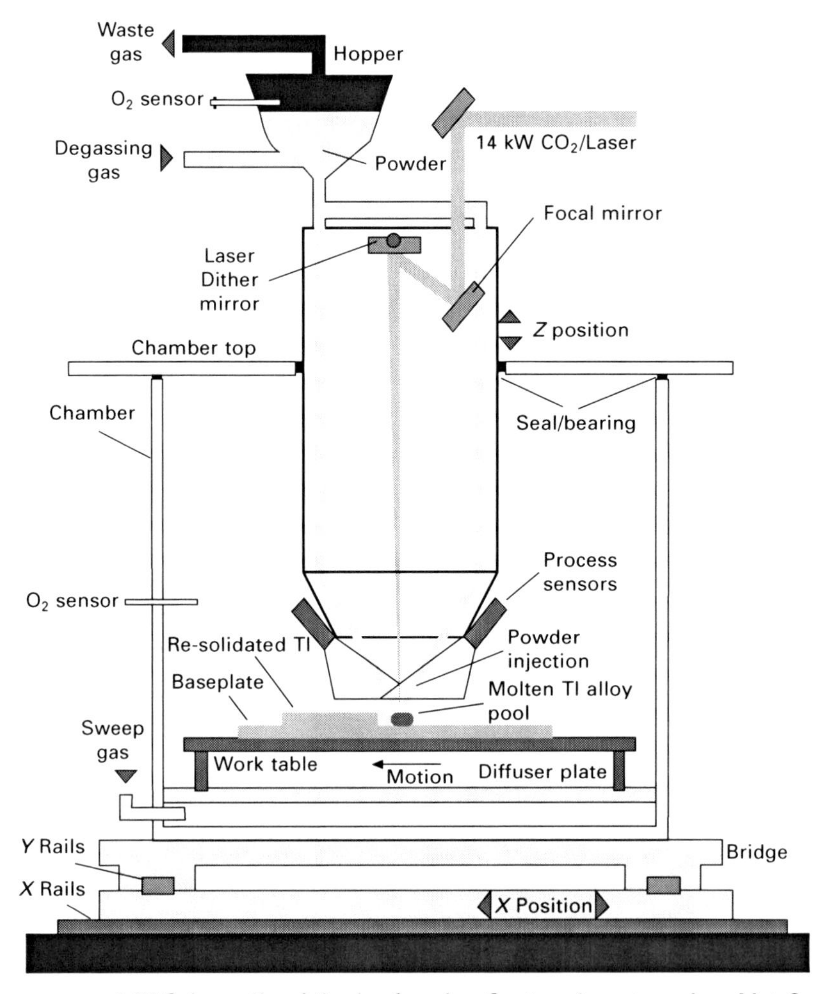

8.26 Schematic of the Lasforming System (courtesy AeroMet Corp.).

available and supplies of certain alloys have limited availability.[29] Typical parts fabricated by ALM are shown in Figs 8.29–8.32.[29] The advantage of attaching features is demonstrated in Figs 8.33 and 8.34, and the capability of using ALM in repair of a part is illustrated in Fig. 8.35.[29]

Metal powder injection molding (PIM) is based upon the injection molding of plastics; the process has been developed for long production runs of small (normally below 400 g) complex-shaped titanium parts in a cost-effective manner. This technique involves melting and pelletization of a mixture of titanium powder and a binder which are injected into a die, the binder removed chemically/thermally, and finally the part is sintered. By increasing the metal (or ceramic) particle content, the process evolved into a process for production of high density metal, intermetallic or ceramic components (Fig. 8.36).[30,31]

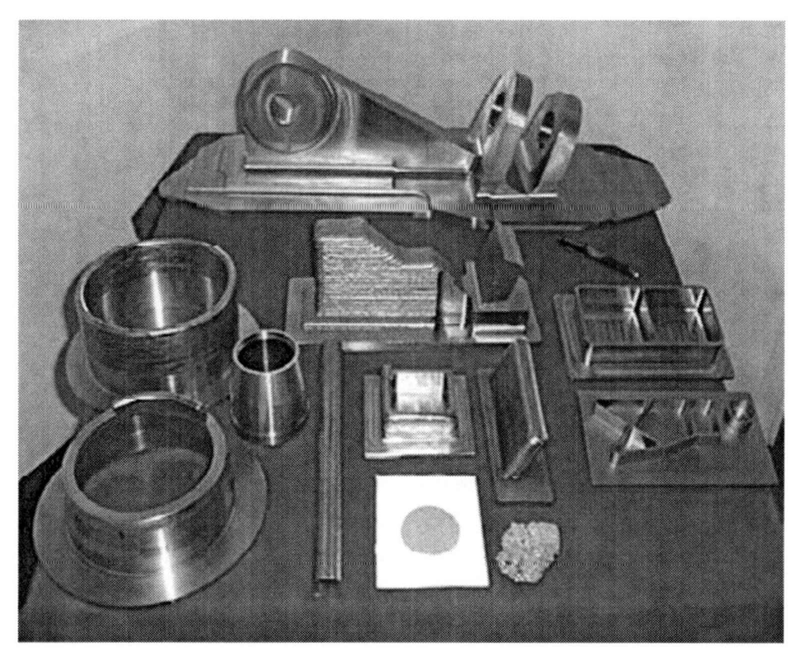

8.27 Examples of Lasform shapes (courtesy AeroMet Corp.). A further example of an ALM component is shown in Fig. 8.28.

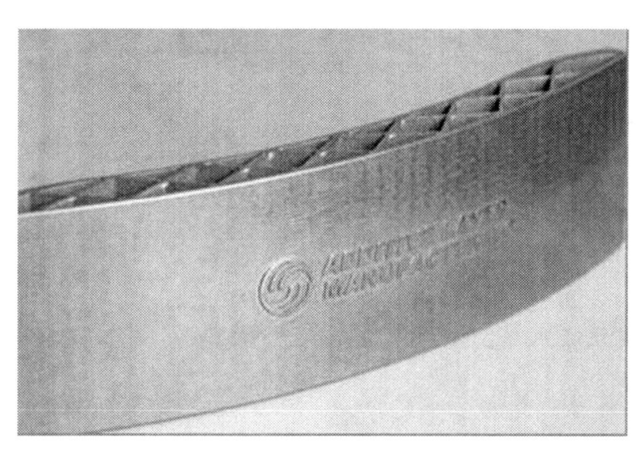

8.28 Additive layer manufacturing. Sandwich structure aerofoil demonstrator produced at the Centre for Additive Layer Manufacturing (CALM) using an electron beam chamber and powder feed.

This method allows the fabrication of components with good mechanical properties provided the chemistry (particularly oxygen) is controlled.[32,33] Typical shapes are shown in Fig. 8.37. By incorporating a porous layer on the surface of body implant parts Praxis are able to cause bone ingrowth and an improved bonding between the implant and the bone.[34]

Aerospace Prototype

Material: Ti6Al4V
Size: ø 180 × 300 mm
Weight: 5.5 kg
Build time: 40 h

8.29 Aerospace application for ALM.

Femoral stem

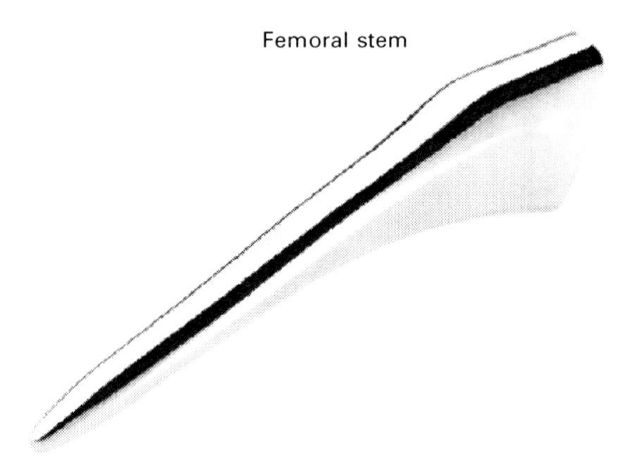

8.30 Medical implant application for ALM.

The majority of the early work on developing a viable titanium PIM process was plagued by the unavailability of suitable powder, inadequate protection of the titanium during elevated temperature processing and less than optimum binders for a material as reactive as titanium.[31] However, some PIM practitioners have now learned what the titanium community has long known – that titanium is the universal solvent and must be treated accordingly.[7–11,35]

Titanium has a high capacity to form interstitial solutions readily with a wide range of commonly encountered elements including carbon, oxygen and nitrogen, which has presented several challenges for titanium PIM development efforts. These interstitial solutions are undesirable since they

8.31 Lock barrel prototype fabricated by ALM.

8.32 Automotive pulley prototype fabricated by ALM.

significantly degrade the ductility of sintered titanium PIM parts. Therefore, it is advantageous to use a binder, which can be completely removed from the green PIM part without leaving these detrimental impurities behind. This is particularly true when fabricating structural aerospace and medical implant parts, which can require oxygen impurity levels below 300 ppm to meet ASTM F 167 Standards.

Unfortunately, unlike conventional ceramic and ferrous alloy PIM processing, there is a significantly narrower processing window between the debinding cycle and the temperature where impurity diffusion becomes significant within titanium. In general this requires that the titanium PIM binder be essentially removed from the green part at temperatures typically below 260°C to prevent introducing impurities into the sintered parts. Additionally, the binder must exhibit high chemical stability and not undergo catalytic decomposition in the presence of titanium metal powder surfaces

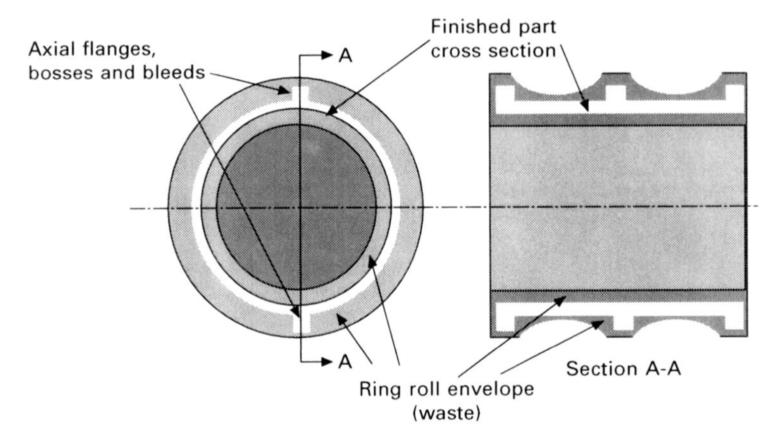

8.33 Material waste in machining features on a forged preform in conventional manufacture. Discrete features like axial flanges and bosses produce a disproportionate increase in forging size and weight.

8.34 Fan case produced by adding features by ALM to a forged preform.

during molding operations, even when held for long isothermal holds within the injection molding machine.

Attempts to adapt conventional ceramic and metal PIM binder systems for titanium processing have met with limited success. This is due to the fact that these systems often employ significant amounts of thermoplastic polymer within their formulations. Unfortunately, even some of the more

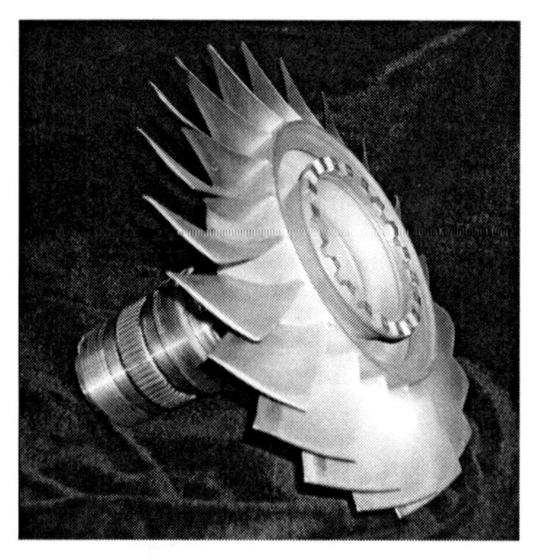

8.35 ALM repair of gas turbine components.

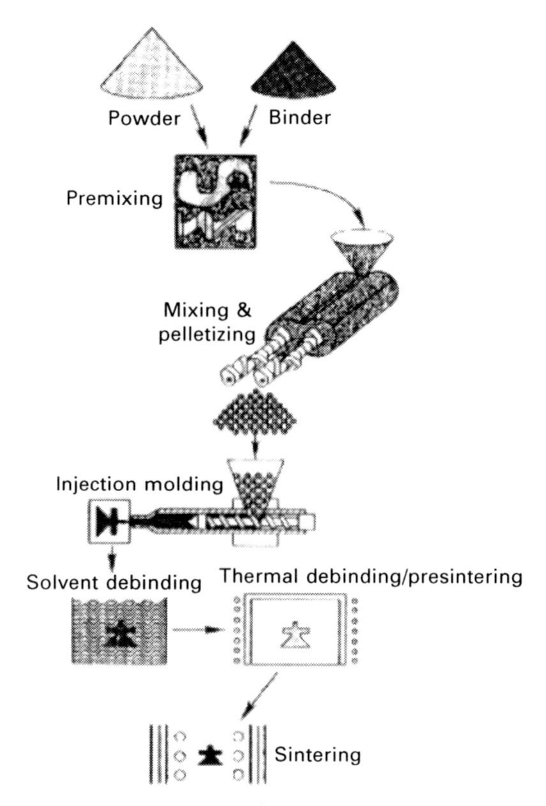

8.36 Schematic of the steps involved in powder injection molding, in which a polymer binder and metal powder are mixed to form the feedstock which is molded, debound and sintered.

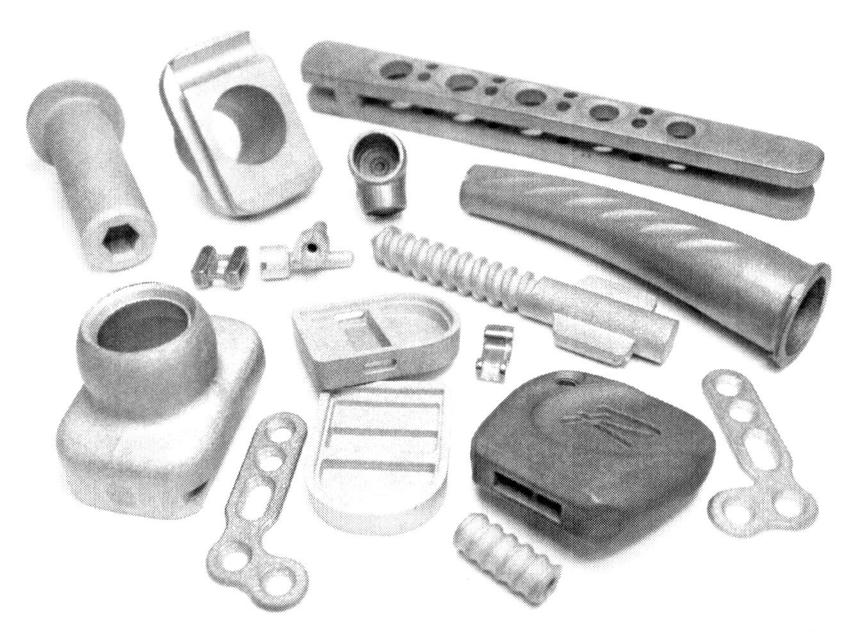

8.37 Titanium MIM components (courtesy of Praxis Technology).

well-known polymer binders known for their ability readily to thermally unzip to their starting monomers (e.g. polymethylmethacrylate, polypropylene carbonate, poly-α-methylstyrene) still tend to introduce impurities into the sintered titanium PIM bodies because their depolymerization occurs close to those temperatures where impurity uptake is initiated. Alternative binder systems based upon catalytic decomposition of polyacetals are promising but require expensive capital equipment to handle the acid vapor catalyst as well as suitable means of eliminating the formaldehyde oligomers that form as polymer decomposition by-products. However, there are a number of binder systems which appear to have the necessary characteristics to be compatible with titanium.[34]

Currently titanium PIM parts run up to a foot in length, but parts over three or four inches (about 50 g weight) are not common, Fig. 8.37. The limiting factors at this time are dimensional reproducibility and chemistry. Owing to the shrinkage, large parts become dimensionally more difficult to make because of loss of shape during shrinkage. If the parts have flat surfaces to rest on the setter they come out fairly consistently. But parts with multiple surfaces that require setters in complex shapes become less practical as the size goes up. Further, large overhanging areas become difficult to control dimensionally owing to the effects of gravity. With increasing experience, the packing density of titanium powder mixes will be increased, especially when new binders become available and the shrinkage can be decreased making the dimensional problems less of a factor.

The current estimate is that worldwide titanium PIM part production is currently at about the 3–5 ton per month level.[31] This market is poised for expansion. What is needed is low cost (less than US$20/lb or US$44/kg) powder of the right size (less than about 40 μm) and good purity (which is maintained throughout the fabrication process). For non-aerospace applications, the purity level of the Ti-6Al-4V alloy can be less stringent; for example, the oxygen level can be up to 0.3 wt% while still exhibiting acceptable ductility levels (the aerospace requires a maximum oxygen level of 0.2 wt%).[11] For the CP grades, oxygen levels can be even higher; up to at least 0.4 wt% (Grade 4 CP titanium has a specification limit of 0.4 wt%).[11] In fact, the Grade 4 CP titanium (UTS 550 MPa (80 ksi)) while having a lower strength than regular Ti-6AL-4V (UTS 930 MPa (135 ksi)) may well be a better choice for many potential PIM parts where cost is of great concern. Grade 4 would allow use of a lower cost starting stock and a higher oxygen content in the final part. Further into the future, the beta alloys with their inherent good ductility (body centered cubic, bcc structure) and the intermetallics with attractive elevated temperature capability are potential candidates for fabrication by PIM. The science, technology and cost now seem to be in the market for titanium PIM to show significant growth.

8.5 Spraying and research-based processes

Spray forming can involve either molten metal [36] or solid powder. Because of its very high reactivity, the challenges associated with molten metal spraying of titanium are quite considerable. However, both spray forming in an inert environment and under reactive conditions have been achieved with appropriately designed equipment.[13] A segmented cold-wall crucible, combined with induction heating and an induction-heated graphite nozzle was used to produce a stream of molten metal suitable for either atomization, to produce powder or spray forming (Fig. 8.38).

Recently there has been increased interest in cold-spray forming involving solid powder particles.[37] Cold spray (<500°C) can produce both monolithic 'chunky' shapes and coated components. In this process, solid powder is introduced into a deLaval-type nozzle and expanded to achieve supersonic flow (Fig. 8.39). Powders are in the range of 1–50 μm, at relatively low temperatures (<500°C) with a velocity in the range of 300–1200 m s^{-1}. Both monolithic 'chunky' shapes and coated components can be produced (Fig. 8.40)[19] and the coatings can be applied even to the inside of tubular components. The density of the sprayed region is less than full density, but this can be increased to 100% density by a subsequent HIP operation.

This technique is also very useful in bonding together normally difficult to bond metals such as titanium and steel (Fig. 8.41).

Rapid solidification (RS), mechanical alloying (MA) and vapor deposition

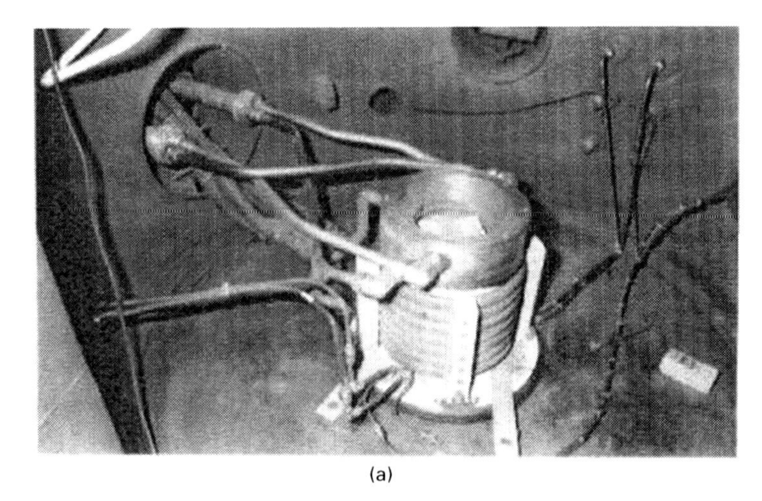

(a)

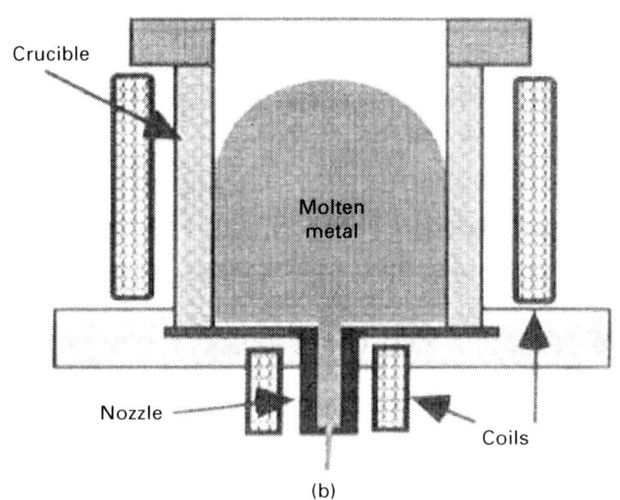

(b)

8.38 (a) Cold-wall induction bottom-pour (CWIBP) crucible installed in a plasma cold hearth furnace, (b) schematic of the CWIBP system.

(VD) all fall in the category of 'far from equilibrium processes'.[13] Novel constitutional (such as extension of solubility levels) and microstructural (in particular microstructural refinement and production of very stable dispersions of second phase particles) effects can be obtained by all three processes, however commercial processes are not on the near horizon. An example of the fine dispersion of second phase particles which can be obtained by RS is shown in Fig. 8.42(a) and nanograined material produced by MA in Fig. 8.42(b).[38,39] These nanograins show surprisingly good stability on exposure to elevated temperatures, especially when yttria particles are dispersed throughout the matrix.

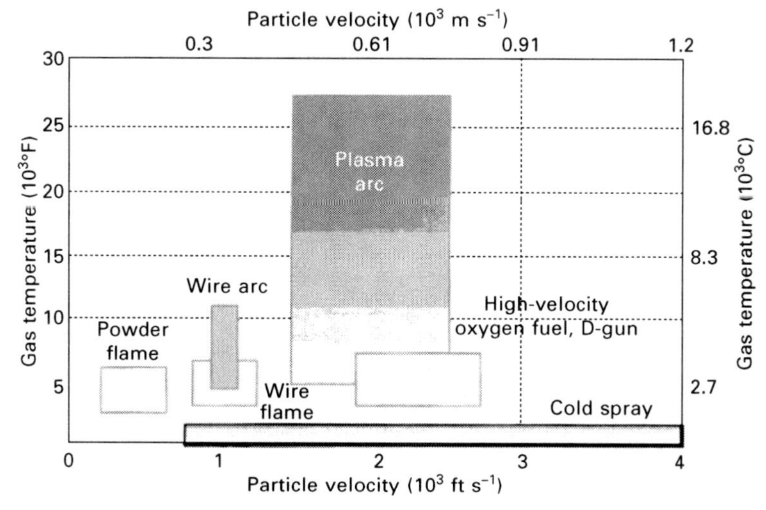

8.39 Temperature versus velocity regimes for common thermal spray processes compared to temperature and velocity in cold spray technology.

8.40 Titanium heat pipe connectors produced by cold spraying (courtesy of M. Jahedi, CSIRO).

The VD approach can be used to alloy together normally virtually immiscible Mg with Ti to create low density alloys akin to Al-Li alloys, and fabricate layered structures at the nano-level (Fig. 8.43).[40]

Also, currently in the research base are thermohydrogen processing (THP)[13,41] and porous structures.[42] By intentionally adding hydrogen to a titanium alloy such as Ti-6Al-4V with a normal PM microstructure, the

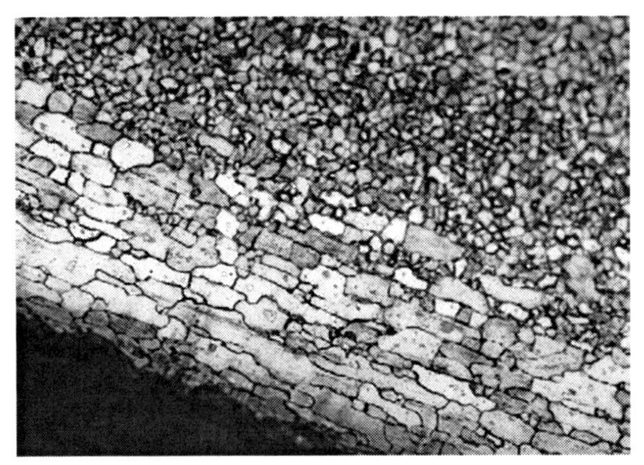

8.41 Optical photomicrograph showing the excellent bonding of titanium to steel (courtesy Ktech).

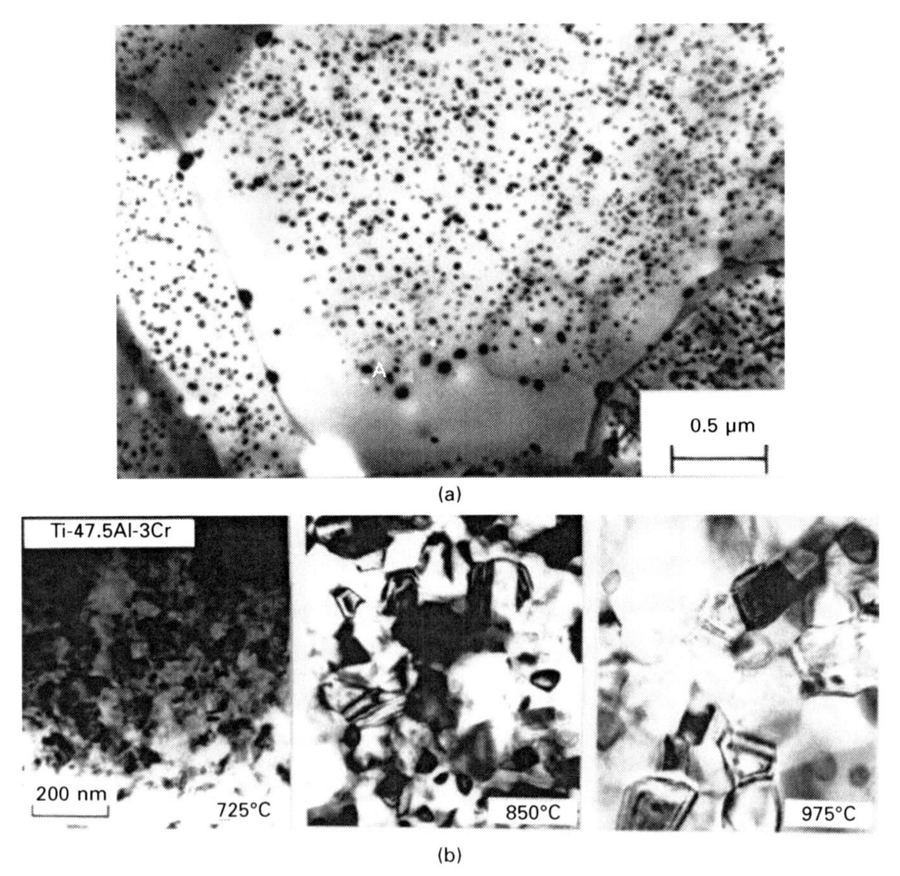

8.42 Titanium aluminide intermetallic alloys exhibiting (a) a fine dispersion of second phase erbia particles (Ti3Al based alloy) and (b) nanograins after HIP at the temperatures indicated (TiAl based alloy).

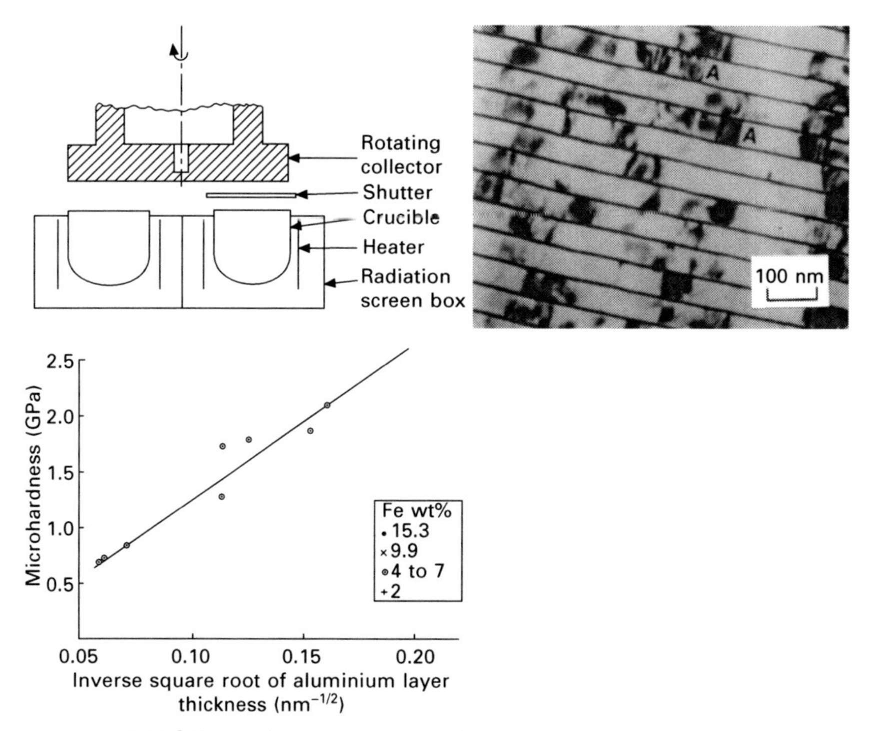

8.43 Schematic cross section of rotating collector used in vapor deposition of layered nanostructured materials (top left). Increase in hardness with decreasing layer spacing (bottom left), and layered nanostructured consisting of layers of Al and Fe (top right).

microstructure can be refined in the dehydrogenated conditions (Fig. 8.44) with an enhancement in mechanical properties.[41]

A novel type of porous low density titanium alloy can be produced by HIP consolidation of alloy powder in the presence of an inert gas such as argon (Fig. 8.45).[13] The tensile strength decreases in a linear manner as the porosity level increases, following the 'rule of mixtures' relationship at least up to 30% porosity level. This material exhibited excellent damping characteristics suggesting a generic area of application. There may also be applications in body implants with the foam integrated in various locations to facilitate growth of bone/flesh into the porous regions promoting a stronger joint. Tensile testing of the bond between foam and dense material indicated a bond strength in excess of 85 MPa well above the Food and Drug Administration (FDA) requirement of 22 MPa for porous coatings on orthopedic implants.

Porous structures with a potential use in honeycomb structures or in sound attenuating or firewall applications are now possible with very precisely controlled porosity levels and architecture using a novel blended metal–plastic approach. (Fig. 8.46).[42]

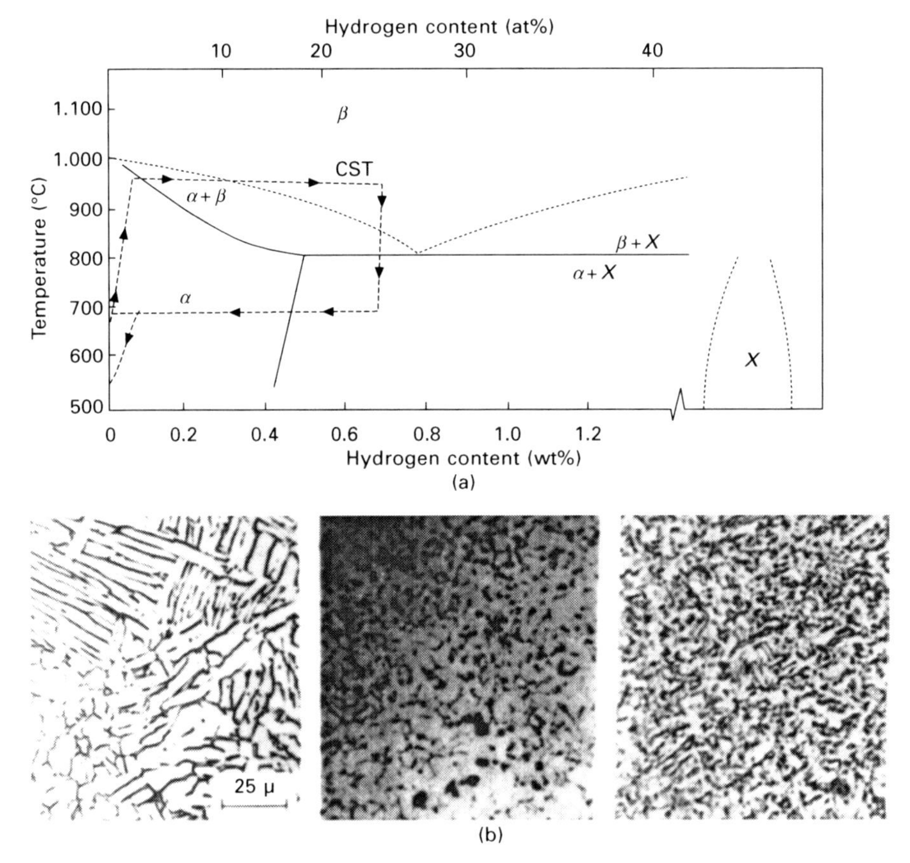

8.44 (a) Pseudo binary phase diagram for Ti-6Al-4V. X represents the hydride phase and CST (constitutional solution treatment) is one possible thermohydrogen processing treatment, and (b) refinement of the microstructure of Ti-6Al-4V powder compact using the thermochemical processing technique, (left) as HIP coarse alpha laths, (center) hydrogenated then compacted, (right) hydrogenated in compacted state. The latter two conditions are after dehydrogenation, both showing a refined alpha microstructure, (center) equiaxed grains, (right) fine alpha laths.

8.6 Future trends

Over the past 30 years, a great deal of money, much of it from US government sources, has been spent in attempting to circumvent the high cost of titanium components for aerospace and terrestrial applications. However, despite a few successes with lower integrity, BE parts and recent advances in PIM, the overall market is small; in total perhaps 20 000 pounds per year maximum worldwide.

However, a variety of high quality, low cost powders should be available

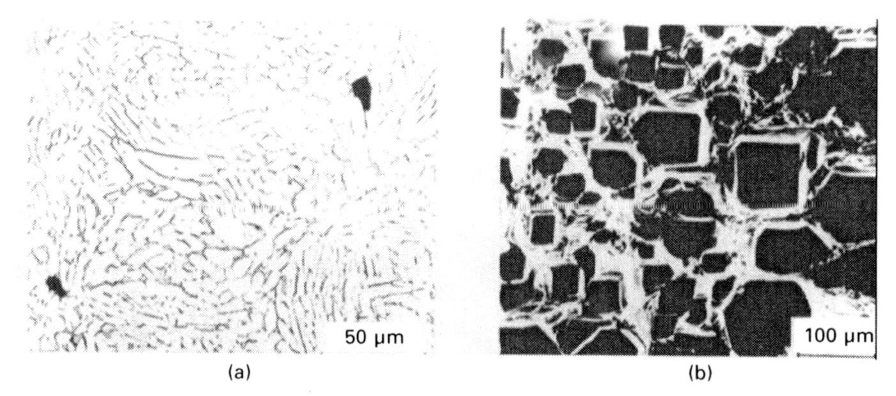

8.45 (a) Optical micrograph of pores in HIP Ti-6Al-4V containing argon after annealing at 700°C (1290°F), and (b) SEM of sample containing up to 40% porosity after an anneal in excess of 1000°C (1830°F).

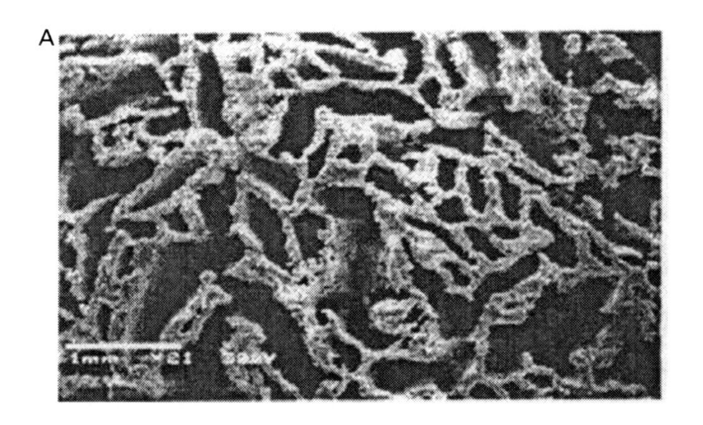

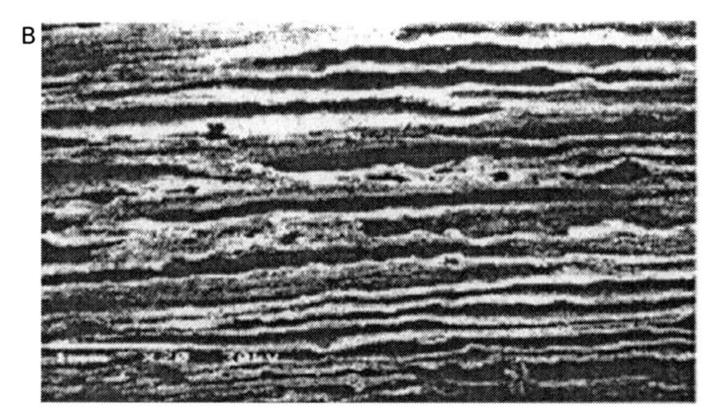

8.46 Micrographs of foams produced by die compaction (A) and extrusion (B). A, cross section is perpendicular to the compaction direction; B, cross section is parallel to the extrusion direction.

8.47 Cold press, sinter and hot rolled preform (weighing 95 kg (210 pounds)) for commanders hatch forging (Bradley Fighting Vehicle) (courtesy of ADMA Products).

soon. There have also been a number of developments, which should lead to a reasonable growth of products produced by the PM method. There appear to be three areas where significant growth can occur: (a) small parts (less than 1 pound in weight) by PIM, (b) larger parts using the BE and PA techniques and (c) generally larger parts using the ALM approach. Early entrants to the PIM marketplace naively largely ignored what every good titanium metallurgist knows – that titanium is the universal solvent. This fact should be clear to current titanium PIM practitioners and growth should be healthy in this area. Blended elemental applications are likely to grow especially with innovative approaches such as use of hydrogenated powder to produce uniformally high densities[20,21] including large armored vehicle parts (Fig. 8.47).

There are barriers to overcome using the PA approach; here the competition is with critical components produced by the ('tried and true') cast-and-wrought or direct casting approach.[43] Recent developments suggest that immediate applications for PA titanium is in complex parts, very difficult to fabricate TiAl intermetallic alloys, and in metal matrix composite concepts.[24,25] The ALM approach should also see significant growth, in competition with the BE and PA techniques. The research-based techniques mentioned in this paper all show promise, but have yet to approach commercial fruition.

8.7 Acknowledgements

The author recognizes useful discussions with Joe Fravel, Serge Grenier, Mitch Godfrey, G. N. Hayek Jr., Andy Hanson, Larry LaVoie, John (Qiang)

Li, Tim McCabe, Steve Miller, John Moll, Vladimir Moxson, Vladimir Duz, Takashi Nishimura, Hiroaki Shiraishi, Tessa Stillman, Yoshy Takada, Bruno Unternährer, Jim Withers and Fred Yolton. We also acknowledge the contribution of Mrs Marlane Martonick for assistance in formatting and typing the text.

8.8 References

1 Froes F H (Sam), Imam M A and Fray D (eds), *Cost Affordable Titanium*, TMS, The Minerals, Metals and Materials Society, Warrendale, Pa, 2004.
2 Gungor M N, Imam M A and Froes F H (Sam) (eds), *Innovations in Titanium Technology*, TMS, Warrendale, Pa. 2007.
3 Imam M A, F H Froes and Dring K F (eds), *Cost-Affordable Titanium III* TMS, Warrendale, Pa. 2010.
4 Congress of the US Office of Technology Assessment, *Advanced Materials by Design* (June 1988).
5 Materials Science and Engineering – *Forging Stronger Links to Users*, NMAB, National Academy Press publication NMAB-492, Washington DC, 1999.
6 Froes F H, Eylon D and Bomberger H (eds) *Titanium Technology: Present Status and Future Trends*, TDA, Dayton, OH, 1985.
7 Froes F H (Sam), Yau T-L and Weidenger H G, 'Titanium, Zirconium and Hafnium', *Materials Science and Technology – Structure and Properties of Nonferrous Alloys*, Chapter 8, KH Matucha (ed.), VCH Weinheim, FRG, 1996, 401.
8 Froes F H (Sam), 'Titanium', *Encyclopedia of Materials Science and Engineering*, Chapters 3.3.5a–3.3.5e, P. Bridenbaugh (ed.) Elsevier, Oxford, UK, 2000.
9 Froes F H (Sam), 'Titanium alloys', *Handbook of Advanced Materials*, Chapter 8, James K. Weasel (ed.), McGraw-Hill Inc., New York, NY, 2000.
10 Froes F H (Sam), 'Titanium metal alloys', *Handbook of Chemical Industry Economics, Inorganic*, Jeff Ellis (ed.), John Wiley and Sons, New York, 2000.
11 Boyer R R, Welsch G and Collings E W (eds), *Materials Properties Handbook: Titanium Alloys*, ASM International, Materials Park, OH, 1994.
12 Froes F H and Eylon D, 'Powder metallurgy of titanium alloys', *International Materials Reviews*, 1990, **35**, 162.
13 Froes F H and Suryanarayana C, 'Powder processing of titanium alloys', *Reviews in Particulate Materials*, A. Bose, R.M. German and A. Lawley (eds), MPIF, Princeton, NJ, 1993, **1**, 223.
14 McCracken C G, 'Manufacture of hydride-dehydride low oxygen Ti-6Al-4V (Ti-6-4) powder incorporating a novel powder de-oxidation step', *Euro P/M Conference 2010.*
15 Abkowitz S M, Abkowitz D, Fisher H, and Main D H (2011), 'Affordable P/M titanium – Microstructure, properties and products', *MPIF Conference*, 2010.
16 McCracken C G, private communication February 14, 2012.
17 Bertolini M, private communication April 21, 2010.
18 Withers J, private communication, May 29, 2011.
19 Barnes J E, private communication, November 7, 2011.
20 Abakumov G I, Duz V A and Moxson V S (2010), 'Titanium alloy manufactured by low cost solid state PM processes for military, aerospace and other critical applications,' ITA Conference.

21 Ivasishin OM, Savvakin DV, Froes FH, Moxson VS and Bondareva KA, *Mater Technol*, 2002, **17**(1), 20–5.
22 Moxson V S, private communication, October 17, 2011.
23 Moxson V S, ADMA Corp., private communication, October 31, 2011.
24 Moll J H and McTiernan B J 'A powder metallurgy approach to fabrication of Ti-Al-type intermetallics', *MPR*, 2000, January, 18–26.
25 Moll J H 'Fabrication of titanium intermetallics via a powder metallurgy approach', JOM, 2003, **52**(5), 32.
26 Clemens H, Schretter P, Wurzwallne K, Bartels A and Koeppe C (2003), *Structural Intermetallics*, R. Darolia (ed) TMS, Warrendale, Pa, 2003, 205.
27 Samarov V, private communication, March 17, 2011.
28 Arcella F and Froes F H (Sam), 'Production of titanium aerospace components from powder using laser forming', *JOM*, 2000, **52**(5), 28.
29 Sears J (2011), reported at PowderMet, San Francisco, May 18–21.
30 German RM, *Powder Metallurgy Science*, 2nd edition, MPIF, Princeton, NJ, 1994, chapter 6, p. 192 *et seq*.
31 Froes F H (Sam) and German R M, *Metal Powder Report*, 2000, **55**(6) 'Titanium powder injection molding (PIM)', 12.
32 German R 'Powder injection molding of titanium components', *Powder Injection Molding Int*, 2009, **3**(4), 21–37.
33 Baril E 'Titanium alloys by powder injection molding', Powder Injection Molding Int, 2010, **4**(4), 22–32.
34 MacNeal T, private communication, May 2, 2011.
35 Froes F H (Sam) 'Advances in titanium metal injection molding', *TMS Symposium on Innovations in Titanium Technology*, Gungor M N, Imam M A and Froes F H (Sam) (eds), Orlando, Fl, February 25–March 1, 2007, 157–66.
36 Froes F H (Sam) (Feb 2000) 'Conference report: Fourth International Conference on Spray Forming', Baltimore, Maryland, October 24–26, 1999, *Light Metals Age*, vol. 58, p. 72.
37 Segal L A E, Papyrin A N, Conway Jr J C and Shapiro D (Sept 1998), 'A review of cold forming of metals', JOM, **50**(9), 52.
38 Srisukhumbowornchai N, Senkov O N, Froes F H, Öveçoglu M L and Hebeisen J (1997), 'Stability of nanocrystalline structures in a Ti-47.5Al-3Cr (at %) alloy produced by mechanical alloying and hot isostatic pressing', *Synthesis/Processing of Lightweight Metallic Materials – II*, CM Ward-Close, F.H. Froes, D.J. Chellman and S.S. Cho (eds), TMS, Warrendale, Pennsylvania, 1997, 243.
39 Trivedi P B, Baburaj E G, Geng A, Ovecoglu M L, Patankar S and Froes F H, 'Grain size control in Ti-48Al-2Cr-2Nb with yttrium additions', *J. Alloys and Compounds*, 2003, March, 100–6.
40 Ward-Close M, private communication, Nov 11, 2003.
41 Froes F H, Senkov O N and Qazi J I, 'Hydrogen as a temporary alloying element in titanium alloys : thermohydrogen processing', *Int. Mater Rev*, 2004, **49**(3–4), 227–245.
42 R. Loutfy (2000), MER Corp., Tucson, AZ, unpublished work.
43 Machiavelli, *Il Principe*, 1513.

9
Metal-based composite powders

N. LLORCA-ISERN and C. ARTIEDA-GUZMÁN,
Universitat de Barcelona, Spain

DOI: 10.1533/9780857098900.2.241

Abstract: Since the early 1990s considerable effort has been devoted to the development of metal-based composite powders (MeCP). Reinforcements in MeCP can vary from intermetallic to ceramic or polymer, depending on composition and can also be microstructured or nanostructured, depending on the size of the constituent materials. Composite powders can be used at the macro- and microscale to produce dense composite objects, composite coatings, to provide a combination of properties in one component or to provide specific properties to withstand extreme conditions in service. In addition to this, technology for the synthesis of nanodevices has also evolved. Metal composite powders are produced by a variety of methods based on solid-, liquid- and gas-phase synthesis and mechanosynthesis. Functionality and design are the current drivers for the development of metal composite powders.

Key words: applications, gas-phase synthesis, liquid-phase synthesis, metal composite powders, solid-phase synthesis.

9.1 Introduction

The objective of this chapter is to provide an overview of metal-based composite powders (MeCPs), including the processes for their synthesis, the most frequently employed composite systems, the linking between composite powders and the common applications of MeCPs. Some future trends will also be discussed.

The main reason for producing composite powders is, of course, to optimise the properties of a product, but they have the added advantage of making it easy to combine dissimilar materials that are difficult to obtain by conventional composite production. Composite powders can also be used to advantage in components where extremes of size are necessary, such as nanotechnology and large dimension objects, for which the need to ensure size-related properties is critical.

From the macroscale and microscale points of view, the objectives when using composite powders are focused on the need to produce a dense composite object, a composite coating, a combination of different properties, or to ensure that certain properties are present even in severe service conditions. In addition to this, the production of nanodevices in nanoscience and nanotechnology can be considered a breakthrough for the composite powder sector.

Solid-phase and liquid-phase sintering are the consolidation processes that are most often used to achieve these objectives. Liquid-phase sintering is particularly common: metals (such as copper) or polymers are used as a wetting reinforcement matrix to produce the metal matrix composites (MMCs) and polymer matrix composites (PMCs). Composite coatings can be made using composite powders as the raw material (a direct composite process) or by mixing the different constituents during the coating process (an indirect composite process).

Composite materials are produced by a variety of methods. Conventional melting and casting techniques are not ideal for producing composites because they are unable to ensure uniformity of dispersoids. It is therefore recommended that powder metallurgy (PM) is used for the production of such materials. PM techniques are known to allow distribution of the reinforcement particles, but without the segregation phenomena typical of the casting processes (Fogagnolo *et al.*, 2004; Khakbiz and Akhlaghi, 2009). Mechanical synthesis (MS), also known as mechanical alloying (MA), is another widely used preparation technique for obtaining composite powders.

Macroparticles, microparticles and nanoparticles are always formed in a medium, which may be in the gas, liquid, or solid phase. Classification that is based on the media in which the particles are formed is perhaps the most important and most widely used. By focusing on the media in which the particles are formed, many issues that are common to all of the techniques within each specific medium are included. The media and the different techniques for particle syntheses are usually classified as gas phase, liquid phase or solid phase.

Gas phase: vapour condensation, vapour reaction (flame synthesis, chemical vapour reactions) and aerosol (spray pyrolysis or vapour pyrolysis, flame-aerosol pyrolysis). These techniques use precursors in order to activate the reactions. Two points must be taken into account: the precursor must be dissolved in the fluid, but not react with it and the product must not dissolve in the fluid, nor react with it. In the aerosol process, the main difference between spray and flame pyrolysis is the source of heat. In spray pyrolysis, the aerosol is carried by the gas into a preheated furnace, where the liquid droplets undergo solvent evaporation, solute precipitation, solute decomposition and oxide sintering in order to gain the final particles. In flame-aerosol pyrolysis, heat is supplied by the burning fuel gases and organic solvents that are in direct contact with the precursor.

Liquid phase: chemical precipitation and coprecipitation, hydrothermal, solvothermal (non-hydrolytic), forced hydrolysis, sol–gel, microwave heating, reduction in solution and electrochemical. The chemical composition of the particles can be tailored in the liquid phase. Surface controlling agents can be added during or after particle formation in order to control the size and to prevent agglomeration.

Solid phase: mechanical milling, mechanochemical processing, cryochemical processing, self-combustion and solid-state synthesis. The kinetic energy from a grinding medium is transferred to a coarse-grained metal, ceramic, or polymeric sample material with the purpose of reducing size or alloying or combining immiscible systems. The severe plastic deformation induced by these high-energy processes leads to the nanostructuration of the components. The self-combustion method has been widely used in the synthesis of ceramic particles.

In all gas-phase and liquid-phase synthesis methods and most solid-phase methods, their most significant and basic part is a chemical reaction. These include, but are not exclusive to: reduction (redox) reactions, oxidation (redox) reactions, precipitation and coprecipitation reactions, hydrogenation or condensation, and addition or displacement reactions. Temperature, pressure, catalysation or a particular atmosphere are important factors. The name of the process is sometimes based on one of these precise parameters, which results in a long list of different terms which can cause confusion when trying to find or decide on a method for producing a particular composite powder. Consequently, there is considerable interest in simplifying the classification; as suggested above, a simple option is the medium in which the formation of the composite particle takes place.

9.2 Metal-based composite powder production

Although there are different techniques for producing metal-based composite powders, three groups are generally recognised. These are mechanosynthesis, sol–gel methods and pyrolysis.

9.2.1 Solid phase: mechanosynthesis

Mechanical alloying (MA) is a low-temperature powder processing method that was developed in the 1960s (Benjamin, 1988), but which reached its full commercial status in the 1980s (Fisher and Haeberle, 1988). Originally, MA was developed to produce oxide dispersion strengthened steels (ODS) and Ni-based superalloys (Benjamin, 1990), but nowadays it is used in the production of aluminium, copper alloys (Fogagnolo *et al.*, 2002), iron or refractory metals and especially intermetallics (Benn *et al.*, 1988). In the last two decades, different mechanical routes have been developed with the aim of producing advanced materials using mechanical energy to achieve chemical reactions or structural changes and involving the milling of the constituents (e.g. pure metals, or compounds or blends) (Chinicas, 2006).

Mechanical milling (MM), mechanical alloying (MA) or mechanosynthesis (MS) and mechanochemical synthesis (MCS) or reactive milling (RM) refer to processes involving the mechanical energy that is produced in vibratory mills, attritor mills, tumbling ball mills or planetary ball mills. With the exception of mechanical milling, in which no chemical changes are produced, the processes involve chemical or metallurgical reactions, which are either direct or induced by the high energy of the process. High energy ball milling chemical reactions are one of the most effective ways of synthesising composites and nanocomposites of various classes of compounds, for example metals, oxides, nitrides, borides, carbides, salts and organic compounds, in reactions involving activation in solids, and displacement or redox reactions between a reactive metal and metal oxide.

Important research was done by Schaffer and McCormick (1989) on the relationship of the diffusion rate to the diffusion path and they demonstrated that the solid-state reaction of copper oxide by calcium with an enthalpy of –473 kJ mol^{-1} of CaO produced pure Cu. On the other hand, the reduction of iron oxide by aluminium for the synthesis of nanomagnetic composites has been studied by Takacs (2002) and the combustion reaction was detected empirically. For example, the presence of fine Al_2O_3 particles in a copper matrix not only increases the hardness of this material, but it also diminishes the grain growth rate at temperatures near the melting point of copper. It is well known that the amount, size and distribution of reinforcing particles plays a critical role in enhancing or limiting the overall properties of the composite. This effect was also observed by other researchers in NiTi/Al_2O_3 nanocomposites (Mousavi *et al.*, 2009). Adding reinforcement particles to the matrix does not always produce a fine and homogenous composite distribution. To achieve a fine distribution of nanosized Al_2O_3 particles, the internal oxidation and MA are often used.

Dispersing fine reinforcement particles in a metallic matrix or a nanostructured matrix is beneficial to the mechanical properties of the composite. However, the former alone is not always recommended when changes in the physical properties may be involved, but fine and well-distributed reinforcement with a nanostructured matrix can overcome this limitation. Different theories have been developed in order to explain the effect of inclusions, for example the effects on the magnetic properties. The chemical precursors typically consist of mixtures of oxides, chlorides and/or metals that react either during milling or during subsequent heat treatment to form a composite consisting of ultrafine particles within a matrix. Since high-energy milling processes significantly reduce the size of the crystals involved, research is now focused on the production of nanocomposite powders. Thus the direct consequences are, first, the possibility of synthesising the particles of the composite system *in situ* or *ex situ* and, second, producing them with microstructural refinement at the nanometre scale. This is why the majority of the literature from the past ten years deals with the production of nanocomposite or nanostructured composite powders.

The capacity of the MA technique to produce composite powders with a uniform distribution of nanoparticles within the matrix alloy has resulted in the synthesis of a variety of nanocomposite systems such as the Mg/carbon nanotube (CNT) (Huang *et al.*, 2007), Mg/Cr_2O_3 (Vijay *et al.*, 2006), Cu/Fe_3C (Correia *et al.*, 2007), Co/Al_2O_3 (Li *et al.*, 2007) Zn/Al_2O_3 (Karimzadeh *et al.*, 2008) or Ni/AlN (Chung *et al.*, 2003).

Mechanochemical processing is a relatively new technique in the preparation of nanosized materials. It has been shown that enhanced reaction rates can be achieved and dynamically maintained during milling as a result of microstructural refinement and mixing processes accompanying repeated fracture, welding and deformation of particles during collision events (Suryanarayana, 2001; Schaffer and McCormick, 1991). MA or MM can be used as an initial step prior to other synthesis techniques and also as the final step following the synthesis method. This will be discussed with examples in the next part of the chapter.

Mechanochemical synthesis (MCS) or reactive milling (RM) reactions fall into two categories as suggested by Botta *et al.*, (2001), namely:

(i) Those which occur during the mechanical activation process and where the reaction enthalpy is highly negative. The adiabatic temperature (at heat release) is T_{ad} = 1300–1800K.
(ii) Those which occur during subsequent thermal treatment and where the reaction enthalpy is only moderate. The adiabatic temperature is T_{ad} < 1300K.

The first type of reaction takes place in two distinct modes of reaction kinetics, that is, either a combustion reaction or a progressive reaction (Takacs, 2002). The latter reaction may extend to a very small volume of powder mixture resulting in a gradual transformation ($T_{adiabatic}$ < 1800K). The former is a self-propagating combustion reaction that can be initiated when the reaction enthalpy is sufficiently high ($T_{adiabatic}$ > 1800K) (Munir, 1988; Botta *et al.*, 2001). The combustion reaction can be ignited by the aforementioned high-energy ball milling. In addition to the energy dissipated by the collision events, the reaction enthalpy will also contribute to the temperature rise during milling and will ultimately cause thermal instability and combustion. Since diffusion and thus reaction rates depend on temperature, the overall rate of reaction is affected by the reaction enthalpy. Schaffer and McCormick (1991) stated that in reactions that may involve the combustion effect, it is required that the adiabatic temperature, T_{ad}, be at least 1800K. The adiabatic temperature is a measure of the local heat generated by the reaction and is often taken as a measure of the driving force. Thus the value of T_{ad} can be used as a suitable criterion for anticipating the occurrence of a self-propagating combustion reaction in ball milling processes. It is worth noting that if the matrix system is diluted, T_{ad} will decrease. The value of

T_{ad} for any reaction, taking into consideration the dilution effect, can be calculated from the reaction enthalpy ΔH_r:

$$\Delta H_r + \Sigma \int_{T_0}^{T_m} C_{p,S} + \Sigma \Delta H_t + \Sigma \int_{T_m}^{T_{ad}} C_{p,L} = 0 \qquad [9.1]$$

where $C_{p,S}$ and $C_{p,L}$ are the mean heat capacities in the solid and liquid phases of the products respectively, T_0, T_m and T_{ad} are the initial, melting and adiabatic temperatures respectively and the ΔH_t is the state transformation enthalpy.

Nanostructured materials inherit relatively large interfacial energy (Gleiter, 2000). When such nanostructures are generated by extensive plastic deformation, for example by high-energy ball milling (HEBM), the stored energy in the material also becomes significant. These factors enhance the chemical reactivity of the milled product and in some cases novel metastable phases form in the course of HEBM (Koch and Whittenberger, 1996; Suryanarayana, 2001; Zhang, 2004).

When mechanochemical synthesis involves a reduction reaction, mechanical milling alone may not be sufficient for the reduction reaction to take place completely, for example if the reaction enthalpy is moderately high. Thus, a combination of mechanical and thermal treatment makes the material system a likely candidate for the synthesis of nanocomposites. Some composite systems involve post thermal treatment in an inert or a reactive atmosphere in order to form the final composite structure; this is a two-step synthesis.

Other composites can be synthesised by a displacement reaction. A reaction can take place thermodynamically at room temperature if its ΔG^0_{298} has a high negative value (it is highly exothermic) and provided the kinetics are fast. In this regard MA can enhance the kinetics of the reaction by creating a high diffusivity path, providing an extensive interface area between reactants and by the dynamic removal of reaction products from the interfaces through repeated fracturing and cold welding of powder particles. MA also increases the number of dislocations and grain boundaries and this shortens the diffusion paths.

The reinforcement particles are incorporated into the matrix by means of a high-energy process (HEM). It allows the production of diverse combinations of metals and reinforcement components and has led to the development of novel materials that cannot be produced by melting metallurgy (Ag/SnO_2, Al/SiC, $MCrAlY/SnO_2$, $NiCrBSi/TiB_2$, and Stellite/VC, among others).

In the effort to produce nanoparticles, some new innovations have been added to these traditional methods, such as introducing reactive materials into the milling process and combining milling with solid-state synthesis. Some new approaches include the extraction of soluble molecules from a uniform molecular solid mixture by acid/base leaching or evaporation in order to leave molecular skeleton residues that become nanoparticles.

When mechanochemical synthesis includes a combustion process that is induced inside a vial, which is similar to thermally ignited self-propagating high-temperature synthesis (SHS), it is called a mechanically induced self-sustaining reaction (MSR) (Takacs, 2002; Lu and Li, 2005; Kim *et al.*, 2006). The synthesis of ternary carbonitride phases that could be employed as master alloys has produced, via different methods such as thermal treatments (Pastor, 1988), self-sustaining high-temperature synthesis (Munir and Ealamloo, 1994; Yeh and Chen, 2005, 2007, Yeh and Liu, 2006), or mechanochemical synthesis (Córdoba *et al.*, 2005, 2007a). Recently, MSR synthesis has been proposed as a reliable and easy method for obtaining high purity quaternary $Ti_yNb_{1-y}C_xN_{1-x}$ carbonitrides (Córdoba *et al.* 2007b, 2009). If a second MA step for mixing them with refractory, iron or Co–Ni alloys is carried out, novel cermet-like powders can be produced.

Limitations of mechanosynthesis

The first potential limitation is the low productivity of the high-energy ball mills, which makes it difficult to introduce mechanochemistry in large-scale technology. The advent of horizontal attritor mills with different types of milling media has overcome this problem to a certain extent. The second limitation is contamination from the milling media and the reaction of the milling media or the milled products with the atmosphere in which the synthesis is performed. The latter shortcoming can be minimised if the time for intensive mechanical activation is sufficiently short and this is possible when mechanocomposites are used as precursors for traditional synthesis methods. In this way, all of the advantages of the mechanochemical approach are conserved, while the limitations may be reduced significantly. The oxidation of pure metals seems to be inevitable during milling. Fortunately, recent research reveals that when the grain sizes of the powders are reduced to the nanometre scale, the bulk sintered part can achieve nearly full densification (Lee and Kim, 1995; Kim and Moon, 1998; Kim *et al.*, 2004a). High-energy ball milling offers a large number of possible routes for synthesising new materials and is a promising method for industrial scale-up synthesis.

9.2.2 Liquid phase: sol–gel technology

In this chemical process, a colloidal solution (sol) gradually evolves towards the formation (gellation) of a gel-like dual phase system, formed by a continuous network of a solid in a continuous network of fluid. It is a wet chemical technique widely used in materials science and more specifically in ceramic engineering. A drying process removes the remaining solvent and a final sintering thermal treatment can be carried out if further polycondensation or densification is needed.

Varma *et al.* (1990) prepared Ag–YBCO superconductors from composite powders via the citrate gel route. The raw materials were cupric nitrate, yttrium nitrate, barium nitrate, silver nitrate and citric acid. The nitrates were dissolved, a stoichiometric amount of citric acid was added, and the pH value was adjusted. The mixture was concentrated to a viscous state in a steam bath and an oven to obtain a citrate gel. The gel was then thermally decomposed in air and the powder was ground, pressed and sintered in a static air atmosphere. Nanostructured Ni–Y_2O_3/ZrO_2 composite powders were synthesised by Grossmann *et al.* (1995) employing the sol–gel method. The characteristics of the powder can be controlled through choice of the oxidation and reduction conditions, and homogeneously dispersed nickel within a ZrO_2 matrix could be generated. Ragunathan *et al.* (1993) synthesised W–Cu composite powders at lower temperatures than those required for the methods that are more commonly used. Greater control of the chemistry and homogeneity are possible with sol–gel synthesis.

9.2.3 Gas phase: pyrolysis

This synthesis route is capable of forming nanoparticles once a sufficient degree of supersaturation of condensable products has been reached in the vapour phase. Once nucleation occurs, the growth of particles is fast and takes place by coalescence; depending on the temperature, spherical or agglomerated particles are produced. The process has been successfully employed in synthesising various nanometre-scale materials including metals, simple and complex metal oxides, carbides, nitrides, rare earth doped oxides and multi-elemental glasses, among others. Laser pyrolysis is a gas-phase synthesis method where a flowing reactive gas is heated rapidly with a laser.

Spray pyrolysis was used to fabricate Ag–TiO_2 composite powders for application in oxide barrier filaments of superconducting multifilamentary tapes. The composite powders were prepared from an aqueous solution of TiO_2 and $AgNO_3$ (Matsumoto *et al.*, 2002). Majumdar *et al.* (1998) synthesised Ag–SiO_2 and Ag–CuO by spray pyrolysis using mixtures of aqueous silver nitrate and colloidal silica, resulting in high purity and a high level of compositionally homogenous composite powders.

9.3 Copper- and aluminium-based composite powder systems

MMCs refer to a kind of material in which rigid ceramic or another high-strength metal or alloy reinforcements are embedded in a ductile metal or alloy matrix. MMCs combine metallic properties, such as ductility and toughness, with characteristics such as high strength, moduli leading to greater strength in shear and compression, and to higher service temperature capabilities.

The attractive physical and mechanical properties that can be obtained with MMCs, such as a high specific modulus, strength, and thermal stability, have been documented extensively (Tjong and Ma, 2000; Flom and Arsenault, 1986; Nardone and Prewo, 1986; Mortensen *et al.*, 1988).

Interest in MMCs for use in the aerospace and automotive industries, as well as other structural applications, has increased over the past 20 years. This is a result of the availability of relatively inexpensive reinforcements and the development of various processing routes which result in reproducible microstructures and properties (Chou *et al.*, 1985). *In situ* MMC powders with a wide range of matrix materials (including aluminium, titanium, copper and iron) and second-phase particles (including borides, carbides, nitrides, oxides and their mixtures) can be produced.

Since the objective of the reinforcement is to increase the stiffness and strength of the matrix, metallic or ceramic particles are used in the synthesis of metal composite powders. With their large elastic modulus and high strength, ceramic particles are ideal reinforcing particles. In principle, many ceramic powders meet these requirements, but there are other restrictions which influence the choice of the reinforcement for a particular composite. Many of the ceramic particles of interest are thermodynamically unstable when they are in contact with pure metals and will react to form compounds at the interface between the particles and the surrounding matrix (Mortensen, 2007, p 490). There are other factors that should be considered, in addition to reaction stability, when choosing an appropriate reinforcement. If the composite is made by MA or milling of solid powders, the ratio of the metal powder size to that of the reinforcing particles is important if a uniform distribution of the reinforcement is to be achieved. In general, if the size ranges of the metal and ceramic particles are similar, the reinforcement distribution will be more uniform. The reinforcing particles in composite materials with metal alloy matrices, metal carbides (SiC, TaC, WC, B_4C, TiC), metal nitrides (TaN, ZrN, Si_3N_4, TiN), metal borides (TaB_2, ZrB_2, TiB_2, WB) and metal oxides (ZrO_2, Al_2O_3, ThO_2) are applied.

In the following sections, a selection of metallic matrices and the most commonly used reinforcements are analysed.

9.3.1 Copper-based composite powders

Copper and copper alloys constitute one of the major groups of commercial materials. They are widely used in engineering because of their excellent properties, for example, outstanding resistance to corrosion, easy production, high conductivity and good fatigue resistance (Davis, 2001, p 4). The main problem with copper alloys is the low intrinsic strength of these materials; for this reason, they need to be alloyed with different insoluble elements to achieve particle strengthening. Potential alloying elements that can be used to reinforce copper include Cr, W, Ta, Nb, Mo and V. On the other hand, copper can also be strengthened by reinforcement with ceramic particles such as TiC and NbC. These reinforcements not only increase the microhardness of copper but they also maintain this microhardness at elevated temperatures (Hussain *et al.*, 2008).

Sheibaina *et al.* (2009) suggested that copper matrix composites were promising candidates for application in electrical sliding contacts where good wear resistance and high thermal and electrical conductivity are needed. For example, the trend in the development of electrode materials for high-current applications is to design copper matrix composites containing high melting temperature components. These can be produced by dispersing hard particles like oxides, carbides or nitrides into the copper matrix, using either liquid-state or solid-state techniques (Tjong and Lau, 2000; Lee and Lee, 1999).

There are some publications focused on the production of pure Cu matrix nanocomposites such as Cu–Al_2O_3 (Wu and Li, 2000; Ying and Zhang, 2000), Cu–MgO (Mulas *et al.*, 1999), Cu–TiB_2 (Dong *et al.*, 2002), Cu–ZnO (Castricum *et al.*, 2001) and Cu–MnO (Sheibani *et al.*, 2009), using an MA method. Very few methods used copper alloy as the matrix (Cu(Mo)), yet it seems that the presence of a nanocrystalline Cu(Mo) alloy as a matrix and the homogenous distribution of Al_2O_3 reinforcements can be helpful in achieving better mechanical properties in a nanocomposite. However, the wettability of the reinforcement from Cu is higher when it is pure.

The mechanochemical reduction of copper oxide with different reductants such as Fe, Al, Ti, Ca, Ni and C has already been studied (Schaffer and McCormick, 1990, 1991; Sheibani *et al.*, 2007). The products were usually a mixture of copper with dispersed oxides particle, that is copper matrix composites. When this process is combined with mechanical milling (reactive milling), nanostructured composites with a uniform distribution of reinforcement particles are synthesised.

Cu–15 wt% Mo/30 vol% Al_2O_3 nanocomposites can be synthesised by a displacement reaction between Al and MoO_3 in a Cu–22.5 wt% MoO_3–8.5 wt% Al powder mixture according to the following reaction (Sabooni *et al.*, 2010):

$$MoO_3 + 2Al \rightarrow Al_2O_3 + Mo \qquad [6.2]$$

$$\Delta G^0_{298} = -915 \text{ kJ mol}^{-1}; \Delta H^0_{298} = -966 \text{ kJ mol}^{-1}$$

The very high negative value of ΔG^0_{298} (−915 kJ mol^{-1}) indicates that this reaction can thermodynamically take place at room temperature. However, the kinetics may delay the reaction. MA can enhance the kinetics of the reaction owing to the repeated fracturing and cold welding of powder particles. Immediate formation of Al_2O_3 is promoted by dynamically formed high Al/MoO_3 interface areas, as well as the short circuit diffusion path that is provided by increasing the number of defects (such as dislocations and grain boundaries) during MA (Heidarpour *et al.*, 2009). It has been found by different researchers that MA induces the dissolution of metallic atoms into the crystal lattice of the accompanying metal. The relative crystallography, size of atoms and affinity will enhance this process or make it more difficult. Also the presence of ceramic particles will influence the diffusion path (Llorca-Isern *et al.*, 2010).

There have also been studies of the *in situ* formation of Cu–MnO nanocomposite powder by mechanochemical reactions between CuO and Mn. Particular attention has been paid to the mechanism of this process and its structural evolution and morphological variation during mechanical milling. Since T_{ad} in this reaction is well above the melting point of Cu, 1358K, it is certain that Cu is completely melted during the combustion reaction. This is in agreement with the observation made by Zhang and Richmond (1999) regarding the spherical morphology of the Cu particles undergoing phase separation from AlO_3.

Pure copper, iron and cobalt together with Al_2O_3 powder were mixed in a planetary ball mill in order to form Cu–Fe–Co/Al_2O_3 composite powders. The reaction stages and resulting magnetic properties were analysed in order to determine the influence of the ceramic powder on the process and properties, compared to the metallic system. Al_2O_3 interacts with the metallic dissolution of Fe and Co atoms in Cu, resulting in a barrier to the progress of the atomic diffusion. Figure 9.1 shows a (CuFeCo) metallic aggregate after mechanical milling. It has a plastic appearance with almost flat surfaces, as welding is the predominant process in these kinds of mechanical alloys. However, in

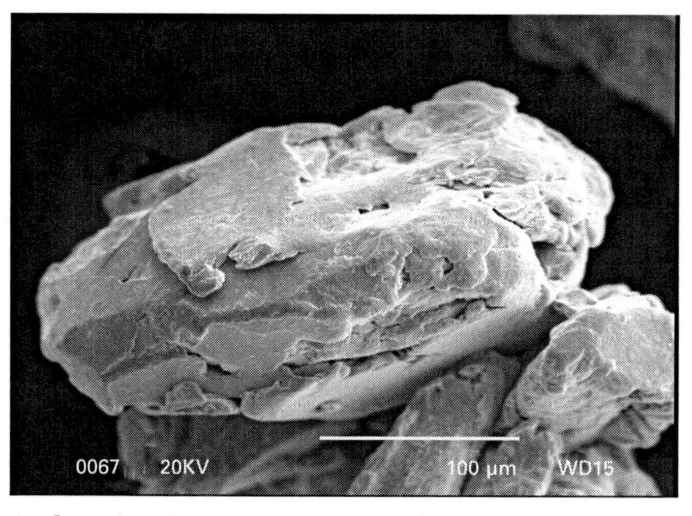

9.1 Scanning electron microscopy (SEM) micrograph of the metallic Cu–Fe–Co aggregate after HEBM showing ductile features of the powder.

Fig. 9.2, the aggregate of a metallic system with 3% Al_2O_3 acting as a barrier to the plastic deformation hampers the mobility of dislocations generated by the mechanical process, resulting in increased fragility of the material so that eventually it will fracture. Consequently the fracture aspect of this composite powder is not plastic but brittle (Llorca-Isern *et al.*, 2010).

An aggregate of composite powder (CuFeCo + 10% Al_2O_3) processed by HEBM is shown in Fig. 9.3(a) and (b). Owing to their hardness and brittleness, ceramic particles are broken and refined during the mechanical process. Al_2O_3 particles are then trapped between deformed metallic particles, hindering the progress of diffusion. In Fig. 9.3(a) and (b), the alumina particles can be seen as dark spots embedded within the metallic matrix; as the milling time favours fracturing, there are different sizes of particles.

Concerning the magnetic properties, the composite powder showed that the presence of alumina increases its coercivity (H_c) and remanence (M_r). The processed metallic powder presents magnetic domains with defined block wall contours showing a dendritic morphology, as can be seen in the Fig. 9.4(a); this was also observed by Zeng *et al.* (2007) who suggested that this morphology was typical for the magnetic domains of metals processed by MA. Looking at the morphology of the magnetic domains in Fig. 9.4(b) (obtained by the authors), corresponding to the composite powders (metallic system with 10% alumina), two features are notable: first, the diffused block wall contours and, second, the different branched forms adopted by the magnetic domains. The magnetic domains of composite powders are less homogenous and present diffused features that are probably caused by

9.2 SEM micrograph of the composite Cu–Fe–Co and 3% alumina (%w) after HEBM showing brittle features of the metal-based composite powder.

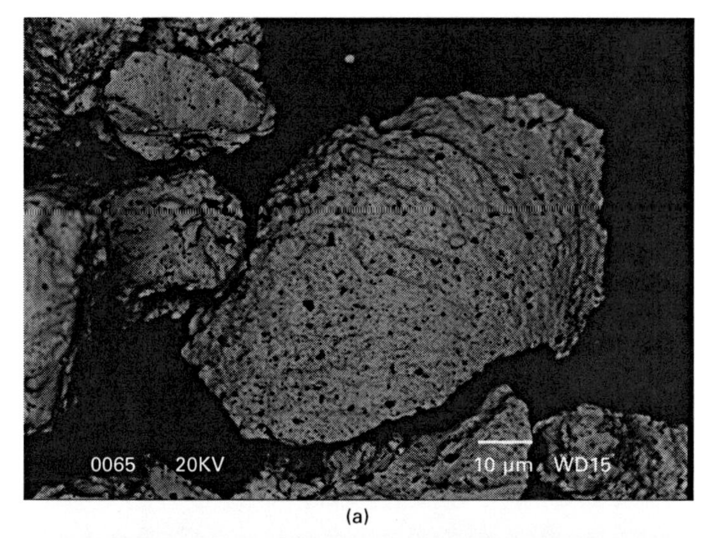

(a)

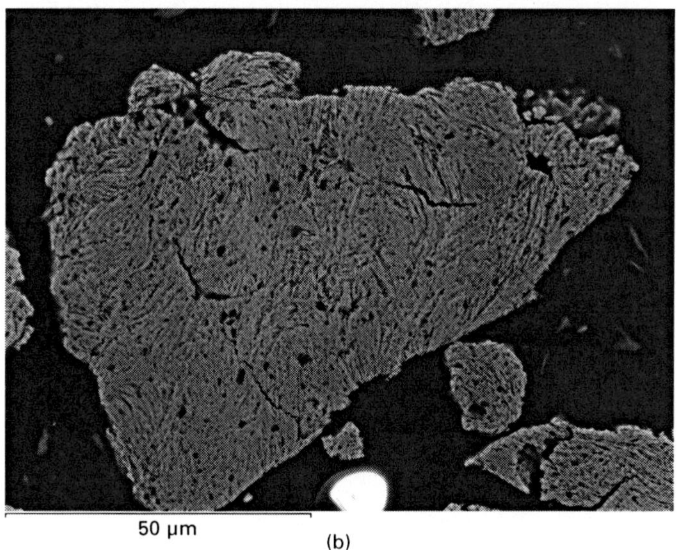

(b)

9.3 (a) SEM-BSE (back scattered electron) micrograph of Cu-Fe-Co and 10% alumina (%w) aggregate after HEBM. The dark spots correspond to the alumina particles embedded in the metallic matrix. (b) SEM-BSE cross-section micrograph of Cu-Fe-Co and 10% alumina (%w) after HEBM. The dark spots correspond to the alumina particles in the metal composite powder.

pinning the alumina particles, hindering the orientation and reorientation of the magnetic dipoles in these domains.

In some studies, (Lee *et al.*, 2001; Korác *et al.*, 2007), Cu–Al_2O_3 nanocomposite powders were synthesised via the thermochemical or

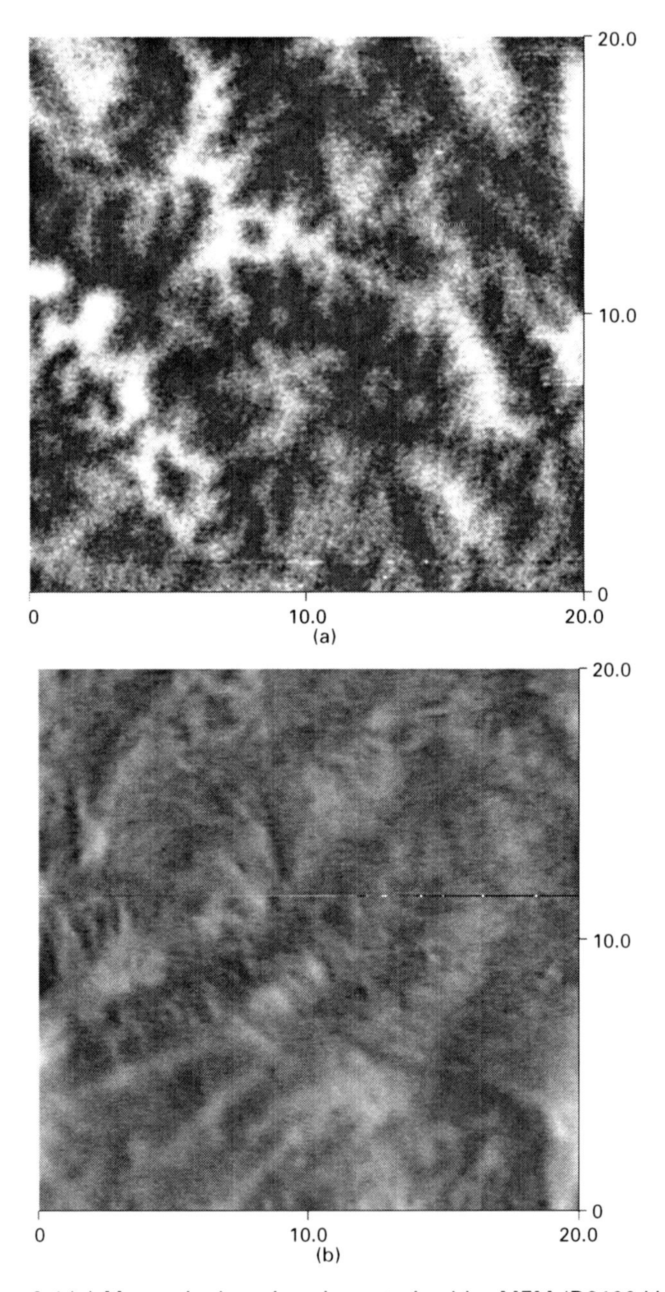

9.4 (a) Magnetic domains characterized by MFM (D3100 Veeco magnetic force microscope) after HEBM in Cu–Fe–Co metallic powder. Image area of 20 μm × 20 μm. (b) Magnetic domains characterized by MFM (D3100 Veeco magnetic force microscope) after HEBM in Cu–Fe–Co and 10% Al_2O_3 (%w) metal composite powder. Image area of 20 μm × 20 μm.

chemical route, in which the input materials are in a liquid state. Owing to the development of contemporary materials with advanced properties, there has been considerable interest in this synthesis method for the production of ultra-fine and nanocomposite powders (Jena *et al.*, 2001).

9.3.2 Aluminium-based composite powders

Aluminium-based matrix composites (AMC) that have been reinforced with hard ceramic particles have received considerable interest because they are relatively easy to process and can offer nearly isotropic properties in comparison to fibre-reinforced composites. In addition, these composites exhibit high strength and stiffness, creep resistance and superior wear resistance, and also provide good electrical and thermal conductivity. All of these properties make particle reinforced AMCs attractive for a wide range of applications in the automotive, aerospace and transport industries. Reinforcement particles used in AMCs include nitrides, borides, carbides and oxides (Everett and Higby, 1991).

The one-step *in situ* production of aluminium matrix composite powders has been successfully realised by means of mechanical alloying. The composite material powders were manufactured by simultaneously mixing the elemental powders in the appropriate percentages to obtain the aluminium alloy matrix and the selected carbides. In this case, the elemental powders used for the aluminium alloy were aluminium, copper, silicon and magnesium. Two different carbides were used as reinforcement for the composite material: VC and TiC (Ruiz-Navas *et al.*, 2006).

Aluminium alloys manufactured by MA show better properties, especially at higher temperatures, than alloys obtained by atomisation or conventional means (Last and Garret, 1996). The reasons for this include the reduction of grain size (Van Meter *et al.*, 1992), the high level of work hardening and the fine dispersion of precipitates (mainly oxides) in the microstructure (Mishra *et al.*, 1992; McCormick and Froes, 1998). Numerous aluminium alloys are manufactured by MA (Li and Lai, 1998), including alloys from the 6xxx (Mukai and Ishikawa, 1995) and 7xxx (Polkin and Borzov, 1995) series. Aluminium matrix composite powders that have been reinforced with SiC and Al_2O_3 (Faure and Brune, 1987; Bhadury *et al.*, 1996; Ravikiran and Surappa, 1997) have been obtained in order to improve the wear behaviour of the aluminium alloy matrix. More recent studies have considered the possibility of adding intermetallics, which is a potential success owing to their hardness and compatibility with the matrix (González-Carrasco *et al.*, 1994; Torralba *et al.*, 1997; Adamiak *et al.*, 2004).

Aluminium alloys have reasonable strength and ductility with corresponding workability. Therefore, these alloys have been widely used for the synthesis of nanocomposites with different nano-sized ceramic particulates such as Al_2O_3 (Arami *et al.*, 2008; Razavi-Hesabi *et al.*, 2006), CNT (Pérez-Bustamante, 2008) and $MoAl_x$ (Maiti and Chakraborty, 2008) by MA. The 6xxx series of Al–Mg–Si alloys (i.e. 6061) is a good choice as the matrix alloy owing to its excellent mechanical properties, good weldability and corrosion resistance.

Even though the *in situ* composites have significant advantages, some synthesis routes may lead to composites with an inhomogeneous microstructure with various unstable and/or undesirable phases (Lü *et al.*, 2000, 2001; Tee *et al.*, 2001). These undesirable phases might drastically reduce the mechanical properties. For example, Tjong *et al.* (2005) have studied the mechanical properties of *in situ* Al–10 wt% TiB_2 and found that the formation of an Al_3Ti phase has a negative impact on the mechanical performance of *in situ* composites. They proposed that the elimination of the intermetallic Al_3Ti phase is a primary task for developing *in situ* Al–TiB_2 composites with superior mechanical performance. Recently, Sadeghian *et al.* (2011) used a two-step MA process and obtained *in situ* formation of TiB_2 particles in the Al matrix with a fine and uniform distribution, thus preventing the formation of undesirable compounds. Consequently, strict control of the process may ensure the synthesis.

Based on the Al–Ce/Al_2O_3 system, Reddy *et al.* (2007) concluded that mechanical milling is not sufficient for the reduction reaction when the reaction enthalpy is moderately high. They suggested that a combination of mechanical and thermal treatment would make the material system a likely candidate for the synthesis of nanocomposites. The chemical redox reaction of this system is:

$$3CeO_2 + 4Al \rightarrow 2Al_2O_3 + 3Ce \qquad [9.3]$$

Boron carbide (B_4C) is one of the hardest and lightest materials and thus a candidate for reinforcing light ductile metals. However, owing to the poor wettability of B_4C with the molten Al alloys, it is difficult to produce dense Al–B_4C composites by liquid-phase approaches. It has been reported that Al requires a temperature as high as 1100°C in order to wet the B_4C surface completely. Processing at such high temperatures leads to the formation of a series of components by chemical reactions between these two phases (Shorowordi *et al.*, 2003; Lee and Kang, 2001). Ye *et al.* (2006b,c) have optimised the cryomilling technique for processing composite powders that consist of particulate B_4C reinforcement in different aluminium matrices. It has been reported that Al–B_4C composites prepared by consolidation of the cryomilled powder particles exhibited better wear resistance (Tang *et al.*, 2008) and improved mechanical properties (Ye *et al.*, 2006a) when compared to the unreinforced alloys. The effect of interfacial debonding on the mechanical properties of Al–B_4C composites consolidated by different thermomechanical techniques was reported in another study (Zhang *et al.*,

2006). However, in the aforementioned studies, cryomilling was used to prepare the Al–B_4C composite powders by using micrometre-sized B_4C particles. Nano-sized boron carbide particles can be produced from commercially available boron carbide particles and mechanically alloyed with Al6061 to synthesise the corresponding nanocomposite powder successfully (Khakbiz and Akhlaghi, 2009).

9.4 Other metal-based composite powders

9.4.1 Iron-based composite powders

The main advantages of the PM route for iron-based systems are raw material savings, low energy costs, higher content of alloying elements and the possibility of ceramic particle additions. However, the materials produced by this technique generally suffer from the problem of contaminated matrix-reinforcement interfaces. Consequently, techniques involving *in situ* generation of the reinforcing phases are preferred as the synthesis route for these materials. As a result, the reinforcement surfaces are likely to be free from gas absorption, oxidation or other detrimental surface reaction contamination, and the interface between the matrix and the reinforcement bond tends to be stronger (Zhang *et al.*, 1999). Some of these technologies include liquid–solid or liquid–liquid reactions, and self-propagation high-temperature synthesis (SHS). *In situ* formation of dual carbides (TiC and other carbides) in ferrous composites has already been studied. For example, Dogan and Hawk (1995) formed *in situ* TiC and $(Cr,Fe)_7C_3$ carbides in Fe-based composites; Farid *et al.* (2007) synthesised *in situ* TiB_2 and TiC in stainless steel matrix composites; Jiang *et al.* (2000) produced *in situ* (TiW) Cp/Fe composites, and Fu and Xu (2010) produced VC as an alternative addition to TiC/Fe matrix composites, producing dense (Ti,V)C/Fe-matrix composites by an *in situ* reaction combined with planetary ball mixing techniques.

The strategy applied for iron-based materials is usually to use reinforcement particles which conform to precipitation in the initial matrix alloy. This can be applied when high-alloyed initial powders are needed, for example if Fe-Cr-B has to be reinforced by CrB_2 (Lampke *et al.*, 2011).

Crisan and Crisan (2011) studied Fe and Fe_3O_4 (magnetite) powders obtained by ball milling. They demonstrate that the nanocomposite powder undergoes an incomplete redox reaction during preparation with formation of FeO. Magnetite and iron powders are gradually transformed via redox reactions and, at the end of the heating stage, wursite is the main phase observed in the sample.

9.4.2 Titanium-based composite powders

Titanium matrix composites (TiMCs) that have been reinforced with ceramic particles have considerable potential for structural applications in the aerospace, transportation and industrial sectors, primarily because of their superior stiffness, toughness, elevated temperature resistance and excellent specific strength. Many reinforcements are used in TiMCs, including SiC, Al_2O_3, TiO_2, TiC, TiB, TiB_2 and rare earth oxides. In addition, elements of rare earth added to the TiMCs can greatly increase the strength of the matrix alloy at high temperatures. Rare earth oxide is produced from the rare earth element and solid solution oxygen in a matrix alloy; it is beneficial for grain refinement, fatigue resistance and thermal stability (Zhang *et al.*, 2004b).

Titanium alloys are the metallic biomaterials that are most commonly used for medical implants owing to their good biocompatibility, high chemical stability in the physiological environment and excellent mechanical properties. For these reasons, effort has been devoted to synthesising titanium alloy–hydroxyapatite (Ti6Al4V/HA) composites. HA has poor mechanical properties that are improved by the Ti alloy. Plasma spraying is the most popular technique for adding a coating of HA to the titanium substrate. The brittle nature of the HA layer often results in wearing of the coating. The development of Ti6Al4V/HA composite powders aims to solve such problems. Thian *et al.* (2001) studied the formation of Ti6Al4V/HA composite powders using a technique called the 'ceramic slurry approach', which was employed to produce Ti6Al4V/HA composite powders; the outer layer consists of a biocompatible HA surface and the inner core consists of mechanically strong Ti6Al4V. Ti6Al4V can also be reinforced with TiC. Ar-atomisation produces a composite powder containing very small TiC particles, as proposed by Hu and Loretto (1994).

The use of innovative reinforcement materials plays an important role in the production of composite materials with outstanding properties. Several studies present nanometre-scale reinforcements as potentially successful materials for the production of metal matrix composites. A few works present titanium MMCs (TiMCs) strengthened with nano-carbon reinforcements. Montealegre *et al.* (2011) studied the behaviour of TiMCs reinforced with nano-carbon materials, CNTs and nanodiamonds (NDs), which were produced via powder metallurgy under different processing conditions. It was shown that TiMCs with a well distributed and small reinforcing phase presented the best flexural strength (1473 MPa for TiMCs with NDs hot pressed at 900°C, compared to 400 MPa for titanium pure matrix (Boyer *et al.*, 1994; Kondoh *et al.*, 2009)).

Research was carried out by Joshi *et al.* (2002). A titanium composite powder (Ti–TiO_2) was produced by *in situ* metal oxidation during MA. Begin-Colin *et al.* (1993) obtained metal–ceramic composite powders of Al_2O_3–Ti by MA using TiO_2 and Al as raw materials. These are both examples of TiMC powders.

9.4.3 Intermetallic-based composite powders

Different groups of intermetallics are considered in this section: primarily Fe_3Al, Ti_3Al, NiAl and Ni_3Al. In general, intermetallics are obtained by self-propagated high-temperature synthesis and then milled. In particular, Fe_3Al-based intermetallics possess attractive properties, including considerable hardness, high melting points, relatively low densities compared with Fe-based and Ni-based alloys, and excellent oxidation and corrosion resistance (Matsuura *et al.*, 2006). As a result, these compounds are useful for many structural applications, including gas metal filters, heating elements, heat treatment fixtures, high-temperature dies and moulds and cutting tools (Knibloe *et al.*, 1992; Yoshimi and Hanada, 1997). Addition of a third alloying element such as Ti or Cr to the Fe_3Al intermetallic compound can lead to an improvement in mechanical properties, including ductility at room temperature and creep resistance at high temperatures, chemical stability and tribological behaviour by solid solution and/or precipitation hardening, as well as grain boundary strengthening (Zhu *et al.*, 2000; Fortnum and Mikkola, 1987; Mendiratta *et al.*, 1987).

It has also been shown that the dispersion of a hard second phase material (e.g. Al_2O_3) within the matrix can enhance the mechanical properties (Whelham, 1998). Intermetallic reinforced alumina composites exhibit several advantages such as high strength, good wear resistance and improved fracture toughness (Schicker *et al.*, 1999; Horvitz *et al.*, 2002). Khodaei *et al.* (2009) synthesised Fe_3Al–Al_2O_3 nanocomposites by MA of an Al and Fe_2O_3 powder mixture and reported that prior to the Fe_2O_3–Al combustion reaction, the powder particles attained a nanocrystalline structure. This promotes the Fe_2O_3–Al reaction by providing high diffusivity paths.

Mechanical alloying of Fe, Al and TiO_2 powder mixtures led to the formation of (Fe, Ti)$_3$Al–Al_2O_3 nanocomposites. Rafiei *et al.* (2009) found that the TiO_2 oxide is gradually reduced by Al during milling. This reaction led to the formation of crystalline Ti and amorphous Al_2O_3:

$$4Al + 3TiO_2 \rightarrow 3[Ti]_{Fe} + 2Al_2O_3 \quad [9.4]$$

The ΔG°_{298} for reaction [9.4] has a very high negative value (–498.22 kJ mol^{-1}) indicating that this reaction can take place thermodynamically at room temperature.

Upon further milling, Ti (reduced from TiO_2) and the remaining Al were dissolved into an Fe lattice and formed an Fe(Al,Ti) solid solution that transformed to an (Fe,Ti)$_3$Al intermetallic compound with a disordered DO_3 structure at longer milling times. Annealing the final structure led to the crystallisation of amorphous Al_2O_3 and ordering of the $(Fe,Ti)_3Al$ matrix. If extra Fe and Al is added to the starting powder mixture to obtain an $(Fe,Ti)_3Al$ matrix, it will dilute the starting mixture and consequently T_{ad} will be lowered. Thus, the reaction mode can change from self-propagation combustion to a gradual or progressive mode of reaction.

There have been several studies into the preparation of titanium aluminide and iron aluminide matrix nanocomposites by MA. Zhang *et al.* (2004a) reported that MA of an Al–TiO_2 powder mixture leads to the formation of a range of Ti-based composites including $Ti(Al,O)/Al_2O_3$, $Ti_xAl_y(O)/Al_2O_3$ and $Ti_3Al/TiAl$. Forouzanmehr *et al.* (2009) synthesised the $TiAl/Al_2O_3$ nanocomposite by MA of an Al and TiO_2 powder mixture and reported that TiO_2 is gradually reduced by Al during ball milling.

TiAl-based intermetallics have received considerable interest regarding various applications in the aerospace, automotive and energy production industries, owing to their relatively low densities, high specific strength, excellent oxidation and corrosion resistance and adequate creep resistance at high temperatures. Therefore, the TiAl-based alloys most commonly studied have a composition of 44–48% Al and contain additions of other elements (such as Nb, Cr, W, Si, B, etc) to improve the creep and/or oxidation properties and tensile ductility. Compounds such as Al_2O_3, TiB_2, Ti_5Si_3 and Ti_2AlC have been identified as compatible and thermochemically stable reinforcing phases for the γ-TiAl matrix (Ward-Close *et al.*, 1996; Ramaseshan *et al.*, 1999).

Yeh and Shen (2009) carried out a study of TiAl–Ti_2AlC *in situ* composite preparation with a broad range of compositions, conducted by self-propagating high-temperature synthesis (SHS) of compressed samples with a mixture of elemental powders. Several carbide NiAl intermetallic composite powders have been produced by simultaneous reaction synthesis of the carbide and the intermetallic, with the objective of enhancing metallic cermets by replacing the metallic binder with NiAl (McCoy and Shaw, 1994).

NiAl has also been reinforced with TiB_2 or Al_2O_3, the latter produced by mixing NiO and Al, and it was mechanically alloyed to produce the composite powder (Cheng, 1994; Oleszak, 2001). Ni_3Al and NiAl intermetallics have been used as reinforcement of Al 2124 aluminium alloy, showing higher thermal stability and better mechanical properties than the same metallic alloy reinforced with common ceramics (Torres *et al.*, 2002).

9.4.4 Refractory-based composite powders

A wide variety of refractory materials are available. These materials can be classified in three general families of refractories, in terms of their chemical composition:

- oxide refractories
- non-oxide refractories
- composite refractories.

Concerning metal-based refractories, silver-based refractory contact materials produced by powder metallurgy are used extensively as contact materials owing to their properties, which include high conductivity, good resistance to welding and corrosion, high melting temperatures and hardness. Ag–65 wt% W composite is widely used in air circuit breakers in the 50–100 Å range (Lee, 1997). Findik and Uzun (2003) studied the effect of graphite additions on electrical conductivity and hardness of a 60 wt% Ag–37 wt% WC–3 wt% C contact material produced by powder metallurgy.

Tungsten and molybdenum are two refractory metals that have attracted great interest for high-temperature applications in industries such as consumer electronics, aerospace, telecommunications, medicine and defence, owing to their excellent heat resistance. The high melting temperatures of the metals (3422°C for tungsten and over 2623°C for molybdenum), however, makes it extremely difficult to process the materials by melting and casting. Conventionally, tungsten and molybdenum-based alloys and composites are produced by a powder metallurgical method, followed by minor machining or grinding, if necessary.

An important refractory metal-based composite is the W–Cu system, which has superior thermal management properties and a high microwave absorption capacity. It is predominantly used for heavy-duty electrical contacts and arcing resistant electrodes. Tungsten–copper alloy powders are used in many fields on account of the high electric and thermal conductivities of copper and the high melting point of tungsten. Good thermal and electrical conductivity and a low thermal expansion coefficient contribute to making these composites advanced engineering materials. However, because the W–Cu system exhibits mutual insolubility, W–Cu powder compacts have a bad sintering capacity, even when using liquid-phase sintering at above the melting point of the Cu phase. For this reason, it has been reported that W–Cu composite powders can be produced by ball milling (Gaffet *et al.*, 1991), oxide co-reduction (Gusmano *et al.*, 2001), freeze-drying technique (Xi *et al.*, 2010), mechano-chemical methods (Kim *et al.*, 2004b) and MA (Xiong *et al.*, 1995; Doré *et al.*, 2004; Alam, 2006). However, only a few reports are available on the synthesis of W–Cu composite powders through chemical routes (Ardestani *et al.*, 2009; Hashempour *et al.*, 2010). Compared to MA processes, chemical synthetic approaches have the advantage of better control of particle size, shape and distribution through adjustment of the reaction parameters.

Sahoo *et al.* (2011) addresses the synthesis of W–Cu nanocomposites by a multivariate route, comprising polyol and thermal decomposition processes.

W–Cu nanocomposite powders were synthesised by simultaneous reduction of $Cu(acac)_2$ by polyethyleneglycol (PEG-200) and decomposition of $W(CO)_6$ in diphenyl ether. The composition of the resultant W–Cu nanocomposites could be easily altered by adjusting the ratio of metal precursors, since both are solid powders.

9.5 Applications

Among the diversity of applications in which Me-based composite powders have been used, we can identify four sectors that continue to be employed and probably will be considered in the immediate and mid-range future. First, the enhanced electrical properties are welcomed for the production of electrodes or parts of SOFCs (solid oxide fuel cells) (Changsheng *et al.*, 2011). Second, the magnetic properties of Fe–Co-based metallic alloys have been successfully applied by combining them with some oxides, particularly alumina, in order to strengthen and stabilise the grain size of the metallic matrix. Another example of where the magnetic properties were the key part of the system was the result of combining Fe and ferrite. Taking into account that MA was used as the production method, the magnetic domains can be highly influenced by the process and thus internal stresses and grain/domain size are the key factors to be controlled. Microelectronics and sensor devices are among the most frequent applications of this type of composite powder. Nanocomposite powders can help to produce higher efficacy in doped semiconductors.

When mixing magnetic metals or alloys with polymers, metal-based polymer composite powders can offer magnetic properties that are strongly related to powder particle size and will produce soft magnetic composites. In particular, Nowosielski (2007) studied high energy milling of amorphous ribbons of metallic $Fe_{78}Si_9B_{13}$ and $Fe_{73.5}Cu_1Nb_3Si_{13.5}B_9$ glasses. After annealing to the point of nanocrystallising them, they were mixed with a polymer and finally consolidated.

Biomedicine is the third sector where objects produced from metal-based composite powders find a market. It is well known that HA-ceramic composite powders are in great demand for bone restructuration, but some attempts have also been carried out to combine Ti–Al–V with HA. Biocompatibility and inert bioreactivity are the objectives for any material to be used in this sector.

Structural properties can be enhanced by Me-composite powder-based materials for applications where these properties are essential; however, other functions may also be crucial, thus affecting material selection. The sintering step for the production of composite material products is nowadays well known, as confirmed by the increasing number of scaled-up laboratory processes. Thus, general structural requirements compose the fourth main group of applications (Neikov *et al.*, 2009).

CNT reinforced metal-matrix consolidated composites are still at the research stage. However, there are several potential applications for these composites. Owing to their high strength, wear resistance and low density, these applications are piston rings, gears and cylinder liners in the automobile industry, and aircraft brakes and landing gear in the aerospace industry. If they also have a low coefficient of thermal expansion, structural radiators in space applications or devices in electronic packaging are also potential applications for these materials. Microelectromechanical systems (MEMS), sensors, batteries and energy storage systems need a high elastic modulus, large surface area, reduced response times and high current density, characteristics that can also be found in some metallic composite powder combinations (Bakshi *et al.*, 2010). CNTs have better strength and stiffness than carbon fibres and hence have the potential to replace carbon-fibre reinforced MMCs in various applications. Overcoming the challenges in processing will result in efficient use of the mechanical properties and will result in the strongest MMCs known to mankind.

9.6 Future trends

It is worth noting that nanocomposites have their own particular route and applications in nanotechnology. Among these applications, supercapacitors, thermoelectric materials, templates, sensors, biosensors and controllers appear to be the future for this family of materials. Much research has been undertaken into utilising CNTs as reinforcement for composite materials since their discovery by Iijima in 1991. However, CNT-reinforced MMCs have received little attention. These composites are being prepared for use in structural applications because of their high specific strength, as well as their use as functional materials owing to their excellent thermal and electrical conductivities.

Since 1970, carbon-fibre reinforced composites have been extensively used for a wide array of applications such as aircraft brakes, space structures, military and commercial planes, lithium batteries, sporting goods and structural reinforcement in construction. In the last decade, a number of studies have been carried out using CNT as reinforcement in different materials, namely polymers, ceramics and metals. Bakshi *et al.* (2010) studied journal articles published on CNT-reinforced composites in the last decade and found that the majority of research had been carried out into reinforcement of polymers by CNT. This can be attributed primarily to the relative ease of polymer processing, which often does not require the high temperatures for consolidation that are needed for metals and ceramic matrixes. This is surprising considering the fact that most of the structural materials used today are metals. Also, the number of publications on different metal matrix-CNT composites from 1997 to 2008 was plotted and it could be observed that there has been an increase in the number of publications on that topic since 2003. In addition, the metal matrixes in CNT-reinforced MMCs that were most used in the last decade were Al, Ni and Cu.

CNT-reinforced metal matrix (MM-CNT) composites are prepared by a variety of processing techniques. Since realising that the most critical issues in the processing of CNT-reinforced MMCs are (i) dispersion of CNTs and (ii) the interfacial bond strength between CNT and the matrix, many researchers have adopted modified steps in their approaches. Powder metallurgy is the most popular and widely applied technique for preparing MM-CNT composites by MA. Esawi and Morsi (2007) used MA for the first time to generate a homogenous distribution of 2 wt% CNT within Al powders. Milling for up to 48 hours led to good dispersion of CNTs but resulted in the formation of large spheres (>1 mm) owing to cold welding. Yang and Schaller (2004) achieved a homogeneous distribution of CNT in a Mg matrix by mechanically mixing the powders in an alcohol and acid mixture, followed by sintering at 823K.

Electrodeposition and electroless deposition are the second most important techniques for the deposition of thin coatings onto MM-CNT composites, as well as for deposition of metals onto CNTs. Walid (2008) researched CNT–Cu composite powders; different CNT volume fractions were prepared in an alkaline citrate bath by electroless copper deposition. The composite powders produced were heat treated and sintered using a spark plasma technique. The CNTs were implanted into the copper particles. To achieve better dispersion in the Cu matrix, the CNTs were electrolessly coated with Ni before hot pressing at 1373K, which ultimately resulted in improved mechanical and wear properties for the composite (Deng *et al.*, 2007).

A novel approach (Xu *et al.*, 2010) was developed to obtain uniformly dispersed CNT reinforcement in Al matrix composite powders. The process involved *in situ* synthesis of carbon nanostructures on the surface of aluminium powders through Friedel–Craft alkylation, and a homogenous dispersion of the CNTs in the aluminium powders was obtained.

The critical issues for mechanical properties in MM–CNT composites are the homogeneous distribution of CNTs in the metal matrix and the interfacial reaction and bonding to the matrix so that they work as effective reinforcement.

9.7 References

This chapter has aimed to reflect the continuing research and effort that has broadened our knowledge and understanding of Me-based composite powders. There are several key books to consult, including: the *Handbook of Non-Ferrous Metal Powders: Technologies and Applications* (Neikov *et al.*, 2009), the *ASM Metals Handbook: Powder Metallurgy* (Eisen *et al.*, 1998), and the *Concise Encyclopaedia of Composite*

Materials (Kelly, 1994 and Mortensen, 2007). The authors would like to thank the valuable contributions of all of the authors and collaborators.

Adamiak M, Fogagnolo JB, Ruiz Navas EM, Dobrzanski LA and Torralba JM (2004), 'Mechanically milled AA6061/(Ti_3Al)P MMC reinforced with intermetallics – the structure and properties', *J Mater Proc Technol*, **155–6**, 2002–6.

Alam SN (2006), 'Synthesis and characterization of W–Cu nanocomposites developed by mechanical alloying', *Mater Sci Eng*, **433**, 161–168.

Arami H, Simchi A and Seyed Reihani SM (2008), 'Mechanical induced reaction in Al–CuO system for *in-situ* fabrication of Al based nanocomposites', *J Alloys Compd*, **465**, 151–6.

Ardestani M, Rezaie HR, Arabi H and Razavizadeh H (2009), 'The effect of sintering temperature on densification of nanoscale dispersed W\20–40wt% Cu composite powders', *Int J Refract Met Hard Mater*, **27**, 862–7.

Bakshi SR, Lahiri D and Agarwal A (2010), 'Carbon nanotube reinforced metal matrix composites – a review', *Int Mater Rev*, **55**, 41–64.

Begin-Colin S, de Araujo Pontes LR, Le Caer G, Pianelli A, Mocellin A and Matteazzi P (1993), 'Synthesis of metal-ceramic composite powders (Al_2O_3-Ti) by mechanical alloying', in *Proceedings International Conference Mechanochemistry*, Tkacova K (ed), Kosice.

Benjamin JS (1988), 'The mechanical alloying process', *Mod Dev Powder Metall*, **21**, 397–414.

Benjamin JS (1990), 'Mechanical alloying – A perspective', *Metal Powder Rep*, **2**, 122–7.

Benn RC, Mirchandani PK and Watwe AS (1988), 'Intermetallic systems produced by mechanical alloying', *Mod Dev Powder Metall*, **21**, 479–93.

Bhadury A, Gopinathan V, Ramakrishnan P and Miodownik AP (1996), 'Microstructural changes in a mechanically alloyed Al –6.2 Zn–2.5 Mg– 1 Cu alloy (7010) with and without particulate SiC reinforcement', *Metall Mater Trans*, **27**, 3718–26.

Botta WJ, Tomasi R, Pallone EMJA and Yavari AR (2001), 'Nanostructured composites obtained by reactive milling', *Scr Mater*, **44**, 1735–40.

Boyer R, Collings EW and Welsch G (1994), *Material Properties Handbook: Titanium Alloys*, ASM International, Ohio.

Castricum HL, Bakker H and Poels EK (2001), 'Oxidation and reduction in copper/zinc oxides by mechanical milling', *Mater Sci Eng*, **304–6**, 418–23.

Changsheng D, Hongfei L, Kazuhisa S and Toshiyuki H (2011), 'Co-precipitation synthesis and characterization of NiO-$Ce_{0.8}Sm_{0.2}O_{1.9}$ nanocomposite powders: effect of precipitation agents', *J Nanosci Nanotechnol*, **11**(3), 2336–43.

Cheng T (1994), 'Mechanical alloying of NiAl-based composites and cold sintering phenomenon', *Scr Metall Mater*, **31**(11), 1599–604.

Chicinas I (2006), 'Soft magnetic nanocrystalline powders produced by mechanical alloying routes', *J Optoelectron Adv Mater*, **8**, 439–48.

Chou TW, Kelly A and Okura A (1985), 'Fibre-reinforced metal matrix composites' *Composites*, **16**, 187–206.

Chung KH, He J, Shin DH and Schoenung JM (2003), 'Mechanisms of microstructure evolution during cryomilling in the presence of hard particles', *Mater Sci Eng, A*, **356**, 23–31.

Córdoba JM, Sayagués MJ, Alcala MD and Gotor FJ (2005), 'Synthesis of titanium carbonitride phases by reactive milling of the elemental mixed powders', *J Am Ceram Soc*, **88**, 1760–4.

Córdoba JM, Sayagués MJ, Alcala MD and Gotor FJ (2007a), 'Monophasic $Ti_yNb_{1-y}C_xN_{1-x}$ nanopowders obtained at room temperature by MSR', *J Mater Chem*, **17**, 650–653.
Córdoba JM, Sayagués MJ, Alcala MD and Gotor FJ (2007b), 'Monophasic nanostructured powders of niobium, tantalum, and hafnium carbonitrides synthesized by a mechanically induced self-propagating reaction', *J Am Ceram Soc*, **90**, 381–7.
Córdoba JM, Avilés MA, Sayagués MJ, Alcalá MD and Gotor FJ (2009), 'Synthesis of complex carbonitride powders $Ti_yMT_{1-y}C_xN_{1-x}$ (MT: Zr, V, Ta, Hf) via a mechanically induced self-sustaining reaction', *J Alloys Comp*, **482**, 349–55.
Correia JB, Marques MT, Carvalho PA and Vilar R (2007), 'Hardening in copper-based nanocomposites', *J Alloys Compd*, **434–5** (31), 301–3.
Crisan O and Crisan AD (2011), 'Phase transformation and exchange bias effects in mechanically alloyed Fe/magnetite powders', *J Alloys Comp*, **509**, 6522–7.
Davis JR (2001), *Copper and Copper Alloys*, ASM International, Ohio.
Deng CF, Zhang XX, Wang DZ and Ma YX (2007), 'Calorimetric study of carbon nanotubes and aluminum', *Mater Lett*, **61**, 3221–3.
Dogan ON and Hawk JA (1995), 'Abrasion resistance of *in-situ* Fe-TiC composites', *Scr Metall*, **33**, 953–8.
Dong SJ, Zhou Y, Shi YW and Chang BH (2002), 'Formation of a TiB_2-reinforced copper-based composite by mechanical alloying and hot pressing', *Metall Mater Trans*, **33A**, 1275–1280.
Doré F, Martin CL and Allibert CH (2004), 'Apparent viscosity of W–Cu powder compacts during sintering', *Mater Sci Eng*, **383**, 390–398.
Eisen WB, Ferguson BL, German RM, Iacocca R, Lee PW, Madan D, Moyer K, Sanderow H and Trudel Y (eds) ASM Handbook volume 7: Powder Metallurgy: Technologies and Applications, (1998), ASM International, The Materials Information Society, Ohio.
Esawi A and Morsi K (2007), 'Dispersion of carbon nanotubes (CNTs) in aluminum powder', *Composites: Part A*, **38**, 646–650.
Everett RK and Higby PL (1991), 'Expansivity of diboride-particulate/aluminium composites', *Scr Metall Mater*, **25**, 625.
Farid A, Guo S, Cui F-e C, Feng P and Lin T (2007), 'TiB_2 and TiC stainless steel matrix composites', *Mater Lett*, **61**, 189–191.
Faure JF and Brune G (1987), 'New PM aluminium alloys for wear and sliding applications', *Metal Powder Rep*, **42**, 101–103.
Findik F and Uzun H (2003), 'Microstructure, hardness and electrical properties of silver-based refractory contact materials', *Mater Des*, **24**, 489–492.
Fischer JJ and Haeberle RM (1988), 'Commercial status of mechanically alloyed materials', *Mod Dev Powder Metall*, **21**, 461–477.
Flom Y and Arsenault RJ (1986), 'Deformation of SiC/Al composites', *J Metal*, **38**, 31–34.
Fogagnolo JB, Amador D, Morales F, Ruiz-Navas E and Torralba JM (2002), 'Powder characterization of Al–Cu produced by mechanical alloying', *Adv Powder Metall Part Mater*, **1**, 191–7.
Fogagnolo JB, Robert MH, Ruiz-Navas EM and Torralba JM (2004), '6061 Al reinforced with zirconium diboride particles processed by conventional powder metallurgy and mechanical alloying', *J Mater Sci*, **39**, 127–32.
Forouzanmehr N, Karimzadeh F and Enayati MH (2009), 'Synthesis and characterization of $TiAl/_{\alpha}$-Al_2O_3 nanocomposite by mechanical alloying', *J Alloys Compd*, **478**, 257–9.
Fortnum RT and Mikkola DE (1987), 'Effects of molybdenum, titanium and silicon

additions on the D03 = B2 transition temperature for alloys near Fe_3Al', *Mater Sci Eng*, **91**, 223–31.
Fu S and Xu H (2010), 'Microstructure and wear behaviour of (Ti,V)C reinforced ferrous composite', *J Mater Eng Perform*, **19**, 825–828.
Gaffet E, Louison C, Harmelin M and Faudot F (1991), 'Metastable phase transformations induced by ball-milling in the Cu–W system', *Mater Sci Eng*, **134**, 1380–1384.
Gleiter H (2000), 'Nanostructured materials: basic concepts and microstructure', *Acta Mater*, **48**, 1–29.
González-Carrasco JL, García-Cano F, Caruana G and Liebich M (1994), 'Aluminum/Ni_3Al composites processed by powder metallurgy', *Mater Sci Eng*, **183**, 5–8.
Grossmann J, Rose K and Sporn D (1995), 'Processing, properties and microstructural design of sol–gel derived nanostructured Ni–Y_2O_3/ZrO_2', *Ceram Trans* **51**, 713–717.
Gusmano G, Bianco A and Polini R (2001), 'Chemical synthesis and sintering behaviour of highly dispersed W/Cu composite powders', *J Mater Sci*, **36**, 901–907.
Hashempour M, Rezaie HR, Razavizadeh H, Salehi MT, Mehrjoo H and Ardestani M (2010), 'Investigation on fabrication of W\Cu nanocomposite via a thermochemical coprecipitation method and its consolidation behavior', *J Nano Res*, **11**, 57–66.
Heidarpour A, Karimzadeh F and Enayati MH (2009), '*In situ* synthesis mechanism of Al_2O_3–Mo nanocomposite by ball milling process', *J Alloys Compd*, **477**, 692–695.
Horvitz D, Gotman I, Gutmanas EY and Claussen N (2002), '*In situ* processing of dense Al_2O_3–Ti aluminide interpenetrating phase composites', *J Eur Ceram Soc*, **22**, 947–954.
Hu D and Loretto MH (1994), 'Microstructural characterisation of a gas atomized Ti6Al4V-TiC composite', *Scr Metall Mater*, **31**(5), 543–548.
Huang ZG, Guo ZP, Calka A, Wexler D and Liu HK (2007), 'Effects of carbon black, graphite and carbon nanotube additives on hydrogen storage properties of magnesium', *J Alloys Compd*, **427**, 94–100.
Hussain Z, Othmana R, Longa BD and Umemotob M (2008), 'Synthesis of copper–niobium carbide composite powder by *in situ* processing', *J Alloys Comp*, **464**, 185–189.
Jena PK, Brocchi EA and Motta MS (2001), '*In situ* formation of Cu–Al_2O_3 nano-scale composites by chemical routes and studies on their microstructures', *Mater Sci Eng, A*, **313**, 180–186.
Jiang WH, Fei J and Han XL (2000), '*In Situ* Synthesis of (TiW)Cp/Fe Composites', *Mater Lett*, **46**, 222–224.
Joshi PB, Marathe GR, Murti NSS, Kaushik VK and Ramakrishnan P (2002), 'Reactive synthesis of titanium matrix composite powders', *Mater Lett*, **56**, 322–328.
Karimzadeh F, Enayati MH and Tavoosi M (2008), 'Synthesis and characterization of Zn/Al_2O_3 nanocomposite by mechanical alloying', *Mater Sci Eng, A*, **486**, 45–48.
Kelly A (1994), Concise Encyclopaedia of Composite Materials, London, Elsevier Science.
Khakbiz M and Akhlaghi F (2009), 'Synthesis and structural characterization of Al–B_4C nano-composite powders by mechanical alloying', *J Alloys Comp*, **479**, 334–1.
Khodaei M, Enayati MH and Karimzadeh F (2009), 'Mechanochemically synthesized Fe_3Al–Al_2O_3 nanocomposite', *J Alloys Compd*, **467**, 159–162.
Kim JC and Moon I (1998), 'Sintering of nanostructured W–Cu alloys prepared by mechanical alloying', *Nanostruct Mater*, **10**, 283–290.
Kim DG, Kim GS, Suk MJ, Oh ST and Kim YD (2004a), 'Effect of heating rate on microstructural homogeneity of sintered W–15wt%Cu nanocomposite fabricated from W–CuO powder mixture', *Scr Mater*, **51**, 677–681.

Kim DG, Kim GS, Oh ST and Kim YD (2004b), 'The initial stage of sintering for the W–Cu nanocomposite powder prepared from W–CuO mixture', *Mater Lett*, **58**, 578–581.

Kim JW, Chung HS, Shim JH, Ahn JP, Cho YW and Oh KH (2006), 'Synthesis and liquid phase sintering of TiN/TiB_2/Fe–Cr–Ni nanocomposite powder', *J Alloy Compd*, **422**, 62–66.

Knibloe JR, Wright RN, Sikka VK, Baldwin RH and Howell CR (1992), 'Elevated temperature behavior of Fe_3Al with chromium additions', *Mater Sci Eng*, **153**, 382–386.

Koch CC and Whittenberger JD (1996), 'Mechanical milling/alloying of intermetallics', *Intemetallics*, **4**, 339–355.

Kondoh K, Threrujirapapong T, Imai H, Umeda J and Fugetsu B (2009), 'Characteristics of powder metallurgy pure titanium matrix composite reinforced with multi-wall carbon nanotubes', *Compos Sci Technol*, **69**, 1077–1081.

Korác M, Andic Z, Tasic M and Kamberovic Z (2007), 'Sintering of Cu–Al_2O_3 powders produced by a thermomechanical route', *J Serbian Chem Soc*, **72**, 1115–25.

Lampke T, Wielage B, Pokhmurska H, Rupprecht C, Schuberth S, Drehmann R and Schreiber F (2011), 'Development of particle-reinforced nanostructured iron-based composite alloys for thermal spraying', *Surf Coat Technol*, **205**, 3671–3676.

Last HR and Garret RK (1996), 'Mechanical behavior and properties of mechanically alloyed aluminum alloys', *Metall Mater Trans*, **27**, 737–745.

Lee PW (1997), Powder metal technologies and applications, ASM International, Ohio.

Lee BS and Kang S (2001), 'Low-temperature processing of B_4C–Al composites via infiltration technique', *Mater Chem Phys*, **67**, 249–255.

Lee JS and Kim TH (1995), 'Densification and microstructure of the nanocomposite W-Cu powders', *Nanostruct Mater*, **6**, 691–694.

Lee YF and Lee SL (1999), 'Effects of Al additive on the mechanical and physical properties of silicon reinforced copper matrix composites', *Scr Mater*, **41**(7), 773–778.

Lee DW, Ha GH and Kim BK (2001), 'Synthesis of Cu–Al_2O_3 nano composite powder', *Scr Mater*, **44**, 2137–2140.

Li L and Lai M (1998), *Mechanical Alloying*, *Kluwer Academic Publishers*, Boston.

Li J, Ni X and Wang G (2007), 'Microstructure and magnetic properties of Co/Al_2O_3 nanocomposite powders', *J Alloys Compd*, **440**, 349–356.

Llorca-Isern N, Artieda-Guzmán C, Porras-Mateu N and Roca-Vallmajor A (2010), 'Structural and magnetic properties of nanocrystalline Cu–Fe–Co–Al_2O_3 composite powders processed by mechanical alloying', *Pulvimetalurgy World Congress Firenze*, Italy, volume **2**, 256.

Lu CJ and Li ZQ (2005), 'Structural evolution of the Ti–Si–C system during mechanical alloying', *J Alloy Compd*, **395**, 88–92.

Lü L, Lai MO and Wang HY (2000), 'Synthesis of titanium diboride TiB_2 and Ti–Al–B metal matrix composites', *J Mater Sci*, **35**, 241–248.

Lü L, Lai MO, Su Y, Teo HL and Feng CF (2001), '*In situ* TiB_2 reinforced Al alloy composites', *Scr Mater*, **45**, 1017–1023.

Maiti R and Chakraborty M (2008), 'Synthesis and characterization of molybdenum aluminide nanoparticles reinforced aluminium matrix composites', *J Alloys Compd*, **458**, 450–456.

Majumdar D, Kodas TT and Glicksman HD (1998), 'Generation of novel silver-silica powders by spray pyrolysis', *World Congress on Particle Technology*, 3, Brighton, UK, 1711–1719.

Matsumoto M, Kaneko K, Yasutomi Y, Ohara S, Fukui T and Ozawa Y (2002), 'Synthesis of TiO_2-Ag composite powder by spray pyrolysis', *J Ceram Soc Jpn*, **110**, 60–62.

Matsuura K, Obara Y and Kudoh M (2006), 'Fabrication of TiB_2 particle dispersed FeAl-based composites by self-propagating high-temperature synthesis', *ISIJ Int*, **46**, 871–874.

McCormick PG and Froes FH (1998), 'The fundamentals of mechanochemical processing', *JOM*, **50**(11), 61–65.

McCoy KP and Shaw KG (1994), 'Fabrication of carbide intermetallic composite powders by reaction synthesis', in *Specialty Materials and Composites. Advances in Powder Metallurgy & Particulate Materials*, Lall C (ed), Metal Powder Industries Federation, Michigan, 179–87.

Mendiratta MG, Ehlers SK and Lipsitt HA (1987), 'DO_3-B2-á phase relations in Fe–Al–Ti alloys', *Metall Trans*, **18**, 509–18.

Mishra RS, Bieler TR and Mukherjee AK (1992), 'On the superplastic behaviour of mechanically alloyed aluminium alloys', *Scr Metall Mater*, **26**, 1605–1608.

Montealegre I, Neubauer E, Angerer P, Danninger H and Torralba JM (2011), 'Influence of nano-reinforcements on the mechanical properties and microstructure of titanium matrix composites', *Compos Sci Technol*, **71**, 1154–1162.

Mortensen A (2007), *Concise Encyclopaedia of Composite Materials*, 2nd edition, Elsevier, Oxford.

Mortensen A, Cornie JA and Flemings MC (1988), 'Solidification processing of metal-matrix composites', *J Met*, **40**, 12–19.

Mousavi T, Karimzadeh F and Abbasi MH (2009), 'Mechanochemical assisted synthesis of NiTi intermetallic based nanocomposite reinforced by Al_2O_3', *J Alloys Compd*, **467**, 173–178.

Mukai T and Ishikawa K (1995), *J Jpn Soc Powder, Powder Metall*, **42**, 180–184.

Mulas G, Varga M, Bertoti I, Molnar A, Cocco G and Szepvolgyi J (1999), '$Cu_{40}Mg_{60}$ and Cu–MgO powders prepared by ball-milling: characterization and catalytic tests', *Mater Sci Eng*, **267**, 193–199.

Munir ZA (1988), 'Synthesis of high temperature materials by self-propagating combustion methods', *Bull Amer Ceram Soc*, **67**, 342–349.

Munir Z and Ealamloo-Grami M, (The Regents of the University of California) 1994. *Synthesis of Transition Metal Carbonitrides*, US patent application 5,314,656.1994-May-24.

Nardone VC and Prewo KW (1986), 'On the strength of discontinuous silicon carbide reinforced aluminum composites', *Scr Metall*, **20**, 43–48.

Neikov O, Naboychenko S, Mourachova I, Gopienko V, Frishberg I and Lotsko D (2009), *Handbook of Non-Ferrous Metal Powders: Technologies and Applications*, Elsevier, Amsterdam.

Nowosielski R (2007), 'Soft magnetic polymer–metal composites consisting of nanostructural Fe-basic powders', *J Achievements Mater Manufactur Eng*, **24**(1), 68–77.

Oleszak D (2001), 'Mechanically alloyed nanocrystalline NiAl–Al_2O_3 composite powders', in *Science of Metastable and Nanocrystalline Alloys: Structure, Properties and Modelling*, Dinesen AR, Eldrup M, Juul Jensen D, Linderoth S, Pedersen TB, Pryds NH, SchroederPedersen A and Wert JA (eds), Risoe National Laboratory, Roskilde, 3–7.

Pastor H (1988), 'Titanium-carbonitride-based hard alloys for cutting tools', *Mater Sci Eng*, **105–6**, 401–409.

Pérez-Bustamante R, Estrada-Guel I, Antúnez-Flores W, Miki-Yoshida M, Ferreira PJ and Martínez-Sánchez R (2008), 'Novel Al-matrix nanocomposites reinforced with multi-walled carbon nanotubes', *J Alloys Compd*, **450**, 323–326.

Polkin IS and Borzov AB (1995), 'New materials produced by mechanical alloying', *Adv Perfor Mater*, **2**, 99–109.

Rafiei M, Enayati MH and Karimzadeh F (2009), 'Mechanochemical synthesis of $(Fe,Ti)_3Al$–Al_2O_3 nanocomposite', *J Alloys Comp*, **488**, 144–147.

Raghunathan S, Allen RJ, Persad C, Bourell DL, Eliezer Z and Marcus HL (1993), 'Tungsten and tungsten-based composites and alloys for high heat flux duty', *Specialty Materials and Composites. Advances in Powder Metallurgy & Particulate Materials*, in Lall C (ed), Metal Powder Industries Federation, Michigan, 175–187.

Ramaseshan R, Kakitsuji A, Seshadri SK, Nair NG, Mabuchi H, Tsuda H, Matsui T and Morii K (1999), 'Microstructure and some properties of TiAl–Ti_2AlC composites produced by reactive processing', *Intermetallics*, **7**, 571–7.

Ravikiran A and Surappa MK (1997), 'Oscillations in coefficent of friction during dry sliding of A356 Al–30% wt SiCp MMC against steel', *Scr Mater*, **36**, 95–98.

Razavi-Hesabi Z, Simchi A and Seyed Reihani SM (2006), 'Structural evolution during mechanical milling of nanometric and micrometric Al_2O_3 reinforced Al matrix composite', *Mater Sci Eng*, **428**, 159–168.

Reddy BSB, Das K, Pabi SK and Das S (2007), 'Mechanical-thermal synthesis of Al-Ce/Al_2O_3 nanocomposite powders', *Mater Sci Eng A*, **445–6**, 341–6.

Ruiz-Navas EM, Fogagnolo JB, Velasco F, Ruiz-Prieto JM and Froyen L (2006), 'One step production of aluminium matrix composite powders by mechanical alloying', *Composites*, **37**, 2114–2120.

Sabooni S, Mousavi T and Karimzadeh F (2010), 'Mechanochemical assisted synthesis of Cu(Mo)/Al_2O_3 nanocomposite', *J Alloys Comp*, **497**, 95–99.

Sadeghian Z, Lotfia B, Enayatib MH and Beiss P (2011), 'Microstructural and mechanical evaluation of Al–TiB_2 nanostructured composite fabricated by mechanical alloying', *J Alloys Comp*, **509**, 7758–7763.

Sahoo PK, Kamal SSK, Premkumar M, Sreedhar B, Srivastava SK and Durai L (2011), 'Synthesis, characterization and densification of W\Cu nanocomposite powders', *Int J Refract Met Hard Mater*, **29**, 547–554.

Schaffer GB and McCormick PG (1989), 'Combustion synthesis by mechanical alloying', *Scr Metall*, **23**, 835–838.

Schaffer GB and McCormick PG (1990), 'Displacement reactions during mechanical alloying', *Metall Trans*, **21**(10), 2789–2794.

Schaffer GB and McCormick PG (1991), 'Anomalous combustion effects during mechanical alloying', *Metall Trans*, **22**(12), 3019–3024.

Schicker S, Garcia DE, Gorlov I, Janssen R and Claussen N (1999), 'Wet milling of Fe/Al/Al_2O_3 and Fe_2O_3/Al/Al_2O_3 powder mixtures', *J Am Ceram Soc*, **82**, 2607–12.

Sheibani S, Ataie A, Heshmati-Manesh S and Khayati GR (2007), 'Structural evolution in nano-crystalline Cu synthesized by high energy ball milling', *Mater Lett*, **61**, 3204–3207.

Sheibani S, Khakbiza M and Omidib M (2009), '*In situ* preparation of Cu–MnO nanocomposite powder through mechanochemical synthesis', *J Alloys Comp*, **477**, 683–687.

Shorowordi KM, Laoui T, Haseeb ASMA, Celis JP and Froyen L (2003), 'Microstructure and interface characteristics of B_4C, SiC and Al_2O_3 reinforced Al matrix composites: a comparative study', *J Mater Process Technol*, **142**, 738–743.

Suryanarayana C (2001), 'Mechanical alloying and milling', *Prog Mater Sci*, **46**, 1–184.

Takacs L (2002), 'Self-sustaining reactions induced by ball milling', *Prog Mater Sci*, **47**, 355–414.

Tang F, Wu XL, Ge SR, Ye JC, Zhu H, Hagiwara M and Schoenung JM (2008), 'Dry sliding friction and wear properties of B_4C particulate-reinforced Al-5083 matrix composites', *Wear*, **264**, 555–561.

Tee KL, Lü L and Lai MO (2001), '*In situ* stir cast Al–TiB_2 composite: processing and mechanical properties', *Mater Sci Technol*, **17**, 201–206.

Thian ES, Khor KA, Loh NH and Tor SB (2001), 'Processing of HA-coated Ti-6Al-4V by a ceramic slurry approach: an *in vitro* study', *Biomaterials*, **22**, 1225–32.

Tjong SC and Lau KC (2000), 'Abrasive wear behavior of TiB_2 particle-reinforced copper matrix composites', *Mater Sci Eng*, **282**, 183–186.

Tjong SC and Ma ZY (2000), 'Microstructural and mechanical characteristics of *in situ* metal matrix composites', *Mater Sci Eng*, **29**, 49–113.

Tjong SC, Wang GS and Mai YW (2005), 'High cycle fatigue response of *in-situ* Al-based composite containing TiB_2 and Al_2O_3 submicron particles', *Compos Sci Technol*, **65**, 1537–1546.

Torralba JM, da Costa CE, Cambronero LEG and Ruiz-Prieto JM (1997), 'PM aluminium composite reinforced with Ni_3Al', *Key Eng Mater*, **127–131**, 929–936.

Torres B, Lieblich M, Ibanez J and Garcia-Escorial A (2002), 'Mechanical properties of some PM aluminide and silicide reinforced 2124 aluminium matrix composites', *Scr Mater*, **47**(1), 45–49.

Van Meter ML, Kampe SL and Lawley A (1992), 'Dispersion Strengthened P/M Al-Mg Alloys' in *Advances in Powder Metallurgy and Particulate Materials*, Campus JM and German RM (eds), Metal Powder Industries Federation, Princeton, 285–301.

Varma HK, Kumar KP, Warrier KGK and Damodaran AD (1990), 'Silver-yttrium barium copper oxide composite derived from citrate gel', *Supercond Sci Technol*, **3**(2), 73–75.

Vijay R, Sundaresan R, Maiya MP and Srinivasa Murthy S (2006), 'Hydrogen storage properties of Mg–Cr_2O_3 nanocomposites: the role of catalyst distribution and grain size', *J Alloys Compd*, **424**, 289–293.

Walid MD (2008), 'Processing and characterization of CNT/Cu nanocomposites by powder technology', *Powder Metall Met Ceram*, **47**, 9–10.

Ward-Close CM, Minor R and Doorbar PJ (1996), 'Intermetallic-matrix composites – a review', *Intermetallics*, **4**, 217–29.

Welham NJ (1998), 'Mechanical activation of the formation of an alumina-titanium trialuminide composite', *Intermetallics*, **6**, 363–368.

Wu JM and Li ZZ (2000), 'Nanostructured composite obtained by mechanically driven reduction reaction of CuO and Al powder mixture', *J Alloys Compd*, **299**, 9–16.

Xi X, Xu X, Nie Z, He S, Wang W, Yi J and Tieyong Z (2010), 'Preparation of W–Cu nano-composite powder using a freeze-drying technique', *Int J Refract Met Hard Mater*, **28**, 301–304.

Xiong CS, Xiong YH, Zhu H and Sun TF (1995), 'Synthesis and structural studies of Cu–W alloys prepared by mechanical alloying', *Nanostruct Mater*, **5**, 425–432.

Xu X, Li Z, Zhang D and Chen Z (2010), '*In situ* synthesis of nanostructured carbon reinforcement in aluminum powders', *Mater Lett*, **64**, 1154–1156.

Yang J and Schaller R (2004), 'Mechanical spectroscopy of Mg reinforced with Al_2O_3 short fibres and C nanotubes', *Mater Sci Eng*, **370**, 512–515.

Ye JC, Han BQ and Schoenung JM (2006a), 'Mechanical behaviour of an Al–matrix composite reinforced with nanocrystalline Al-coated B_4C particulates', *Philos Mag Lett*, **86**, 721–732.

Ye JC, He JH and Schoenung JM (2006b), 'Cryomilling for the fabrication of a particulate B_4C reinforced Al nanocomposite: Part I. Effects of process conditions on structure', *Met Mater Trans*, **37**, 3099–3109.

Ye JC, Lee Z, Ahn B, He JH, Nutt SR and Schoenung JM (2006c), 'Cryomilling for the fabrication of a particulate B_4C reinforced Al nanocomposite: Part II. Mechanisms for microstructural evolution', *Met Mater Trans A*, **37**, 3111–3117.

Yeh CL and Chen YD (2005), 'Synthesis of niobium carbonitride by self-propagating combustion of Nb–C system in nitrogen', *Ceram Int*, **31**, 1031–9.

Yeh CL and Chen YD (2007), 'Combustion synthesis of vanadium carbonitride from V–C powder compacts under nitrogen pressure', *Ceram Int*, **33**, 365–371.

Yeh CL and Liu EW (2006), 'Preparation of tantalum carbonitride by self-propagating high-temperature synthesis of Ta–C system in nitrogen', *Ceram Int*, **32**, 653–658.

Yeh CL and Shen YG (2009), 'Formation of TiAl–Ti_2AlC *in situ* composites by combustion synthesis', *Intermetallics*, **17**, 169–173.

Ying DY and Zhang DL (2000), 'Processing of Cu–Al_2O_3 metal matrix nanocomposite materials by using high energy ball milling', *Mater Sci Eng*, **286**, 152–156.

Yoshimi K and Hanada S (1997), 'The strength properties of iron aluminides', *JOM*, **49**(8), 46–49.

Zeng Q, Baker I, McCreary V and Yan Z (2007), 'Soft ferromagnetism in nanostructured mechanical alloying FeCo-based powders', *J Magn Magn Mater*, **318**, 28–38.

Zhang DL (2004), 'Processing of advanced materials using high-energy mechanical milling', *Prog Mater Sci*, **49**, 537–560.

Zhang DL and Richmond JJ (1999), 'Microstructural evolution during combustion reaction between CuO and Al induced by high energy ball milling', *J Mater Sci*, **34**, 701–706.

Zhang X, Lu W, Zhang D and Wu R (1999), '*In situ* technique for synthesizing (TiB+TiC)/Ti composites', *Scr Mater*, **41**, 39–46.

Zhang DL, Cai ZH and Adam G (2004a), 'Mechanical milling of Al/TiO_2 composite powders', *JOM*, **56**, 53–56.

Zhang GJ, Ando M, Yang JF, Ohji T and Kanzaki S (2004b), 'Boron carbide and nitride as reactants for *in situ* synthesis of boride-containing ceramic composites', *J Eur Ceram Soc*, **24**, 171–178.

Zhang H, Chen MW, Ramesh KT, Ye J, Schoenung JM and Chin ESC (2006), 'Tensile behavior and dynamic failure of aluminum 6092/B_4C composites', *Mater Sci Eng*, **433**, 70–82.

Zhu SM, Tamura M, Sakamoto K and Iwasaki K (2000), 'Characterization of Fe_3Al-based intermetallic alloys fabricated by mechanical alloying and HIP consolidation', *Mater Sci Eng*, **292**, 83–89.

10
Porous metals: foams and sponges

R. GOODALL, The University of Sheffield, UK

DOI: 10.1533/9780857098900.2.273

Abstract: This chapter describes the processing and properties of metals containing significant fractions of porosity, processed using powders. The basic concepts used in porous materials research are introduced and the different types of processing techniques that have been explored are surveyed. The reported property data for different foams are collated and used to illustrate the range of properties that have been achieved and methods to predict the properties of porous metals from elementary knowledge about their structure are discussed. Finally, the outlook for porous metals research and some likely future directions of fruitful enquiry are suggested.

Key words: material properties, metal foams, processing.

10.1 Introduction

A distinct class of materials with their own properties and behaviours, porous metals (called metal foams or sponges) have been the subject of research for some time and are starting to be applied in engineering situations. This chapter looks at why these materials are so interesting, how they are processed, what properties they display and the underlying rules that they follow. The emphasis is on porous metals processed by a solid state powder route and the literature on these materials is reported. However, in general the same relationships between the porous structure and the base metal properties, and the overall foam behaviour are found in all foamed materials.

Porous metals are interesting for research and engineering for two principal reasons: they are able to achieve unusual properties in combination and these properties can be tailored to suit the needs of a particular application:

Property combinations: Porous metals can be thought of as composites and can display some of the properties of metals (such as high electrical and thermal conductivity, ductile failure at high stress, etc) with some of the properties of a porous structure (such as permeability to fluids, low density, etc). These can be combinations that are impossible to achieve with other materials; for example, a highly thermally conducting material that is permeable to fluids would be suitable as a heat exchanger.

Tailoring of properties: The properties of a porous metal depend on both the properties of the metal in dense (i.e. non-porous) form, and also on the porous structure (essentially the amount, shape, size and distribution of porosity). As a result, small structural changes can be used to adjust the properties (within certain limits), allowing the precise properties required to be obtained. This could, for example, be of use in a protective application (against crushing or impact), where the strength could be adjusted to ensure maximum energy absorption, while ensuring that the protected component did not exceed the safe load or impulse.

10.1.1 Porous metals

Any metal with a fluid phase (liquid or, most usually, gas) distributed throughout it could be classified as a porous metal, although in practice the study of these materials is generally confined to those showing significant levels of porosity (greater than 50%) and where the second phase regions are uniformly distributed throughout the material (unlike the case of a cast part with a large internal pore, for example). In these structures, the regions of free space are called pores or cells and the solid phase is called the dense or parent metal, with the individual structural elements being termed struts, if they are thin relative to the pores, and being referred to as cell walls if they are larger. Figure 10.1 shows an example image of a porous structure with some of these features indicated.

Another important structural change can be discovered if it is imagined

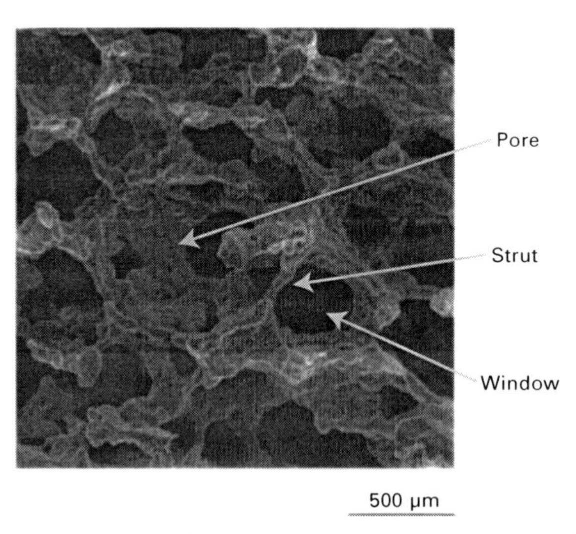

10.1 Example of a porous metal, indicating some of the typical features. The image (taken from Zhao and Sun, 2001) shows an aluminium foam processed by the space holder method. Foam micrograph reprinted from *Scripta Materialia*, Copyright (2001), with permission from Elsevier.

how the structure will change going from a metal with a low amount of porosity to one with much higher levels. With small amounts of a fluid phase, the pores will be isolated, existing as bubbles within the matrix. This is a closed cell structure and one that might be correctly called a metal foam. If we increase the amount of porosity, it is clear that at some point the pores will start to intersect and are likely to join, making a network that connects together and communicates with the external environment (this will occur at a volume fraction porosity of 0.64 (Arzt, 1982)). This is an open celled structure, and could be called a metal sponge, although the term metal foam is also applied.

10.1.2 Characteristics

In determining the properties of a porous metal, the two most important factors are the metal it is made from and the porous structure. The composition of the solid and any microstructure are easily determined by standard analysis techniques (e.g. energy dispersive X-ray spectrometry and optical microscopy), but the porous structure can be quite complex and difficult to quantify simply. The easiest identification to make is if the foam has open cells (which interconnect and are accessible by the outside environment) or closed cells (which are individually isolated and enclosed by metal, even if it is only a thin cell wall). Nevertheless, to interpret the properties it is normally necessary to quantify the structure and for this reason a number of parameters are commonly identified, Table 10.1.

In addition to these values, which would normally be given as the average for the whole specimen, the distribution of each of them will be important. In particular, for porous metals processed from powders, it may be found that the pore size distribution is bimodal. One peak may represent the size of the main pores, with a further peak at a size similar to the size of the powder particles, which represents residual porosity from incomplete sintering. If the

Table 10.1 Parameters frequently measured to characterise the structure of a porous metal

Parameter		Units	Measurement method
Volume fraction solid	Porosity	–	By measuring mass and volume, knowing density of base metal is required to calculate the porosity
	Density	kg m^{-3}	
Pore size (diameter)		m (mm)	Using image analysis on 2D sections (macrographs, optical or electron microscopy), for grain size/shape, using software or methods such as linear intercept (Higginson and Sellars, 2003). Measurements in 3D made using X-ray tomography (e.g. Tuncer *et al.*, 2011)
Pore aspect ratio		–	
Window size		m (mm)	

data is assimilated into a single average value, this underlying complexity can be missed.

10.1.3 Further information

Research has been carried out on metal foams since the 1960s and has been at a high level of intensity since around the 1980s. As a result there are a larger number of general reference works and reviews of the subject as a whole than will be covered in this chapter. Here the focus will be on metal foams and sponges processed using techniques based on powder metallurgy, but the wider subject can be accessed through books and review articles (Ashby *et al.*, 2000; Banhart, 2001; Dunand, 2004; Conde *et al.*, 2006; Colombo and Degischer, 2010; Goodall and Mortensen, 2013), and the proceedings of conferences such as the biennial MetFoam conference (Banhart *et al.*, 2003; Nakajima and Kanetake, 2005; Lefebvre *et al.*, 2007).

10.2 Powder processing: partial sintering and space holders

10.2.1 Partial sintering

When uniform powders are placed together, they will pack with space in between them which varies as a function of their coordination number (Arzt, 1982). A random packing of monosized spheres is usually taken to fill 64% of the available volume, although this will be more if the powder is pressed to form a compact. When sintered, the particles will bond and the pore fraction will be reduced. The simplest way to produce a porous metal is to interrupt this process at some point before full densification has been reached, Fig. 10.2. This was one of the first approaches used to produce porous metals from powder, for example with copper in the 1960s (Goodstein *et al.*, 1966) and some of the early methods are examined by Liu and Liang (2001). Sintering of powders is used commercially to produce porous metals by Schunk Sinter Metals (http://www.sintermetalltechnik.com). With larger pores, these can be used to control air flow as filters or silencers, and with smaller pores they can form self-lubricating bearings.

In more recent work, control of the final porosity over a relatively wide range (5–37%) has been achieved in titanium (Oh *et al.*, 2003), although the limit of the packing of particles (filling 64% of space) is once again roughly respected. Porous steel has been produced in this manner (Arockiasamy *et al.*, 2010) and titanium-based alloys (including NiTi with shape memory behaviour) have also been produced using hot pressing to accelerate bonding (Lagoudas and Vandygriff, 2002; Nomura *et al.*, 2005), although this may reduce the porosity in the final sample.

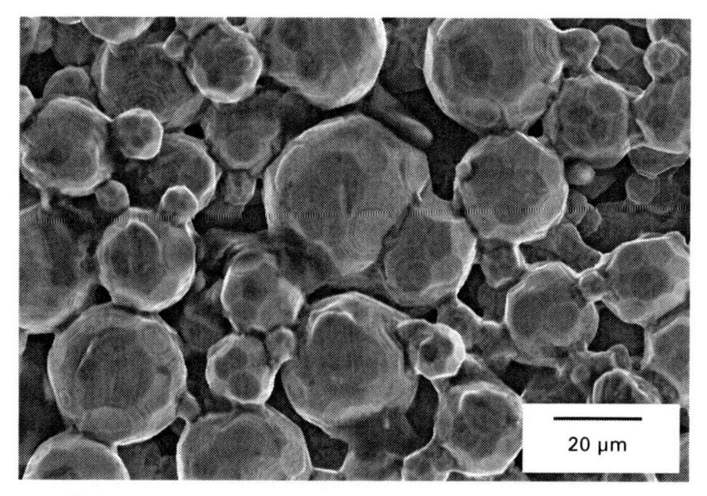

10.2 Porous titanium produced by partial sintering of gas atomised powder. Scanning Electron Microscope image courtesy of Mr Mark Taylor, the University of Sheffield.

With some metals, such as porous titanium, large particles (180–2000 μm diameter) have been used to increase the amount of porosity retained (Thieme *et al.*, 2001). As the driving force for sintering is less, liquid phase sintering was used to accelerate the formation of bonds. In this work, a more complex structure was made by combining together blocks of different structures at the green body stage, giving an overall part with a graded structure. Liquid phase sintering by niobium additions has also been used to accelerate consolidation in NiTi with NaCl space holders (Bansiddhi and Dunand, 2009).

The heat for the process need not always be provided externally. Where porous alloys or intermetallics are required (possibly to achieve a foam with a high melting point), it may be possible to produce them through reaction between different powders. These reactions can be exothermic and, because of the high surface area, have a tendency to enter positive feedback, releasing more heat and speeding up as the reaction progresses throughout the sample. This has been investigated as a specific technique, often termed self-propagating, high-temperature synthesis, SHS (Kobashi and Kanetake, 2002; Li *et al.*, 2002; Kim *et al.*, 2004; Biswas, 2005; Kanetake and Kobashi, 2006), typically used for NiTi and NiAl. Although based on a runaway reaction, meaning the process may seem difficult to control, with careful preparation and selection of the appropriate starting conditions (e.g powder size) and the inclusion of additional compounds that either enhance (e.g. B_4C) or suppress (e.g. TiC) the reaction, porous metals with controlled structures can be produced (Kobashi and Kanetake, 2002; Biswas, 2005; Kobashi *et al.*, 2006a, 2006b). It is even possible to include dense metal faceplates in the process to create sandwich panels directly (Nabavi and

Khaki, 2011). What causes the porosity to be developed in these alloys (which may be from around 10–85% (Kobashi and Kanetake, 2002)) is not fully understood, but may be due to the speed of the reaction limiting the time for porosity contained within the compact to be removed by evolution of the structure and because of the effect of hot gases generated during the process (Kanetake and Kobashi, 2006).

Even if there is no exothermic reaction, when powders of different elements or alloys are sintered together, the Kirkendall effect can act to increase porosity. The effect normally produces very fine scale porosity when the interdiffusion rates of the two metals are not equal and has been observed to occur at a sufficient level to generate a porous material with NiTi and Ni_3Al (He *et al.*, 2007; Dong *et al.*, 2009; Wen *et al.*, 2010; Liang *et al.*, 2011), but has also been used as a means of increasing porosity by adding to the pore fraction (Wang *et al.*, 1998) or by expanding a percolating network of particles (Ismail *et al.*, 2011).

Sintering of shapes other than powders has also been employed. While this includes exotic shapes such as hollow spheres, produced by coating expanded polystyrene spheres with metal powder and sintering (Nagel *et al.*, 1997; Uslu *et al.*, 1997; Lim *et al.*, 2002; Andersen *et al.*, 2000; Roy *et al.*, 2011), simpler shape changes will still show an effect. With a deviation from spherical particles, the percentage of space filled is generally reduced (Cumberland and Crawford, 1987; Yu and Standish, 1993), as these shapes do not pack as efficiently and the overall porosity of the powder compact can be increased by a substantial amount if extreme shapes such as fibres are used (Fedorchenko, 1979; Ducheyne *et al.*, 1978; Markaki *et al.*, 2003; Delannay, 2005; Clyne *et al.*, 2006; Tan *et al.*, 2006). With the lower number of particle to particle contacts in such systems, a route such as liquid phase sintering ensures good bond strength between fibres and enhances the properties of the material produced (Markaki *et al.*, 2003).

10.2.2 Space holders

Commonly, simply sintering powders does not produce materials with enough porosity, or a large enough pore size for many applications and workers in the field wish to enhance these and obtain some degree of control over the process. This is often done by including some other phase in the process, which occupies the space that will be required for the pores in the final sample. Although this can be a second phase which is in itself porous, forming what is called a *syntactic* foam (e.g. Xue and Zhao, 2011), more often it will be removed by some treatment at a later stage; this removable phase is called the *space holder*. The generic process is shown schematically in Fig. 10.3, where metal powder and space holder particles are blended, compacted and the space holder is then removed before sintering. The space

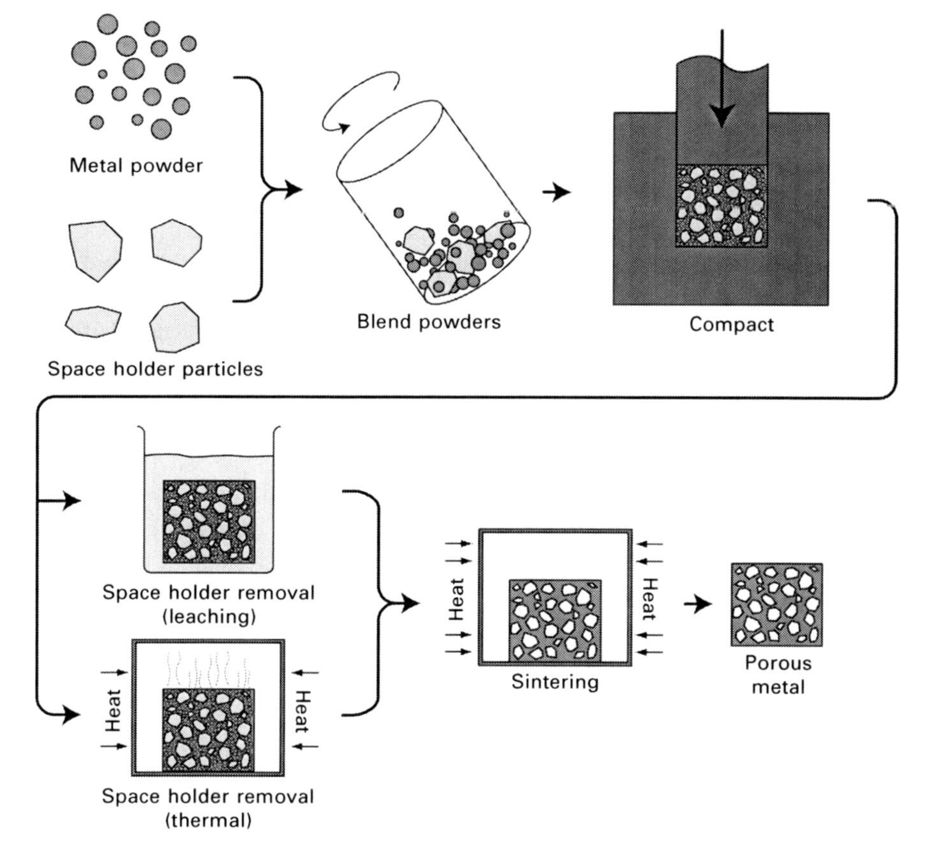

10.3 Schematic diagram of a space holder route for the production of porous metal.

holder also enhances green body strength (Laptev *et al.*, 2004, 2005), as does the degree of compaction, as would be expected. This can be sufficient that machining can be carried out on the green body, to avoid potentially challenging machining of the porous part (Laptev *et al.*, 2004).

Variants of the process have been used for all of the main metals and alloys from which foams and sponges are commonly made, including aluminium (Zhao and Sun, 2001; Wen *et al.*, 2003; Sun and Zhao, 2003; Zhao *et al.*, 2004; Jiang *et al.*, 2005b; Sun and Zhao, 2005a; Hakamada *et al.*, 2005a; Bin *et al.*, 2007), titanium (Bram *et al.*, 2000; Wen *et al.*, 2001; Rak *et al.*, 2003; Laptev *et al.*, 2004; Dunand, 2004; Rak and Walter, 2006; Esen and Bor, 2007; Ye and Dunand, 2010) and a variety of steels (Bram *et al.*, 2000; Bakan, 2006). As processes have been developed separately, different space holders tend to be preferred for different materials. For example, for aluminium foams, NaCl is frequently used, for titanium and its alloys, it may also be NaCl (where the metal is sintered with the space holder in the solid state (Ye and Dunand, 2010)), carbamide (essentially urea pellets, used as fertilizer) (Tuncer *et al.*, 2011), magnesium particles (Esen and Bor, 2007, 2011) or a polymer binder phase.

Sintering–dissolution process

One of the first space holder methods developed to a significant extent (sometimes termed the sintering–dissolution process, SDP) used aluminium and NaCl (Zhao and Sun, 2001). Powders were blended together and compacted and the compact sintered below both the melting point of aluminium (660°C) and salt (801°C). To increase the rate, the temperature can be taken 10–20°C above the melting point of aluminium; at temperatures higher than this, the aluminium will begin to be excluded from the compact, as liquid aluminium does not wet salt (Zhao *et al.*, 2004). Finally, the space holder NaCl phase can be removed by leaching in water, leading to the formation of a foam of the type seen in Fig. 10.1. The removal of salt is normally successful, although for porosities below 60% some residual space holder may remain (Zhao and Sun, 2001), which is in agreement with models developed for dissolution from a multi-phase compact (Zhao, 2003). Addition of magnesium (Sun and Zhao, 2003, 2005a) may be made as a sintering aid, acting to reduce the oxide layer, although it has sometimes been found to be less effective than a combined addition of magnesium and tin (Zhao *et al.*, 2004). Where they work, these additions are helpful, as the pressure that can be applied without damaging the space holder particles (which are usually more fragile than the metals being sintered) prevents there being as high green densities as would be recommended with dense components in powder metallurgy (Zhao *et al.*, 2004). The SDP for aluminium is highly developed, with simulations of the compact being used to predict the spatial distribution of heat over time, the sintering time required (Sun and Zhao, 2005b) and modelling of the dissolution behaviour of a leachable space holder material being performed (Zhao, 2003; Sun and Zhao, 2005b).

Some workers have substituted a spark plasma sintering (SPS) step for conventional sintering in the NaCl space holder method (Wen *et al.*, 2003; Hakamada *et al.*, 2005a), including for NiTi (Fu *et al.*, 2006), where very fine grained metal powders (50 nm) were used. This produces a particularly fine grain sized microstructure, as the shortened processing time and time at elevated temperature experienced by the material means that grain growth cannot occur (Wen *et al.*, 2003). To further increase the porosity, SPS has been used with polymer spheres coated with a nickel alloy (Song and Kishimoto, 2006), although as the foams are closed celled (Hakamada *et al.*, 2005b), the polymer or its remnants can be trapped inside.

Sintering with other space holders

NaCl is by no means the only space holder used. Carbamide (urea) and ammonium hydrogen carbonate are frequently employed. These are not removed by dissolution as NaCl is, but can be removed by treatments at high temperature (either before sintering, or as an initial step in the sintering heat treatment). Methods using these space holders were developed at around the same time as the SDP, principally for producing higher melting point metals with porosity. The same procedures of blending (often using a solvent) and pressing powders, substituting ammonium hydrogen carbonate and carbamide for NaCl, have been used for titanium and magnesium foams (Bram *et al.*, 2000; Wen *et al.*, 2001, 2002a, 2002b, 2004; Zhuang *et al.*, 2008; Niu *et al.*, 2009; Nouri *et al.*, 2010; Tuncer *et al.*, 2011), superalloys (Bram *et al.*, 2000; Mi *et al.*, 2009), stainless steel (Gulsoy and German, 2008) and copper foams (Hakamada *et al.*, 2007), and examples have also been reported of polypropylene carbonate (PPC) (Hong *et al.*, 2008) and starch being used for titanium (Mansourighasri *et al.*, 2012). All of these are removed by a thermal treatment before sintering, which is effective in retaining the porous shape, even though they decompose at relatively low temperatures (around 200°C).

Carbamide is also a popular space holder for aluminium (Jiang *et al.*, 2005a) and has been used for stainless steels, not being removed thermally, but by water leaching (Bakan, 2006). Carbamide particles are available in different shapes, as either rough spheres or high aspect ratio flakes. The use of different forms of carbamide allows the pore shape to be controlled as this shape is preserved in the porous material (Bram *et al.*, 2000; Jiang *et al.*, 2005b).

Another important space holder is potassium carbonate, K_2CO_3 (Zhao *et al.*, 2005) and its use is sometimes called the lost carbonate process. This has the advantage that it is thermally decomposed, so an additional leaching treatment is not required. It also is not removed until high temperatures (891°C, when it melts and decomposes simultaneously (Zhao *et al.*, 2005)), meaning that it can contribute to structural integrity up to high temperatures. This has been used for higher melting point metals, such as copper (Zhao *et al.*, 2005; Thewsey and Zhao, 2008; El-Hadek and Kaytbay, 2008) and iron (Ma *et al.*, 2006).

Porous intermetallics and alloys

There is an interest in porous alloys and intermetallic materials for a range of applications, including high temperature use. Shape memory alloys of NiTi have been made by blending and cold pressing elemental powders on their own (Yuan *et al.*, 2004), or with NaCl (Zhao *et al.*, 2009), followed by sintering where reaction between them takes place. The process has also been applied to a ferritic stainless steel containing Cr and Mo, using NaCl space holders (Scott and Dunand, 2010). With just elemental powders used, the pore size tends to be small (a few hundred micrometers) and the porosity low (around 30–40%); although these may be controlled by the use of a hot isostatic pressing (HIP) treatment (Yuan *et al.*, 2008). The inclusion of space holders permits high porosities, with a bimodal pore size distribution, where the larger pores are introduced by the salt particles, the smaller by the Kirkendall effect or residual porosity from the powder compact.

NiTi has also been processed from prealloyed powder, using NaCl (Bansiddhi and Dunand, 2008, 2009) or NaF as a space holder (Bansiddhi and Dunand, 2007). In both cases HIP was used to cause densification, although the high temperatures used necessitate either the use of a high melting point space holder (such as NaF, which is removed by a treatment in water of several weeks' duration (Bansiddhi and Dunand, 2007)) or by carrying out the process rapidly so that the space holder does not melt (as is done with NaCl (Bansiddhi and Dunand, 2008)). The advantages of the latter process stem mainly from the more easily obtained and handled space holder and improved process speed, and the reduced corrosion, cost and toxicity risks that result. The same process has been used with NaCl for another shape memory CuAlMn alloy (Gong *et al.*, 2011a, 2011b). As this is processed at high temperature (930°C), the salt does not have to be dissolved before the process is begun, as it will evaporate during the sintering process.

Metal injection moulding with space holders

Polymer particles have been used as space holders (Li and Lu, 2011) and organic binders may also be used to hold the powder in a specific shape to help with processing a specific component; this can play a dual role in preserving free space until a later stage of sintering (Rak and Walter, 2006). With this method, it is not possible to control pore size and shape externally, as the space holder has no rigid shape of its own, but this limitation is balanced by the advantage of not having to add a specific space holder phase, as the organic binder can be used to generate parts with complex shape (allowing net shaping or near net shaping of components) by methods like metal injection moulding (MIM).

Metal injection moulding (MIM) is another technique for forming powders into a shape for sintering and can often result in residual porosity (e.g. Guo *et al.*, 2009), suggesting that it may be exploited for porous metal production. The process can be used with elemental powders to produce a complex shape, which can then be processed further by a procedure such as SHS (Guoxin *et al.*, 2008), or sintered to allow reaction (Köhl *et al.*, 2009; Ismail *et al.*, 2011). MIM can also be used with elemental or prealloyed powders (Bram *et al.*, 2011), using partial sintering to produce a porous material, or by including a space holder material in the mix (Köhl *et al.*, 2011). Examples have been demonstrated for Ti-6Al-4V with small (<50 μm) poly(methylmethacrylate) (PMMA) particles as space holders (Engin *et al.*, 2011), where the small size of the space holders avoid difficulties with material flow during injection, and for 316L stainless steel with slightly larger PMMA particles (around 75 μm) (Manonukul *et al.*, 2010). The advantage of PMMA is that it breaks down thermally and is easily removed without leaving a residue that could contaminate the metal parts or impair sintering, although the small pore size and early removal of PMMA in the sintering process mean that the pore fractions obtained are low. Alternatively, NaCl can be used as the space holder for titanium (Chen *et al.*, 2009; Bram *et al.*, 2011) and NiTi production (Köhl *et al.*, 2009; Bram *et al.*, 2011), which has been demonstrated for pore sizes of around 300–400 μm. This space holder material is removed by dissolution in water before heat treatment, which means that no specific treatments during the sintering are required and that the pores are all open.

Freeze casting

A different route from adding preformed space holders to a powder mix, or using a space holder that has no rigid shape, is to form space holder particles *in situ*. The freeze casting process, originally developed for ceramic powders, has been applied to titanium powder (Chino and Dunand, 2008; Fife *et al.*, 2009; Li and Dunand, 2011) and recently stainless steel (Driscoll *et al.*, 2011). In the process, a slurry is formed of water with powder in suspension (Chino and Dunand, 2008) and this is frozen in a directional manner. As the ice crystals form, they displace the powder particles, forming a space holder phase *in situ*, concentrating the powder particles in certain regions and removing liquid from around the particles. The powder is therefore brought into close contact, so that weak bonds may form. In the next stage, the space holder is removed by sublimation, leaving pores in the structure of powder particles, which a short sintering step can give the required cohesive strength. Due to directional solidification the ice crystals tend to become aligned along the direction of heat extraction and the pores are highly anisotropic (Chino and Dunand, 2008).

As the freezing is directional (and ice crystals tend to be highly planar (Fife *et al.*, 2009)) and the microstructure therefore not uniform, porous materials processed in this way can have highly anisotropic properties (Driscoll *et al.*, 2011). It is not usually possible to obtain metal powder particles as small as the ceramic powders used in freeze casting (which are often submicrometre in size), but the processing is improved by having the smallest particles possible (as they are more easily moved by the ice crystals as they form (Chino and Dunand, 2008)), as is the final material (as smaller particles sinter more rapidly, forming stronger bonds) (Fife *et al.*, 2009).

Template methods

Other methods use a polymer foam as a template to define the porosity (in a sense, the polymer foam holds the space, although this is not done solely by occupying it as with other space holders). A different method that has been used for very low density nickel and stainless steel foams is by placing the powder particles in a slurry rather than a binder and coating an open cell carbon foam (which is itself formed from a polyurethane foam) (Queheillalt *et al.*, 2004). In this method the carbon foam is retained at the core of the struts in the foam. Titanium alloy (Ti-6Al-4V) has also been produced in foam form by using a polyurethane template dipped in slurry containing metal powder and a polymer binder, with the sintering step for the powder also removing the polymeric binder and foam (Li *et al.*, 2005). Perhaps more ambitious is the combination of the formation of the foam scaffold and inclusion of the metal powder in one process: the dispersal of large quantities of metal powder (titanium, iron and copper have been used) in one of the components of a foamable polymer system, such as polyurethane, the processing of the polymer, causing it to foam, and the subsequent heat treatment to remove the polymer and sinter the metal sufficiently to provide cohesive strength (Jee *et al.*, 2000; Guo *et al.*, 2000; Xie and Evans, 2004). Although the polymer foam is closed celled, the metal foams produced will be open celled (Jee *et al.*, 2000), with high fractions of porosity, typically over 90% (Jee *et al.*, 2000; Guo *et al.*, 2000). A related process uses a slurry of metal powder containing phosphoric acid (Angel *et al.*, 2005). The reaction between the powder and the acid liberates hydrogen, which foams the slurry forming a green body which can be sintered.

10.3 Powder processing: gas entrapment and additive layer manufacturing

10.3.1 Gas entrapment

To attempt to access different pore structures, workers have sought to use gas pressure to provide additional expansion of bodies formed from powder. This has some limitations, as for the effect of the gas pressure to be felt, the temperature needs to be high and the material needs to be in a highly deformable state (Kearns *et al.*, 1988) – in some cases, metals with superplastic properties have been employed (Dunand and Teisen, 1998) – making titanium a particularly suitable metal, although other materials, notably NiTi with superelastic properties, have been produced (Oppenheimer *et al.*, 2004; Greiner *et al.*, 2005). Also, the effect of the gas will only be felt while the pores remain closed. As pores start to intersect each other and joining therefore becomes more likely at porosities between 50 and 64%, the upper limit of the porosity that is achieved by these methods tends to be around this point.

The most common version of the method sees a powder being compacted under an inert gas environment, such as argon, so that quantities of this gas become entrapped within the compact (Schwartz *et al.*, 1998). A similar goal may be achieved by cryogenic milling of the powder in the presence of argon (VanLeeuwen *et al.*, 2011), which results in very fine scale porosity. The compact is then sintered under vacuum, which causes the metal particles to bond, but at the same time expands the pores owing to the pressure of the trapped gas (Schwartz *et al.*, 1998; Dunand and Teisen, 1998; Davis *et al.*, 2001; Murray and Dunand, 2003, 2004; Murray *et al.*, 2003; Oppenheimer *et al.*, 2004; Greiner *et al.*, 2005). Figure 10.4 shows a schematic diagram of the process. Control of the process can control the foam produced. Applying a uniaxial stress during expansion can constrain the foam and cause the

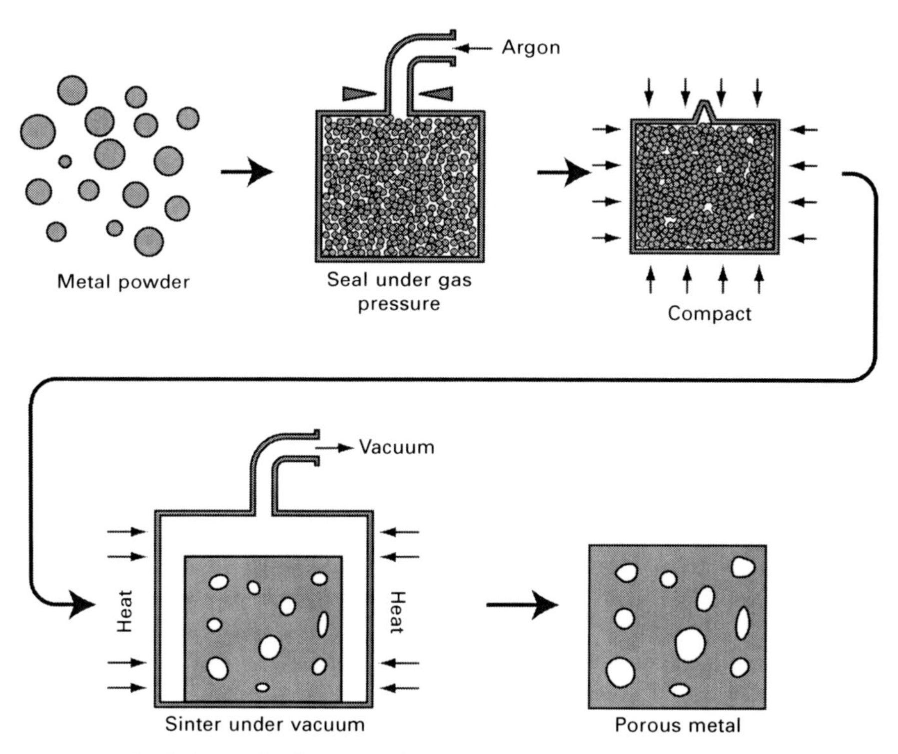

10.4 Schematic diagram of the gas entrapment process for porous metal creation.

pores to become aligned (Davis *et al.*, 2001), while repeated heating and cooling cycles can result in the membranes between the cells rupturing and the foam becoming much more open (Murray and Dunand, 2003). Workers have attempted to increase the porosity by including chemical agents that break down at high temperature to release gas, for example TiH_2 (Ricceri and Matteazzi, 2003) although success has been limited, at least in part by the high solubility in metal of the hydrogen gas released.

10.3.2 Additive layer manufacturing-based techniques

With current demands for efficiency in material and energy usage, new manufacturing techniques such as additive layer manufacturing (ALM) processes (also known as powder bed methods) are the subject of great current research interest. In this chapter it will not be attempted to give a full account of all these methods, but rather to focus on their use to produce porous metals.

For the majority of the materials discussed in the previous sections, the porosity contained within them is random in nature, being in pores of unpredictable size, shape and location (even if we can define global measures to characterise some of these properties, see earlier). The materials that can be made through ALM processes are very different in nature. The essential principle of the process translates a 3D image held in a computer into a metallic part, with very low levels of wastage or final machining required. This is carried out in specialised equipment, such as the systems sold commercially by Arcam AB (http://www.arcam.com/), EOS Electro Optical Systems (http://www.eos.info/en/home.html), Renishaw (http://www.renishaw.com) and others.

The processes have arisen from rapid prototyping, where a computer-aided design (CAD) model is created and formed into a solid part by adding layers (hence *additive layer manufacturing*, ALM). These methods were first applied to polymers, although some were adapted to produce metal parts; for example, 3D printing has been used to deposit shapes made from a slurry containing Ti-6Al-4V powder (Li *et al.*, 2006). This succeeded in making simple structures, with a minimum spatial resolution of around 500 μm.

Currently these rapid net shape manufacturing techniques are also available for metals. In one version, laser direct metal deposition (LDMD), metal powder is fed into a laser beam, which causes it to melt and be deposited on a substrate. By moving the powder injection and laser beam around the substrate, a 3D shape can be built up. Porous lattices have been achieved in Ti-6Al-4V and 316L stainless steel (Ahsan *et al.*, 2011), although the limited overhangs possible probably mean that the porosity will be limited to relatively low levels. A more widely used system for porous parts is shown schematically in Fig. 10.5. This uses a deposited layer of metal powder

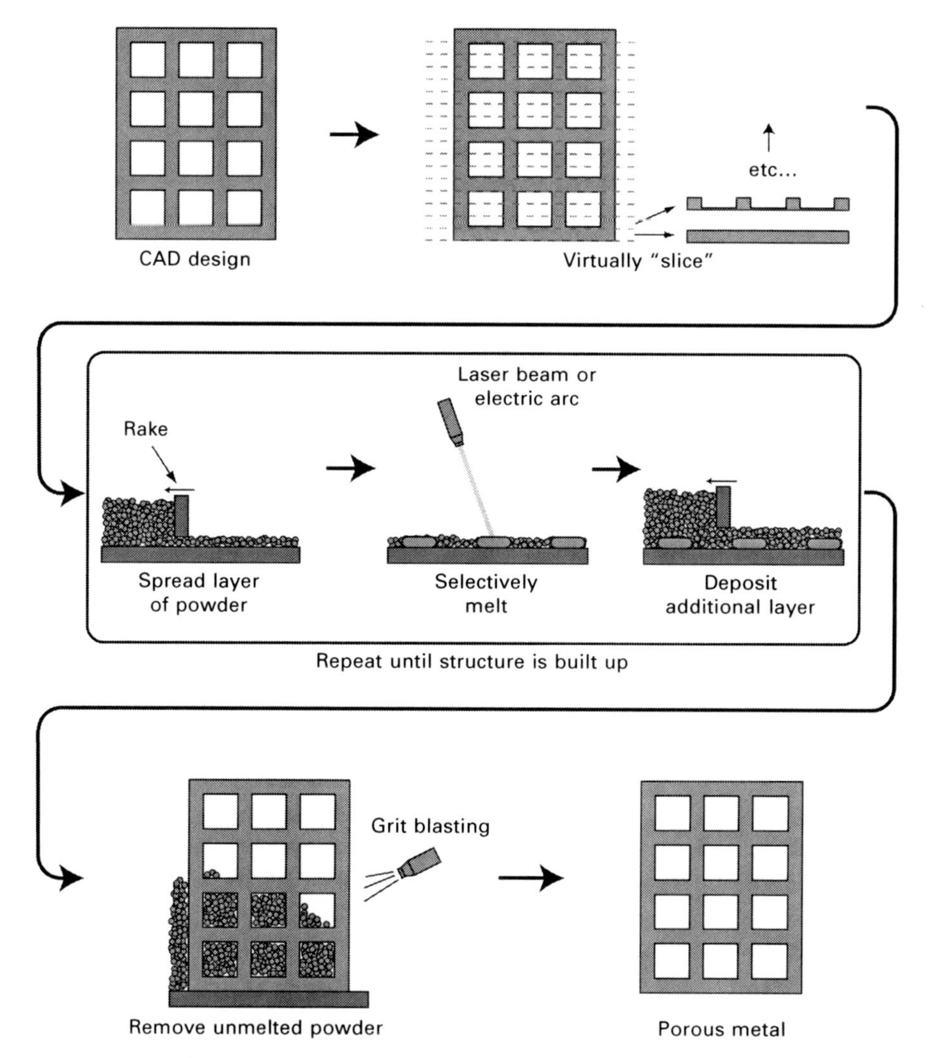

10.5 Schematic diagram representing the production of a metallic lattice using an ALM method.

(typically spherical particles are preferred owing to improved flowability). Once the layer has been spread across a build surface, a focussed heat source, such as a laser or electron beam is used to melt powder selectively in areas where the CAD model requires solid. In other areas it is left unheated. After treatment, another layer is deposited and the process is repeated until a 3D shape, encased and permeated by unmelted powder, is obtained. To produce a porous shape, this unmelted powder will have to be removed by grit blasting, for example. The powder can be recycled back into the process.

Systems where the metal powder is melted by an electron beam (often termed electron beam melting, EBM) have most commonly been used to make metallic lattices, often using titanium or Ti-6Al-4V, as these high value materials are common choices for the near net shaping offered by the machine. The most common choice of structure is a simple cubic lattice, or a lattice following the bond structure in diamond (Heinl *et al.*, 2007, 2008b; Cansizoglu *et al.*, 2008; Murr *et al.*, 2010a, 2010b; Li *et al.*, 2012). However, much more complex designs are possible, including lattices with re-entrant struts that are able to show negative Poissons ratios (auxetic behaviour) (Schwerdtfeger *et al.*, 2010). Porous shapes have also been made in a similar technique using a laser (selective laser melting, SLM), both in regular cubic-based lattices and structures replicating those of cancellous bone (Pattanayak *et al.*, 2011).

The main advantage of techniques like EBM is the great flexibility in the structures that can be produced, offering the possibility of having highly tailored structures. As well as regular lattices, random structures replicating regular foams (using tomography images of other foam types as the input) have been made (Murr *et al.*, 2010b, 2011; Ramirez *et al.*, 2011). The flexibility is such that the struts can be made both dense and hollow (Murr *et al.*, 2010b).

The nature of the processing will affect the material produced. The angle of struts relative to the build direction (the normal to the plane of the layers) can have an effect on the properties displayed by the lattice (Cansizoglu *et al.*, 2008) and the amount of energy input will affect the strut thickness and hence density (Heinl *et al.*, 2008a). In mechanical tests on the lattices it has been found that they have a relatively low fatigue resistance, partly due to the structure (as porous metals have low fatigue resistance generally and also because the microstructural condition of the Ti-6Al-4V used was not optimal for fatigue resistance (Li *et al.*, 2012)).

Because of interest from biomedical applications, lattices have also been produced from Co–Cr alloys (Murr *et al.*, 2011) and from copper for applications that require thermal or electrical flow (Ramirez *et al.*, 2011). For biomedical use the material condition should be suitable, as the alloys are well known and many researchers are looking at implantable parts made using such techniques (vanNoort, 2012), but the conductivity may be limited, as oxides, formed during the processing, were found throughout the copper lattices produced (Ramirez *et al.*, 2011).

10.4 Properties of porous metals

In the broadest sense, the properties displayed by a porous material will follow the same trends as seen in composites, with the properties of the whole being a blend of the properties of the individual constituents. An important simplification can be made in that one of the limits is the properties of free space, which for most properties can be taken as zero.

10.4.1 Mechanical properties

Data reported in the literature for the Young's modulus and compressive strength (the most common properties measured) of porous metals processed from powder are summarised in Table 10.2 and the variation in relative Young's modulus (the Young's modulus of the foam divided by that of the metal from which it is made) with porosity is shown in Fig. 10.6 for different processing techniques and metals. It is clearly seen that there is a fall off in both of these properties as the porosity increases, with a very wide spread in the properties displayed. This is quite usual for foams (see the results summarised in Ashby *et al.* (2000), for example), owing to the random nature of the structure and the potentially severe effect of defects. With sintered powders, the potential for defects is even larger, enhancing the spread.

Although in foams generally the pore size does not affect the mechanical response (with some exceptions where the surface plays a significant role (Diologent *et al.*, 2009)), there is some evidence that larger pore size foams processed by powders have improved mechanical properties (Bin *et al.*, 2007), which was attributed to a change in the aspect ratio of the walls, although it is not clear why this may arise. In similar tests the same effect was found and in X-ray tomography investigations it was found that a larger pore size gives rise to larger struts (Tuncer *et al.*, 2011), although intriguingly, the consolidation within them is reduced, with more pore wall porosity, smaller interconnection size and a lower strut density. Nevertheless it appears that the strut size is the dominant effect.

Metal foams are often highlighted for applications where their energy absorbing qualities may be employed. Care needs to be taken with experimental investigations, as it has been found that the energy absorbed in a dynamic mode of loading for aluminium foams is less than that absorbed in static tests (Sun and Zhao, 2003). Observations on similar materials show that the plateau stress is increased at higher strain rates (Hakamada *et al.*, 2005a) and so the reduced energy absorption must be due to densification occurring at a lower strain. In NiTi by contrast, it has been found that the Young's modulus is higher in dynamic testing than in quasistatic testing, attributed to the superplastic behaviour of the base metal in this case (Greiner *et al.*, 2005). It has been found that NiTi has good energy absorbing and damping characteristics, attributed to the hysteresis behaviour of the dense metal (Köhl *et al.*, 2011). This has been employed to make dense materials for energy absorption, by infiltrating a porous NiTi network with magnesium (Li *et al.*, 2010). Investigation of the behaviour and mechanisms operating

Table 10.2 Literature data for the mechanical properties of porous metals processed from powders

Processing method	Metal	Pore size (μm)	Pore fraction	E (GPa)	σ_y (MPa)	Ref.
Partial sintering	Ti		0.04–0.36	7.8–88.8	–	(Oh *et al.*, 2003)
		200–1000	0.24–0.57	14.1–57.7	–	(Thieme *et al.*, 2001)
Self-propagating High-temperature Synthesis	NiTi	100–280	0.6	0.12–0.2	–	(Kim *et al.*, 2004)
Space holder	Ti	200–500	0.78	5.3	–	(Wen *et al.*, 2001)
		525	0.45–0.7	0.42–8.8	15–116	(Esen and Bor, 2007)
		100–500	0.45–0.52	1.7–2.8	99–132	(Hong *et al.*, 2008)
		200–500	0.8	2.9	–	(Wen *et al.*, 2002a)
		200–500	0.35–0.8	2.9–10.3	25–478	(Wen *et al.*, 2002b)
		100–300	0.64–0.79	1.9–3.7	–	(Mansourighasri *et al.*, 2012)
		355–500	0.42–0.51	23–39	–	(Ye and Dunand, 2010)
		50–300	0.42–0.72	0.28–3.03	17.5–316	(Chen *et al.*, 2009)
		140–1800	0.64	3.8–6.1	45–87	(Tuncer *et al.*, 2011)
	Al		0.51	–	2.8	(Jiang *et al.*, 2005a)
	Mg	200–500	0.5	0.35	–	(Wen *et al.*, 2001)
		70–400	0.35–0.55	0.7–1.8	11–17	(Wen *et al.*, 2004)
		200–400	0.36–0.43	3.6–18.1	15–31	(Zhuang *et al.*, 2008)
	Cu	300–425	0.02–0.78	1.26–14.5	3.5–258	(Hakamada *et al.*, 2007)
		120–500	0.5–0.85	8.1–33.7	–	(El-Hadek and Kaytbay, 2008)
	Stainless Steel 316L	750–1000	0.7	0.27–1.4	–	(Bakan, 2006)
	Stainless Steel	750–1000	0.4–0.6	46–55	832–1211	(Gulsoy and German, 2008)
MIM with space holders	Ti	10–41	0.27–0.56	0.8–2.2	653–1244	(Engin *et al.*, 2011)
		400	0.51	22.5	140	(Bram *et al.*, 2011)
Trapped gas expansion	Ti	200	0.22–0.41	39–60	120–200	(Davis *et al.*, 2001)
Powder on foam template	Ni alloy	8000	0.89–0.93	0.77–1.87	1.3–3.5	(Queheillalt *et al.*, 2004)
	Ti-6Al-4V	400–700	0.9	0.8	10.3	(Li *et al.*, 2005)

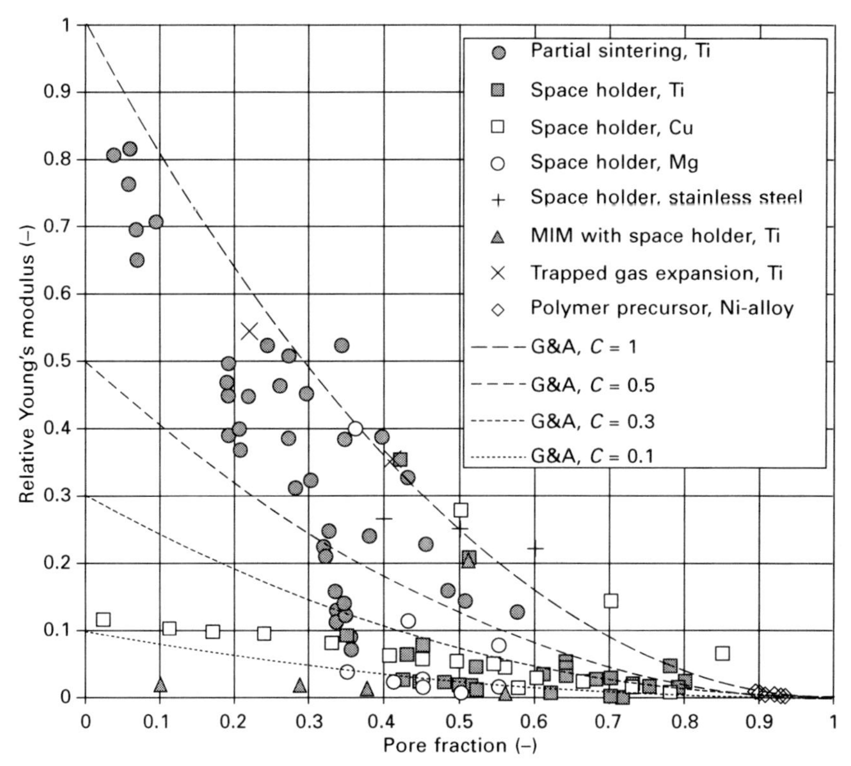

10.6 Literature data for the relative Young's modulus (E_{foam}/E_{metal}) of various porous metals produced from powder, plotted against pore fraction. Data for partial sintering with Ti are from Thieme *et al.* (2001) and Oh *et al.* (2003), for space holder with Ti from Wen *et al.* (2001, 2002a, 2002b), Esen and Bor (2007), Hong *et al.* (2008), Chen *et al.* (2009), Ye and Dunand (2010), Tuncer *et al.* (2011) and Mansourighasri *et al.* (2012), for space holder with Cu from Hakamada *et al.* (2007) and El-Hadek and Kaytbay (2008), for space holder with Mg from Wen *et al.* (2004) and Zhuang *et al.* (2008), for space holder with stainless steel from Bakan (2006) and Gulsoy and German (2008) for MIM with space holder from Engin *et al.* (2011) and Bram *et al.* (2011), for trapped gas expansion from Davis *et al.* (2001) and for polymer precursor from Queheillalt *et al.* (2004). Note that a wide variety of pore sizes are represented by the data. Also shown are the predictions of the Gibson–Ashby (G&A) model (Table 10.4) for open celled foams, with various values of the constant, *C*.

at different temperatures has shown that the trends in behaviour of the dense metal are conserved (Hakamada *et al.*, 2005a; Scott and Dunand, 2010). Foams processed from powders have also been investigated for creep (Scott and Dunand, 2010), which can be related to the dense metal properties by relatively accessible models (e.g. Hodge and Dunand, 2003; Mueller *et al.*, 2007).

10.4.2 Conduction properties

At temperatures within a few hundred degrees of ambient conditions, heat transport by radiation is low and at cell sizes below 4 mm, convection in the gas phase can be ignored (Skochdopole, 1961), leaving conduction through the solid phase the only significant mechanism of thermal transport. This being the case, parallels can be drawn between electrical and thermal conductivity and the same trends are observed. For most situations in most porous metals, the two properties can be assimilated into one although for some powder processed porous metals a difference has been found, attributed to defects in the structure (Thewsey and Zhao, 2008). For engineering use this is fortunate, as there have been relatively few measurements of either type of conductivity in powder processed metal foams. Some examples, in fact reporting values of electrical conductivity, are given in Table 10.3 and the variation in relative conductivity with porosity is shown in the graph in Fig. 10.7. As with mechanical behaviour, the trend is for the conductivity of a foam to be decreased as the porosity increases, with this reduction more rapid with the initial appearance of pores. Thermal conductivity follows similar trends and data for copper foams with different pore size and density have been reported (Thewsey and Zhao, 2008).

It has been found that the electrical conductivity of porous titanium processed by a space holder method decreases as pore size is reduced. This is, however, traced to differences in the pore shape (the aspect ratio) introduced when the pores are compacted in a uniaxial pressing step (Li and Lu, 2011). Differences in electrical conductivity with pore size have also been observed in copper and iron (Ma *et al.*, 2006). The conductivity is also higher for larger pores, which is suggested to be because of better bonding being achieved in the cell walls, which also have larger dimensions. This is, however, contrary to X-ray tomography results obtained on other types

Table 10.3 Literature data for the conduction properties of porous metals processed from powders. Where multiple data points are given, the minimum and maximum values are quoted in the table

Processing method	Metal	Pore size (μm)	Pore fraction	Conductivity (S m^{-1}) × 10^{6}	Ref.
	Cu	250–>1000	0.7–0.85	0.26–6.26	(Ma *et al.*, 2006)
		120–500	0.5–0.85	0.13–0.31	(El-Hadek and Kaytbay, 2008)
Space holder	Ti	150–400	0.1–0.67	0.33–1.41	(Li and Lu, 2011)
		136–403	0.31–0.64	0.13–0.69	(Li and Zhu, 2005)
	Fe	425–1500	0.68–0.76	0.95–1.45	(Ma *et al.*, 2006)
	Al	452–>2000	0.63–0.83	0.62–4.26	(Ma *et al.*, 2005)

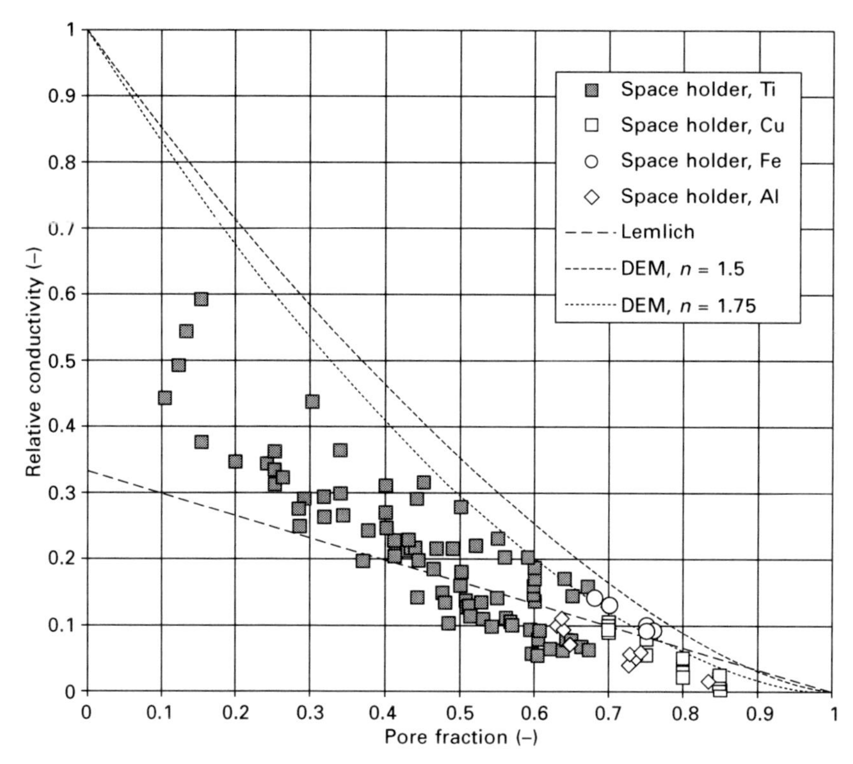

10.7 Literature data for the relative conductivity (S_{foam}/S_{metal}) of various porous metals produced from powder, with porosity. Data for space holder with Ti are from Li and Zhu (2005) and Li and Lu (2011), for Cu and Fe from Ma *et al.* (2006) and for Al from Ma *et al.* (2005). Also shown are the predictions of equations [10.1] and [10.2], the Lemlich and DEM models.

of porous metal processed from powder, where better consolidation is seen with smaller pores and higher pore densities, although larger struts are also produced, which appears to be the dominant effect for mechanical properties (Tuncer *et al.*, 2011).

In general, the conductivity of foams processed by powder methods is below that of other foams, as shown in Fig. 10.7. This is likely to be due to a greater content of impurities, such as entrained oxides, originating from the powder surfaces (Goodall *et al.*, 2006), a conclusion which is reinforced in the work of Li *et al.* (Li and Lu, 2011), who found the conductivity of dense commercial purity titanium processed by the same method as the porous metal in their work was 1.7×10^6 S m^{-1}, where the reference value for (commercial purity) CP Ti is 2.38×10^6 S m^{-1} (Lide, 2001), representing a relative conductivity of a little over 70% when the dense material is produced by that particular powder route.

10.4.3 Biological properties

Foams made from titanium, magnesium and NiTi are often suggested for use as biomaterial implants even if no biological testing is performed (Thieme *et al.*, 2001; Wen *et al.*, 2002a; Oh *et al.*, 2003; Dunand, 2004; Nomura *et al.*, 2005; Nouri *et al.*, 2010) and sufficiently high purities for use in the body have been achieved (Laptev *et al.*, 2004; Hong *et al.*, 2008). This is largely as the dense versions of these metals are currently used, or are leading candidates for actual implants, and the bioinertness is known.

Titanium foams have been pretreated to create a surface suitable for implants (Wen *et al.*, 2002b). Porous titanium has been exposed to simulated body fluid after surface treatments to provide a desirable surface for cell attachment (Hong *et al.*, 2008). While this gave a fine porous texture to the surface of the foam, which should perform well, no cell tests have yet been reported. Highly porous (90%) Ti-6Al-4V alloy has been subjected to cell culture tests, by seeding MC3T3-E1 osteoblast-like cells on samples of foam in cell culture media, with culture times of up to 3 days (Li *et al.*, 2005). It was found that cells attached to the material surface and spread and that there was evidence of extracellular matrix production.

Magnesium can degrade in the body environment, producing compounds that can be excreted through biological processes and is therefore attractive for degradable implants. In tests in a saline solution representing the body environment, breakdown of porous magnesium has been observed with, as expected, the dissolution rate being increased by increasing the porosity (Zhuang *et al.*, 2008).

Porous NiTi is particularly attractive for implants into bone, owing to the similarities in the stress–strain curve possible between the two materials (like bone, NiTi can display hysteresis within the superelastic region) (Bansiddhi *et al.*, 2008). Foams of NiTi have been produced and implanted into animal models (rabbits) (Kim *et al.*, 2004). In these tests, no inflammation was observed and after six weeks bone ingrowth occurred in all specimens examined. However, in the absence of a control test on other porous materials it is difficult to say how this behaviour ranks with other types. As more candidate porous biomaterials are produced from metal powders, a more comprehensive testing programme will be required.

10.5 Prediction of porous metal properties

10.5.1 Models for mechanical behaviour

The mechanical properties of foams are successfully modelled in many cases by simple relationships derived by considering a regular cubic structure (Fig. 10.8) which deforms by beam bending (Gibson and Ashby, 1997). Some of the simple predictions of this approach are given in Table 10.4, further

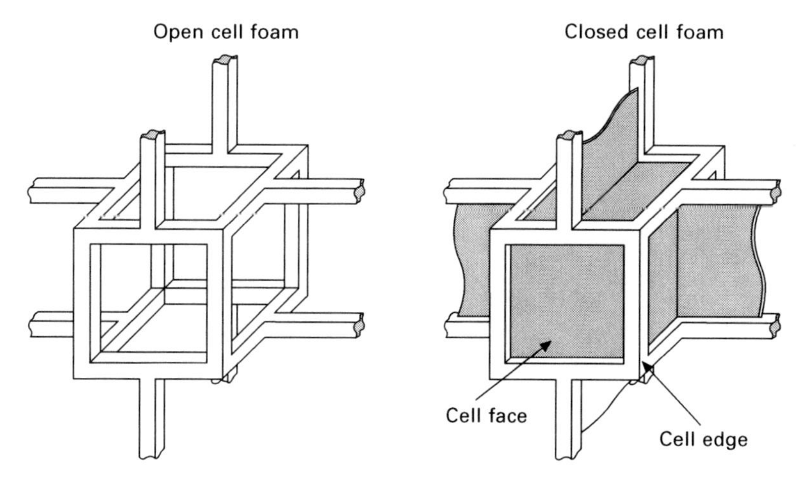

10.8 Regular structures considered by Gibson and Ashby (1997) in deriving their expressions for the properties of foams.

Table 10.4 Expressions for some foam properties derived by Gibson and Ashby (1997). In the table, ρ represents the density, C terms are constants and ϕ is the proportion of the solid in the cell edges as opposed to the faces, with other terms as defined in the table. The superscript * indicates that the property refers to the foam, the subscript s indicates that it refers to the dense metal

Mechanism	Open cell foam	Closed cell foam
Young's modulus, E	$\frac{E^*}{E_s} = C_1 \left(\frac{\rho^*}{\rho_s}\right)^2$	$\frac{E^*}{E_s} = C_1\phi^2 \left(\frac{\rho^*}{\rho_s}\right)^2 + C_1(1-\phi)\frac{\rho^*}{\rho_s}$
Yield strength, σ_y (compression)	$\frac{\sigma_y^*}{\sigma_{ys}} = C_2 \left(\frac{\rho^*}{\rho_s}\right)^{3/2}$	$\frac{\sigma_y^*}{\sigma_{ys}} = C_2 \left(\phi\frac{\rho^*}{\rho_s}\right)^{3/2} + C_2(1-\phi)\frac{\rho^*}{\rho_s}$
Fracture strength, σ_f (compression)	$\frac{\sigma_f^*}{\sigma_{fs}} = C_3 \left(\frac{\rho^*}{\rho_s}\right)^{3/2}$	$\frac{\sigma_f^*}{\sigma_{fs}} = C_3 \left(\phi\frac{\rho^*}{\rho_s}\right)^{3/2} + C_3(1-\phi)\frac{\rho^*}{\rho_s}$

results can be found in Gibson and Ashby (1997). The predictions of the equation for Young's modulus of open cell foams are shown compared to the data in Fig. 10.6.

The Gibson – Ashby model has frequently been used to understand the mechanical response of foams processed from metal powders, for example by Davis *et al.* (2001), Wen *et al.* (2002a), Hakamada *et al.* (2005, 2007) Jiang *et al.* (2005a), Esen and Bor (2007), Chino and Dunand (2008), Niu *et al.* (2009) and Bram *et al.* (2011). However, it does have certain limitations. While it broadly captures the trends, the constants in the equation vary with the precise foam structure and material, and hence have to be measured for each set of experiments, acting like a fitting parameter. For example, many metal foams show mechanical properties that are below the simple Gibson–Ashby predictions (with $C = 1$) and require this constant to be reduced (sometimes this is called the 'knock-down' factor and frequently has a value around 0.3–0.5 (San-Marchi and Mortensen, 2001; Despois *et al.*, 2006)). The reason for this knock down is sometimes attributed to defects and attempts have been made to model their effects (Kepets *et al.*, 2007). For Fig. 10.6, it is clearly seen that some of the porous metals processed from powder, notably the partially sintered powders, follow the model with $C = 1$ reasonably well. Others, particularly some of the space holder examples, require the constant to be nearer 0.3 for a good match.

To go beyond the Gibson–Ashby approach, some workers have used models that may be more specific, but capture the behaviour of a particular foam type more precisely. Other relatively simple analytical models, based on Gibson–Ashby or similar approaches, have been used with some success in situations where they can capture the mechanisms operating in the microstructure (e.g. Li and Dunand, 2011). For sintered fibres, models have been developed for both the thermal behaviour and the stiffness (Clyne *et al.*, 2006), taking the bending of the fibre segment between bonds and accounting for the statistical distribution of these elements at different angles throughout the material. In sintered powders, calculations are sometimes made using the smallest amount of solid cross sectional area normal to the direction of the applied load; this will typically be the necks between the particles (Esen and Bor, 2011). Equations developed using this approach are called minimum surface area models and can work well when the density is high and there are no long ligaments of material that will deform by bending or twisting; this type of behaviour would be better captured by the Gibson–Ashby approach.

Other models have been examined and found to work in specific situations, including Eshelby's theory for isolated inclusions in a matrix (Greiner *et al.*, 2005), which evidently can be applied when the volume fraction porosity is low (around 20% in this case). Others have applied Mori–Tanaka models for the elastic properties of materials containing inclusions (El-Hadek and Kaytbay, 2008) with good correlations, and an extended version of these models indicates that a small amount of microporosity (> 5%), which is frequently present in samples produced from powder, reduces the anisotropy in properties seen when the pores are not uniform in shape and are aligned (as can also happen in materials produced from powders with space holders when a uniaxial pressing step is used) (Gong *et al.*, 2011b).

Another approach that has been shown to work well with foams processed from molten metal in various ways is the use of continuum mechanics approaches developed for composites (treating the porous material as a composite of metal and air) for elastic behaviour and a modified secant modulus method (where the relationship between the elastic behaviour of the foam and the dense metal is used to gauge how the plastic behaviour of the foam will relate to that of the dense metal) for the plastic flow stress (Despois *et al.*, 2006). This has as yet not been tested against porous metals processed from powder.

10.5.2 Models for conduction

In work on electrical conductivity, it has been found that simple models can predict the variation in foam conductivity with density, if the conductivity of the solid is known (the evidence shows that the pore size and shape have little effect, although if the pores are not equiaxed there may be a departure from predictions) (Goodall *et al.*, 2006). The best performing models were those of Lemlich (1978), which was particularly effective for low density foams, with long thin struts and the differential effective medium (DEM) model (Torquato, 2002), which worked well for foams with closed or semi-enclosed pores. The equations for these models are:

Lemlich: $$\frac{K^*}{K_s} = \frac{(1 - V_p)}{3} \qquad [10.1]$$

DEM: $$\frac{K^*}{K_s} = (1 - V_p)^n \qquad [10.2]$$

where K is the conductivity, V_p is the volume fraction porosity and n is an exponent relating to pore shape. The superscript * indicates that the property refers to the foam, the subscript s indicates that it refers to the dense metal. In the DEM, if the pores are spherical, n takes a value 1.5. If the pores are non-spherical, the value will be >1.5. with 1.75 being a typical value for slightly ellipsoidal pores (Goodall *et al.*, 2006).

The predictions of these equations are shown on the graph in Fig. 10.7. Porous metals processed from powders fall below the conductivity predicted by the DEM model, which tends to better represent foams produced by other means with liquid metals as a starting point. The suggested explanation for this is that a powder metallurgical route is likely to include a certain amount of oxide from the surface of the powder, reducing the effective conductivity of the dense metal (Goodall *et al.*, 2006). The conductivity of powder metallurgical foams is somewhat closer to the Lemlich model at high porosities, although it climbs above this when the porosity falls to low values and the pores are isolated; this is unsurprising as the Lemlich model considers high porosity open celled structures with long straight struts.

10.5.3 Simulations

With the structure being such an important factor in determining the properties shown by a foam, and the fact that the structure has a complex, stochastic nature, research has included the use of simulations (such as finite element methods) on real and artificial foam structures to attempt to understand the behaviour. Finite element simulations of mechanical properties on a global and local scale have been performed for titanium foams processed by argon gas trapping (Shen *et al.*, 2006), using as inputs the results of simulations of the growth of pores in 3D. These structures are arrived at from real samples early in the process, where 2D sections are used to capture the distribution of the initially nucleated pores and as such should be superior to artificially generated porous structures. Overall the comparison to experimental stress–strain curves is good (Shen *et al.*, 2006). Finite element models of idealised lattices have also been used to study the effects of defects in the structure (Kepets *et al.*, 2007).

10.6 Future perspectives

Research in metal foams is still highly active, with groups in the UK, Germany, the USA, Japan and elsewhere using powder metallurgical techniques to produce samples, with the same goal of developing new materials and processes, and assisting the transition from research into applications. Recently this research has taken on developments in powder bed rapid manufacturing technologies and has started to produce more of the regular lattice structures that can be easily fabricated using these methods. It is likely that this trend will continue (particularly in view of the rapid developments being made in the processing technology), as an ordered structure is more highly predictable and the possibilities for tailoring the structure to an unprecedented degree of precision are attractive for optimisation of the materials for particular properties and applications. This is starting to include the concept of non-uniform lattices, where the structure varies with location in the component, allowing different behaviours to be prioritised in different parts.

Another factor that is likely to be an important future development is the creation of porous materials made from multiple materials. By using two or more materials, it is possible to engineer structures with unusual properties, such as the zero or negative thermal expansion coefficient structures proposed by Lakes (1996, 2007). The creation of systems with such behaviour on a scale small enough to be considered engineered materials, rather than structures, would open up a new range of exciting possibilities for materials scientists to work with. This will require new advances in the processing areas described here, along with the exploration of new techniques and, importantly, how they may be translated into industrial processes. A thorough understanding of the processes behind the behaviour is also essential, so that the material design process can proceed with knowledge of the affect of the many available variables.

10.7 References

Ahsan, M. N., Paul, C. P., Kukreja, L. M. and Pinkerton, A. J. (2011) 'Porous structures fabrication by continuous and pulsed laser metal deposition for biomedical applications; modelling and experimental investigation'. *J. Mater. Process. Technol.*, **211**, 602–9.

Andersen, O., Waag, U., Schneider, L., Stephani, G. and Kieback, B. (2000) 'Novel metallic hollow sphere structures'. *Adv. Eng. Mat.*, **2**, 192–5.

Angel, S., Beck, W., Harksen, S. and Scholz, P. F. (2005) 'Functional and structural characterisitics of metallic foams produced by the slip reaction foam sintering (SRFS)-process'. *Mat. Sci. Forum*, **492–3**, 39–45.

Arockiasamy, A., Park, S. J. and German, R. M. (2010) 'Viscoelastic behaviour of porous sintered steels compact'. *Powder Metall.*, **53**, 107–11.

Arzt, E. (1982) 'The influence of an increasing particle coordination on the densification of spherical powders'. *Acta Metall.*, **30**, 1883–90.

Ashby, M. F., Evans, A. G., Fleck, N. A., Gibson, L. J., Hutchinson, J. W. and Wadley, H. N. G. (2000) *Metal Foams: A Design Guide*, Butterworth-Heinemann, Boston.

Bakan, H. I. (2006) A novel water leaching and sintering process for manufacturing highly porous stainless steel. *Scr. Mater.*, **55**, 203–206.

Banhart, J. (2001) Manufacture, characterisation and application of cellular metals and metal foams. *Prog. Mat. Sci.*, **46**, 599–632.

Banhart, J., Fleck, N. and Mortensen, A. (2003) Cellular metals: manufacture, properties, applications, *Proceedings Conference Metfoam 2003*, Berlin, 23–25 June 2003, Verlag MIT Publishing, Berlin, Germany.

Bansiddhi, A. and Dunand, D. C. (2007) Shape-memory NiTi foams produced by solid-state replication with NaF. *Intermetallics*, **15**, 1612–1622.

Bansiddhi, A. and Dunand, D. C. (2008) Shape-memory NiTi foams produced by replication of NaCl space-holders. *Acta Biomater.*, **4**, 1996–2007.

Bansiddhi, A. and Dunand, D. C. (2009) Shape-memory NiTi–Nb foams. *J. Mater. Res.*, **24**, 2107–2117.

Bansiddhi, A., Sargeant, T. D., Stupp, S. I. and Dunand, D. C. (2008) Porous NiTi for bone implants: a review. *Acta Biomater.*, **4**, 773–782.

Bin, J., Zejun, W. and Naiqin, Z. (2007) Effect of pore size and relative density on the mechanical properties of open cell aluminium foams. *Scr. Mater.*, **56**, 169–172.

Biswas, A. (2005) Porous NiTi by Thermal Explosion Mode of SHS: Processing, Mechanism and Generation of Single Phase Microstructure. *Acta Mater.*, **53**, 1415–1425.

Bram, M., Stiller, C., Buchkremer, H. P., Stover, D. and Baur, H. (2000) High-porosity titanium, stainless steel and superalloy parts. *Adv. Eng. Mat.*, **2**, 196–199.

Bram, M., Kohl, M., Buchkremer, H. P. and Stover, D. (2011) Mechanical properties of highly porous NiTi alloys. *J. Mater. Eng. Perform.*, **24**, 522–528.

Cansizoglu, O., Harrysson, O., Cormier, D., H. West and Mahale, T. (2008) 'Properties of Ti–6Al–4V non-stochastic lattice structures fabricated via electron beam melting'. *Mater. Sci. Eng. A*, **492**, 468–74.

Chen, L. J., Li, T., Li, Y. M., He, H. and Hu, Y. H. (2009) Porous titanium implants

fabricated by metal injection molding. *Trans. Nonferrous Met. Soc. China*, **19**, 1174–1179.
Chino, Y. and Dunand, D. C. (2008) Directionally freeze-cast titanium foam with aligned, elongated pores. *Acta Mater.*, **56**, 105–113.
Clyne, T. W., Golosnoy, I. O., Tan, J. C. and Markaki, A. E. (2006) Porous materials for thermal management under extreme conditions. *Phil. Trans. R. Soc. A*, **364**, 125–146.
Colombo, P. and Degischer, H. P. (2010) Highly porous metals and ceramics. *Mater. Sci. Technol.*, **26**, 1145–1158.
Conde, Y., Despois, J.-F., Goodall, R., Marmottant, A., Salvo, L., Marchi, C. S. and Mortensen, A. (2006) Replication processing of highly porous materials. *Adv. Eng. Mat.*, **8**, 795–803.
Cumberland, D. J. and Crawford, R. J. (1987) *The Packing of Particles*, Elsevier, Amsterdam.
Davis, N. G., Teisen, J., Schuh, C. and Dunand, D. C. (2001) Solid state foaming of titanium by superplastic expansion of argon-filled pores. *J. Mater. Res.*, **16**, 1508–1519.
Delannay, F. (2005) Elastic model of an entangled network of interconnected fibres accounting for negative Poisson ratio behaviour and random triangulation. *Int. J. Solids Struct.*, **42**, 2265.
Despois, J.-F., Mueller, R. and Mortensen, A. (2006) Uniaxial deformation of microcellular metals. *Acta Mater.*, **54**, 4129–4142.
Diologent, F., Goodall, R. and Mortensen, A. (2009) Surface oxide in replicated microcellular aluminium and its influence on the plasticity size effect. *Acta Mater.*, **57**, 286–94.
Dong, H. X., Jiang, Y., He, Y. H., Song, M., Zou, J., Xu, N. P., Huang, B. Y., Liu, C. T. and Liaw, P. K. (2009) *J. Alloys Comp.*, **484**, 907.
Driscoll, D., Weisenstein, A. J. and Sofie, S. W. (2011) 'Electrical and flexural anisotropy in freeze tape cast stainless steel porous substrates'. *Mater. Lett.*, **65**, 3433–5.
Ducheyne, P., Aernoudt, E. and De_Meester, P. (1978) The mechanical behaviour of porous austenitic stainless steel fibre structures. *J. Mater. Sci.*, **13**, 2650–2658.
Dunand, D. C. (2004) Processing of titanium foams. *Adv. Eng. Mat.*, **6**, 369–376.
Dunand, D. C. and Teisen, J. (1998) Superplastic foaming of titanium and Ti-6Al-4V. *Porous and Cellular Materials for Structural Applications*. In Schwartz, D. S., Shih, D. H., Evans, A. G. and Wadley, H. G. (eds) San Francisco, CA, USA, Materials Research Society Symposium Proceedings, Vol. 521, Warrendale PA, USA.
El-Hadek, M. A. and Kaytbay, S. (2008) Mechanical and physical characterization of copper foam. *Int. J. Mech. Mater. Des.*, **4**, 63–69.
Engin, G., Aydemir, B. and Gülsoy, H. Ö. (2011) Injection molding of micro-porous titanium alloy with space holder technique. *Rare Metals*, **30**, 565–571.
Esen, Z. and Bor, S. (2007) Processing of titanium foams using magnesium spacer particles. *Scr. Mater.*, **56**, 341–344.
Esen, Z. and Bor, S. (2011) Characterization of Ti–6Al–4V alloy foams synthesized by space holder technique. *Mater. Sci. Eng. A*, 528, 3200–3209.
Fedorchenko, I. M. (1979) Progress in work in the field of high-porosity materials from powders and fibers. *Soviet Powder Metall. Met. Ceram.*, **18**, 615–622.
Fife, J. L., Li, J. C., Dunand, D. C. and Voorhees, P. W. (2009) Morphological analysis of pores in directionally freeze-cast titanium foams. *J. Mater. Res.*, **24**, 117–124.
Fu, Y. Q., Gu, Y. W., Shearwood, C., Luo, J. K., Flewitt, A. J. and Milne, W. I. (2006) Spark plasma sintering of TiNi nano-powders for biological application. *Nanotechology*, **17**, 5293–5298.

Gibson, L. J. and Ashby, M. F. (1997) *Cellular Solids*, Cambridge University Press, Cambridge.
Gong, S., Li, Z., Xu, G. Y., Liu, N., Zhao, Y. Y. and Liang, S. Q. (2011a) 'Fabrication, microstructure and property of cellular CuAlMn shape memory alloys produced by sintering–evaporation process'. *J. Alloys Comp.*, **509**, 2924–8.
Gong, S., Li, Z. and Zhao, Y. Y. (2011b) An extended Mori–Tanaka model for the elastic moduli of porous materials of finite size. *Acta Mater.*, **59**, 6820–6830.
Goodall, R. and Mortensen, A. (2013) Porous Metals. In *Physical Metallurgy*, 5th Edition. Laughlin, D. and Hono, K. (eds), Netherlands Elsevier, Amsterdam.
Goodall, R., Weber, L. and Mortensen, A. (2006) The electrical conductivity of microcellular metals. *J. Appl. Phys.*, **100**, 044912.
Goodstein, D. L., Mccormick, W. D. and Dash, J. G. (1966) Sintered copper sponges for use at low temperature. *Cryogenics*, **6**, 167–168.
Greiner, C., Oppenheimer, S. M. and Dunand, D. C. (2005) High strength, low stiffness, porous NiTi with superelastic properties. *Acta Biomaterialia*, **1**, 705–716.
Gulsoy, H. O. and German, R. M. (2008) Sintered foams from precipitation hardened stainless steel powder. *Powder Metall.*, **51**, 350–353.
Guo, Z. X., Jee, C. S. Y., Ozguven, N. and Evans, J. R. G. (2000) Novel polymer-metal based method for open cell metal foams production. *Mat. Sci. Tech.*, **16**, 776–780.
Guo, S., Duan, B., He, X. and Qu, X. (2009) Powder injection molding of pure Titanium. *Rare Metals*, **28**, 261–265.
Guoxin, H., Lixiang, Z., Yunliang, F. and Yanhong, L. (2008) Fabrication of high porous NiTi shape memory alloy by metal injection molding. *J. Mater. Process. Technol.*, **206**, 395–399.
Hakamada, M., Nomura, T., Yamada, Y., Chino, Y., Hosokawa, H., Nakajima, T., Chen, Y., Kusuda, H. and Mabuchi, M. (2005a) 'Compressive properties at elevated temperatures of porous aluminium processed by the spacer method'. *J. Mater. Res.*, **20**, 3385–90.
Hakamada, M., Yamada, Y., Nomura, T., Kusuda, H., Chen, Y. and Mabuchi, M. (2005b) Effect of sintering temperature on compressive properties of porous aluminum produced by spark plasma sintering. *Mater. Trans.*, **46**, 186–188.
Hakamada, M., Asao, Y., Kuromura, T., Chen, Y., Kusuda, H. and Mabuchi, M. (2007) Density dependence of the compressive properties of porous copper over a wide denisty range. *Acta Mater.*, **55**, 2291–2299.
He, Y., Jiang, Y., Xu, N., Zou, J., Huang, B., Liu, C. T. and Law, P. K. (2007) Fabrication of Ti-Al micro/nanometer-sized porous alloys through the Kirkendall effect. *Adv. Mater.*, **19**, 2102–2106.
Heinl, P., Rottmair, A., Körner, C. and Singer, R. F. (2007) Cellular titanium by selective electron beam melting. *Adv. Eng. Mater.*, **9**, 360–364.
Heinl, P., Körner, C. and Singer, R. F. (2008a) Selective electron beam melting of cellular titanium. *Adv. Eng. Mater.*, **10**, 882–888.
Heinl, P., Muller, L., Korner, C., Singer, R. F. and Muller, F. A. (2008b) Cellular Ti–6Al–4V structures with interconnected macro porosity for bone implants fabricated by selective electron beam melting. *Acta Biomater.*, **4**, 1536–1544.
Higginson, R. and Sellars, M. (2003) *Quantitative Metallography*, Maney, London.
Hodge, A. M. and Dunand, D. C. (2003) Measuring and modelling of creep in open-cell NiAl foams. *Met. Mater. Trans. A*, **34**, 2353–2363.
Hong, T. F., Guo, Z. X. and Yang, R. (2008) Fabrication of porous titanium scaffold materials by a fugitive filler method. *J. Mater. Sci: Mater. Med.*, **19**, 3489–3495.

Ismail, M. H., Goodall, R., Davies, H. A. and Todd, I. (2011) Porous NiTi alloy by metal injection moulding/sintering of elemental powders: Effect of sintering temperature. *Mater. Lett.*, **70**, 142–145.

Jee, C. S. Y., Ozguven, N., Guo, Z. X. and Evans, J. R. G. (2000) Preparation of high porosity metal foams. *Met. Mater. Trans. B*, **31**, 1345–1352.

Jiang, B., Zhao, N. Q., Shi, C. S., Du, X. W., Li, J. J. and Man, H. C. (2005a) 'A novel method for making open cell aluminium foams by powder sintering process'. *Mater. Lett.*, **59**, 3333–6.

Jiang, B., Zhao, N. Q., Shi, C. S. and Li, J. J. (2005b) Processing of open cell aluminium foams with tailored porous morphology. *Scr. Mater.*, **53**, 781–785.

Kanetake, N. and Kobashi, M. (2006) Innovative processing of porous and cellular materials by chemical reaction. *Scr. Mater.*, **54**, 521–525.

Kearns, M. W., Blenkinsop, P. A., Barber, A. C. and Farthing, T. W. (1988) Manufacture of a novel porous metal. *Int. J. Powder Metall.*, **24**, 59–64.

Kepets, M., Lu, T. J. and Dowling, A. P. (2007) Modeling of the role of defects in sintered FeCrAlY foams. *Acta Mech. Sin*, **23**, 511–529.

Kim, J. S., Kang, J. H., Kang, S. B., Yoon, K. S. and Kwon, Y. S. (2004) Porous TiNi biomaterial by self-propagating high-temperature synthesis. *Adv. Eng. Mat.*, **6**, 403–406.

Kobashi, M. and Kanetake, N. (2002) Processing of intermetallic foam by combustion reaction. *Adv. Eng. Mat.*, **4**, 745–747.

Kobashi, M., Kuze, K. and Kanetake, N. (2006a) Cell structure control of porous titanium composite synthesised by combustion reaction. *Adv. Eng. Mat.*, **8**, 836–840.

Kobashi, M., Wang, R. X., Inagaki, Y. and Kanetake, N. (2006b) Effects of processing parameters on pore morphology of combusion synthesised Al-Ni foams. *Mater. Trans.*, **47**, 2172–2177.

Köhl, M., Habijan, T., Bram, M., Buchkremer, H. P., Stover, D. and Koller, M. (2009) Powder metallurgical near-net-shape fabrication of porous NiTi shape memory alloys for use as long-term implants by the combination of the metal injection molding process with the space-holder technique. *Adv. Eng. Mater.*, **11**, 959–968.

Köhl, M., Bram, M., Moser, A., Buchkremer, H. P., Beck, T. and Stöver, D. (2011) Characterization of porous, net-shaped NiTi alloy regarding its damping and energy-absorbing capacity. *Mater. Sci. Eng. A*, **528**, 2454–2462.

Lagoudas, D. C. and Vandygriff, E. L. (2002) 'Processing and characterization of NiTi porous SMA by elevated pressure sintering. *J. Intell. Mater. Syst. Struct.*, **13**, 837–850.

Lakes, R. (1996) Cellular solid structures with unbounded thermal expansion. *J. Mater. Sci. Lett.*, **15**, 475–477.

Lakes, R. (2007) Cellular solids with tunable positive or negative thermal expansion of unbounded magnitude. *Appl. Phys. Lett.*, **90**, 221905.

Laptev, A., Bram, M., Buchkremer, H. P. and Stover, D. (2004) Study of production route for titanium parts combining very high porosity and complex shapes. *Powder Metall.*, **47**, 85–92.

Laptev, A., Vyal, O., Bram, M., Buchkremer, H. P. and Stover, D. (2005) Green strength of powder compacts provided for production of highly porous titanium parts. *Powder Metall.*, **48**, 358–364.

Lefebvre, L. P., Banhart, J. and Dunand, D. C. (2007) 'Porous metals and metallic foams', *Proceedings Conference Metfoam 2007*. Montreal, Sept. 5–7, 2007, DESTech Publications, Lancaster, PA.

Lemlich, R. (1978) A theory for the limiting conductivity of polyhedral foam at low density. *J. Colloid Interface Sci.*, **64**, 107–110.
Li, J. C. and Dunand, D. C. (2011) 'Mechanical properties of directionally freeze-cast titanium foams'. *Acta Mater.*, **59**, 146–58.
Li, B. Q. and Lu, X. (2011) The effect of pore structure on the electrical conductivity of Ti. *Trans. Porous Media*, **87**, 179–189.
Li, C.-F. and Zhu, Z.-G. (2005) Apparent electrical conductivity of porous titanium prepared by the powder metallurgy method. *Chinese Phys. Lett.*, **22**, 2647–2650.
Li, Y. H., Rong, L. J. and Li, Y. Y. (2002) Compressive property of porous NiTi alloy synthesized by combustion synthesis. *J. Alloys Comp*, **345**, 271–274.
Li, J. P., Li, S. H., Blitterswijk, C. A. V. and Groot, K. D. (2005) A novel porous Ti6Al4V: Characterization and cell attachment. *J. Biomed. Mater. Res. A*, **73**, 223–233.
Li, J. P., Wijn, J. R. D., Blitterswijk, C. A. V. and Groot, K. D. (2006) Porous Ti6Al4V scaffold directly fabricating by rapid prototyping: preparation and *in vitro* experiment. *Biomaterials*, **27**, 1223–1235.
Li, D. S., Zhang, X. P., Xiong, Z. P. and Mai, Y. W. (2010) Lightweight NiTi shape memory alloy based composites with high damping capacity and high strength. *J. Alloys. Comp.*, **490**, L15–L19.
Li, S. J., Murr, L. E., Cheng, X. Y., Zhang, Z. B., Hao, Y. L., Yang, R., Medina, F. and Wicker, R. B. (2012) Compression fatigue behavior of Ti–6Al–4V mesh arrays fabricated by electron beam melting. *Acta Mater*, **60**, 793–802.
Liang, W., Jiang, Y., Hongxing, D., He, Y., Xu, N., Zou, J., Huang, B. and Liu, C. T. (2011) The corrosion behavior of porous Ni3Al intermetallic materials in strong alkali solution. *Intermetallics*, **19**, 1759–1765.
Lide, D. R. (2001) *CRC Handbook of Chemistry and Physics*, CRC Press, Boca Raton, FL.
Lim, T. J., Smith, B. and Mcdowell, D. L. (2002) Behaviour of a random hollow sphere metal foam. *Acta Mater.*, **50**, 2867–2879.
Liu, P. S. and Liang, K. M. (2001) Functional materials of porous metals made by P/M, electroplating and some other techniques. *J. Mater. Sci.*, **36**, 5059–5072.
Ma, X., Peyton, A. and Zhao, Y. (2005) Measurement of the electrical conductivity of open-celled aluminium foam using non-contact eddy current techniques. *NDTandE Int.*, **38**, 359–367.
Ma, X., Peyton, A. J. and Zhao, Y. Y. (2006) Eddy current measurements of electrical conductivity and magnetic permeability of porous metals. *NDTandE Int.*, **39**, 562–568.
Manonukul, A., Muenya, N., Léaux, F. and Amaranan, S. (2010) 'Effects of replacing metal powder with powder space holder on metal foam produced by metal injection moulding'. *J. Mater. Process. Technol.*, **210**, 529–35.
Mansourighasri, A., Muhamad, N. and Sulong, A. B. (2012) Processing titanium foams using tapioca starch as a space holder. *J. Mater. Process. Technol.*, **121**, 83–89.
Markaki, A. E., Gergely, V., Cockburn, A. and Clyne, T. W. (2003) Production of a highly porous material by liquid phase sintering of short ferritic stainless steel fibres and a preliminary study of its mechanical behaviour. *Compos. Sci. Technol.*, **63**, 2345–2351.
Mi, G. F., Li, H. Y., Liu, X. Y. and Wang, K. F. (2009) Preparation, Structure and Mechanical Properties of Nickel Based Porous Spherical Superalloy. *J. Iron Steel Res., Int.*, **16**, 92–96.
Mueller, R., Soubielle, S., Goodall, R., Diologent, F. and Mortensen, A. (2007) On the steady state creep of microcellular metals. *Scr. Mater.*, **57**, 33–6.

Murr, L. E., Gaytan, S. M., Medina, F., Lopez, H., Martinez, E., Machado, B. I., Hernandez, D. H., Martinez, L., Lopez, M. I., Wicker, R. B. and Bracke, J. (2010a) Next-generation biomedical implants using additive manufacturing of complex, cellular and functional mesh arrays. *Phil. Trans. R. Soc. A*, **368**, 1999–2032.

Murr, L. E., Gaytan, S. M., Medina, F., Martinez, E., Martinez, J. L., Hernandez, D. H., Machado, B. I., Ramirez, D. A. and Wicker, R. B. (2010b) Characterization of Ti–6Al–4V open cellular foams fabricated by additive manufacturing using electron beam melting. *Mater. Sci. Eng. A*, **527**, 1861–1868.

Murr, L. E., Amato, K. N., Li, S. J., Tian, Y. X., Cheng, X. Y., Gaytan, S. M., Martinez, E., Shindo, P. W., Medina, F. and Wicker, R. B. (2011) Microstructure and mechanical properties of open-cellular biomaterials prototypes for total knee replacement implants fabricated by electron beam melting. *J. Mech. Behav. Biomed. Mater.*, **4**, 1396–1411.

Murray, N. G. D. and Dunand, D. C. (2003) 'Microstructure evolution during solid-state foaming of titanium'. *Comp. Sci. Tech.*, **63**, 2311–16.

Murray, N. G. D. and Dunand, D. C. (2004) Effect of thermal history on the superplastic expansion of argon-filled pores in titanium: Part I Kinetics and microstructure. *Acta Mater.*, **52**, 2269–2278.

Murray, N. G. D., Schuh, C. A. and Dunand, D. C. (2003) Solid-state foaming of titanium by hydrogen-induced internal-stress superplasticity. *Scr. Mater.*, **49**, 879–883.

Nabavi, A. and Khaki, J. V. (2011) Manufacturing of aluminum foam sandwich panels: comparison of a novel method with two different conventional methods. *J. Sandwich Struct. Mater.*, **13**, 177–187.

Nagel, A. R., Uslu, C., Lee, K. J., Cochran, J. K. and Sanders, T. H. (1997) Steel closed cell foams from direct oxide reduction. In *Synthesis/Processing of Lightweight Metallic Materials II*. Ward-Close, C. M., Froes, F. H., Chellman, D. J. and Cho, S. S. (eds) The Minerals, Metals and Materials Soc., Warrendale, PA, USA.

Nakajima, H. and Kanetake, N. (2005) Porous metals and metal foaming technology, *Proceedings Conference MetFoam* 2005 Kyoto, Japan, Sept. 21-23rd 2005, Japan Institute of Metals, Sendai, Japan.

Niu, W., Bai, C., Qiu, G., Wang, Q., Wen, L., Chen, D. and Dong, L. (2009) Preparation and characterization of porous titanium using space-holder technique. *Rare Metals*, **28**, 338–342.

Nomura, N., Kohama, T., Oh, I. H., Hanada, S., Chiba, A., Kanehira, M. and Sasaki, K. (2005) Mechanical properties of porous Ti–15Mo–5Zr–3Al compacts prepared by powder sintering. *Mater. Sci. Eng. C*, **25**, 330–335.

Nouri, A., Hodgson, P. D. and Wen, C. E. (2010) Effect of process control agent on the porous structure and mechanical properties of a biomedical Ti–Sn–Nb alloy produced by powder metallurgy. *Acta Biomater.*, **6**, 1630–1639.

Oh, I. K., Nomura, N., Masahashi, N. and Hanada, S. (2003) Mechanical properties of porous titanium compacts prepared by powder sintering. *Scr. Mater.*, **49**, 1197–1202.

Oppenheimer, S. M., O'Dwyer, J. G. and Dunand, D. C. (2004) 'Porous, superelastic NiTi produced by powder metallurgy'. *TMS Lett.*, **1**, 93–4.

Pattanayak, D. K., Fukuda, A., Matsushita, T., Takemoto, M., Fujibayashi, S., Sasaki, K., Nishida, N., Nakamura, T. and Kokubo, T. (2011) Bioactive Ti metal analogous to human cancellous bone: Fabrication by selective laser melting and chemical. *Acta Biomater.*, **7**, 1398–1406.

Queheillalt, D. T., Katsumura, Y. and Wadley, H. N. G. (2004) Synthesis of stochastic open cell Ni-based foams. *Scr. Mater.*, **50**, 313–317.

Rak, Z. S. and Walter, J. (2006) Porous titanium foil by tape casting technique. *J. Mater. Proc. Tech.*, **175**, 358–363.

Rak, Z., Berkeveld, L. D. and Snijders, G. (2003) *Method for producing a porous titanium material article*. International Patent WO 03/092933A1.

Ramirez, D. A., Murr, L. E., Li, S. J., Tian, Y. X., Martinez, E., Martinez, J. L., Machado, B. I., Gaytan, S. M., Medina, F. and Wicker, R. B. (2011) Open-cellular copper structures fabricated by additive manufacturing using electron beam melting. *Mater. Sci. Eng. A*, **528**, 5379–5386.

Ricceri, R. and Matteazzi, P. (2003) PM processing of cellular titanium. *Int. J. Powder Metall.*, **39**, 53–61.

Roy, S., Wanner, A., Beck, T., Studnitzky, T. and Stephani, G. (2011) Mechanical properties of cellular solids produced from hollow stainless steel spheres. *J. Mater. Sci.*, **46**, 5519–5526.

San-Marchi, C. and Mortensen, A. (2001) Deformation of open-cell aluminium foam. *Acta Mater.*, **49**, 3959–3969.

Schwartz, D. S., Shih, D. S., Lederich, R. J., Martin, R. L. and Deuser, D. A. (1998) Development and scale-up of the low density core process for Ti-64. In *Porous and Cellular Materials for Structural Applications*. Schwartz, D. S., Shih, D. H., Evans, A. G. and Wadley, H. G. (eds) San Francisco, CA, USA, Materials Research Society Symposium Proceedings Vol. 521, Warrendale PA, USA.

Schwerdtfeger, J., Heinl, P., Singer, R. F. and Körner, C. (2010) Auxetic cellular structures through selective electron-beam melting. *Phys. Status Solidi*, **247**, 269–272.

Scott, J. A. and Dunand, D. C. (2010) 'Processing and mechanical properties of porous Fe–26Cr–1Mo for solid oxide fuel cell interconnects'. *Acta Mater.*, **58**, 6125–33.

Shen, H., Oppenheimer, S. M., Dunand, D. C. and Brinson, L. C. (2006) Numerical modelling of pore size and distribution in foamed titanium. *Mech. Mater.*, **38**, 933–944.

Skochdopole, R. E. (1961) The thermal conductivity of foamed plastics. *Chem. Eng. Prog.*, **57**, 55–59.

Song, Z. and Kishimoto, S. (2006) The cell size effect of closed cellular materials fabricated by pulse current assisted hot isostatic pressing on the compressive behaviour. *Scr. Mater.*, **54**, 1531–1535.

Sun, D. X. and Zhao, Y. Y. (2003) Static and dynamic energy absorption of Al foams produced by the sintering and dissolution process. *Metall. Mater. Trans. B*, **34**, 69–74.

Sun, D. X. and Zhao, Y. Y. (2005a) Phase changes in sintering of Al/Mg/NaCl compacts of manufacturing Al foams by the sintering and dissolution process. *Mat. Lett.*, **59**, 6–10.

Sun, D. X. and Zhao, Y. Y. (2005b) Simulation of thermal diffusivity of Al/NaCl powder compacts in producing Al foams by the sintering and dissolution process. *J. Mater. Process Tech.*, **169**, 83–88.

Tan, J. C., Elliott, J. A. and Clyne, T. W. (2006) Analysis of tomography images of bonded fibre networks to measure distributions of fibre segment length and fibre orientation. *Adv. Eng. Mater.*, **8**, 495–500.

Thewsey, D. J. and Zhao, Y. Y. (2008) Thermal conductivity of porous copper manufactured by the lost carbonate sintering process. *Phys. Status Solid A*, **205**, 1126–1131.

Thieme, M., Wieters, K. P., Bergner, F., Scharnweber, D., Worch, H., Ndop, J., Kim, T. J. and Grill, W. (2001) Titanium powder sintering for preparation of a porous functionally graded material destined for orthopaedic implants. *J. Mater. Sci.: Mater. Med.*, **12**, 225–231.

Torquato, S. (2002) *Random Heterogeneous Media*, Springer, New York.
Tuncer, N., Arslan, G., Maire, E. and Salvo, L. (2011) Investigation of spacer size effect on architecture and mechanical properties of porous titanium. *Mater. Sci. Eng. A*, **530**, 633–642.
Uslu, C., Lee, K. J., Sanders, T. H. and Cochran, J. K. (1997) 'Ti-6Al-4V hollow sphere foams'. In *Synthesis/Processing of Lightweight Metallic Materials II*. Ward-Close, C. M., Froes, F. H., Chellman, D. J. and Cho, S. S. (eds) The Minerals, Metals and Materials Soc., Warrendale, PA, USA.
Vanleeuwen, B. K., Darling, K. A., Koch, C. C. and Scattergood, R. O. (2011) Novel technique for the synthesis of ultra-fine porosity metal foam via the inclusion of condensed argon through cryogenic mechanical alloying. *Mater. Sci. Eng. A*, **528**, 2192–2195.
Vannoort, R. (2012) The future of dental devices is digital. *Dental Mater.*, **28**, 3–12.
Wang, N., Starke, E. A. and Wadley, H. N. G. (1998) Porous Al alloys by local melting and diffusion of metal powders. In *Porous and Cellular Materials for Structural Applications*. Schwartz, D. S., Shih, D. H., Evans, A. G. and Wadley, H. G. (eds) San Francisco, CA, USA, Materials Research Society Symposium Proceedings Vol. 521, Warrendale PA, USA.
Wen, C. E., Mabuchi, M., Yamada, Y., Shimojima, K., Chino, Y. and Asahina, T. (2001) Processing of biocompatible porous Ti and Mg. *Scr. Mater.*, **45**, 1147–1153.
Wen, C. E., Mabuchi, M., Yamada, Y., Shimojima, K., Chino, Y., Hosokawa, H. and Asahina, T. (2003) Processing of fine-grained aluminium foam by spark plasma sintering. *J. Mat. Sci. Lett.*, **22**, 1407–1409.
Wen, C. E., Yamada, Y., Shimojima, K., Chino, Y., Asahina, T. and Mabuchi, M. (2002a) Processing and mechanical properties of autogenous titanium implant materials. *J. Mater. Sci.: Mater. Med.*, **13**, 397–401.
Wen, C. E., Yamada, Y., Shimojima, K., Chino, Y., Hosokawa, H. and Mabuchi, M. (2002b) Novel titanium foam for bone tissue engineering. *J. Mater. Res.*, **17**, 2633–2639.
Wen, C. E., Yamada, Y., Shimojima, K., Chino, Y., Hosokawa, H. and Mabuchi, M. (2004) Compressibility of porous magnesium foam: dependency on porosity and pore size. *Mater. Lett.*, **58**, 357–360.
Wen, C. E., Xiong, J. Y., Li, Y. C. and Hodgson, P. D. (2010) 'Porous shape memory scaffolds for biomedical applications: a review', *Phys. Scr.*, T139, 014070.
Xie, S. and Evans, J. R. G. (2004) High porosity copper foam. *J. Mat. Sci.*, **39**, 5877–5880.
Xue, X. B. and Zhao, Y. Y. (2011) 'Ti matrix syntactic foam fabricated by powder metallurgy: particle breakage and elastic modulus'. *JOM*, **63**, 43–7.
Ye, B. and Dunand, D. C. (2010) Titanium foams produced by solid-state replication of NaCl powders. *Mater. Sci. Eng. A*, **528**, 691–697.
Yu, A. B. and Standish, N. (1993) Characterisation of non-spherical particles from their packing behaviour. *Powder Technol.*, **74**, 205–213.
Yuan, B., Chung, C. Y. and Zhu, M. (2004) Microstructure and matrensitic transformation behaviour of porous NiTi shape memory alloy prepared by hot isostatic pressing processing. *Mat. Sci. Eng. A*, **382**, 181–187.
Yuan, B., Zhu, M., Gao, Y., Li, X. and Chung, C. Y. (2008) Forming and control of pores by capsule-free hot isostatic pressing in NiTi shape memory alloys. *Smart. Mater. Struct.*, **17**, 025013.
Zhao, Y. Y. (2003) Stochastic modelling of removability of NaCl in sintering and dissolution process to produce Al foams. *J. Porous Mater*, **10**, 105–111.

Zhao, Y. Y. and Sun, D. X. (2001) A novel sintering-dissolution process for manufacturing Al foams. *Scr. Mater.*, **44**, 105–110.
Zhao, Y. Y., Han, F. and Fung, T. (2004) Optimisation of compaction and liquid-state sintering in sintering and dissolution process for manufacturing Al foams. *Mat. Sci. Eng. A*, **364**, 117–125.
Zhao, Y. Y., Fung, T., Zhang, L. P. and Zhang, F. L. (2005) Lost carbonate sintering process for manufacturing metal foams. *Scr. Mater.*, **52**, 295–298.
Zhao, X., Sun, H., Lan, L., Huang, J., Zhang, H. and Wang, Y. (2009) Pore structures of high-porosity NiTi alloys made from elemental powders with NaCl temporary space-holders. *Mater. Lett.*, **63**, 2402–2404.
Zhuang, H., Han, Y. and Feng, A. (2008) Preparation, mechanical properties and in vitro biodegradation of porous magnesium scaffolds. *Mater. Sci. Eng. C*, **28**, 1462–1466.

11
Evolution of microstructure in ferrous and non-ferrous materials

H. DANNINGER, C. GIERL-MAYER and S. STROBL, Vienna University of Technology, Austria

DOI: 10.1533/9780857098900.2.308

Abstract: The most prominent features of powder metallurgy (PM) materials are their fine and regular microstructure and in many cases their porosity. Here, it is shown how the porosity changes with manufacturing parameters in sintered materials and how preparation has to be done to avoid artefacts. The matrix microstructures, with regard to the alloying technique and resulting element distribution, and the microstructural development during sintering of powder injection moulded products are described. The fine homogeneous microstructure is a typical feature of fully dense PM materials as shown for tool steels and hard metals. The pronounced effect of doping elements on microstructural stability is presented for PM refractory metals.

Key words: homogeneity, phases, porosity, powder metallurgy microstructures, sintering.

11.1 Introduction

Powder metallurgy (PM) products have a different manufacturing route from standard metallic components and this has a strong impact on the microstructure and resulting properties. The fact that in PM processing, none or, at best, only a small fraction of liquid phase is present – in contrast to ingot metallurgy which starts from fully liquid materials – results in usually finer and much more homogeneous microstructures; by the powder route, immiscible components such as W and Cu or Al and Al_2O_3 can be combined. However, this tendency to lack of segregation also means that PM materials are more sensitive to impurities, particles from slag, ceramic linings and dust which are not removed by gravity segregation, as in the case of classical ingot metallurgy, but remain within the powder. Therefore, cleanliness is an essential precondition in PM manufacturing and this is evident when entering a PM factory in comparison to a foundry.

The huge variety of starting powders available and alloying variants for PM multicomponent systems means that even for a given overall composition a multitude of different microstructures can be obtained. In many PM products, porosity is a further microstructural component feature, in part as an inevitable, but frequently unwelcome consequence of the manufacturing process; in part, however, porosity is the *raison d'etre* for PM materials as, for example, in self-lubricating bearings or sintered filters. In particular the presence of the porosity also has a strong impact on the metallographic preparation of PM materials; specific care has to be taken to depict the porosity metallographically realistically since both unintentional closing and enlarging of the pores is easily done during preparation, as will be discussed below.

11.2 Metallographic preparation techniques for powder metallurgy products

In general, the metallographic preparation techniques used for the preparation of conventional metals can also be applied to powder metallurgy products; there are however several differences and peculiarities that have to be considered. The main difference from cast and wrought products is once more the porosity of many PM materials. In contrast to pores present in particular in cast products, the pores in PM materials – except in filters – are fine and mostly very regularly distributed (Figs 11.1 and 11.2). In metallographic preparation, there is a pronounced risk of artefacts, the pores being partially or completely closed during grinding and polishing.[1,2] This risk is the more pronounced the higher the ductility of such materials. Very easily sintered plain iron can appear fully dense after metallographic preparation (Fig. 11.3(a)) although density measurement through water displacement (Archimedes) method shows that there is considerable porosity. Frequently the presence of pores can be recognized by the emergence of wormlike lines in the seemingly dense material. One remedy for this problem is the use of mainly abrasively acting media; extended polishing with 3 μm diamond grit is preferable here compared to SiC, Al_2O_3 or Cr_2O_3. Diamond polishing is also recommended if X-ray diffraction (XRD) is to be done on the sections since in particular the softer polishing media also result in compressive residual stresses that adversely affect the precision of the XRD measurement.

The second way to reveal the pores precisely is by combining polishing with etching (Fig. 11.3(b)). By appropriate combining of both, a realistic image of the pores can be obtained, the metal layers covering the pores that have been generated by polishing being etched off. This is of particular importance if the properties of a sintered material – mechanical, magnetic or electrical – are to be related to the pore geometry, which happens frequently. However, there is also a risk of overetching, that is opening the pores too much. To obtain a correct image of the pores, a commonly used measure is to compare the porosity obtained by water displacement with that obtained from the sections by quantitative metallography; if both are in good agreement it can be assumed that the section has been prepared correctly.[2]

A third method is impregnation of the pores – after 6 μm diamond polishing

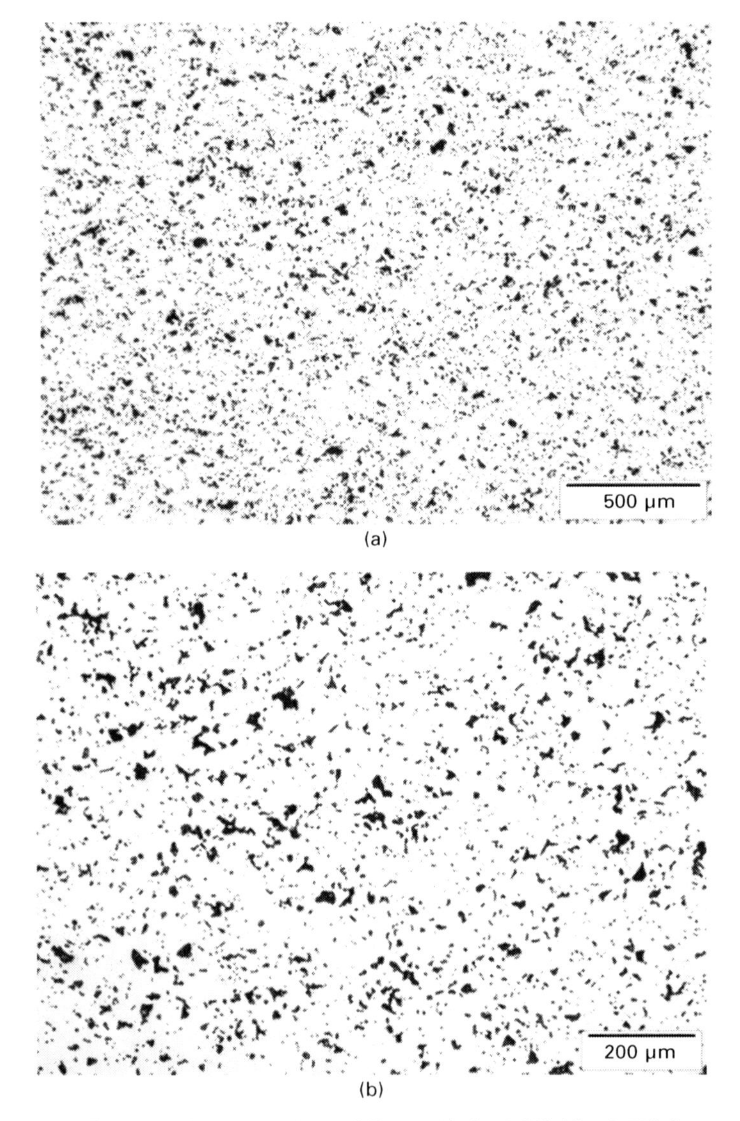

11.1 Porosity in sintered steel Fe–1.5% Cr–0.2% Mo–0.5% C (compacted at 700 MPa, sintered 60 min at 1280°C): (a) overview, (b) higher magnification.

– with low-viscosity resins that are cured afterwards, either thermally or catalytically, by the presence of metal surfaces and the absence of air. If this is done properly, pore closing can be effectively prevented.

Another feature of many PM products that affects the metallographic preparation is the widely varying hardness of the microstructural constituents. In many types of sintered steels, chemically heterogeneous microstructures

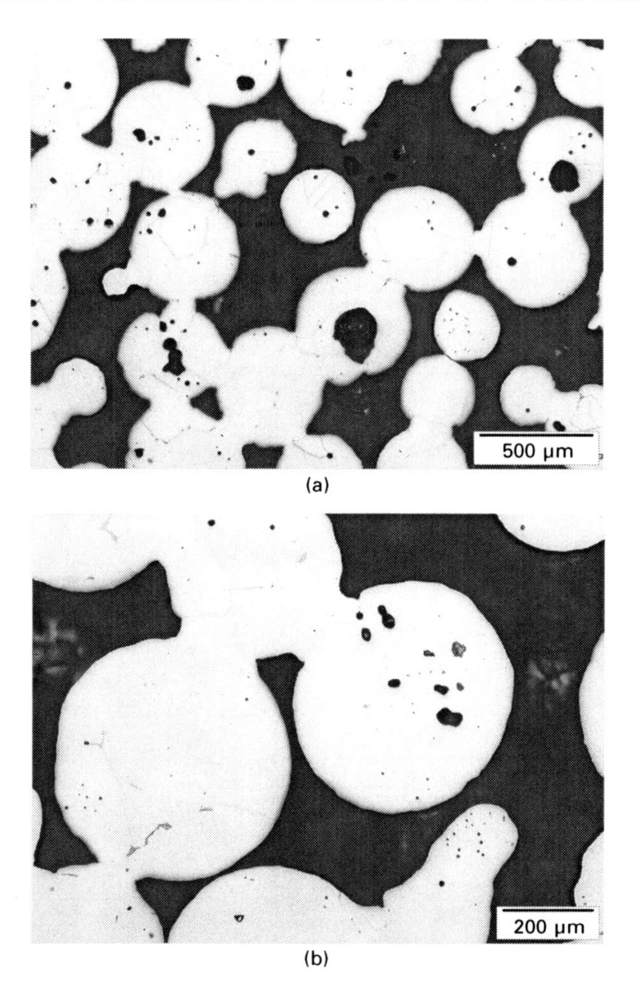

11.2 Porosity in sintered bronze filter (pre-alloyed powder Cu–11Sn, sintered for 60 min at 810°C in H_2): (a) overview, (b) detail.

are observed and, as a consequence, ferrite and retained austenite are present, as well as martensite,[3–7] and the transition may be within a few micrometres. In hard metals, metallic Co and WC form the microstructure, the size of the phases being frequently << 1 μm.[8] This means that polishing techniques have to be applied that prevent preferential removal of the softer phase; once more, using diamond grit for polishing is recommended.

When etching, the presence of open porosity may be a nuisance since the etchant enters the pores and then gradually tends to exude, resulting in overetched stains around the pores. This is particularly common with heat treated, martensitic or bainitic steels (Fig. 11.4). Here, once more impregnation of the section with low-viscosity resins is an effective measure for sealing

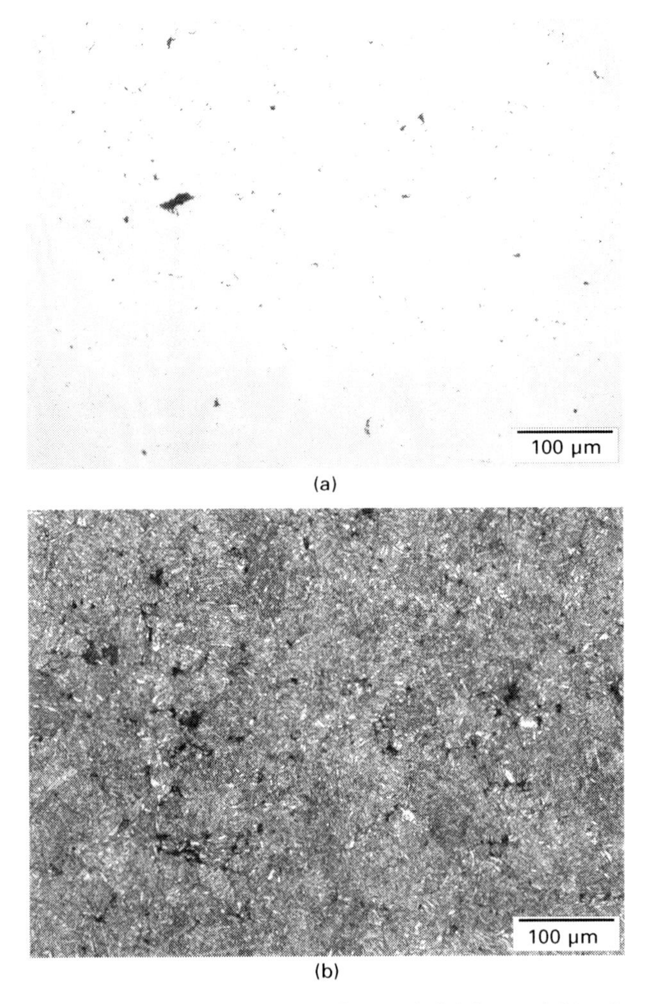

11.3 Sintered steel Fe–1.8% Cr–0.5% C (sintered 60 min at 1250°C in N_2): (a) pores partially closed by polishing, (b) same as (3a), Nital etched.

the pores. Sometimes swabbing produces better results compared with immersion.

Overetching may, however, be useful for multiphase materials. If one of the phases can be selectively removed, better insight into the three-dimensional structure may be obtained. This is shown in Fig. 11.5 for tungsten heavy alloys:[9] while in the plane section the distribution of the phases can only be estimated, after etching away the tungsten phase a much more clear image of the structure can be gained, in particular with respect to the connectivity of the W phase.

11.4 Image of (a) overetched (Nital etched, 60 s) and (b) properly etched (Nital etched, 12 s) sintered steel (Fe–0.85% Mo–0.7% C, sintered 1 h at 1150°C in N_2).

11.2.1 Fractographic techniques

The fracture modes encountered in PM materials are basically the same as those in ingot metallurgy materials; these are ductile rupture, transgranular (cleavage) fracture and intergranular failure, examples being shown in Fig. 11.6. However, both the wide range of compositions and the combination of widely different constituents results in a large variety of appearance.

(a)

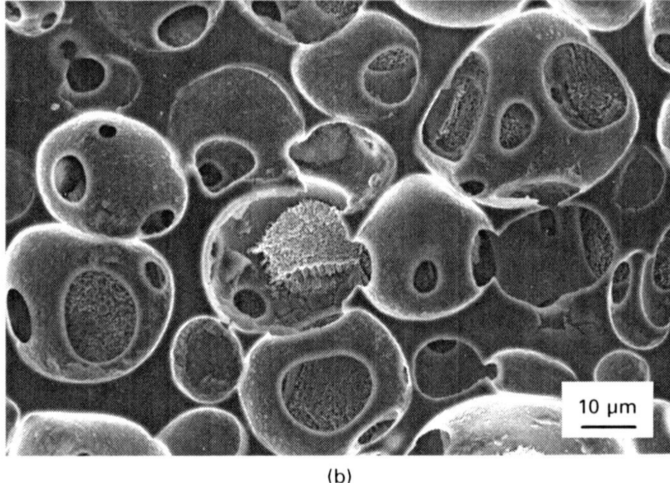

(b)

11.5 Tungsten heavy alloy W–6.7% Ni–3.3% Fe, liquid phase sintered at 1475°C: (a) metallographic section, (b) section, long time etched with saturated $Cu(NH_3).SO_4$ solution.

For multiphase materials, which are very common in PM, the interfaces are of crucial importance for the mechanical behaviour; low interfacial strength is usually linked to inadequate mechanical properties, as shown for W–Ni–Fe heavy alloys in Fig. 11.7: in the brittle materials the Ni–Fe–W binder is easily separated from the W spheres while in the ductile one excellent adherence is observed.[9]

Furthermore, the presence of pores may strongly affect the fracture mode. Here, in particular, the presence of interconnected pores plays a major role, since this type of porosity implies isolated sintering contacts or

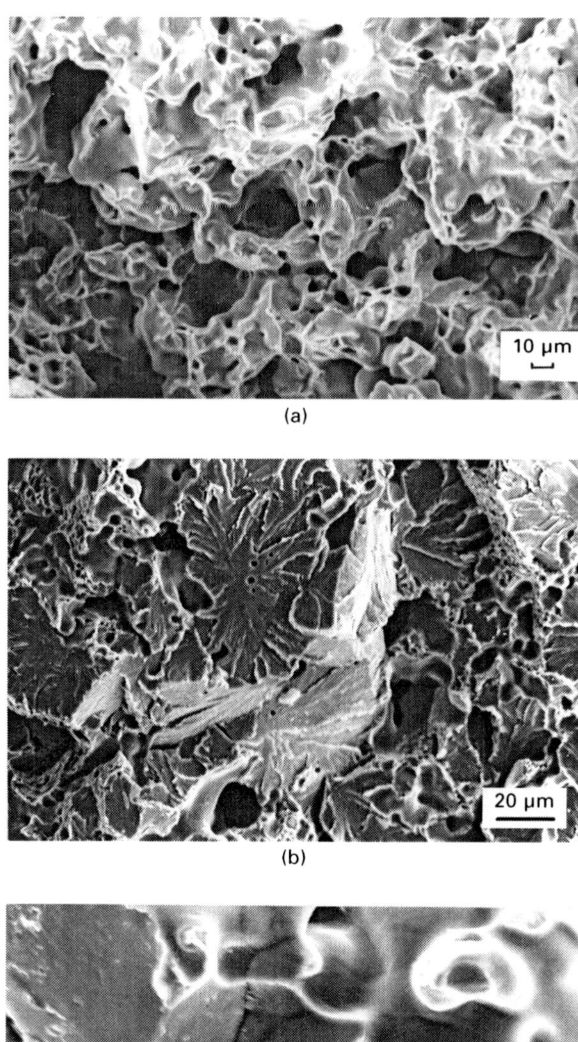

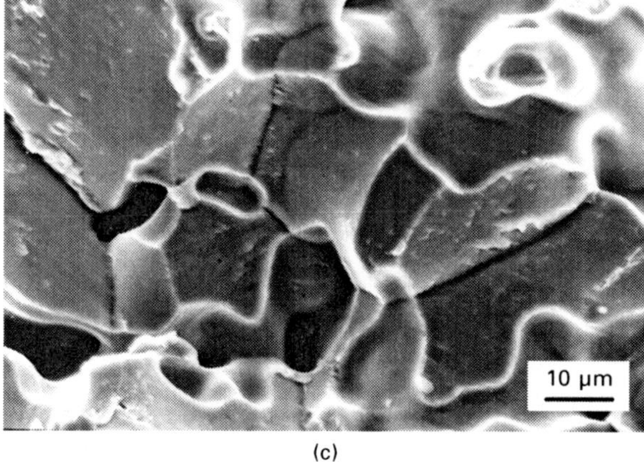

(c)

11.6 Different fracture modes in ferrous PM materials: (a) ductile rupture of Fe–3% Cu, (b) mixed cleavage–ductile rupture in sintered Fe–Mo–C, (c) intergranular failure in Fe–Mo–C–P.

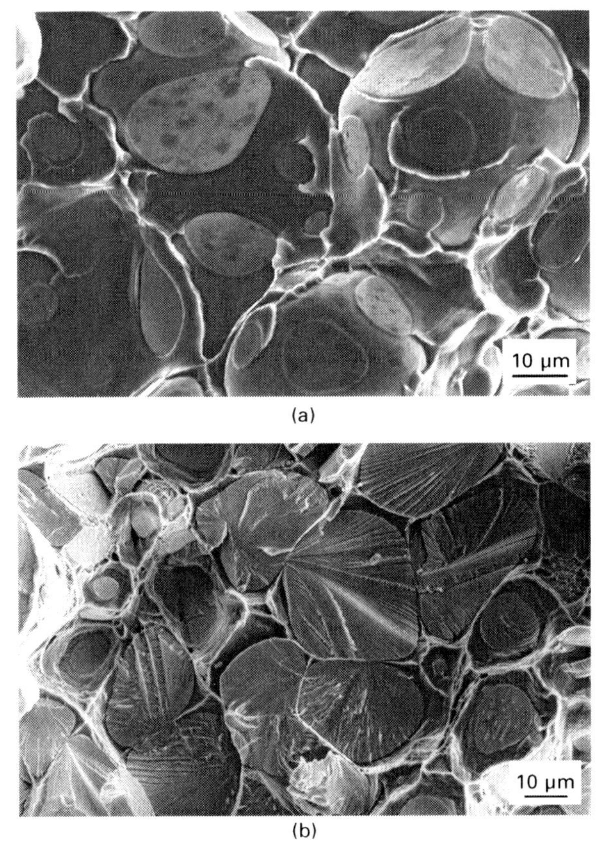

11.7 W heavy alloys with low and high interfacial strength and resulting widely different ductility.[9] (a) low interfacial strength, brittle, (b) high interfacial strength, ductile.

'necks'. It has been shown quite early[10,11] that such isolated necks exhibit a different behaviour compared to bulk material. For fully dense (Armco) or high density sintered plain iron, fracturing at 77K, at the temperature of liquid nitrogen for example, results in pronounced cleavage, as shown in Fig. 11.8(a). If however sintered iron with a porosity of 12–15%, which is fully interconnected, is tested at the same temperature, ductile rupture of the individual necks occurs despite the low temperature (Fig. 11.8(b)). Slesar[10, 11] explained this by the absence of dislocation pile-up effects in the very small cross sections of the necks, dislocation sliding also being possible at low temperatures.

Since in porous sintered materials, for example sintered steels or Al alloys, in reality the sintering contacts are of relevance for the properties and not the pores – pores do not bear any load nor conduct heat or electricity – measuring the dimensions of the contacts is an important task. Since the pore structure,

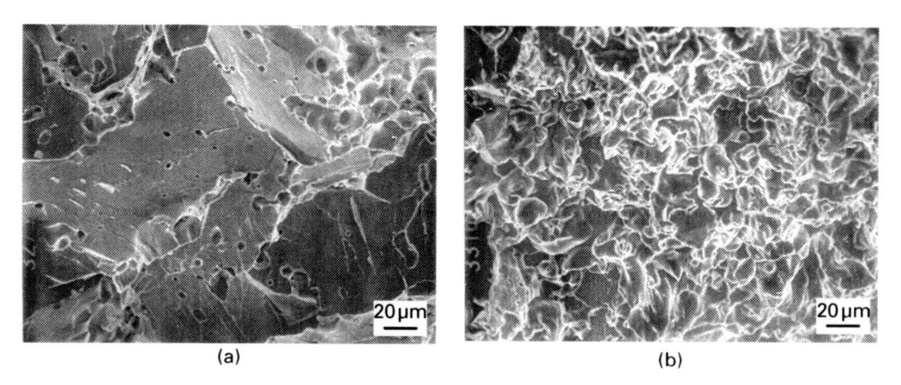

11.8 Fracture surfaces of sintered plain iron with different porosity, broken at 77K. Total porosity, (a) 3.8%, (b) 13.3%.

at least for interconnected porosity, is a three-dimensional pore network, metallographic techniques, which depict only a two-dimensional section, are not very helpful. Fractographic approaches are more suitable; from fractographs the 'effective load-bearing cross section' A_c (or its opposite, the 'plane porosity P'_x) can be measured by quantitative image analysis.[11,12] Of course an image that is taken about 90° to the general plane of the fracture surface will not yield the real area of contacts that have fractured in an angle to this plane but only a projection of this area into the plane. However, this projection of the real contact area gives the correct figure when correlating the load-bearing cross section to properties that are measured by loading perpendicular to the plane, that is tensile or push-pull fatigue loading.

In any case it should be considered that it is usually the size of the sintered, undestroyed, contact that is of interest. Therefore fracture surfaces that have been generated with significant deformation before fracture tend to yield erroneous results for the load-bearing cross section A_c. The same holds if failure occurs predominantly through the particle cores and not through the sintering necks, as in the case of pronounced cleavage. While plastic deformation tends to result in too low values for A_c, cleavage results in too high values. Therefore, the way that fracture surfaces are generated is of crucial importance. With many materials, impact loading at 77K is a suitable way to generate low-deformation impact fracture, although with some risk of cleavage. Fatigue fracture surfaces are usually the best choice, in particular for failure after high cycle fatigue loading, that is closely above the fatigue endurance strength, since the slow crack growth results in preferential propagation through the weakest areas, the contacts, without any marked plastic deformation.

If such fracture surfaces are obtained, the area of the broken necks can be measured on fractographic images using image analyzing software packages. Fully automatic software is frequently difficult to use since the contrast between the broken necks and the internal pore surfaces may be too low to detect reliably in automatic systems. If the failed necks exhibit a finely rugged structure, as for example with some heat treated sintered steels, the grey scale gradient can be used,[13] since the internal pore surfaces are rather uniformly grey, with only slight gradients. Fracture surfaces thus evaluated are shown in Fig. 11.9; in Fig. 11.9(b) which relates to Fig. 11.9(a) and in Fig. 11.9(d) which relates to Fig. 11.9(c) the detected area of the failed neck is shown in white. If however the contrast between broken neck and pore is visible only to the operator but not to the system, semi-automatic systems have to be used (e.g. Dlapka *et al.*[14]), an approach that at least alleviates the contrasting problem but tends to bring about a considerable 'operator effect'.

Recently it has been stated that the load-bearing cross section can also be obtained from quantitative metallography data,[15] that is from sections, but it has still to be checked how reliable this approach is over a wide range of porosity and materials. Of course the sensitivity to metallographic preparation is also very pronounced here.

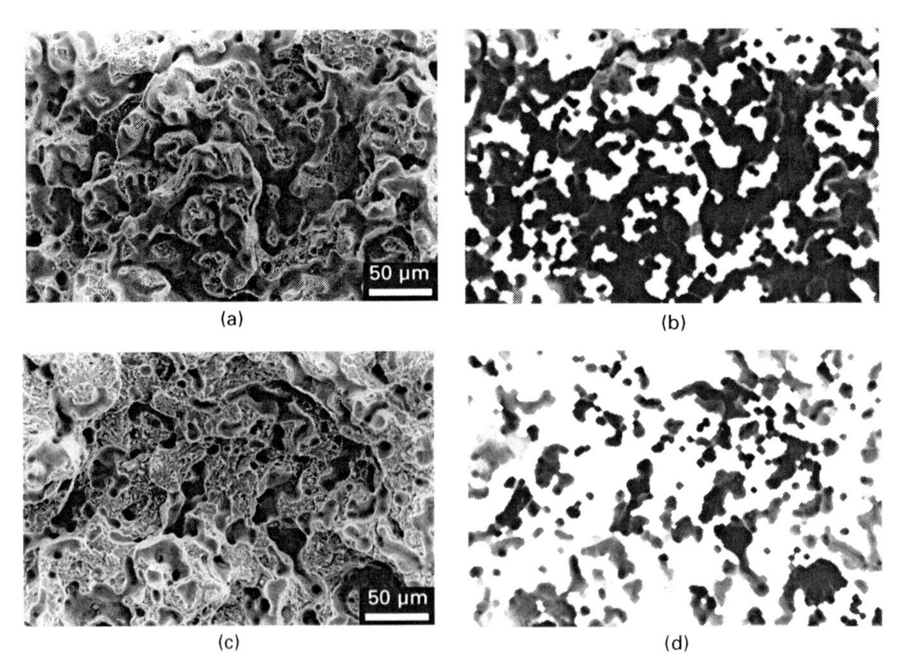

11.9 Fracture surfaces of sintered steel Fe–1.5% Mo–0.7% C, quenched and tempered, and impact tested at 77K:[13] (a) total porosity 12.3%, (b) as (a), broken contacts marked white, (c) total porosity 6.2%, (d) as (c), broken contacts marked white.

11.3 Microstructures of ferrous powder metallurgy materials

11.3.1 Sintered steels: evolution of pore structure

In sintered steels, the porosity depends on one hand on the compacting pressure, which, combined with the compactibility of the starting powder mix, defines the total porosity, and on the other hand on the sintering process which defines the pore morphology, usually represented by shape factors (e.g. Blanco *et al.*[16]) and, within a certain porosity range, the connectivity as well. Typical images of sintered steels with different total porosity but sintered in the same way are shown in Fig. 11.10; images of steels with the same porosity but differently sintered are shown in Fig. 11.11.

As stated above, in the porosity range common for sintered steel parts, the pores are virtually interconnected and open to the surface, as they are in the initial state, for example, in a dewaxed green compact. At higher density levels, however, the pore channels linking the triple junctions tend to become smaller and smaller, and if intense sintering is applied, these channels are closed and the triple junctions remain as isolated pores. The transformation from interconnected to isolated pores is difficult to record, in principle only low deformation fracture studies as described above being suited to reveal this process. Since, however, interconnected porosity can be regarded to be roughly equivalent to open porosity and isolated porosity to closed porosity, measuring the closed porosity through He pycnometry is a reasonably good approach to determine the isolated porosity (although it must be considered that there may be closed pores that are interconnected). Furthermore, for numerous secondary operations it is the open porosity that counts, for example for gas carburizing or electroplating.

He pycnometry studies[14] have shown that the transition from open to closed porosity is mainly a function of the density, that is the total porosity, but in addition both the sintering parameters and the steel composition play a major role. Higher sintering temperature and a longer time results in pore closing at lower density levels and, equally, Cr prealloyed sintered steels show earlier pore closing than Mo prealloyed ones (Fig. 11.12).

So far it has been assumed that all pores present are left over from the green compact and have just had their morphology changed by the sintering process. However, there is also formation of new pores during sintering, the so-called secondary porosity (Fig. 11.13). These pores are generated from alloy element particles if transient liquid phase is formed during sintering.[17] The particles melt, either by congruent melting (as in the case of Cu) or by eutectic reaction with the matrix[18] and the liquid is rapidly distributed in the steel matrix by capillary forces, the driving force being the fast formation of solid solution, in other words it is an entropy-driven process. Depending on the way these pores are generated, their size is correlated to that of the

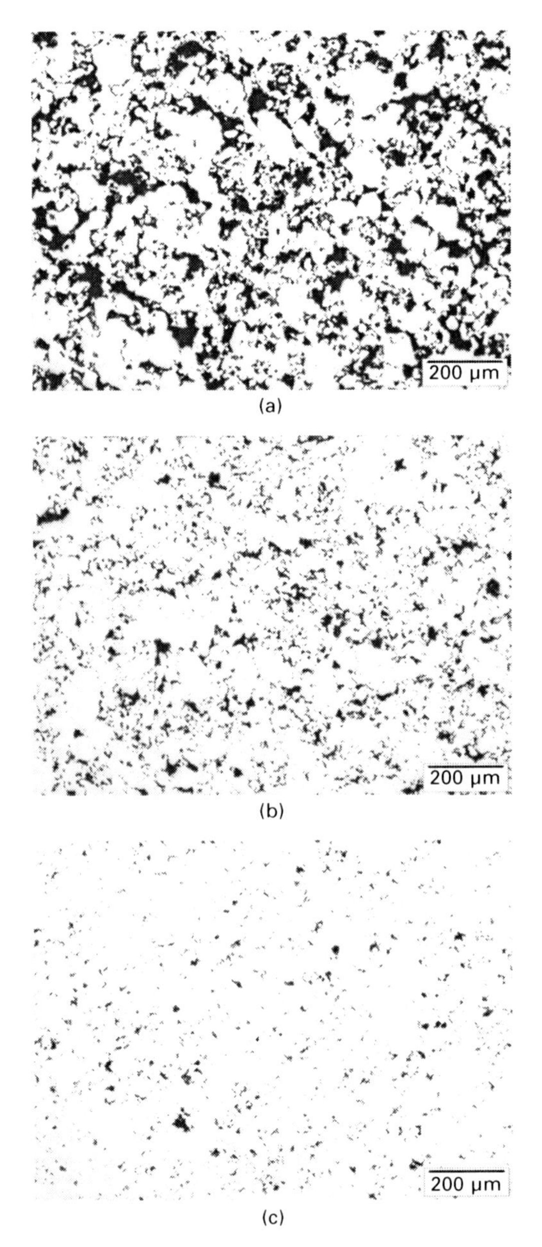

11.10 Metallographic sections of sintered steel Fe–0.85% Mo–0.3% C with varying porosity, sintered 30 min 1120°C, unetched: (a) 25% total porosity, (b) 9% total porosity, (c) 5% total porosity.

initial alloy element particle in different ways: for congruent melting, the size of the pore is virtually the same as that of the original particle, while for eutectic melting, as for example for Mo particles in an Fe-C matrix, the

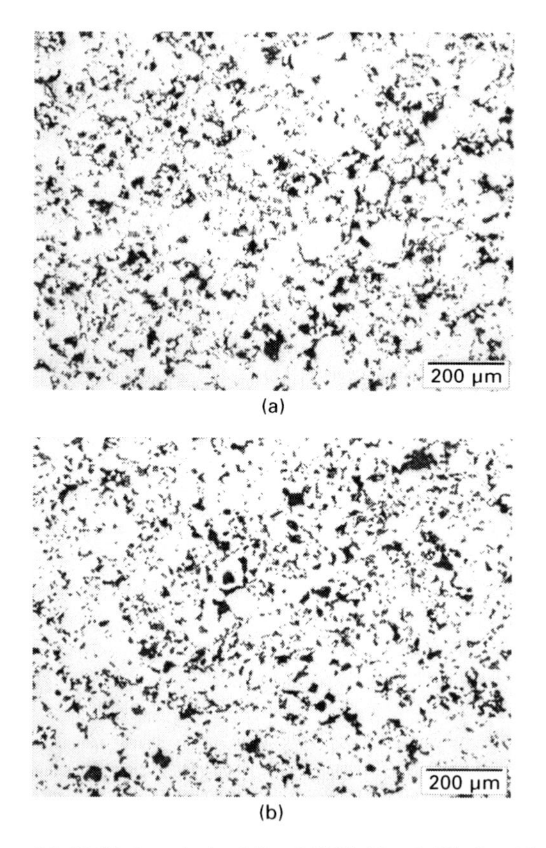

11.11 Sintered steel Fe–0.85% Mo–0.3% C with 15% porosity, unetched: (a) sintered 30 min 1120°C, (b) sintered 60 min 1280°C.

pore may be double the size of the original particle, since most of the melt consists of matrix material. These secondary pores may have an adverse effect on the properties, in particular on the fatigue endurance strength, since they act as crack initiation sites.[19,20] Therefore, both careful selection of the alloy powder grade as well as proper mixing, to avoid formation of alloy particle agglomerates, are required.

11.3.2 Sintered steels: austenite grain size

Compared to wrought steels, sintered steels are 'heat treated' (= sintered) at very high temperatures. Compared to the standard austenitizing temperature of a structural steel, the common sintering temperatures are extremely high. Therefore, excessive austenite grain growth would be expected, which is, however, not the case (see Fig. 11.14).

The reason for this is the pinning effect of the pores. As has been shown by Dlapka *et al.*,[21] that even fairly low volume fractions of pores effectively

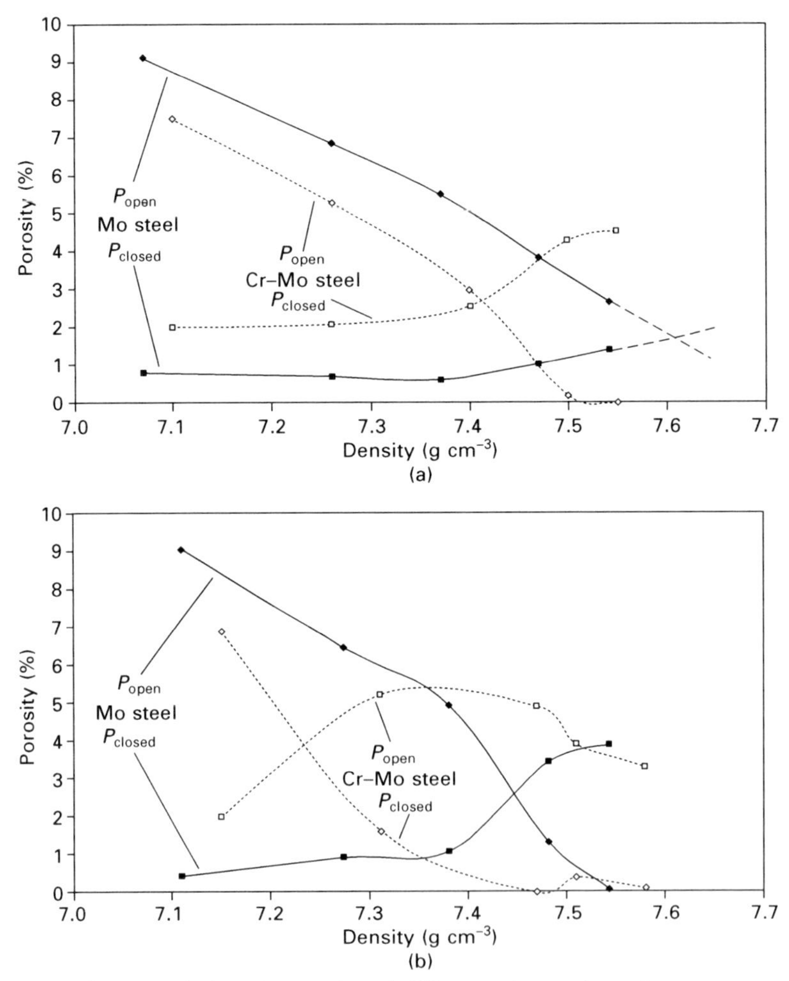

11.12 Open and closed porosity of different sintered steels as a function of the density and the sintering parameters.[14] (a) Sintered at 1120°C, (b) sintered at 1250°C.

prevent grain growth during sintering as well as during heat treatment, meaning that sintered steels are insensitive to overheating. The maximum austenite grain size is the size of the original powder particles. Only in the case of very high relative density, >7.6 g cm^{-3}, combined with intense sintering, has significant grain growth been recorded; under normal conditions, austenite grain growth can be safely neglected with sintered steels.

11.3.3 Sintered steels: carbon dissolution

As with wrought steels, with sintered steels the most common alloy element is carbon. It is usually introduced by admixing fine natural graphite grades,

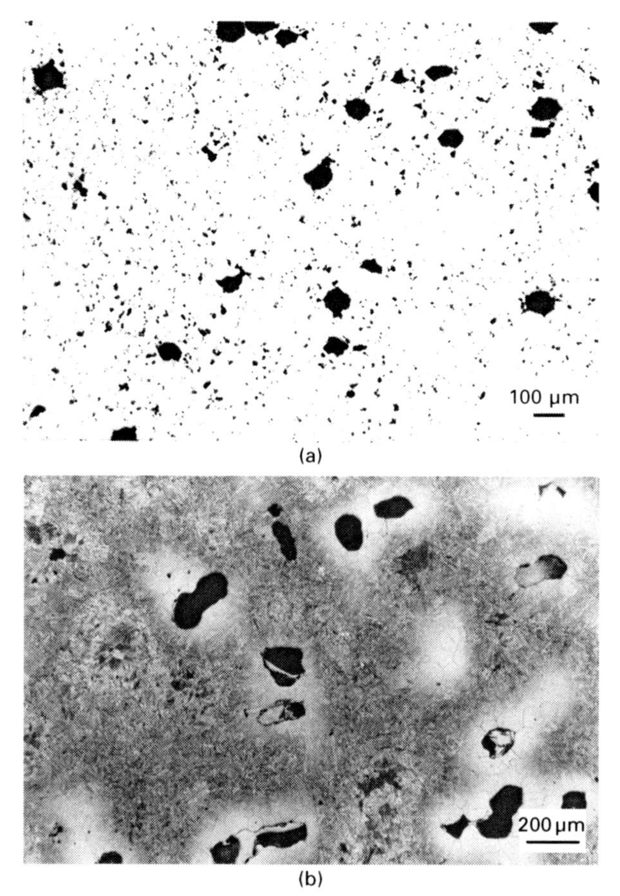

11.13 Secondary porosity in different sintered steels prepared from mixed powders: (a) Fe–3% Cu, unetched, (b) Fe–1.5% Mo–0.7% C, Nital etched.

the graphite being dissolved during sintering. It has been observed that this dissolution does not occur immediately when the eutectoid temperature is attained but that considerable overheating is necessary; in plain carbon steels prepared from atomized iron powder, the graphite is dissolved in the temperature range 900–1000°C, as indicated both by metallographic analysis, considering the amount of pearlite formed, see Fig. 11.15 (from Momeni[22], and analysis of the free carbon (Fig. 11.16). The traditional notion is that natural graphite grades are more readily dissolved in iron than artificial ones; however, today very suitable artificial grades are commercially available.[23,24]

11.3.4 Sintered steels: effect of alloying techniques

One of the main parameters affecting the microstructure of sintered alloy steels is the alloying technique (see also Chapter 7). This is something unknown

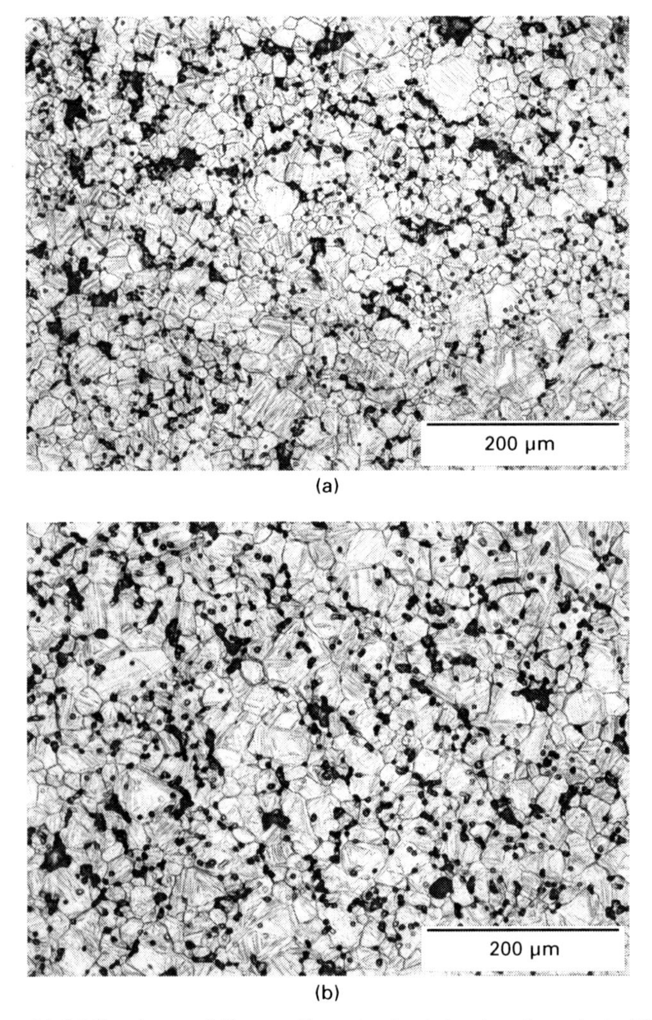

11.14 Sections of Cr prealloy steel, sinter hardened at different temperatures and simultaneously thermally etched: (a) 1120°C, (b) 1250°C.

with wrought steels, but for sintered steels the same nominal composition may result in completely different microstructures, depending on the starting powder used, in particular if chemically homogeneous (prealloyed) or heterogeneous (mixed or diffusion bonded) grades are employed. In Fig. 11.17 different types of sintered steels are shown: the quite regular upper bainitic structure of an Mo prealloy steel is visible compared to a similar steel produced from a mixed powder; in the mixed powder, after sintering at moderate temperature a mixed microstructure is formed containing pearlite, bainite and some martensite, depending on the local Mo content. Undissolved Mo

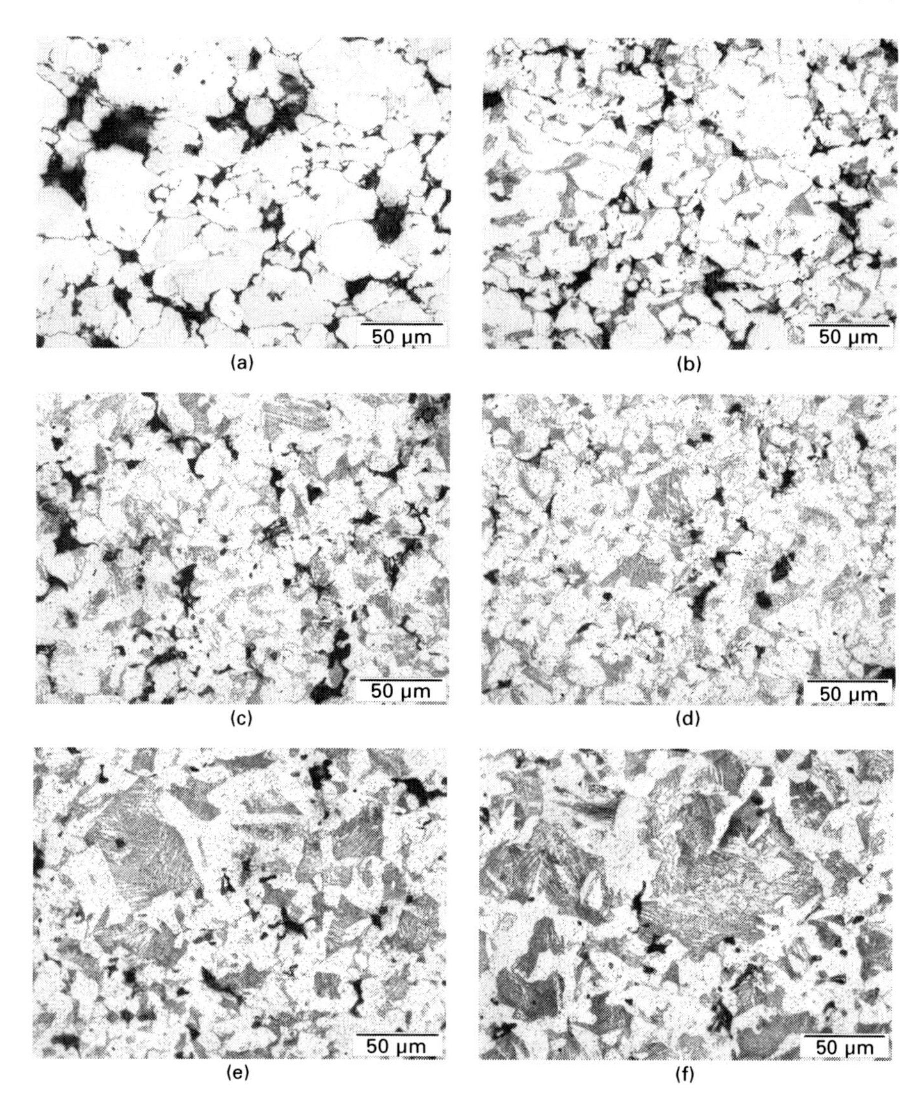

11.15 Fe–0–5% C after sintering for 60 min at different temperatures:[22] (a) 700°C, (b) 800°C, (c) 900°C, (d) 1000°C, (e) 1100°C, (f) 1200°C.

is also visible which has, however, been transformed into carbide. Typically, a prealloyed steel will exhibit virtually the same microstructure regardless of the sintering temperature while changing the sintering temperature may have a dramatic effect on the microstructure of a mixed or diffusion bonded steel, as visible when comparing Figs. 11.17(b) and 11.17(d).

Diffusion bonded Ni–Cu–Mo steels result in the typical 'Distaloy' microstructure (after the trade name of a major manufacturer), as shown

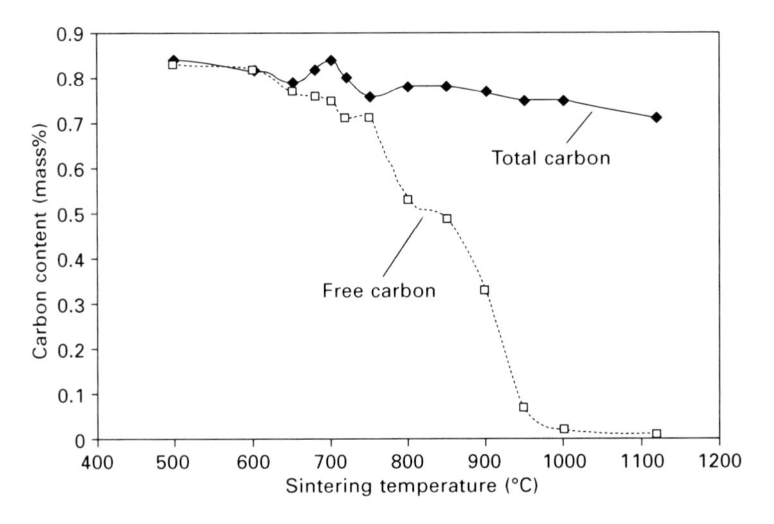

11.16 Free graphite in Fe–0.8% C as a function of the sintering temperature. Starting materials are water atomized iron powder and natural graphite.[23]

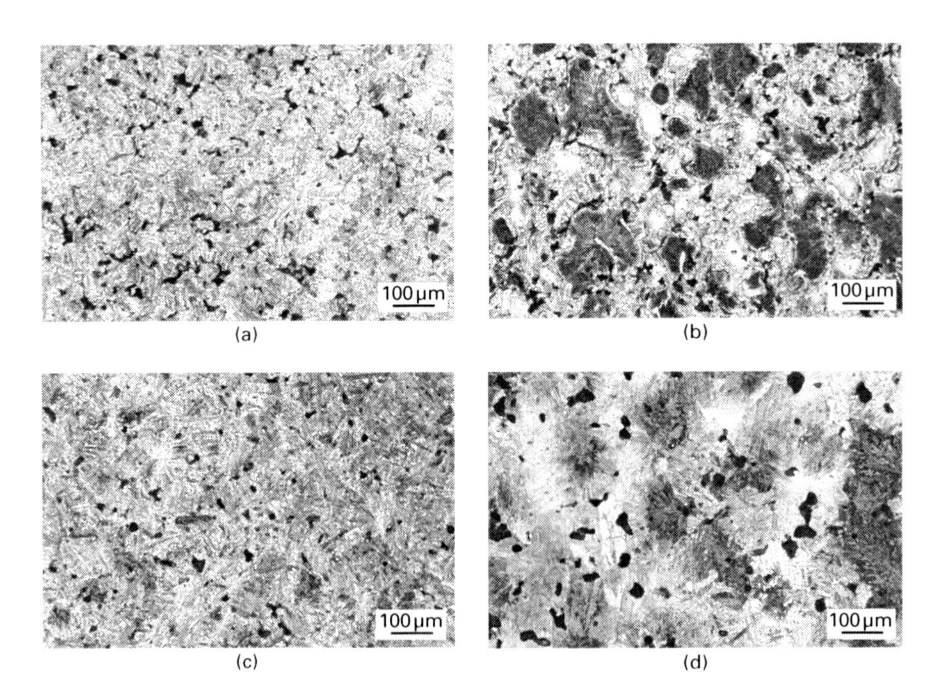

11.17 Fe–1.5% Mo–0.7% C, prealloyed vs. mixed, sintered 1200°C vs. 1320°C: (a) prealloyed, 1200°C, (b) mixed, 1200°C, (c) prealloyed, 1320°C, (d) mixed, 1320°C.

in Fig. 11.18(a). Here, widely varying microstructural constituents such as ferrite, pearlite, bainite, martensite and retained austenite are found closely adjacent to one another, as a consequence of the heterogeneous distribution of the alloy elements.[25] There are however also 'hybrid' variants, which combine prealloying and diffusion bonding; most commonly they are based on Mo prealloyed steel grades and contain Ni or Cu or both as diffusion bonded alloy elements. Such steel grades are widely used as sinter hardening grades; typical microstructures are shown in Fig. 11.18(b).

An attractive way to produce sintered steels with regular heterogeneous microstructure is by coating. Iron or steel powders can be coated with

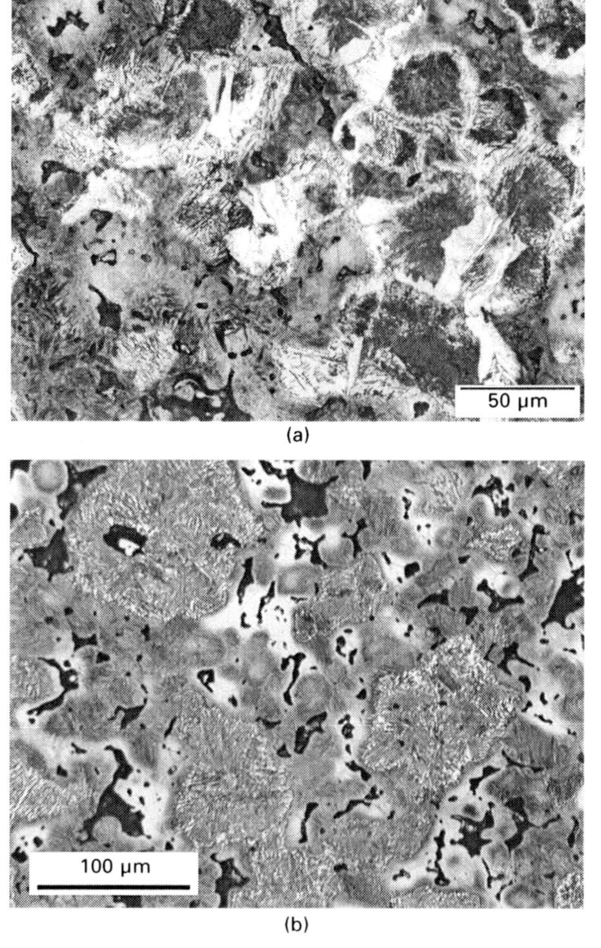

11.18 Microstructure of different diffusion alloyed and hybrid alloyed (prealloy + diffusion bonding) sintered steels (photos: Höganäs AB): (a) diffusion bonded Ni–Cu–Mo alloy steel, (b) hybrid alloy steel (Fe–Mo)–Ni–Cu.

Cu by cementation (Fig. 11.19(a)); such alloying results in more regular microstructure,[26] in particular less free Cu (Fig. 11.19(b) and (c)). Coating with other elements, such as Ni, is more difficult and must be done, for example, by electroplating, since the common electroless 'Ni' deposits contain significant amounts of P and thus cannot be regarded as real Ni layers.[27] Ni electroplated iron powder can be processed to steels with 'micrográdient' structure, that is a fairly regular but heterogeneous distribution of Ni, with resulting graded transition between the microstructural constituents (Fig. 11.19(d)).

11.3.5 Sintered steels: effect of heat treatment

Sintered steels can be heat treated in the same way as wrought steels when considering their special features. The open pores cause problems when quenching in an oil or salt bath, being filled with the quench media, and during

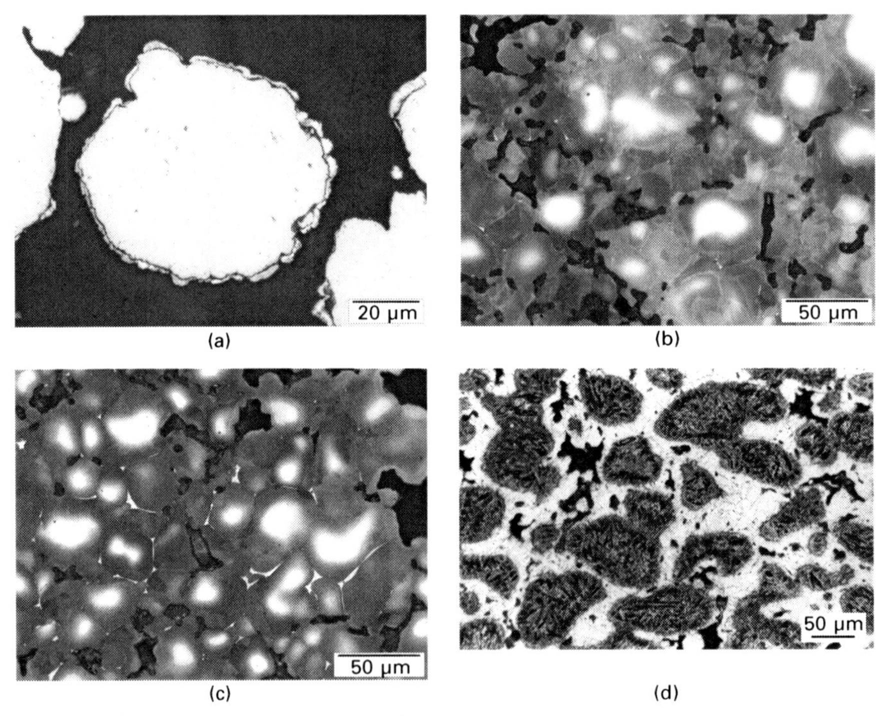

11.19 Microstructures of sintered steels manufactured from coated starting powders: (a) Cu coated Fe powder (Fe–8% Cu), (b) Cu alloy steel (Fe–8% Cu) from coated powder (dark areas: Fe–Cu solid solution; white: unalloyed iron), (c) Cu alloy steel (Fe–8% Cu) from mixed powders, bright grain boundary phase: free copper (d) sintered steel prepared from Ni coated Fe powder (dark areas: pearlite/bainite; light areas: martensite and retained austenite).

gas carburizing or nitriding, resulting in a tendency to 'through treatment', that is carburizing or nitriding of the cores. On the other hand, overheating is not a problem owing to the grain growth inhibiting effect of the pores.

Sintered steel parts are frequently sinter hardened, by blowing cold gas onto them immediately upon leaving the sintering zone of the furnace. If the alloy composition is suitably selected, fully martensitic microstructures are obtained, as shown in Fig. 11.20. Since the cooling rates obtained with sinter

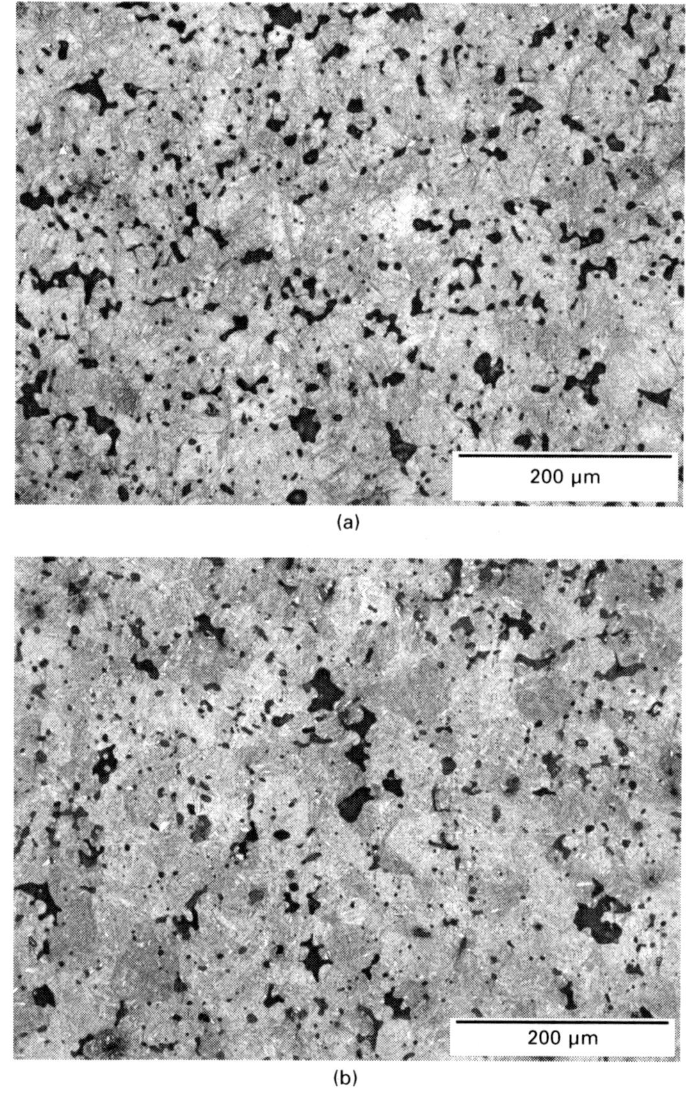

11.20 Microstructures of sinter hardened Cr–Mo prealloyed steels:[28] (a) as gas quenched, (b) gas quenched and tempered at 180°C.

hardening are barely >3 K s^{-1}, air hardening behaviour is required, which means sufficient amounts of Cr, Mo, Ni and Cu, with the combination of Mo+Cu or Cr+Cu being particularly effective. Tempering or at least stress relieving is usually done after sinter hardening. Generally, sinter hardening is more economical than separate heat treatment and also results in cleaner parts without oil; the penalty is the higher alloy element content necessary.

Quench and temper treatments are frequently done as induction hardening of the surface and subsurface zone, which is the standard procedure, for example for sprockets: In Fig. 11.21, for a typical sintered steel, the hardened martensitic surface zone is shown compared to the non-hardened ferritic–pearlitic core. Rapid austenitizing is followed by emulsion quenching, which

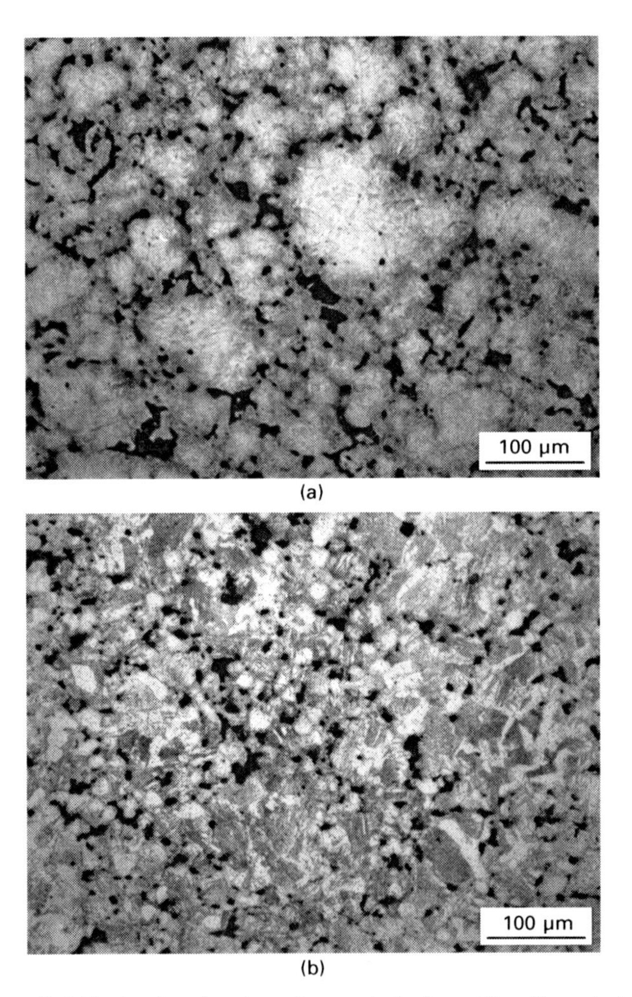

11.21 Induction hardened sprockets from Cu alloyed sintered steel: (a) induction hardened surface zone, (b) base material.

results in hard surfaces and reasonably tough cores, although the process has to be done properly to avoid quench cracks.

The most popular way of obtaining hard surfaces and tough cores is, however, thermochemical treatment by carburizing, carbonitriding or nitriding. In case of parts with homogeneous porosity, the problem of through carburizing can at least be alleviated by adapting the parameters; for surface densified parts, it is however extremely difficult to obtain well carburized densified areas without overcarburizing the non-densified surfaces. Here, either plasma carburizing or low pressure carburizing combined with high pressure gas quenching are alternatives, in particular LP carburizing has been shown to be suited to surface densified PM gears.[29] Since both processes use oxygen-free carbon carriers – CH_4 and C_2H_2 or C_3H_8, respectively – they are also applicable for Cr alloyed sintered steels and do not suffer from the oxidation problems encountered with gas carburizing using $CO–H_2$ mixes. Microstructures of carburized parts are shown in Fig. 11.22.

For nitriding, standard gas nitriding tends to result in through nitriding, with pronounced expansion; therefore, plasma nitriding is the method of choice here, which limits the nitriding effect on the surface. Mo prealloyed steels are particularly well suited here, and more recently Cr and Cr–Mo prealloyed steels have also been used.[30] In general, chemically homogeneous prealloyed steel grades are better suited to thermochemical treatments than heterogeneous ones prepared from mixed or diffusion bonded powders.

11.3.6 Sintered steels: microstructures of joints

As explained in Chapter 7, joining PM parts to each other or to wrought steel parts is a very common procedure, but the processes applied differ from those used in standard steel metallurgy. Fusion welding is done mainly through laser or electron beam welding, with a very narrow joint (see Fig. 11.23(a) and (b)), to avoid pore agglomeration and the formation of large voids. Otherwise, projection welding is the preferred process, once more offering a very limited weld zone. Frequently this is done as capacitor discharge welding, for which a typical joint is shown in Fig. 11.24.

Brazing PM parts is usually done during sintering, as sinter brazing. Here the effect of open pores that tend to wick rapidly the liquid filler has to be considered, leaving an empty gap in the joint. Therefore, reactive fillers have to be used that solidify in contact with iron through peritectic reactions,[31] thus blocking the pores and keeping the filler in the joint. Typically, Ni–Cu–Mn base fillers are used, which leave a highly hardenable joint, as indicated in Fig. 11.25(a). Recently[32] it has been shown that plain Fe–C fillers with a eutectic composition work well if the sintering temperature is sufficiently high; the advantage is that the composition is very similar to that of the

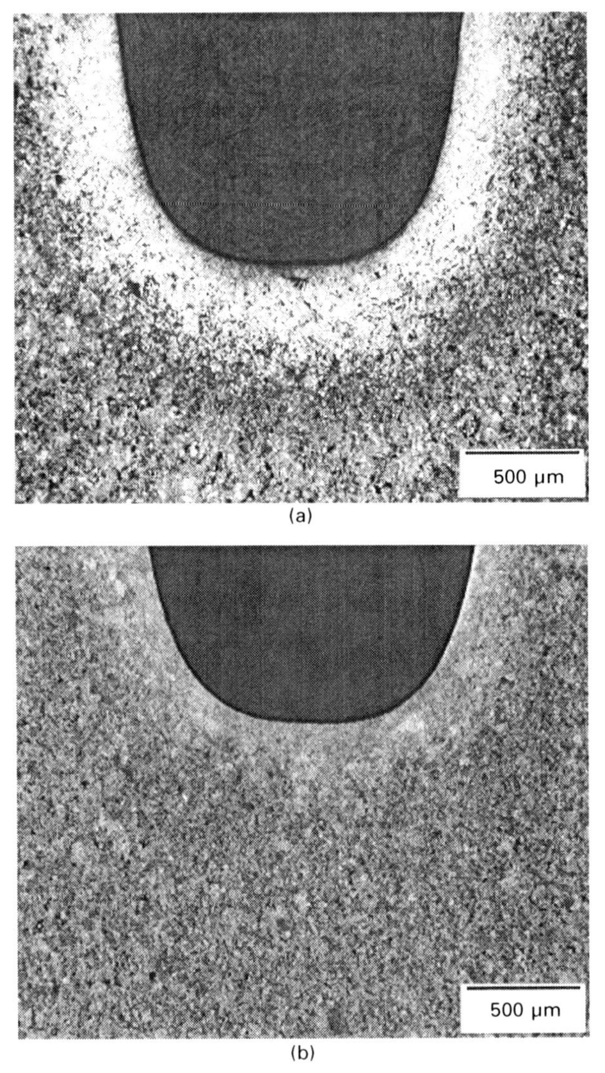

11.22 Microstructures of surface densified sintered steel gears, differently carburized: (a) low pressure carburized, (b) standard gas carburized.

parts to be joined and thus the mechanical and electrochemical behaviour is similar (Fig. 11.25(b)).

'Sinter bonding', a diffusion bonding that uses the different dimensional behaviour, for example of Fe–Cu and Fe–C during sintering, is also very common in PM parts production. If, in a concentric joint, the inner part swells and the outer one shrinks, a solid metallic bond is generated through diffusion processes that give excellent strength without any additional manufacturing steps being necessary.

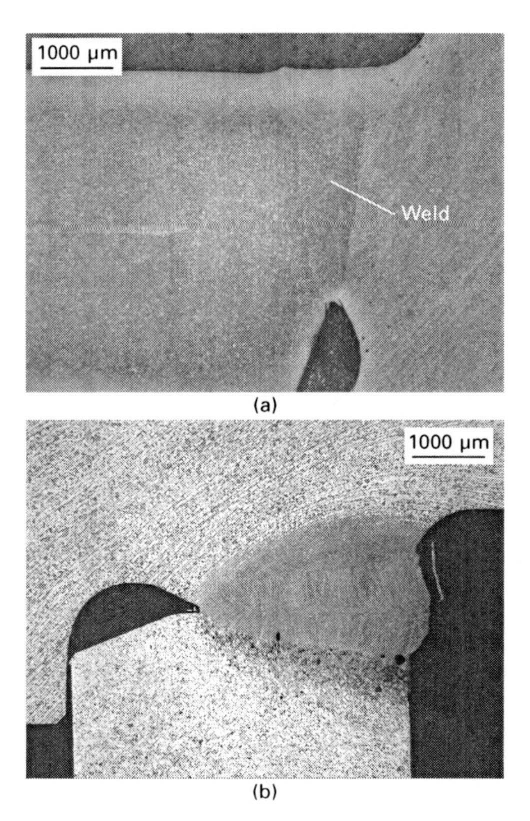

11.23 Joint between sintered steel and wrought steel parts generated by special fusion welding techniques (photos: MIBA): (a) laser welding, (b) electron beam welding.

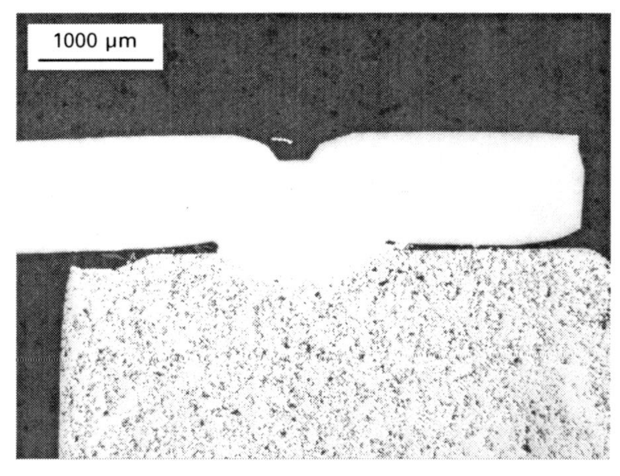

11.24 Joint between sintered steel and wrought steel parts generated by capacitor discharge welding (photo: MIBA).

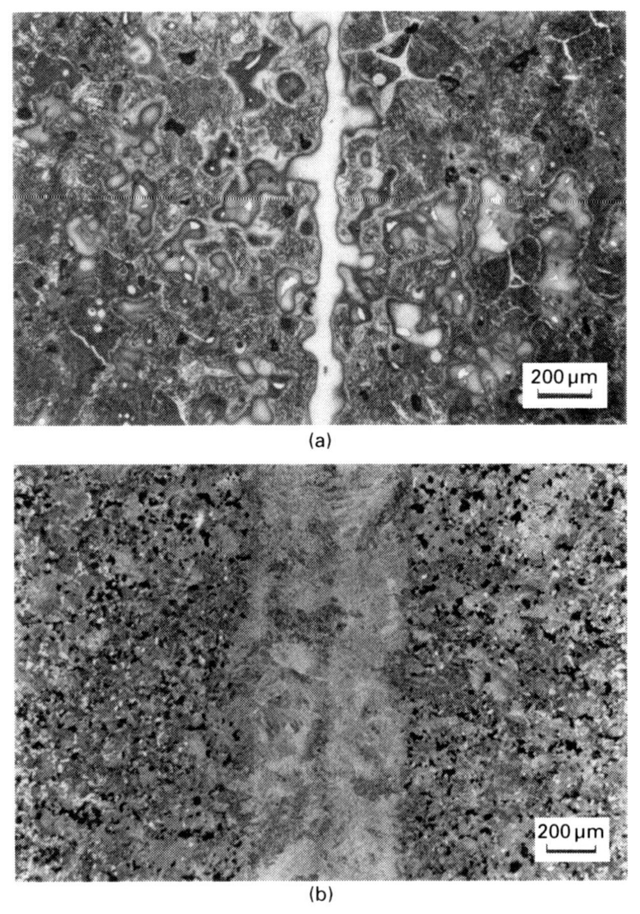

11.25 Sintered steel parts sinter brazed with reactive fillers: (a) Ni–Cu–Mn–Fe–B filller, (b) near-eutectic Fe–C filler.

11.3.7 Metal injection moulding (MIM) ferrous materials

The most characteristic feature of metal injection moulding (MIM) parts is the pronounced change in the porosity during sintering. While a debinded MIM part usually has a porosity in the range of >40%, after sintering the porosity is virtually always <5% and frequently close to zero. Furthermore, the pore connectivity changes: in a 'brown' (debinded) part the porosity is fully interconnected, to enable removal of the 'backbone' binder component in the first stage of sintering. Subsequently, the pores not only decrease in size but also become isolated, forming more or less spherical voids, that is during sintering the structure changes from a 'sponge-type', with open and interconnected porosity, to a 'swiss cheese' one, with isolated and well rounded pores. This change in the structure is easily visible from

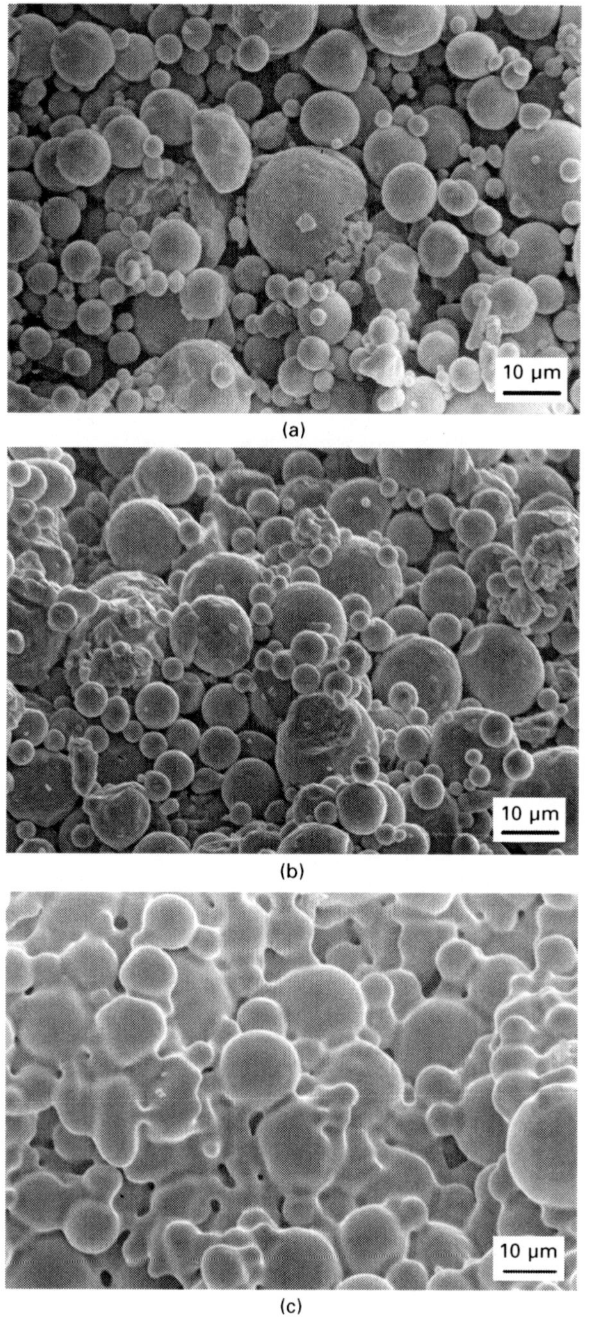

11.26 Stainless steel MIM specimens in different manufacturing states, from powder to virtually dense product (photos: Fraunhofer-IFAM Bremen): (a) gas atomized MIM powder, (b) thermally debinded compact, (c) sintered 5 min at 1180°C, (d) sintered 5 min at 1340°C, (e) sintered 1 h at 1300°C, (f) sintered 1 h at 1350°C.

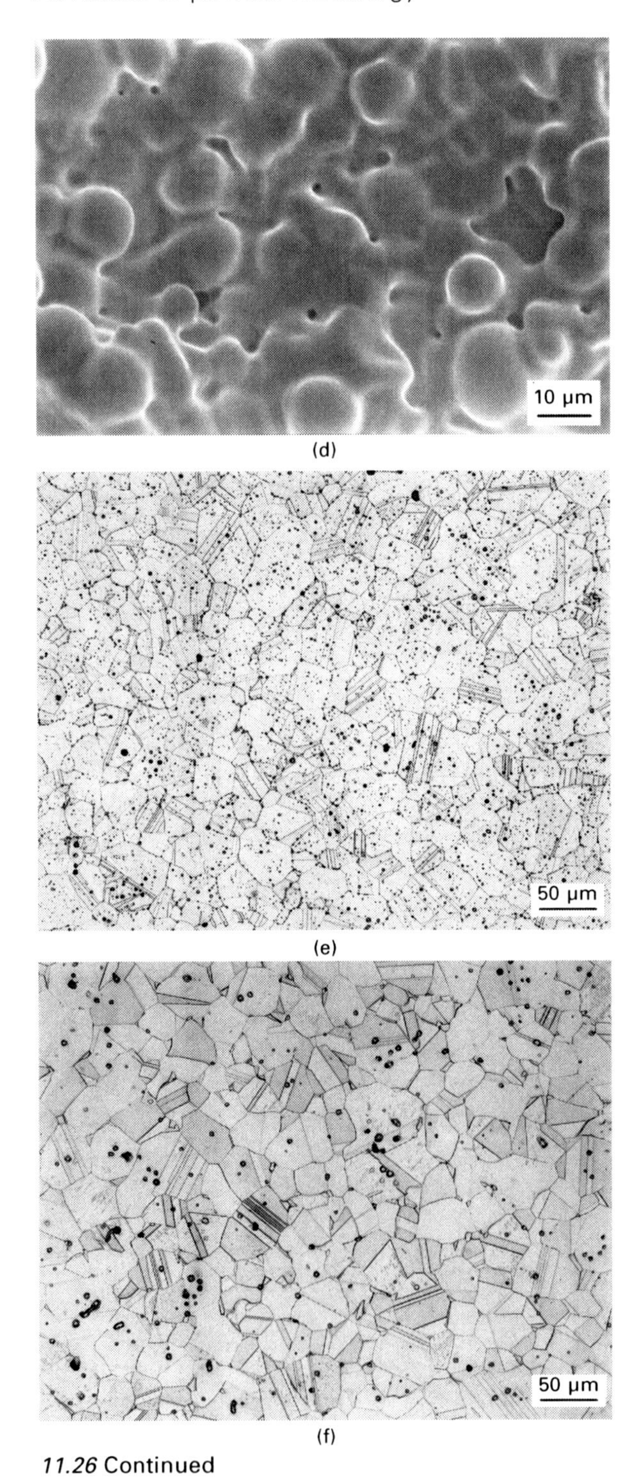

11.26 Continued

Fig. 11.26(a)–(f), depicting the different manufacturing state of a stainless steel part, from the powder to an almost fully dense material with only a few small pores.

11.3.8 Microstructures of powder metallurgy tool steels

As stated in Chapter 7, one of the major advantages of powder metallurgy tool steels compared to ingot metallurgy (IM) ones is the fine and isotropic microstructure attainable via the powder metallurgy route. This holds for virtually all grades of tool steels: cold and hot work tool steels, high speed steels and injection moulding grades. This means that compared to IM steels which exhibit a pronounced microstructural orientation, caused by the hot working necessary to disintegrate the brittle eutectic networks, the PM grades have the same microstructure in all directions and the same properties, which is very helpful for the designer of a tool.

In Fig. 11.27, the microstructure of an ingot metallurgy cold work tool steel (1.2379/AISI D2) is shown perpendicular and parallel to the tooling direction; the pronounced anisotropy is clearly visible. In Fig. 11.28, a powder metallurgy grade is shown; here, there is a finer and much more regular microstructure. It is also evident that the carbides are more rounded. The same holds for high speed steels (HSS); here the fine and rounded carbides dominate. This rounded shape may slightly lower the cutting performance compared to the coarse, angular carbides in IM HSS, but the toughness, hot and cold workability and also the grindability are very much improved.

In PM HSS, the composition is qualitatively the same as in IM HSS, and the carbide types are therefore also identical, mostly MC and M_6C. However, the content of carbide-forming elements may be markedly higher, up to >15%, and the carbide content is also higher accordingly. The finer microstructure of the PM steels can accommodate the high carbide contents without compromising the toughness, which would not be possible with IM steels. For cold work tool steels, PM steels are primarily VC based, that is the microstructure is dominated by MC carbides and not by M_7C_3 as in the case of IM steels. This gives very high wear resistance, in particular to abrasion, combined with attractive toughness.

As stated in Chapter 7, another approach for producing tool steels through powders is pressing and sintering, using water atomized powders. Here maintaining the optimum sintering temperature is of crucial importance; the optimum temperature between undersintering, with resulting porosity, and oversintering, which causes grain growth, has to be safely met.

The press-and-sinter route offers the advantage of additional hard phase reinforcement. This was already found in the 1930s when the so-called 'ferro-TiC', in fact steel-TiC composites, were invented. These materials are

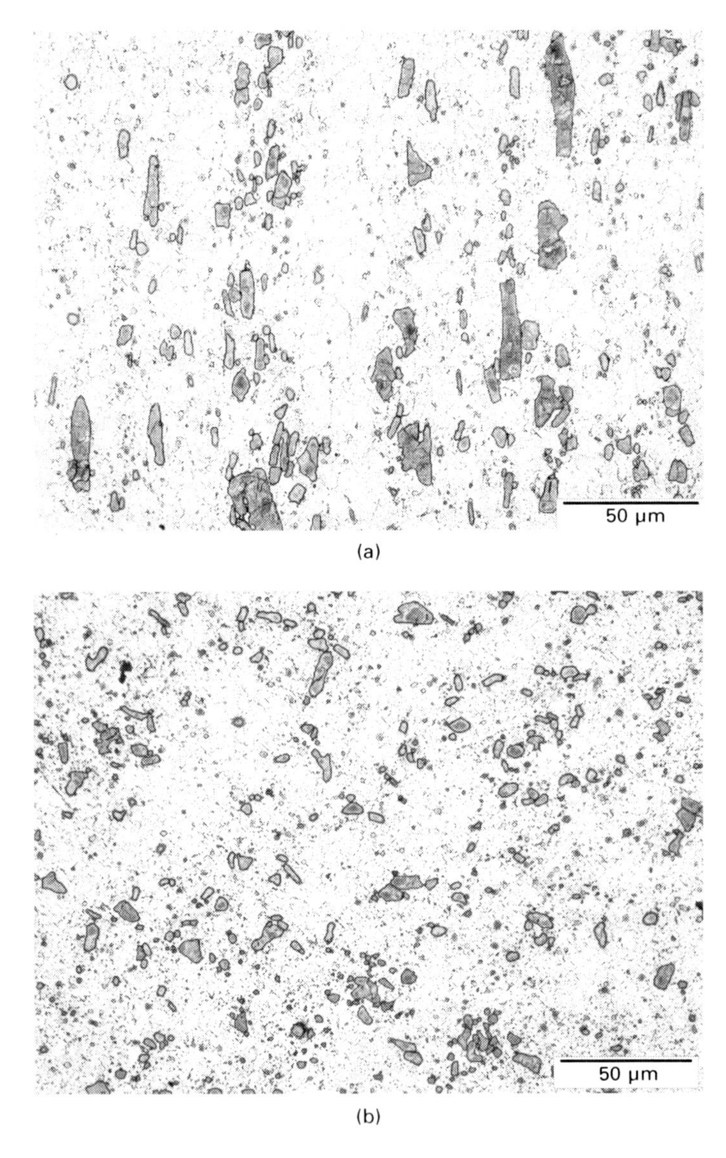

11.27 Wrought cold work tool steel (AISI D2 / 1.2379), longitudinal and cross sections:[33] (a) longitudinal section, (b) cross section.

commercially available today under the brand name 'Ferro-Titanit' (DEW) and, by choosing the right matrix material, can be tailored to different property profiles, the chance to combine high abrasion resistance with equally high corrosion resistance being particularly attractive, for example in polymer processing.

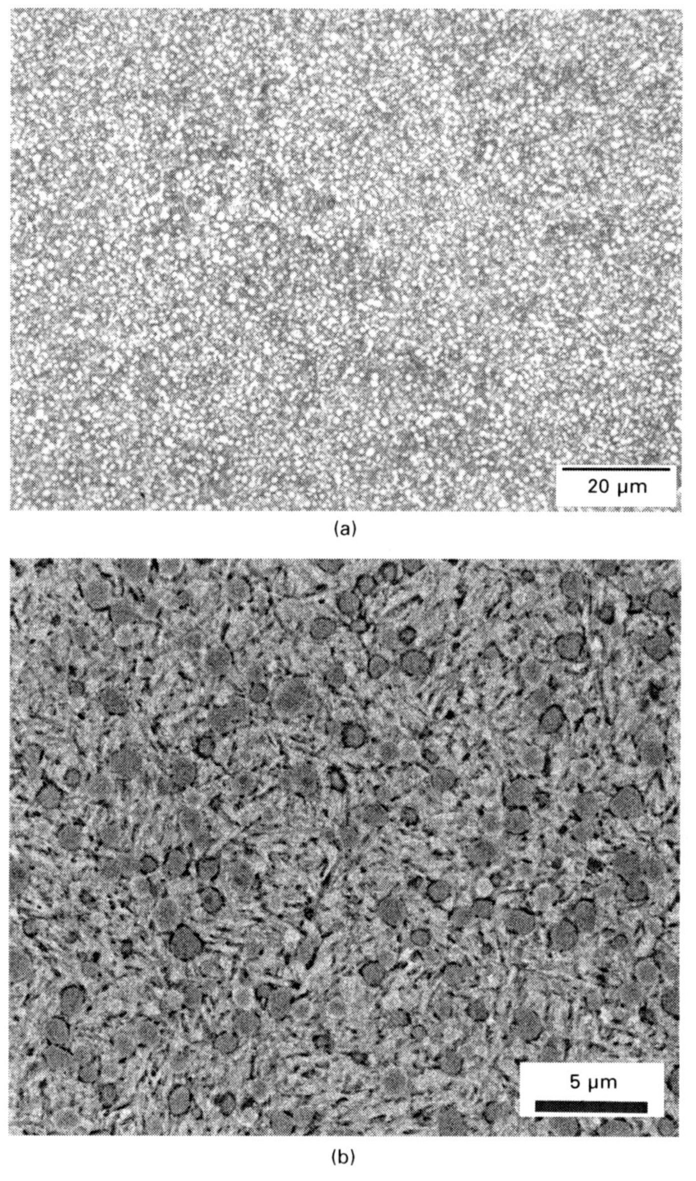

11.28 PM cold work tool steel Böhler K390:[33] (a) light optical image, (b) SEM image.

11.4 Non-ferrous materials

11.4.1 Cu-based sintered materials

From the microstructural viewpoint, Cu-based sintered materials follow similar lines to Fe-based ones. The pores present in the green compact ('primary pores') are mostly retained during sintering and tend to inhibit grain growth. Secondary pores may be generated in compacts from mixed powder grades in the presence of low-melting alloy elements, typically Sn, if too coarse Sn powders are used or if agglomerates are formed during mixing. In particular when sintering bronze from prealloyed powders, for example for filter purposes, the different diffusion coefficients of Cu and Sn in the Cu lattice cause enrichment of Sn in the sintering contacts, which can be identified by a lighter colour there compared to the standard yellow bronze colour (see Fig. 11.29).

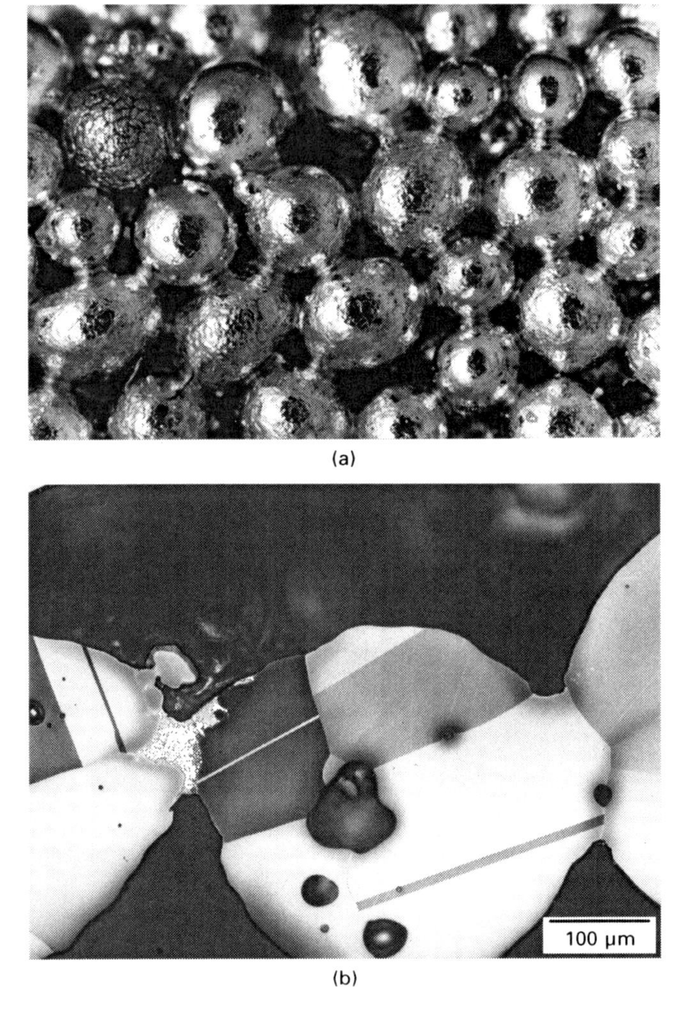

11.29 Gravity sintered bronze filter (prealloyed powder Cu–11Sn, sintered at 800°C, 30 min, H_2): (a) gravity sintered bronze filter, optical image, (b) as (a), microstructure.

Brass is much more tricky to sinter than bronze owing to the high vapour pressure of Zn – the boiling point of Zn is in the range of the usual sintering temperatures – and therefore measures have to be taken to avoid excessive Zn loss, such as sintering in semi-closed containers; using prealloyed powders is also helpful here. In case of proper sintering, regular microstructures can be obtained, however with a considerable effect of the sintering temperature, as evident from Fig. 11.30.

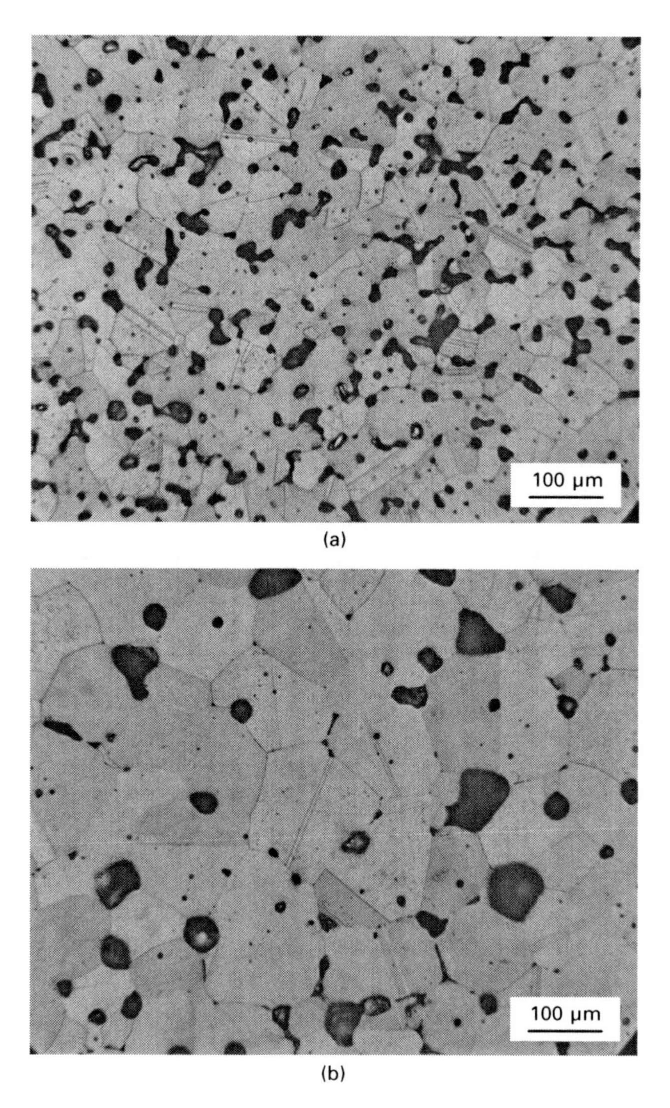

11.30 Microstructures of brass sintered at different temperatures:[34] (a) 910°C, (b) 950°C.

For electrical contact materials, for example in medium voltage interrupters, Cu–Cr is an attractive option; owing to the mutual insolubility, this is a typical two-phase pseudoalloy, Cr particles being embedded in a continuous Cu matrix (Fig. 11.31).

11.4.2 Powder metallurgy light alloys

Powder metallurgy light alloys can be structured into two groups: pressed and sintered parts and fully dense high performance materials, which are available mainly as semi-finished products.

For Al-based precision parts the main difference from other pressed and sintered materials is the presence of stable oxide layers covering the Al particles. These skins prevent sintering and have to be penetrated during the sintering process, which is done by liquid phase sintering in pure N_2,[35–40] usually combined with chemical attack by a strongly reducing metal, typically Mg. Therefore, Al sintered alloys contain elements that form a liquid phase in the range of the typical sintering temperatures (550–620°C), either by eutectic reaction with Al, as in the case of Cu,[36] or by formation of a supersolidus liquid phase, as with Al-Si-*x*-*y* alloys,[38] the latter sintering process resulting in considerable shrinkage while the former is stable dimensionally if properly done. Addition of trace elements such as Sn is also reported to assist sintering.[41]

In the former case, the as-sintered microstructure exhibits Al–Cu phases; after heat treatment, usually a T6 treatment with solutionizing, quenching and artificial ageing, these phases have disappeared, being transformed into

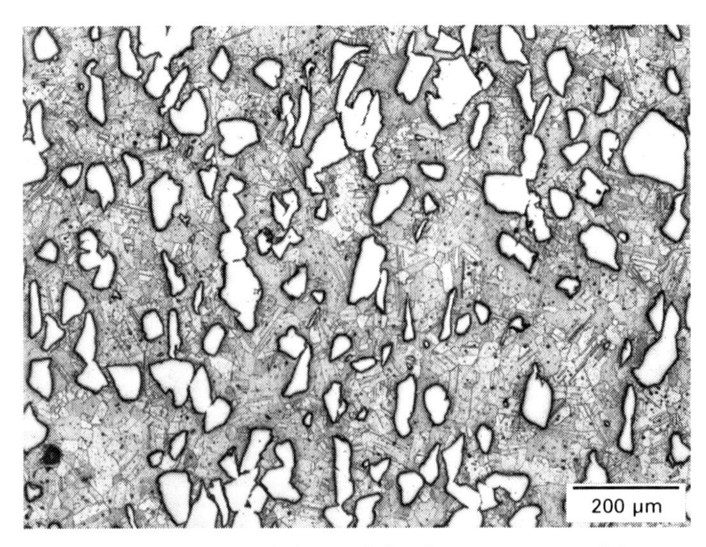

11.31 Microstructure of sintered Cu–Cr contact material.

nanosize precipitates, which results in a fairly homogeneous microstructure. The fragments of the oxide skins remain in the material; usually they are too small to be found in metallographic sections, but they act as microdefects, resulting in dimple formation and lowering the ductility. In fracture surfaces, these oxide fragments are occasionally visible (Fig. 11.32).

The microstructures of hypereutectic Al–Si alloys, in contrast, look quite different: these materials are produced from a mix of plain Al powder and a high-Si masteralloy which forms the supersolidus liquid phase, and this is discernible in the sintered microstructure. This 'hetero-supersolidus sintering' is done at temperatures at which still heterogeneous microstructures are obtained (Fig. 11.33(a)); the dark grey areas are Si phases which give the material attractive wear resistance, while the low porosity, as a consequence of the pronounced shrinkage, results in improved mechanical properties. Sintering at higher temperatures results in homogeneous but markedly coarser microstructures (Fig. 11.33(b)) and inferior mechanical properties. More recently, Al–Zn–Mg–Cu alloys have been prepared following this special 'hetero-supersolidus sintering' route.

Fully dense Al PM products may be compared to PM tool steels: they are expensive but offer superior performance. The main benefits offered by the PM route here are on the one hand the combination of immiscible components, for composite materials such as Al–SiC or Al–Al_2O_3 (Fig. 11.34: extruded Al–Al_2O_3 powder mix) or for dispersion strengthened materials that contain insoluble nanosize phases which strengthen the material in a similar way as precipitates but up to much higher temperatures. The second benefit is the very high cooling rate obtained with particulate materials: very fine and in part supersaturated structures can be obtained, as shown in Fig. 11.35 for Al–Si alloys produced by PM compared to standard IM. For the

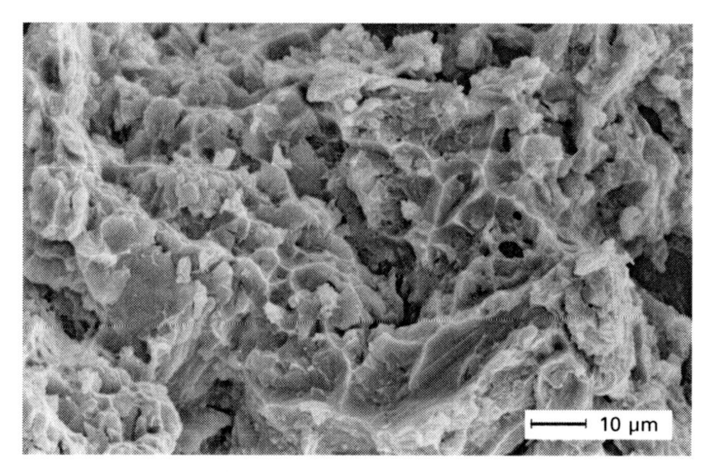

11.32 Fracture surface of sintered Al–Cu–Mg–Si (AA 2014).

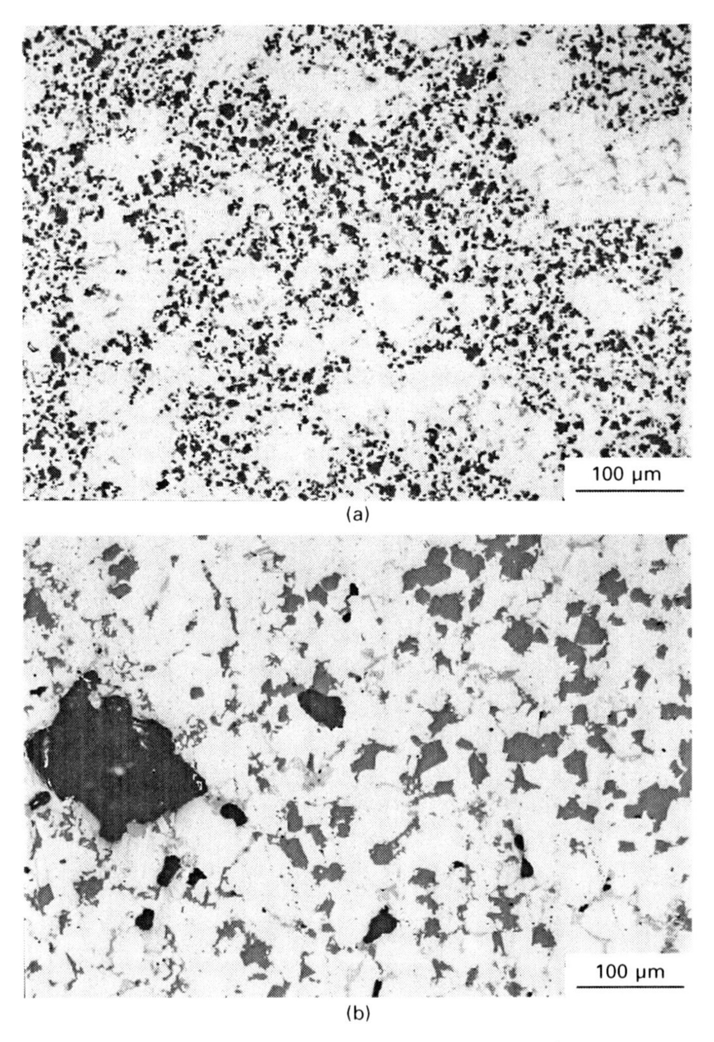

11.33 Hypereutectic Al–Si–Mg–Cu sintered at optimum temperature, 555°C (a) and oversintered, 580°C (b) (photos: Ecka Granulate GmbH).

so-called 'RS' (rapidly solidified) materials, the PM route is the most viable way; here, alloy elements that are insoluble in solid Al are introduced, for example by atomization of a suitable melt, and during processing ultrafine Al–Fe phases are precipitated which have a strengthening effect but with much lower tendency to overageing at higher temperatures than Al–Cu or Zn–Mg based precipitates. Therefore, high strength and creep resistance up to >300°C can be obtained.

PM Ti alloys are produced predominantly by the powder injection moulding route, for example for medical applications; the main features given for

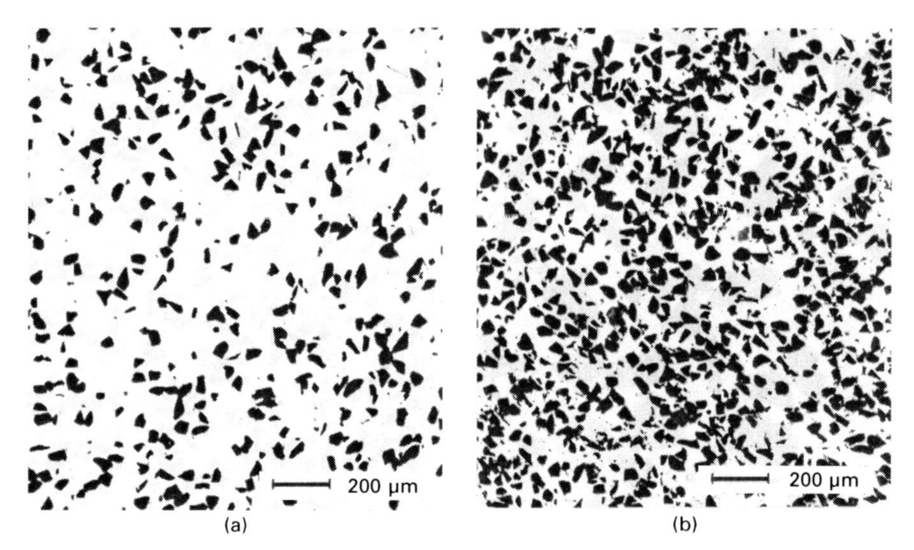

11.34 Microstructures of Al–Al_2O_3 prepared by extrusion of powder compacts: (a) Al – 20 wt% (14.7 vol%) Al_2O_3, (b) Al – 40 wt% (31.5 vol%) Al_2O_3.

MIM steel products also hold here.[42] Typical microstructures are shown in Fig. 11.36; once more the very fine and regular microstructure and the low porosity are evident.

11.4.3 Powder metallurgy refractory metals

Refractory metals have been among the first products of 'modern' powder metallurgy,[43] starting with the metal filament lamp invented by Carl Auer von Welsbach, who used osmium, and the commercialized tungsten filament lamp developed in industrial scale by Coolidge.[44] From the microstructural viewpoint, important aspects are on one hand the importance of attaining high, at best full, density, which is achieved by high temperature sintering with subsequent hot working, and on the other hand the tendency of refractory metals to grain coarsening at the very high service temperatures common for these metals, which results in embrittlement.

The problem of embrittlement has been solved in different ways: for W filaments, the so-called 'non-sag' grades are used in which formation of coarse, elongated grains with high creep resistance is enforced by rows of K bubbles formed at service temperature parallel to the wire axis.[43] Stabilization of the grain structure or forcing grain growth to proceed in a defined direction can also be attained by dispersoids. For W, ThO_2 has been a common additive, for example in welding electrodes; today, Th is regarded as unfavourable and other stable oxides such as La_2O_5 are introduced that give

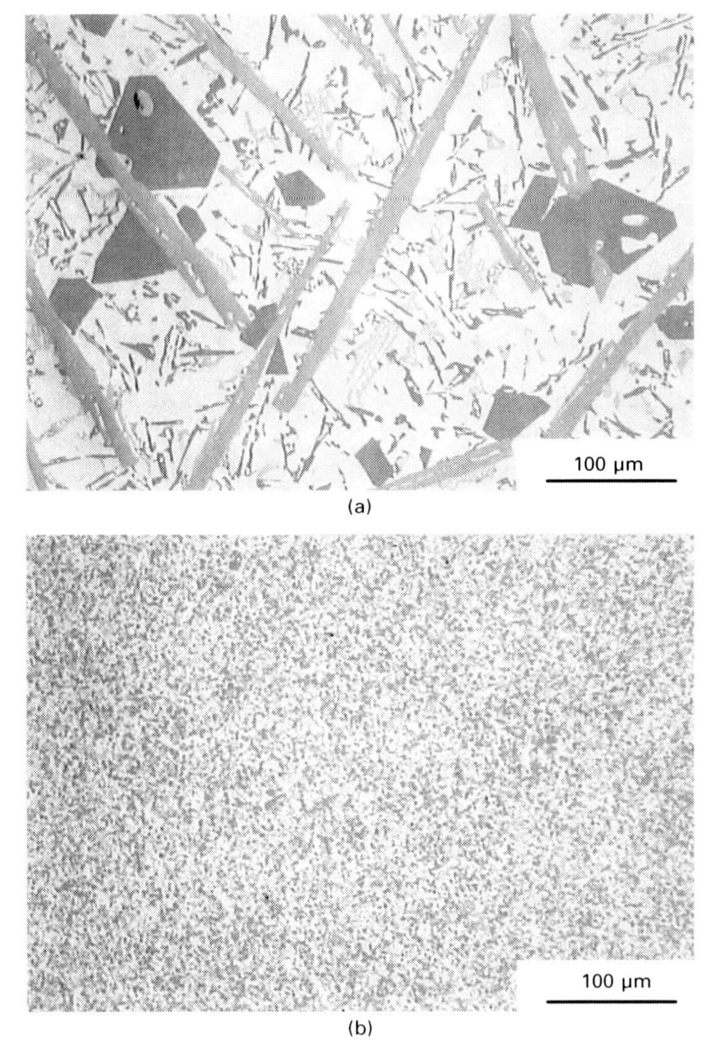

11.35 Hypereutectic Al–Si alloy, differently manufactured (photos: Powder Light Metals GmbH): (a) prepared by die casting, (b) prepared by PM/melt spinning.

fine microstructures (compare Fig. 11.37(a) and (b)). For tungsten, alloying with Re is an effective, though expensive, way to improve the mechanical properties (Fig. 11.37(c)).

For Mo, stabilization by fine carbides of Ti and Zr has proved to be effective, the respective grades are termed 'TZM'; here also, rare earth element oxides are used as dispersoids, for example for lighting purposes. Microstructures are shown in Fig. 11.37(d)–(f). In Fig. 11.37(g) and (h),

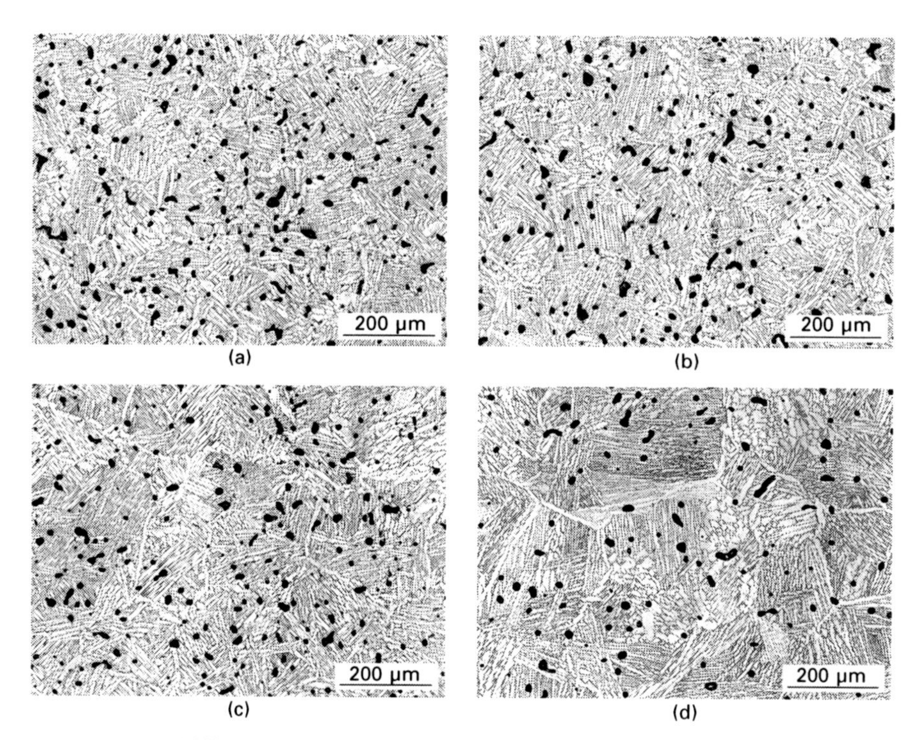

11.36 Microstructure of MIM Ti-6Al-4V, sintered at different temperatures (photos: GKSS): (a) 1250°C, (b) 1300°C, (c) 1350°C, (d)1400°C.

the microstructures of the highly corrosion resistant metals Ta and Nb are shown.

There is also a group of W-based, and to a lesser degree also Mo-based, two-phase materials, so-called 'pseudoalloys', with W and Mo as the main constituent and a binder phase formed of other metals. One example is the group of materials known as 'tungsten heavy alloys' which consist of W spheres embedded in an austenitic Ni–Cu or Ni–Fe(X) matrix. These materials are mixed from the starting powders, pressed and liquid phase sintered; the resulting microstructure, called 'heavy alloy structure', is shown above in Fig. 11.5(a); this combination of hard but fairly brittle W and soft but ductile Ni base matrix results in high strength and ductility, being one of the rare examples in which the optimum properties of different materials are effectively combined.

The other group of two-phase materials is formed by combination of refractory metals with Cu (occasionally also Ag). This aims at combining the high conductivity of Cu and Ag with the high arc resistance of W, for high current switches, or the low thermal expansion of W and Mo, for heat sink applications in electronics. Traditionally these materials have been

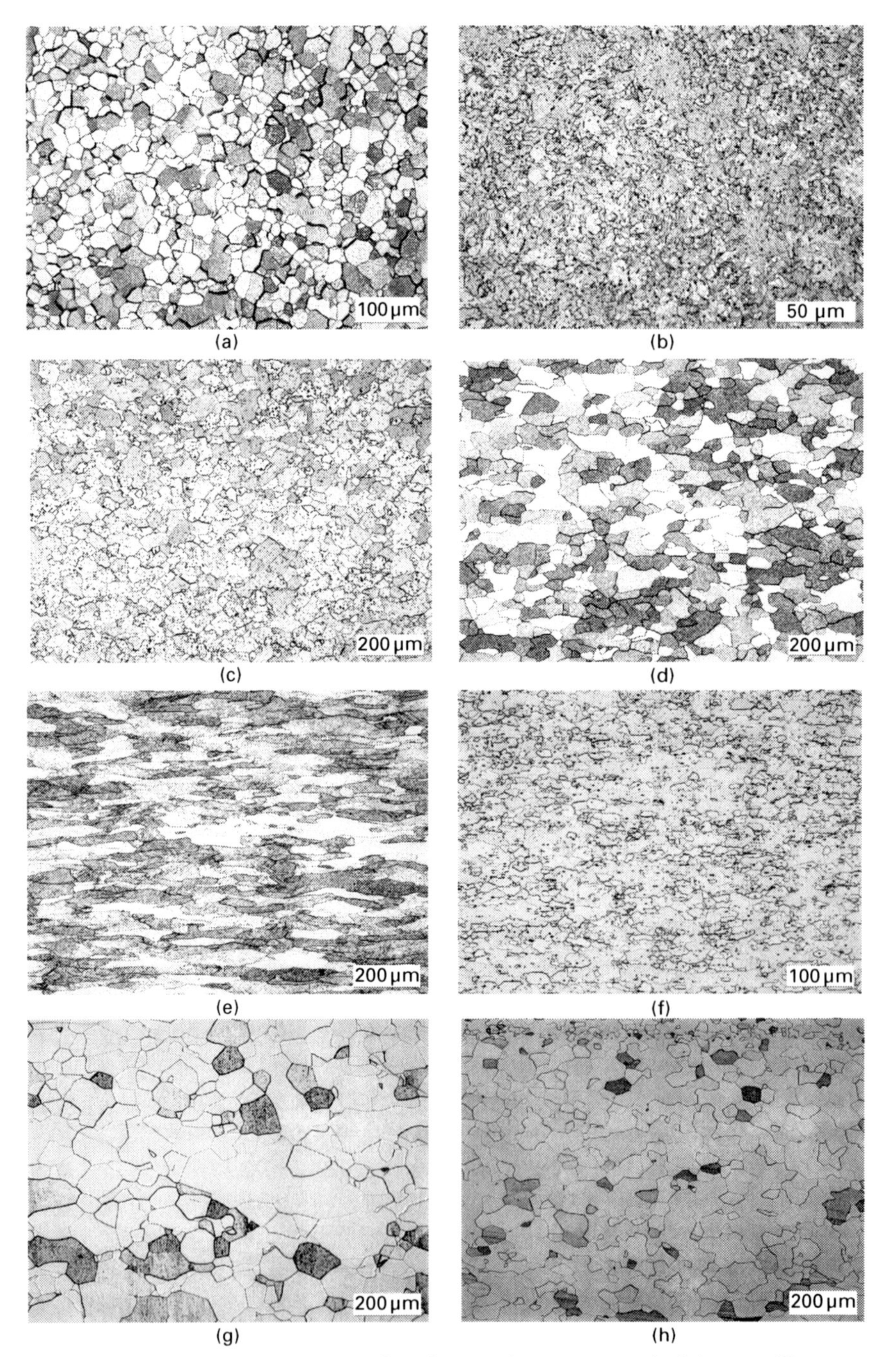

11.37 Microstructures of various refractory metals (photos: Plansee SE): (a) plain W, recrystallized, (b) W–La_2O_3, bar, (c) W–Re, sintered, (d) Mo, recrystallized, (e) Mo rolled, (f) Mo–Y_2O_3–CeO_2, wire, (g) plain Ta, sintered, (h) plain Nb, recrystallized.

manufactured by infiltrating a porous skeleton with liquid Cu or Ag; by using composite powders or ultrafine W powder coated with Cu, the press-and-sinter route can be used which for standard powder mixes yields poor densification during sintering. In Fig. 11.38 the microstructure of a W–20% Cu pseudoalloy is shown in the rolled condition.

11.4.5 Microstructures of hard metals

For hard metals, there are two major groups to be distinguished regarding composition and microstructure: the WC-based ones – the 'traditional' hard metals – and the TiCN-based grades, also known as 'cermets'. One principal difference is that WC is a compound with a precisely defined composition while TiCN may vary in composition: on one hand there is a complete phase field from plain TiC to plain TiN, and on the other hand, metallic elements may be taken into the lattice. Therefore, cermets are less critical regarding composition, in particular carbon content, than WC-based grades and they offer a wider variety of microstructures. In Fig. 11.39, the two microstructural types are shown.

For WC–Co hard metals, the WC grain size is a very important criterion, as is the binder content. For metal cutting, fine and ultrafine grades with low to moderate binder content are mostly used (Fig. 11.40(a) and (b)) which may contain cubic carbides such as TiC, NbC and TaC, while for rock drilling or chipless forming operations, coarser microstructures and higher binder contents are preferred (Fig. 11.40(c) and (d)). For ultrafine grades not only do correspondingly fine WC and Co powders have to be employed, but the

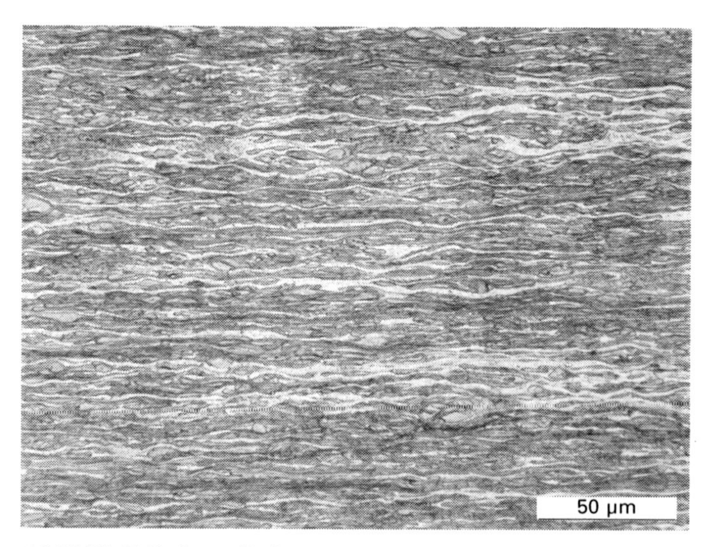

11.38 W–20% Cu, rolled.

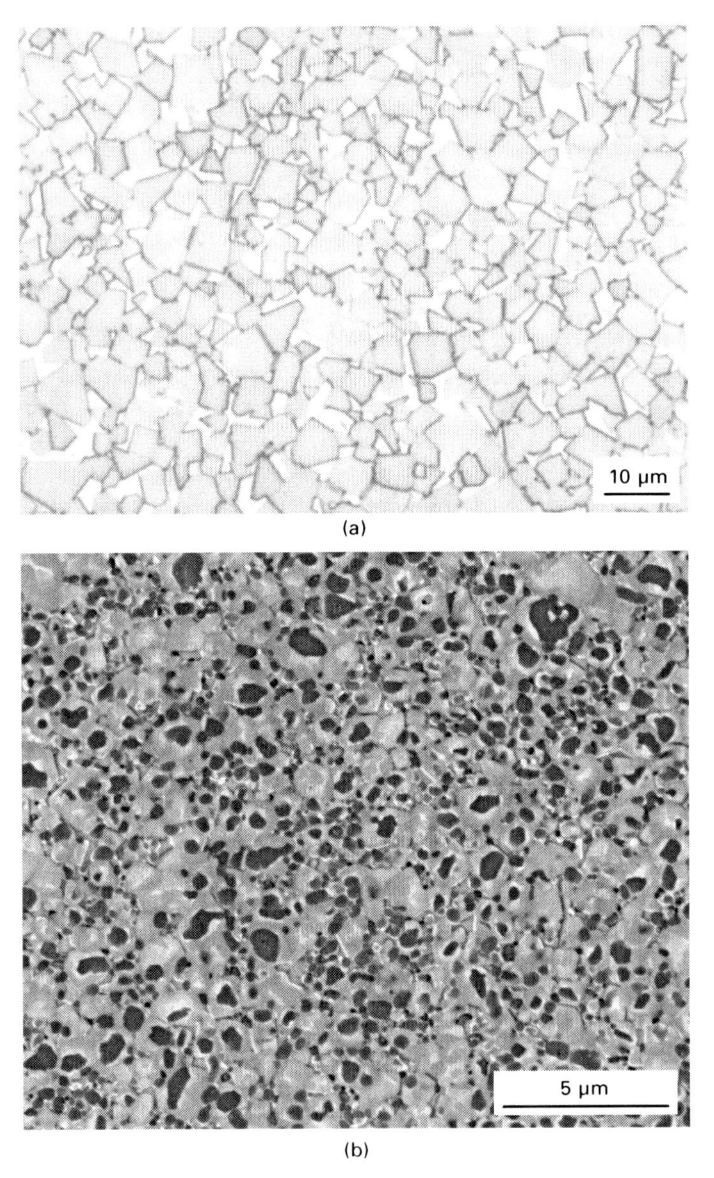

11.39 Microstructures of classical WC–Co hard metal and of cermet: (a) WC–Co hard metal, (b) TiCN base cermet.

WC grain growth during sintering, which is a problem with ultrafine grades although not with standard hard metals, has to be prevented by addition of grain refining additives such as VC or Cr_3C_2. A most critical effect is the formation of isolated very large WC grains in an otherwise ultrafine microstructure which greatly lowers the mechanical properties; this effect

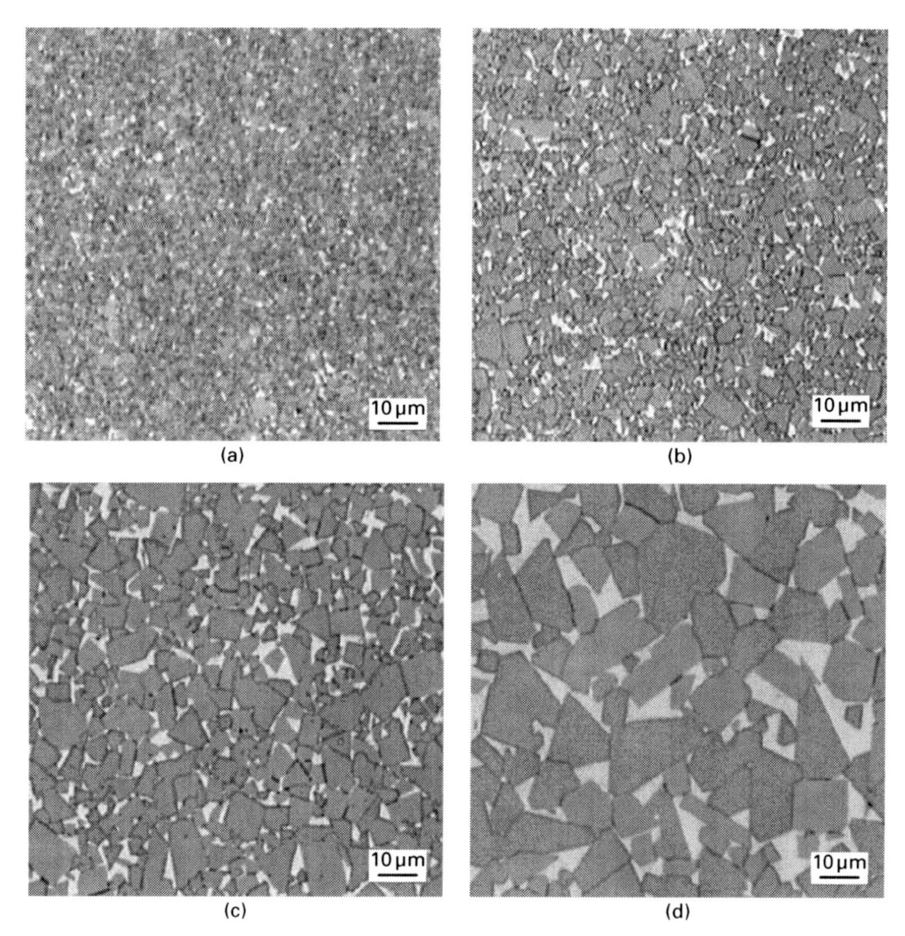

11.40 Microstructure of WC–10 mass% Co hard metal with identical binder content but different WC grain sizes: (a) sub micrometre, (b) fine, (c) coarse, (d) extra coarse.

must be prevented by very careful control of processing and in particular of the chemistry.[45]

By suitable selection of the binder composition, also Ni–Co and Fe–Ni–Co binders being used, different microstructures with their resulting properties are obtained, for example for improved corrosion resistance. Non-conventional WC morphologies are also accessible, for example platelets, as shown in Fig. 11.41; such hard metal grades may offer improved toughness properties. Even hard metals with rounded WC grains can be obtained (Fig. 11.42).

The 'cermet' type hard metals offer the unique chance to produce graded microstructures,[46,47] by changing the C–N ratio in the main hard phase TiCN. By modifying the atmosphere during sintering, either N can be removed from the surface, resulting in plain TiC there, or enriched, forming a high-TiN

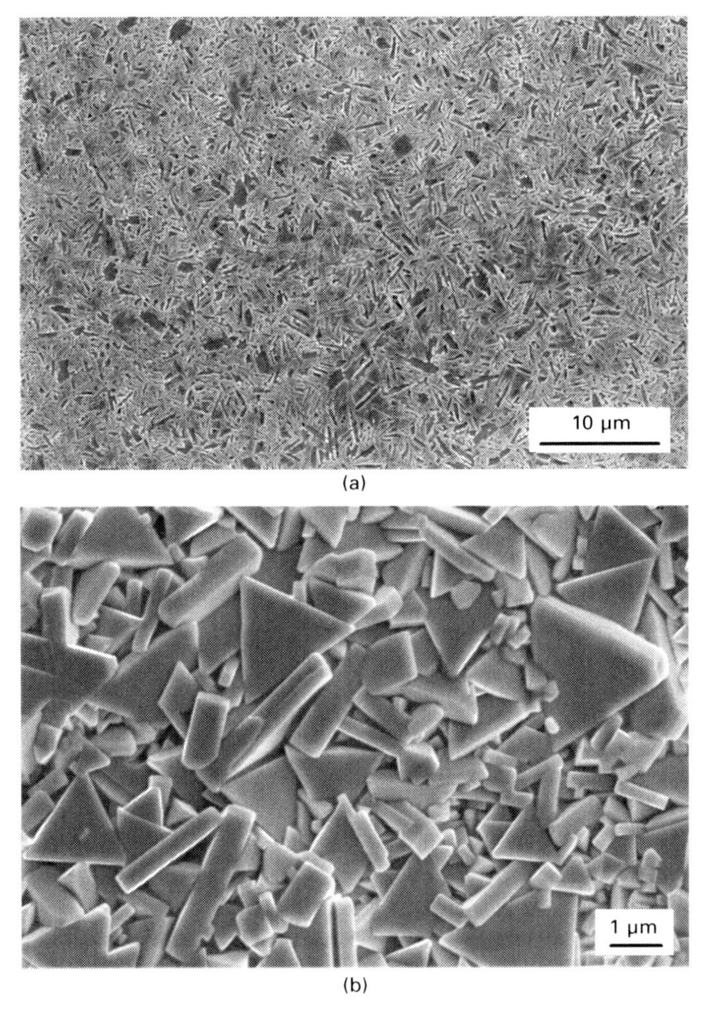

11.41 Microstructure of WC–Co hard metal with platelets: (a) metallographic section, (b) sintered surface.

surface region that exhibits cutting properties similar to those of TiN-coated hard metals but without the problem of a sharp transition between substrate and coating. Recently it has been shown that this technique can also be applied to WC–Co hard metals that contain some Ti.[48]

11.5 Trends in microstructures of powder metallurgy products

For ferrous PM materials one major direction is lowering the total porosity in order to improve the mechanical properties which strongly depend on the

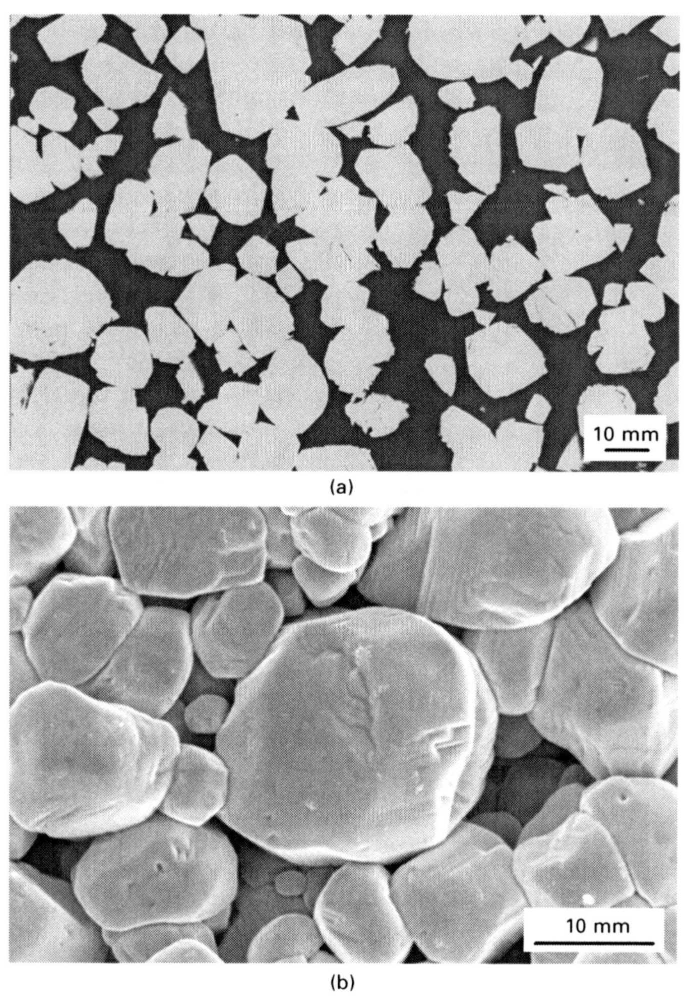

11.42 Microstructure of WC–Co hard metal with rounded WC: (a) metallographic section, (b) section etched with HCl over a long period.

density/total porosity. In particular, combining high relative density with intense sintering can be expected to result in fully closed pores, which, as shown above, significantly improves the mechanical behaviour, increasing the effective load-bearing cross section. For injection moulded products, the pores are isolated anyhow; here, further lowering the porosity, even fully eliminating them by a hot isostatic pressing (HIP) treatment after sintering, is attractive.

Another promising approach is to utilize fully the capabilities of the powder route with regard to the matrix, for example towards manufacturing materials with defined compositional, and thus microstructural, inhomogeneity, concentrating alloy elements and resulting microstructural constituents in those areas where they are most effective, for example at the sintering contacts. This will result in more effective utilization of the mostly expensive alloy elements.

For the fully dense PM materials, finer, more regular microstructures are aimed at, regardless of whether the products are high strength aluminium alloys, tool steels or hard metals. In the latter case, varying the morphology of the hard phases offers significant potential. A further requirement is a decrease in singular defects such as non-metallic inclusions, pore clusters, large carbide grains, both regarding size and frequency, since these are the more probable to act as the sites of crack initiation the higher is the basic performance of the material. Therefore, the full potential of many PM materials can only be exploited in the case of a clean and regular microstructure. On the other hand, the use of PM routes to manufacture composites is still to be fully exploited, regardless of whether metal–ceramic or metal–polymer composites are produced. In general, despite the fairly long history of powder metallurgy, tailoring of specific microstructures still offers numerous opportunities for high performance structural and functional materials.

11.6 Acknowledgements

The authors wish to thank for supplying information and illustrations: W.D.Schubert, W.Lengauer, TU Wien, Vienna, Austria; B.Lindqvist, Höganäs AB, Höganäs, Sweden; M.Dlapka, G.Stetina, MIBA Sinter Austria GmbH, Vorchdorf, Austria; F.Petzoldt, Fraunhofer-IFAM, Bremen, Germany; M.Azadbeh, Sahand Univ., Tabriz, Iran; T.Ebel, GKSS, Geesthacht, Germany; K.Hummert, Powder Light Metals GmbH, Gladbeck, Germany; H.-C.Neubing, formerly Ecka Granulate GmbH, Germany; L.Sigl, Plansee SE, Reutte, Austria; J.L.Garcia, Sandvik Machining Solutions Sverige AB, Stockholm, Sweden.

11.7 Further reading

Books

ASM Handbook Vol.7: Powder Metal Technologies and Applications. ASM, Materials Park OH, 1998.

ASM Handbook Vol.9: Metallography and Microstructures. ASM, Materials Park OH, 2004.

Metals Handbook Vol.12: Fractography. 9th edn, ASM, Materials Park OH, 1987.

W.J.Huppmann and K.Dalal. *Metallographic Atlas of Powder Metallurgy*, Verlag Schmid, Freiburg, Germany, 1986.

P.Beiss, K.Dalal and R.Peters. *International Atlas of Powder Metallurgical Microstructures*, Metal Powder Industries Federation, 2002.

Höganäs Handbook for Sintered Components Vol.6 Metallography, Höganäs, 1999.

Metallographic Preparation of Powder Metallurgy Parts, www.struers.de, know how, application notes.

Journals

Praktische Metallographie / Practical Metallography, Hanser-Verlag (bilingual English–German), 12 issues/year.

11.8 References

1. W. Schatt, K.P. Wieters and B. Kieback, *Pulvermetallurgie*, 2nd edn, Springer, Berlin-Heidelberg-New York, 2007.
2. M. Drozda and W.A. Kaysser, *Pract. Metallogr.*, 1979, **16**, 578–82.
3. U. Engström, in, *Powder Metallurgy–State of the Art*, W.J. Huppmann, W.A. Kaysser and G. Petzow (eds), Verlag Schmid, Freiburg, 1986, 41–70.
4. M.W. Wu, K.S. Hwang and H.S. Huang, *Met. Mater. Trans.*, 2007, **38A**, 1598–1607.
5. F. Castro, S. Sainz, B. Lindsley, K.S. Narasimhan and W.B. James, *Proceedings EuroPM 2011*, Barcelona, EPMA, Shrewsbury, 2011, **1**, 47–53.
6. W. Garcia, S. Sainz, A. Karuppannagounder and F. Castro, *Advances in Powder Metallurgy and Particulate Materials* 2008 (Proceedings Powder Metallurgy World Congress 2008, Washington DC), compiled by R. Lawcock, A. Lawley, P.J. McGeehan, MPIF, Princeton NJ (2008) Part 5, 139–149.
7. F. Bernier, P. Plamondon, J.-P. Baillon and G. L'Esperance, *Powder Metall*, 2011, **54**(5), 559–65.
8. K.J.A. Brookes, *Hardmetals and Other Hard Materials*. International Carbide Data, East Barnet UK, 1992.
9. H. Danninger, M. Schreiner, G. Jangg, B. Lux, *Pract. Met.* 1983, **20**, 64–73.
10. M. Slesar, *Proceedings 7th International Powder Metallurgy Conference*, Dresden, 1981, **1**, 83–103.
11. M. Slesar, E. Dudrova, L. Parilak, M. Besterci, E. Rudnayova, *Sci. Sintering*, 1987, **19**, 17–30.
12. H. Danninger, D. Spoljaric, G. Jangg, B. Weiss, R. Stickler, *Prakt. Metallogr.*, 1994, **31**(2), 56–69.
13. H. Danninger, U. Sonntag, B. Kuhnert, R. Ratzi, *Prakt. Metallogr.*, 2002, **39**(8), 414–25.
14. M. Dlapka, H. Danninger, C. Gierl, B. Lindqvist, *Met. Powder Rep.*, 2010, **65**(2), 30–33.
15. G. Straffelini, A. Molinari, *La Metallurgia Italiana*, 2002, **10**, 31–36.
16. L. Blanco, M. Campos, J.M. Torralba, D. Klint, *Powder Metall.*, 2005, **48**(4), 315–22.

17. B. Kieback, W. Schatt, *Planseeber. Pulvermet.*, 1980, **28**, 204–15.
18. H. Danninger, *Powder Metall. Int.*, 1988, **20**(1), 21–5.
19. D. Spoljaric, H. Danninger, D. Chen, B. Weiss, R. Stickler, *Proceedings PM'94 Powder Metallurgy World Congress 1994 Paris*, SF2M, EPMA, Paris (1994) Vol. II, 827–30.
20. H. Danninger, D. Spoljaric, B. Weiss, *Int. J. Powder Metall.*, 1997, **33**(4), 43–53.
21. M. Dlapka, S. Strobl, H. Danninger, C. Gierl, *Prakt. Metallogr.*, 2010, **47**(12), 686–99.
22. M. Momeni, *PhD thesis*, Vienna University of Technology, 2010.
23. H. Danninger, G. Frauendienst, K.-D. Streb and R. Ratzi, *Mater. Chem. Phys.*, 2001, **67**, 72–7.
24. L. Alzati, R. Gilardi, G. Pozzi, S. Fontana, *Advances in Powder Metallurgy and Particulate Materials* 2011 (Proceedings Powder Met 2011, San Francisco), compiled by I.E. Anderson, T.W. Pelletiers, MPIF, Princeton NJ, 2011, **7**, 11–18.
25. *Höganäs Handbook for Sintered Components Vol.6 Metallography*, Höganäs (1999).
26. M. Kupkova, M. Kupka, S. Strobl, C. Gierl, J. Wagesreither, *Powder Metall. Progr. 7*, 2007, **1**, 35–43.
27. H. Danninger, S. Strobl, R. Guertenhofer, E. Dudrova, *Proceedings 2000 Powder Metallurgy World Congress*, Kyoto, K. Kosuge and H. Nagai (eds), The Japan Society of Powder and Powder Metallurgy, 2001, Part I, 394–397.
28. M. Dlapka, *PhD Thesis*, Vienna University of Technology, 2011.
29. M. Dlapka, C. Gierl, H. Danninger, H. Altena, P. Orth, G. Stetina, *Berg- und Hüttenmaenn. Mh.*, 2009, **154**, 200–204.
30. E. Santuliana, C. Menapace, S. Libardi, G. Lorenzi, A. Molinari, *Int. J. Powder Metall.*, 2011, **47**(6), 38–45.
31. W.V. Knopp, *Advances in Powder Metallurgy and Particulate Materials* 1996 (Proceedings Powder Metallurgy World Congress 1996, Washington DC) T.M. Cadle, K.S. Narasimhan eds., MPIF, Princeton NJ, 1996, **11**, 176–170.
32. H. Danninger, J.M. Garmendia, R. Ratzi, *Powder Metall. Progr.*, 2010, **10**(3), 121–132.
33. C. Sohar, *Lifetime Controlling Defects in Tool Steels*. Springer, Heidelberg-Dordrecht-London-New York, 2011.
34. M. Azadbeh, H. Danninger, C. Gierl, *Proceedings EuroPM2011 Barcelona*, EPMA, Shrewsbury, 2011, **3**, 99–104.
35. S. Storchheim, *Progr. Powder Metall.*, 1962 **18**, 124–130.
36. W. Kehl, H.F. Fischmeister, *Powder Metall.*, 1980, **23**(3), 113–119.
37. H.-C. Neubing, G. Jangg, *Met. Powder Rep.*, 1987, **42**, 354–358.
38. H.-C. Neubing, in, *Pulvermetallurgie in Wissenschaft und Praxis Bd.20*, H. Kolaska (ed), Fachverband Pulvermetallurgie, Hagen, 2004, 3–29.
39. T. Schubert, T. Pieczonka, S. Baunack, B. Kieback, *Proceedings EuroPM2005 Prague*, EPMA, Shrewsbury, 2005, **1**, 3–8.
40. D. Kent, J. Drennan, G. Schaffer, *Acta Mater.*, 2011, **59**, 2469–80.
41. G.B. Sercombe, G.B. Schaffer, *Acta Mater.*, 1999, **47**, 689–97.
42. G.C. Obasi, O.M. Ferri, T. Ebel, R. Bormann, *Mat. Sci. Eng. A*, 2010, **527**, 3929–35.
43. E. Lassner, W.D. Schubert, *Tungsten*. Kluwer Academic/Plenum Publishers, New York, 1999.

44. P.K. Johnson, *Int. J. Powder Metall.*, 2008, **44**(4), 43–8.
45. M. Sommer, W.D. Schubert, E. Zobetz, P. Warbichler, *Int. J. Refract. Met. Hard Mater.*, 2002, **320**, 41–50.
46. W. Lengauer, K. Dreyer, *J. Alloys Comp.*, 2002, 1–2 **338**, 193–212.
47. W. Lengauer, K. Dreyer, *Int. J. Refract. Met. Hard Mater.*, 2006, **24**, 155–161.
48. J.L. Garcia, *Int. J. Refract. Met. Hard Mater.*, 2011, **29**, 306–311.

Part III

Manufacturing and densification of powder metallurgy components

12
Microwave sintering of metal powders

D. AGRAWAL, Pennsylvania State University, USA

DOI: 10.1533/9780857098900.3.361

Abstract: This chapter deals with an overview and current status of the application of microwave energy to the processing of metallic materials for various applications including steel making. The chapter especially focuses on the sintering aspect of some important selected metal powders. Microwave energy has emerged as the most versatile form of energy applicable to numerous diverse fields: communication, chemistry, rubber vulcanization, drying, food processing, medical treatment and diagnosis, and a variety of materials processing fields. The latest application of microwave energy has been to the sintering of metallic powders very effectively. In the last few years many researchers have reported sintering, melting, joining and brazing of metallic materials.

Key words: energy savings, hard metals, metals, microwave sintering, steel making.

12.1 Introduction and background

Since the first application of microwave energy for radar in World War II, it has emerged as the most versatile form of energy to find highly advantageous utilization in many diverse fields such as communications, food processing, rubber vulcanization, textile and wood products, medicine, chemical reactions, drying and sintering ceramic powder, and so on. This has been driven primarily because microwave materials processing offers many advantages over the conventional heating methods; these include substantial enhancements in the reaction and diffusion kinetics, relatively much shorter cycle times, finer microstructures leading to better quality products, substantial energy savings and eco-friendliness.

Microwaves are electromagnetic radiation with wavelengths ranging from about 1 mm to 1 m in free space and frequencies between 300 GHz and 300 MHz, respectively. However, only very few frequency bands in this range are allowed for research and industrial applications to avoid interference with communications which is its most widespread application worldwide. The most common microwave frequencies used for research and industrial applications are 2.45 GHz, 28–30 GHz and 915 MHz. The same frequency at which the kitchen microwaves also operate, 2.45 GHz, is the most common frequency for research and industrial applications.

Microwave processing of materials was mostly confined until the end of the last century to ceramics, semiconductors, inorganic and polymeric materials. Until then, there had been very few reports on microwave processing of metals. The main reason for this lack of work in microwave heating/sintering of metals was due to the misconception that all metals reflect microwave and/or cause plasma formation and hence cannot be heated, except for exhibiting surface heating owing to limited penetration of the microwave radiation. This observation is evident from the conventional view shown in Fig. 12.1 where a plot between microwave energy absorption in solid materials and electrical conductivity[1] is presented. It is evident from this plot that only semiconductors are good microwave absorbers, ceramics/insulators are transparent in microwaves, and the metals should reflect microwaves. However, the researchers did not notice that this relation is valid only for sintered or bulk materials at room temperature and not for powdered materials and/or at higher temperatures. It has been now proved that all metallic materials in powder form do absorb microwaves and even bulk metals if they are pre-heated to a temperature of at least 400°C, also start coupling in the microwave field and can be heated very efficiently to their melting points.

At 2.45 GHz, it is observed that the skin depth in the bulk metals is very low (of the order of a few micrometers) and hence very little penetration of microwaves takes place. However, in the case of fine metal powders, rapid heating can occur. A theoretical model predicted that if the metal powder particle size is less than 100 μm, it will absorb microwaves at 2.45 GHz (unpublished data). It was further observed that the degree of microwave absorption depends upon the electrical conductivity, temperature and the frequency. In magnetic materials, other manifestations of microwave coupling

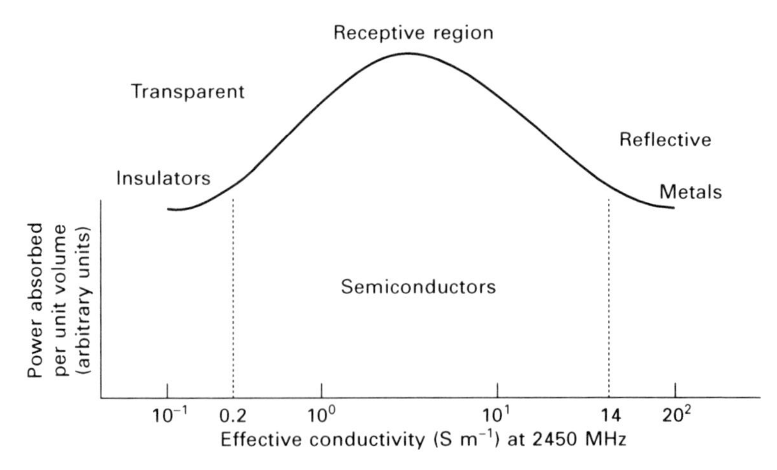

12.1 Microwave energy absorption as a function of electrical conductivity.

include hysteresis losses, dimensional resonances and magnetic resonances caused by unpaired electrons.[2]

The earliest work of microwave interaction with metallic powders is reported in 1979 by Nishitani,[3] who reported that by adding a few percent of electrically conducting powders such as aluminum, the heating rate of the refractory ceramics is considerably enhanced. Walkiewicz *et al.*[4] likewise exposed a range of materials, including six metals, to a 2.45 GHz field and reported modest heating (but not sintering) in the range from 120°C (Mg) to 768°C (Fe). Whittaker and Mingos[5] used the high exothermic reaction rates of metal powders with sulfur for the microwave-induced synthesis of metal sulfides. Sheinberg *et al.*[6] heated Cu powders coated with CuO to 650°C but did not report any sintering of them. Narsimhan *et al.*[7] succeeded in heating Fe alloys in a microwave oven only up to 370°C in 30 min. But in all these studies no sintering of pure metal or alloy powders was reported. It was only in 1998 that the first attempt at microwave sintering of powder metals took place at Pennsylvania State University (Penn State)[8] and since then many other researchers have reported successful sintering of many metallic materials.[9–12]

Figure 12.2 illustrates a schematic of a typical microwave system that is generally used for the processing of metallic materials in the laboratory and is self-explanatory.[13] The system is capable of operating in control atmosphere and temperatures up to 1600°C can be achieved.

This chapter primarily deals with the application of microwave energy to sintering of metal powders, brazing/joining/melting of bulk metals and to steel making. In the following sections recent results of sintering of

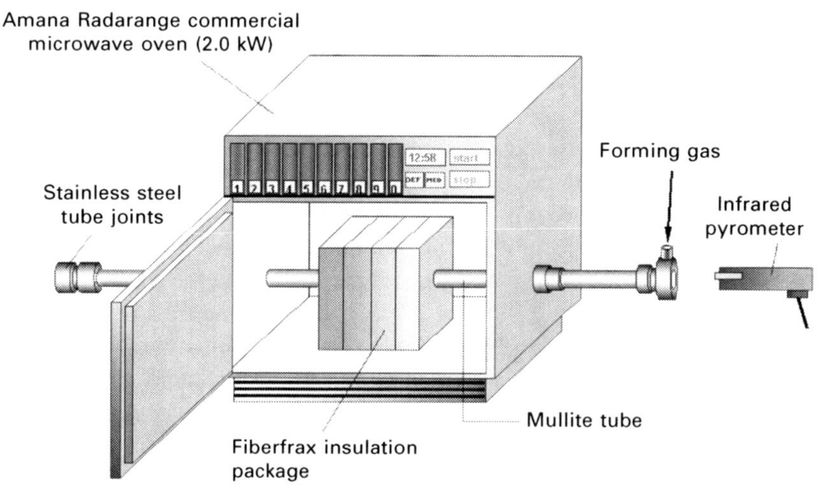

12.2 Schematic of a multimode 2 kW and 2.45 GHz microwave processing system for metallic materials.

metallic powders including pure metals, alloys and steel compositions have been presented. An attempt has also been made to explain the mechanism of microwave heating of powder metals in a separate section. The chapter concludes by providing some opinions about future trends in the field and suggested further reading on the subject.

12.2 Sintering of metallic powders

Microwave sintering of powder metallurgy (PM) green bodies comprising various metals, steels and metallic alloys, in general, produces highly sintered bodies in a very short period of time. In the last decade various metal/alloy powders such as Fe, Cu, Al, Ni, Mo, Co, Ti, W, WC, WHAs, Sn, brass, have been sintered using microwave successfully by various researchers; only the results of some representative metal powders are presented hereunder.

12.2.1 Steel powders

Microwave sintering of various commercial steel compositions was accomplished in a multimode, 2.45 GHz control atmosphere furnace with a typical cycle time of about 90 min, sintering temperature ranges between 1100°C and 1300°C and soaking time of 5–60 min.[13] The mechanical properties, such as the modulus of rupture (MOR) and hardness of microwave-processed samples, were found to be higher than the conventional samples. As an example, microwave sintering of copper steel (MPIF FC-0208 composition) produced good sintered density, hardness, flexural strength, and near net dimensions, thus yielding equivalent or even sometimes superior mechanical properties to conventional sintering (Fig. 12.3).

An important distinction in the microstructures of conventional and microwave-sintered samples observed was that the pores in the microwave-sintered samples (FC208 steel) had more rounded edges than the conventional sample and hence improved ductility and strength.[14] It was found that the conventionally sintered part failed at a load of 320 lb (145 kg) and microwave sintered part at 430 lb (195 kg), indicating an increase of about 30% in the strength. But a more important feature observed was the manner in which the parts failed after applying the maximum load. The conventional part broke into four pieces which is very typical of the standard PM parts. On the other hand, the microwave processed part broke into two flat pieces, indicating a higher ductility. The explanation of this was found in the nature of the porosity. The microwave-sintered samples exhibited round edges in contrast to the typical sharp-edged porosity found in the conventionally sintered samples (Fig. 12.4).

Figure 12.5 shows some commercial steel products for the automotive industry sintered in a microwave field.

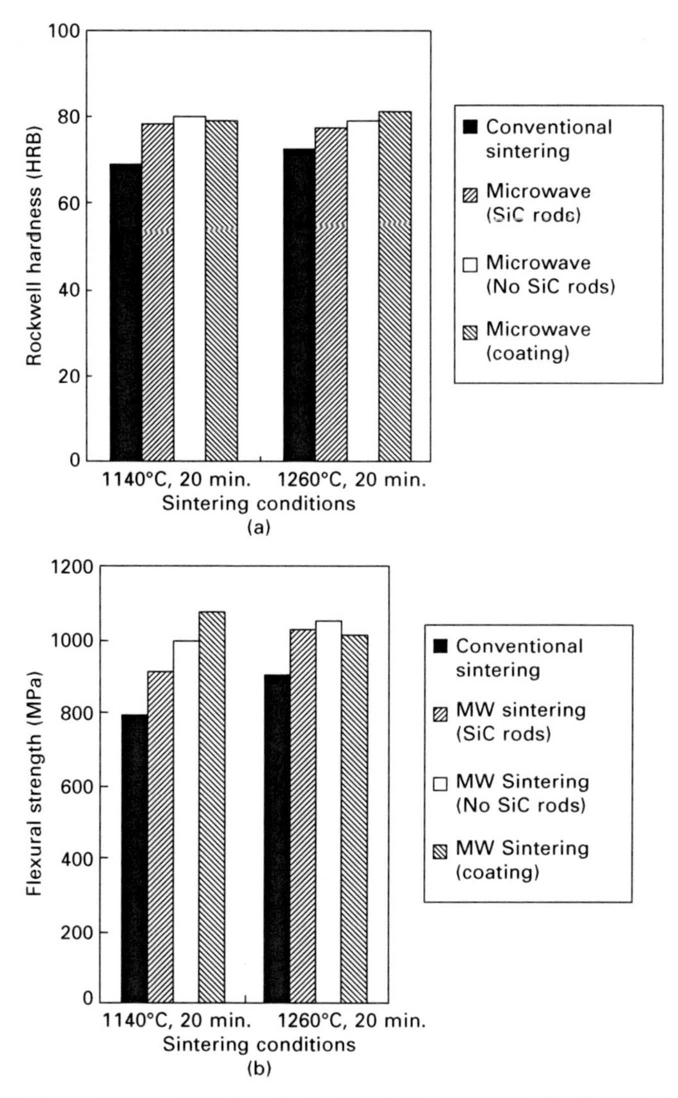

12.3 Comparison of (a) Rockwell hardness and (b) flexural strength of microwave and conventionally sintered FC208 samples of copper steel (FC 208: 2% Cu, 0.08% C).

12.2.2 Refractory metals and their alloys (W, Mo, Re, WHAs)

Refractory metals and alloys are well known for their high mechanical properties which make them useful for a wide range of high temperature applications. However, owing to their refractoriness, it is very difficult to consolidate them under moderate conditions. Conventional heating methods

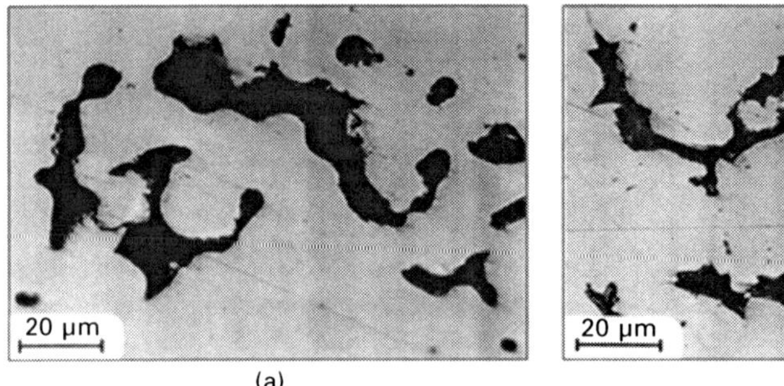

(a) (b)

12.4 Microstructure of Cu steel samples showing differences in the pore shape (black areas) in (a) microwave sintered (round edges) compared with (b) conventionally sintered (sharp edges) samples.

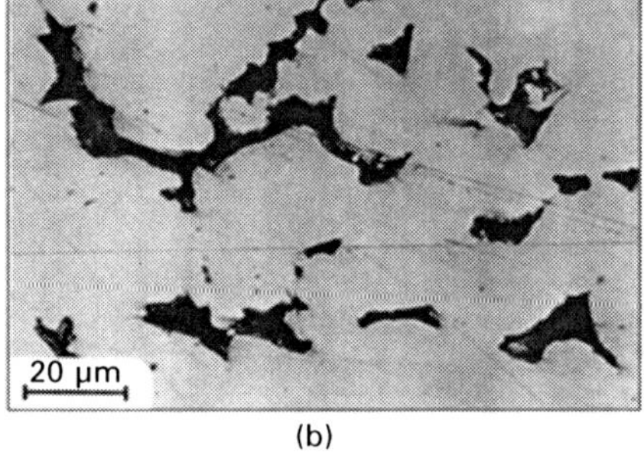

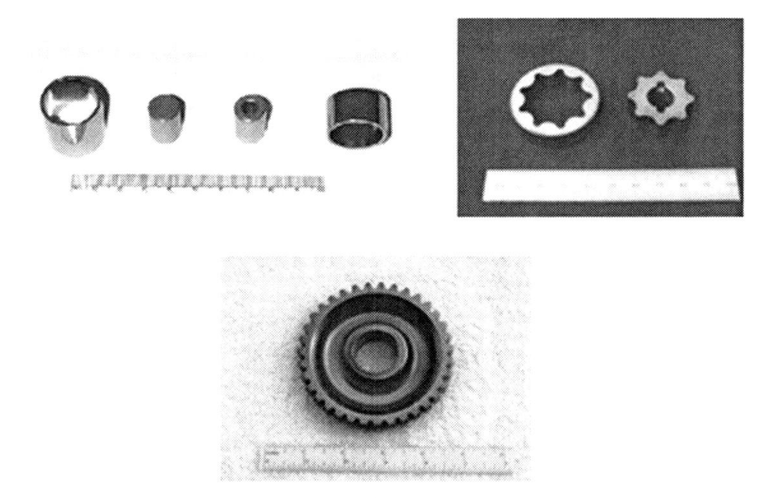

12.5 Various metal/steel parts sintered by a microwave process.

require very high temperatures and long sintering times to achieve high density in these materials. Under such conditions undesired grain growth takes place and as a result their mechanical properties are adversely affected. Microwaves in general lower the sintering temperature and time substantially, leading to fine microstructures. Some refractory metals such as W, Re, Mo, and W-based heavy alloys were microwave sintered at much lower temperatures and sintering times than normally used in a conventional process.[15] Figure 12.6 exhibits microstructures of microwave sintered nano-W powders which have been doped either with Y_2O_3 or with HfO_2 as grain growth inhibitors.[16,17] It is to be noted that microwave sintering at 1400°C for 20 min produced submicrometer size microstructures and densities in the order of 95+%.

Figure 12.7 exhibits a typical pure W dome-shaped sample sintered in microwave to near theoretical density with highly uniform shrinkage.

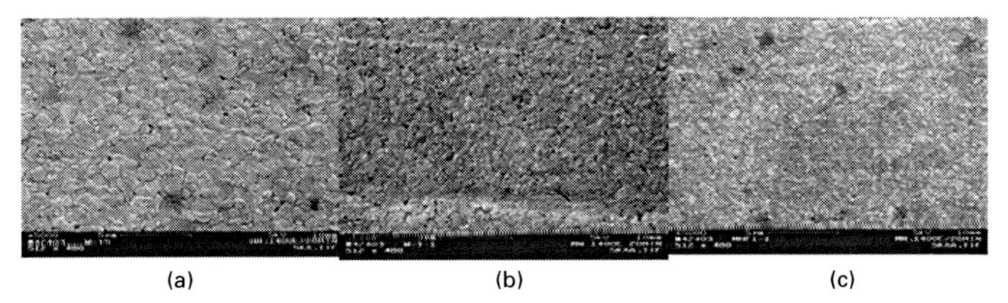

(a) (b) (c)

12.6 Microwave sintered nano-W powders at 1400°C/20 min (a) undoped (1–3 μm), (b) doped with Y_2O_3 (0.5–2 μm) and (c) doped with HfO_2 (0.5–0.75 μm).

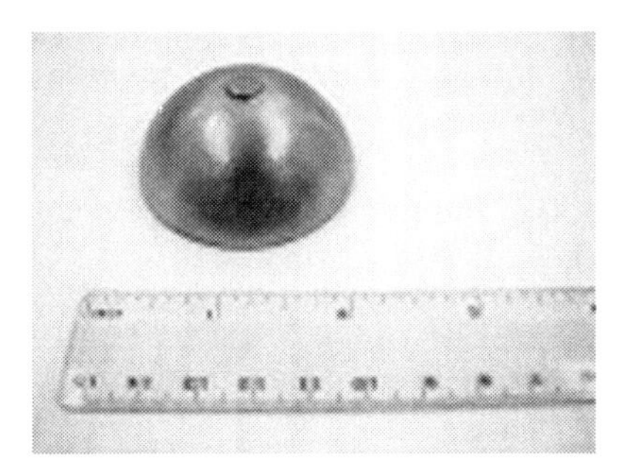

12.7 Pure W dome part sintered in microwave at 1800°C for 30 mins in pure hydrogen.

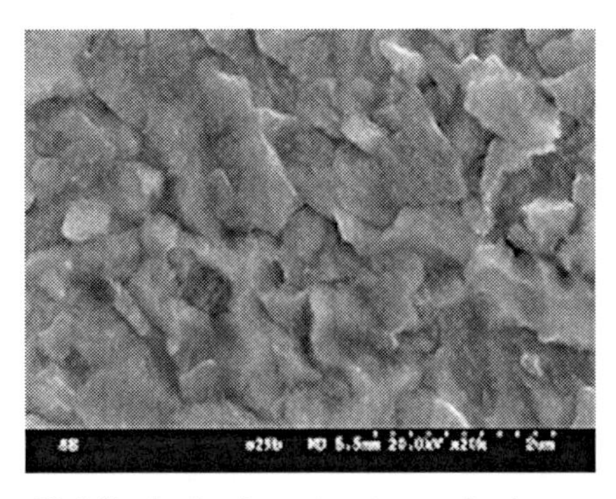

12.8 Typical microstructure of microwave sintered nano-Mo powder (1600°C for 1 min, H_2) showing average grain size of ~0.55 μm.

Figure 12.8 shows a typical microstructure of an Mo sample sintered in microwave at 1600°C for 1 min. The average grain size in this sample is also submicrometer and density about 98+%.[18]

Figure 12.9 exhibits typical microstructures of tungsten heavy alloy, WHA (W-Ni-Fe) samples sintered in microwave and conventional methods

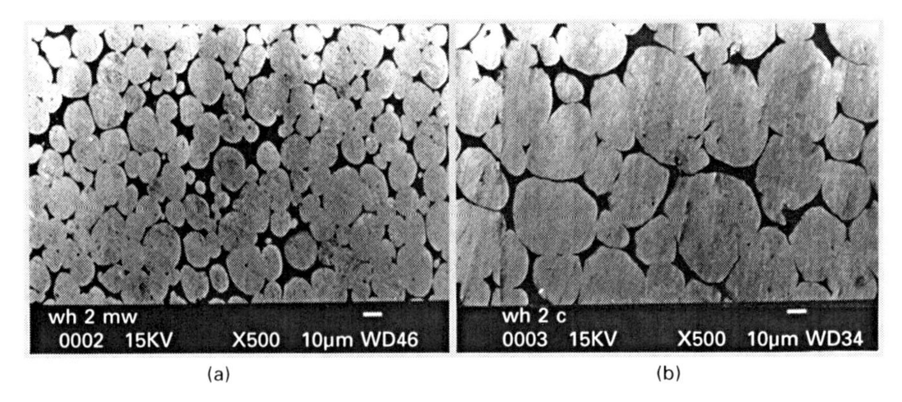

12.9 Microstructures of WHA (92.5W–6.4Ni–1.1Fe) samples sintered at 1500°C in (a) a microwave and (b) a conventional method showing much finer grains in the microwave sample.

at 1500°C, respectively, indicating that the microwave sintered sample has much smaller grain size than in conventional heating.[15]

12.2.3 Sintering of Ti, Al and Cu

Takayama *et al.*[19] used green sample compacts of Ti and Mg metal powders surrounded by BN powder and sintered them in a microwave furnace successfully. They reported higher tensile strength in the microwave-sintered products than the conventional-sintered sample. A comparative study of the sintering behavior of Cu-12Sn bronze system[20] reported that bronze was microwave sintered in significantly less time, resulting in higher density and a more uniform microstructure. Also hardness of the microwave sintered samples compacted at 300 MPa was 50% higher than the conventionally sintered samples. Gupta and Wong compared the properties of pure aluminum sintered using conventional and microwave heating[21] and found that microwave sintered material exhibited improved properties including higher hardness and ultimate tensile strength. In a recent study aluminum alloy was also successfully sintered in a multimode microwave system at 2.45 GHz.[22]

12.2.4 Sintering hard metal (WC–Co)

Hard metal (WC–Co)-based composites (also known as cemented carbides) owing to their unique combination of hardness, toughness and strength are universally used for cutting tools, machining of wear resistant metals, grinding, mining and geothermal oil and gas drilling operations. Conventional methods for sintering WC with Co as a binder phase involve high temperature (up to 1500°C) and lengthy thermal cycles (~24 h) in order to achieve a high degree of densification. These conditions usually favor undesirable WC grain growth in the presence of Co melts. Consequently, the mechanical strength and hardness of the tools are diminished. Finer microstructures provide superior mechanical properties and a longer life for the sintered product. Often, additives such as metal carbides (TiC, VC, TaC) are used to suppress the grain growth, but unfortunately such additives deleteriously affect the mechanical properties of the tools and add substantially to the overall cost of the product. Since microwave heating requires very little time to obtain nearly full densification, the grain growth is relatively suppressed and a finer microstructure is generally obtained without using any grain growth inhibitors.

In 1991, J. Cheng[23] first showed that WC/Co composites could be sintered in a microwave field. Gerdes and Willert-Porada[24] also reported the sintering of similar WC objects from normal size powders, but they followed a reactive sintering route using a mixture of pure W, C and Co instead of normal sintering. In a parallel study at Penn State fully sintered WC/Co commercial green bodies were also achieved[25] in a multimode microwave system, and it was observed that microwave processed WC/Co bodies exhibited much better mechanical properties than the conventional parts, fine and uniform microstructure with little grain growth, and nearly full density without adding any grain-growth inhibitors when sintered at 1250–1320°C for only 10–30 mins.[26–28] The microstructural examination of the microwave sintered WC/Co samples, in general, exhibited a smaller average grain size than the conventionally sintered sample. Microwave sintered parts also showed significant property improvements without varying the composition of the components and without the addition of grain growth inhibitors. The WC/Co part produced by the microwave sintering process exhibited an unprecedented improvement in abrasion resistance (15–30% better), erosion resistance (22% better) and corrosion resistance in 15% HNO_3 (20 × better) without any noticeable loss in hardness or fracture toughness when compared to the conventionally sintered materials. These improvements in the properties are believed to be due to the fine microstructure, uniform cobalt phase distribution and pure Co phase at the grain boundaries in microwave-sintered samples.[29] Figure 12.10 illustrates some commercial WC/Co parts which have been fabricated very successfully using microwave technology.

12.3 Bulk metal processing

12.3.1 Brazing/joining/melting of bulk metals

The application of microwaves to metallic materials has also been extended from sintering to melting, brazing, joining and metal coating of bulk metals.[30–33] Figure 12.11 shows bulk metal pieces of aluminum, copper and stainless steel that have been melted in a microwave field using a special

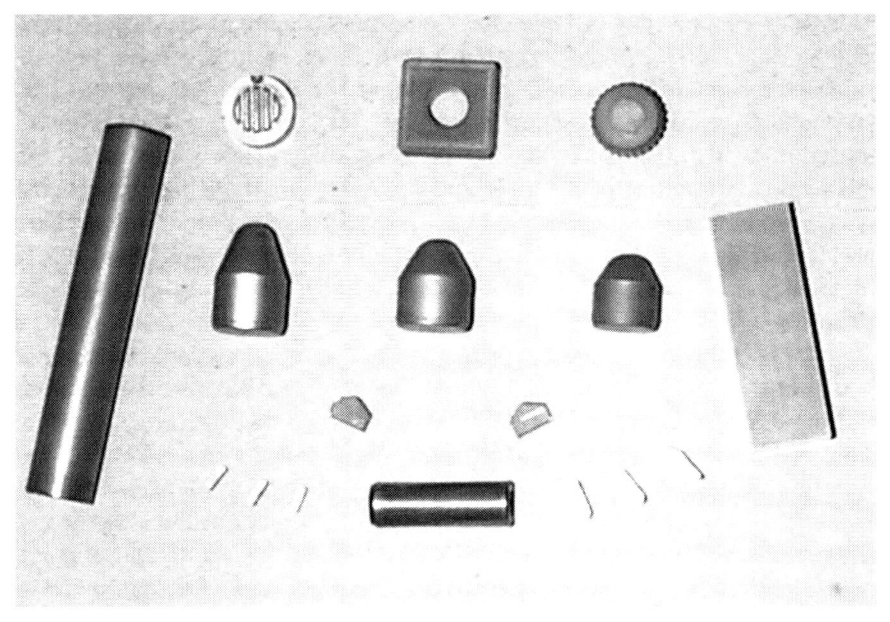

12.10 Some typical WC–Co hard metal commercial parts sintered in a microwave.

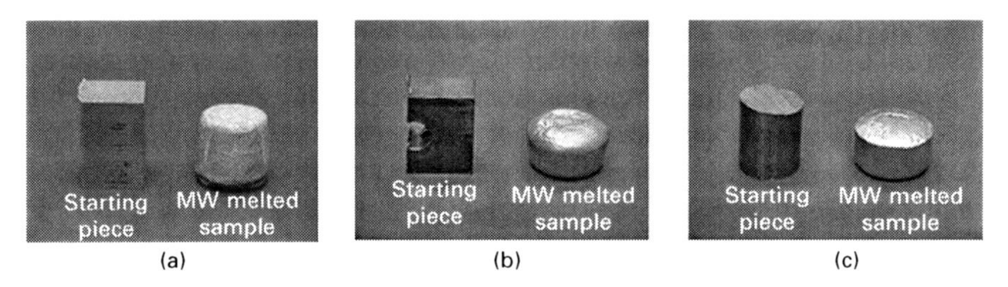

12.11 Some typical bulk metals (a) aluminum, (b) copper and (c) stainless steel before and after melting in a microwave.

insulation package with susceptors. It took less than 30 mins to melt these bulk metal samples.

Microwave heating is a material property and only those materials that couple in the microwave field will be heated. Because of this special feature, it has been found that microwaves selectively heat powder metals and reflect bulk metals at room temperature. This phenomenon has been exploited to braze and join bulk metals using powdered metal/alloy braze mixtures. Bulk steel pieces and tungsten bulk metals have been successfully joined using a powdered metal alloy as the joining medium. An example is shown in Fig. 12.12 in which regular steel and cast iron parts have been joined in the microwave field in 2–3 mins using a braze powder. The joint is almost perfect, as indicated by the microstructural examination of the sample.

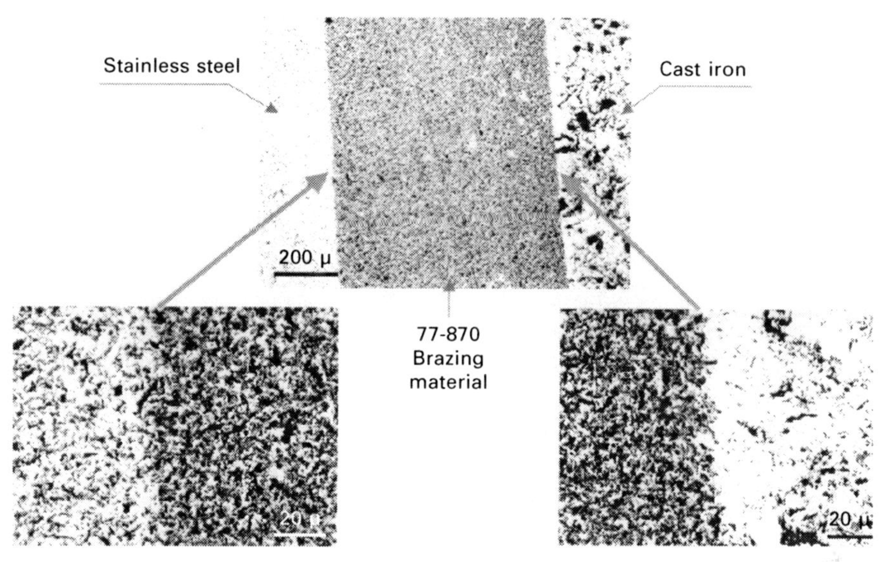

12.12 Microwave joining of stainless steel and cast iron using braze powder.

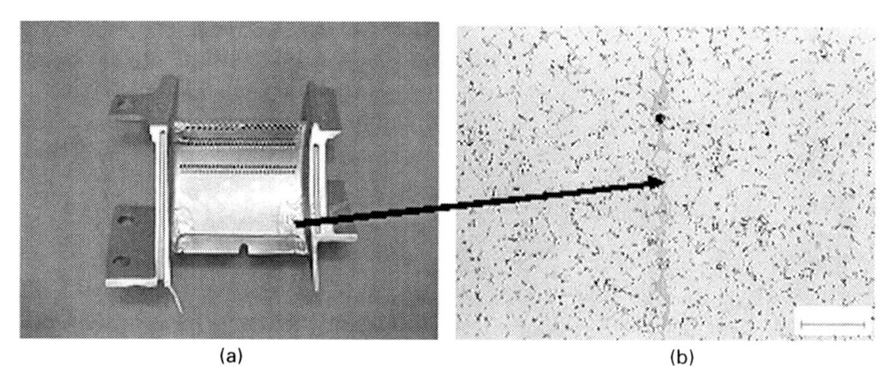

12.13 Microwave repair of damaged aircraft engine super-alloy turbine blade. (a) Brazing powder was pasted on to the damaged (cracked) area and microwave heated to infiltrate the cracks. (b) The microstructure image shows that the cracked area was fully filled with the brazing material.

Repairing/brazing damaged metal parts has also been accomplished at Penn State's microwave processing center. In a systematic investigation, repair of super-alloy based turbine blades used in aircraft engines has been achieved due to microwave's selective heating feature; it is possible to heat only the braze powder placed on the damaged area in the form of a slurry. Heating is very rapid and the brazed material melts in 2–3 mins resulting in a very strong and effective braze/repair, as shown in Fig. 12.13. Although the bonding strength of the two bulk pieces was not measured physically, looking at the microstructures the join appears quite strong.

12.3.2 Metal coatings on metallic substrates

The microwave sintering of metallic coatings deposited by the high velocity particles consolidation (HVPC) process (also known as cold spray) was achieved successfully.[34] The coating materials were nickel, cobalt and 316L stainless steel, deposited on a steel substrate. After microwave sintering, the coating materials were assessed in terms of hardness, microstructure, porosity and interfacial diffusion. The sintered Ni and stainless steel samples showed a relatively large reduction in hardness when compared to as-sprayed samples as a result of stress relief and recrystallization, which was confirmed by the microstructure images. Sintered Co samples showed a slight decrease in hardness owing to the presence of a metastable phase. Grain coarsening was also observed as the sintering temperature increases. The lower temperature sintered samples showed a higher percentage porosity than the as-sprayed sample, which would result from pore coalescence and incomplete sintering; however, the porosity tended to decrease as the sintering temperature increases. Energy dispersive spectroscopy (EDS) results showed that there was inter-diffusion occurring across the coating/substrate interface in the sintered samples, which is desired to improve the coating adhesion. This work shows that coated powdered metals on bulk metal substrates can be successfully sintered in microwaves resulting in good sinterability and adhesion.

12.3.3 Steel making

The Kyoto Protocol 2005 mandates all the signatory member nations to cut down carbon emission for all future technologies. Steel is a basic and highly important material needed for all infrastructure-related components and is a critical material for the economic growth of any country. Conventional steel making methods involve multi-step processes such as raw mixture pelletization, sintering and melting. Each of these steps is highly time and energy consuming. Additionally, they also contribute substantially to the CO_2 build-up in the greenhouse effect. Therefore, any new process for making steel with less CO_2 emission and elimination of pelletization and sintering is always sought after. Recently, Hwang *et al.*[35] succeeded in combining microwave energy with the electric arc furnace to develop a new clean and green steel-making technology in which the CO_2 has been substantially reduced and energy consumption has been cut down by 25% over the conventional basic oxygen furnace (BOF) technology. The unique aspect of this technology is that it utilizes the advantages of rapid volumetric heating, high energy efficiency and enhanced reaction kinetics that microwaves offer.

In another study, Nagata and co-workers[36,37] have succeeded in developing pure microwave-based technology for the direct manufacture of steel with at least a 50% reduction in CO_2 emission. They succeeded in producing highly pure pig iron in a multimode microwave reactor from powdered iron ores, with carbon as the reducing agent, in the nitrogen atmosphere. Iron ore contains magnetic materials like Fe_2O_3, Fe_3O_4 and FeO. Magnetite, Fe_3O_4, particularly has strong ferromagnetic properties and is heated by the magnetic field component of microwaves with great efficiency. In this process carbon powder is used only to reduce the iron ore into Fe and is heated strongly by the electric field component of microwaves. Therefore, the process for making steel (pig iron) in microwaves using a mixture of iron ore (Fe_3O_4) and carbon contains three steps: (i) step one is oxide reduction at very low temperatures, under 600°C. This is the most characteristic process of microwaves in which the direct reaction between the solid–solid phase, that is the solid carbon powder and solid oxide iron ores takes place. (ii) step two is a solid–gas/plasma reaction in the temperature range 600–1100°C, at which the reduction and heating of the mixture continues. Visible light spectroscopy revealed that line emissions originated from Fe atoms and carbon nitride molecules. (iii) step three is complete reduction of ore to produce pure pig iron at around 1100–1380°C. It is estimated that this process suppresses CO_2 emission by about 50% over the conventional process as is explained in the following reactions steps:

microwave steel making process:

$$(Fe_3O_4 + 2C) + MW\ energy = 3Fe + \underline{2}CO_2$$

conventional blast oxygen furnace for steel making:

$$(Fe_3O_4 + 2C) + (2C + 2O_2) = 3Fe + \underline{4}CO_2$$

The microwave-produced pig iron is highly pure and contains impurities of Mg, S, Si, P and Ti of the order of 1/20~1/10th of what is obtained in pig iron produced by modern conventional blast furnaces. The amount of carbon needed was two-thirds of that used in the conventional process to produce the unit weight of steel, if we applied renewable energy or nuclear power to the microwave excitations. Figure 12.14 displays microwave-produced pig iron samples. Since in this process no lime or other chemicals are added, there is no slag formation.

12.4 Microwave–metal interaction: mechanism(s)

In microwave sintering of ceramics or non-conducting materials, there are two issues: rapid heating and rapid material diffusion and/or enhancement of reaction kinetics. The heating part has been widely studied and explained

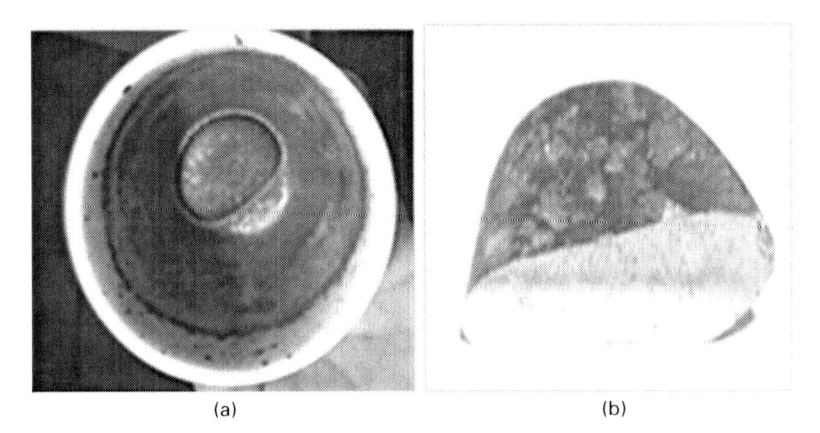

(a) (b)

12.14 Microwave produced pig iron without any slag. (a) No slag, (b) pure pig iron.

in terms of dielectric loss interactions, but the rapid material diffusion and enhancement in reaction kinetics is not yet fully understood and is generally explained by the so-called 'non-thermal' effects responsible for material diffusion. Many different physical phenomena are involved in the microwave processing of materials. One of the recent key discoveries has emphasized that the interaction between microwaves and matter takes place through both the electric field vector and the magnetic field vector.[38–41] Classically, there are various absorption mechanisms identified with ceramics processing that are always associated with the E electric field. Some of them are dipole reorientation, space and ionic charge and crystal defects, which are primarily found in insulators or dielectric materials.

However, in case of metal powders the interaction will be mainly based on the skin depth, electric conduction, scattering, eddy currents and/or magnetic losses as well.[42,43] However, there has not been any successful attempt yet to explain fully the exact causes behind unusual enhancements in reaction and sintering kinetics in a microwave field. Classical sintering equations based on thermal mechanisms cannot fully explain the material diffusion in a microwave field.

The microwave power absorbed per unit volume (P in W m^{-3}) is expressed by the equation:

$$P = 2\pi f_0 (\varepsilon_0 \varepsilon'' E^2 + \mu_0 \mu'' H^2) \quad [12.1]$$

where E and H are the electric and magnetic fields, f_0 is the frequency, ε'' and μ'' are dielectric and magnetic loss factors, respectively. So it can easily be explained that the microwave power absorbed (or the resultant heat) is directly proportional to the dielectric and magnetic losses and, of course, field intensity and frequency. There is another factor responsible for providing uniform and volumetric heating: penetration depth, D, which is the distance in the direction of penetration at which the incident power is reduced to half of its initial value and is a strong function of the loss and frequency of the field:

$$D = 3\lambda_0/8.686\pi \tan\delta(\varepsilon')^{1/2} \quad [12.2]$$

where λ_0 is the wavelength of the microwaves. In bulk metallic materials, the microwave penetration is rather low at room temperature and is commonly described by a quantity known as skin depth δ, given by:

$$\delta = 1/(\pi f_0 \mu \sigma)^{1/2} \quad [12.3]$$

where f_0 is microwave frequency, μ is the permeability, and σ is the electrical conductivity. From this equation it is noted that a bulk metallic material at room temperature would have only a few micrometers of skin depth and would reflect most of the incident power. However, the situation in case of powdered metallic materials is entirely different. They would be easily heated more effectively; smaller metallic particles, especially if they are in the range of their skin depth would be most effective in microwaves.

Cherradi *et al.*[38] claimed that in most ceramics the dielectric loss mechanism was a minor contribution to the power absorbed compared to the induction losses caused by eddy currents. These authors also attributed the heating of metals to eddy current losses from the *H* field. Their evidence, obtained using different size and shape samples in different fields, is rather convincing of the role of such eddy current losses as a major contributor to heating of metals. In a comprehensive work,[39] a survey of a variety of samples of metals, ceramics and composites, showed remarkable differences in their heating behavior depending upon whether they were exposed to the *E* field or *H* field at microwave frequencies. All powder metals were very effectively and rapidly heated in the *H* field but not in the *E* field; the reverse was true in case of insulator or ceramics. Further only powder metals could be heated; a solid metallic rod was not heated at all in either field. All these experiments were done in a single mode cavity where the *E* and *H* field can be separated.

12.5 Future trends

The implications of the application of microwave energy and its advantages for processing metallic materials as shown above are obvious in the field of metallurgy. Metal powders are used in industry for diversity of products and applications. The challenging demands for new and improved processes and materials of high integrity for advanced engineering applications require innovation and newer technologies. Finer microstructures and near theoretical densities in special PM components are still elusive and challenging.

Increasing cost is also a concern of the industry. Researchers are looking for newer technologies and processes to meet these demands. The developments reported herein using microwave processing may offer a new method of meeting these demands for producing better microstructures and properties in powder metal products. A new philosophy of steel making for the 21st century using microwave energy may change the entire metal industry.

However, considering all aspects of microwave technology for high temperature materials processing, including its inherent limitations and reluctance of the industry to adopt new technology for fear of losing the capital investment in the existing conventional technology, it is believed that the future of microwave technology for processing metallic materials is quite bright. New microwave high temperature system manufacturers in China, Japan and India are expected to dominate and change the scene in the next 5–10 years. In fact all the successes in scaling up of microwave technology so far achieved have been with the continuous microwave processing systems for specialty materials such as cemented carbides, ferrites, and varistors, metal products, and so on. And the coming decade will witness more of such systems being built for many other materials and products.

12.6 Further reading

1. *Microwave and Metals*, M. Gupta and E. Wong Wai Leong, John Wiley & Sons, Singapore, 2007.
2. *Microwave and Radio Frequency Applications*, D. C. Folz, J. H. Booske, D. E. Clark, and J. F. Gerling (eds), American Ceramic Society Publishers, Westerville, OH, USA, 2003.
3. *Microwave Theory and Applications in Materials Processing V* (*Ceramic Transactions*, 111) D. E. Clark, G. P. Binner and D. A. Lewis (eds), American Ceramic Society Publishers, Westerville, OH, USA, 2001.
4. *Microwave Theory and Applications in Materials Processing IV* (*Ceramic Transactions*, 80) D. E. Clark, W. H. Sutton and D. A. Lewis (eds), American Ceramic Society Publishers, Westerville, OH, USA, 1997.
5. *Microwave Theory and Applications in Materials Processing III* (*Ceramic Transactions*, 59), D. E. Clark, D. C. Folz, S. J. Oda and R. Silberglitt (eds), American Ceramic Society Publishers, Westerville, OH, USA. 1995.
6. *Advances in Microwave and Radio Frequency Processing (AMPERE)* W. Paroda (ed), Springer, Verlag, Berlin/Heidelberg, 2006.
7. *Microwave and Radio Frequency Applications*, R.L. Schulz and D.C. Folz (eds), Microwave Working Group, Arnold, MD, USA, 2005.
8. *Proceedings of 10th International Conference on Microwave and High Frequency Heating*, Modena, Italy, Sept.12–17, 2005, C. Leonelli and P. Veronesi (eds), Microwave Application Group, Cavriago, Italy, 2005.

9. *Foundations of Electroheat: A Unified Approach*, A.C. Metaxas, John Wiley & Sons, 1996.

12.7 References

1. Barnsley, B. P., 'Microwave processing of materials', *Met. Mater.*, 1989, **5**(11), 633.
2. Newnham, R. E., Jang, S. J., Xu, M. and Jones, F., *Ceram. Trans.*, 1991, **21**, 51.
3. Nishitani, T., *Method for Sintering Refractories and an Apparatus Therefor*, US Patent no. 4,147,911 (April 3, 1979).
4. Walkiewicz, J. W., Kazonich, G. and McGill, S. L., 'Microwave heating characteristics of selected minerals and compounds', *Min. Metall. Process*, 1988, 39–42.
5. Whittaker, A. G. and Mingos, D. M., 'Microwave-assisted solid-state reactions involving metal powders', *J. Chem. Soc. Dalton Trans.*, 1995, 2073–9.
6. Sheinberg, H., Meek, T. and Blake, R., *Microwaving of Normally Opaque and Semi-opaque Substances*, US Patent no. 4,942,278 (17 July, 1990).
7. Narsimhan K. S. V. L., J. Arvidsson, Rutz, G. H. and Porter, W. J., *Methods and Apparatus for Heating Metal Powders*, US Patent no. 5,397,530 (14 March, 1995).
8. Roy, R., Agrawal, D., Cheng, J. and Gedevanishvili, S., 'Full sintering of powdered metals parts in microwaves', *Nature*, 1999, 399,664, (June 17).
9. Takayama, S., Saiton, Y., Sato, M., Nagasaka, T., Muroga, T. and Ninomiya, Y., 'Microwave sintering for metal powders in the air by non-thermal effect', *9th International Conference on Microwave and High Frequency Heating*, Loughborough University, UK, September 2003, 369–72.
10. Sethi, G., Upadhyaya, A., Agrawal, D. and Roy, R., 'Microwave and conventional sintering of pre-mixed and prealloyed Cu-12Sn bronze', *Sci. Sintering*, 2003, **35**, 49–65.
11. Gupta, M. and Wong, W. L. E., 'Enhancing overall mechanical performance of metallic materials using two-dimensional microwave assisted rapid sintering', *Scr. Mater.*, 2005, **52**, 479–83.
12. Panda, S. S., Singh, V., Upadhyaya, A. and Agrawal, D., 'Sintering response of austenitic (316L) and ferritic (434L) stainless steel consolidated in conventional and microwave furnaces', *Scr. Mater.*, 2006, **54**, 2179–83.
13. Anklekar, R. M., Agrawal, D. and Roy, R., 'Microwave sintering and mechanical properties of P/M steel', *Powder Metal.*, 2001, **44**, 355.
14. Anklekar, R. M., Bauer, K., Agrawal, D. and Roy, R., 'Improved mechanical properties and microstructural development of microwave sintered copper and nickel steel PM parts', *Powder Metall.*, 2005, **44**, 39–46.
15. Mondal, A., Agrawal, D. and Upadhyaya, A., 'Microwave sintering of refractory metals/alloys: W, Mo, Re, W–Cu, W–Ni–Cu and W–Ni–Fe alloys, *J Microwave Power Electromag. Energy*, 2010, **44**, 28–44.
16. Jain, M., Skandan, G., Martin, K., Kapoor, D., Cho, K., Klotz, B., Dowing, R., Agrawal, D. and Cheng, J., 'Microwave sintering: a new approach to fine-grain tungsten – I', *Intl. J. Powder Metall.*, 2006, **5**(2), 45–50.
17. Jain, M., Skandan, G., Martin, K., Kapoor, D., Cho, K., Klotz, B., Dowing, R., Agrawal, D. and Cheng, J., 'Microwave sintering: a new approach to fine-grain tungsten – II,' *Intl. J. Powder Metall.*, 2006, **5**(2), 53–57.

18. Chhillar, P., Agrawal, D. and Adair, J. H., 'Sintering of molybdenum metal powder using microwave energy', *Powder Metall.*, 2008, **51**(2), 182–7.
19. Takayama, S., Saiton, Y., Sato, M., Nagasaka, T., Muroga, T. & Ninomiya, Y., 'Microwave sintering for metal powders in the air by non-thermal effect', *9th International Conference on Microwave and High Frequency Heating*, Loughborough University, UK, 2003, 369–372.
20. Sethi, G., Upadhyaya, A., Agrawal, D., 'Microwave and conventional sintering of pre-mixed and prealloyed Cu–12Sn bronze', *Sci. Sintering*, 2003, **35**(49), 49–65.
21. Gupta, M. and Wong, W.L.E. 'Enhancing overall mechanical performance of metallic materials using two-directional microwave assisted rapid sintering', *Scr. Mater.*, 2005, **52**, 479–83.
22. Padmavathia, C., Upadhyaya, A. and Agrawal, D., 'Effect of microwave and conventional heating on sintering behavior and properties of Al–Mg–Si–Cu alloy,' *Mater. Chem. Phys.*, 2011, **130**, 449–457.
23. Cheng, J., *Study on Microwave Sintering Technique of Ceramics Materials*, PhD Thesis, Wuhan University of Technology, China, 1991.
24. Gerdes, T. and Willert-Porada, M. 'Microwave sintering of metal-ceramic and ceramic-ceramic composites', *Mat. Res. Soc. Symp. Proc.*, 1994, **347**, 531–537.
25. Cheng, J., Agrawal, D., Komarneni, S., Mathis, M. and Roy, R., 'Microwave processing of WC-Co composites and ferroic titanates', *Mater. Res. Innovations*, 1997, **1**, 44–52.
26. Breval, E., Cheng, J., Agrawal, D., Gigl, P., Dennis, M., Roy, R. and Papworth, A.J., 'Comparison between microwave and conventional sintering of WC/Co Composites', *Mater. Sci. Eng.*, 2005, **A391**, 285–295.
27. Roy, R., Agrawal, D.K. and Cheng, J. 1999, *An Improved Process and Apparatus for the Preparation of Particulate or Solid Parts*, US Patent no. 6,004,505, December 21, 1999.
28. Agrawal, D., Cheng, J., Seegopaul, P. and Gao, L., 'Grain growth control in microwave sintering of ultrafine WC-Co composite powder compacts', *Proceedings Euro PM'99 Conference* (held in Turin, Italy, Nov. 1999), 2000, 151–8.
29. Agrawal, D., Papworth, A.J., Cheng, J., Jain, H. and Williams, D.B., 'Microstructural examination by TEM of C/Co composites prepared by conventional and microwave processes', *Proceedings 15th International Plansee Seminar*, G. Kneringer, P. Rodhammer and H. Wildner (eds), Plansee Holding AG, Reutte, 2001, volume 2, 677–684.
30. Barmatz, M., Jackson, H. and Radtke, R., *Microwave Technique for Brazing Materials*, US Patent no. 6,054,693 (20 April, 2000).
31. Ripley, E. B., Eggleston, P. A. and White, T. L., in *Proceedings of the Third World Congress on Microwave and Radio Frequency Applications*, D.C. Folz, J. Booske, D. Clark and J. Gerling (eds), ACS Publishers, Westerville, OH, USA, 2003, 241.
32. Moore, A. F., Schechter, D. F. and Morrow, M.S., *Method and Apparatus for Melting Metals*, US Patent Appl. US2003/0089481 (A1, 15 May, 2003).
33. Gedevanishvili, S., Agrawal, D., Roy, R. and Vaidhyanathan, B., *Microwave Processing Using Highly Microwave Absorbing Powdered Material Layers*, US Patent no. 6,512,216 (January 28, 2003).
34. Chanthapan, S., Shoffner, B. W., Eden, T. J. and Agrawal, D., 'Investigation of microwave sintering on high velocity particle consolidation coatings', in *Proceedings 1st Global Congress of Microwave Energy Applications August 2008*, Japan, 2008, 727–30.

35. Hwang J-Y., Huang X., Shi S., 'Microwave heating method and apparatus to reduce iron oxide', in *Materials Processing Under the Influence of External Fields*, Q Han, G Ludka and Q Zhai (eds), TMS, Hoboka, NJ, USA, 2007, 225–34.
36. Ishizaki K., Nagata K., Hayashi T., 'Production of pig iron from magnetite ore-coal composite pellets by microwave heating', *ISIJ Inte. (Iron and Steel Institute of Japan)*, 2006, **46**, 1403.
37. Nagata, K., Ishizaki, K., Kanazawa, M., Hayashi, T., Sato, M. Matsubara, A., Takayama, S., Agrawal, D. and Roy, R., 'A concept of microwave furnace for steel making on industrial scale', in *Proceedings of 11th International Conference on Microwave and High Frequency Heating*, Oradea, Romania, Sept. 2–6, 2007, A. M. Silaghi and I. M. Gordan (eds), 2007, 87–90.
38. Cherradi, A., Desgardin, G., Provost, J. and Raveau, B., 'Electric and magnetic field contribution to the microwave sintering of ceramics', in *Electroceramics IV*, R. Wasner, S. Hoffmann, D. Bonnenberg and C. Hoffmann (eds), RWTN, Aachen, 1994, Vol. II, 1219.
39. Cheng, J., Roy, R. and Agrawal, D., 'Radically different effects on materials by separated microwave electric and magnetic fields,' *Mater. Res. Innovations*, 2002, **5**, 170.
40. Roy, R., Ramesh, P. D., Cheng, J., Grimes, C. and Agrawal, D., 'Major phase transformations and magnetic property changes caused by electromagnetic fields at microwave frequencies', *J. Mat. Res.* 2002, **17**, 3008.
41. Roy, R., Ramesh, P. D., Hurtt, L., Cheng, J. and Agrawal, D., 'Definitive experimental evidence for microwave effects: radically new effects of separated E and H fields, such as decrystallization of oxides in seconds', *Mat. Res. Innovations* 2002, **6**, 128.
42. Ma, J., Diehl, J. F., Johnson, E. J., Martin, K. R., Miskovsky, N. M., Smith, C. T., Weissel, G. J., Weiss B. L. and Zimmerman, D. T., 'Systematic study of microwave absorption, heating, and microstructure evolution of porous copper powder metal compacts.', *J. Appl. Phys.*, 2007, **101**, 074906, 1–8.
43. Yoshikawa, N, 'Fundamentals and applications of microwave heating of metals', *J. Microwave Power Electromag. Energy*, 2010, **44**(1), 4–13.

13
Joining processes for powder metallurgy parts

C. SELCUK, Brunel Innovation Centre, UK

DOI: 10.1533/9780857098900.3.380

Abstract: Powder metallurgy (PM) processes are ideal for rapid production of near net shape parts with complex geometries from a range of materials that would often not be possible to combine otherwise. Certain alloy combinations, chemical compositions in powder form are more favourable for synthesis of materials. This allows PM to be extremely versatile, maximising material utilisation. The net shape capability directly reduces or eliminates secondary operations like machining. Despite this obvious advantage of PM processes, joining materials synthesised from powders has been associated with difficulties related to their inherent characteristics, like porosity, contamination and inclusions, at levels which tend to influence the properties of a welded joint. This chapter presents an overview of joining PM components. It seeks to identify preferred joining processes and identify apparent technology gaps, with an emphasis on offering solutions to welding problems. It also highlights developing approaches.

Key words: joining, near net-shape processing, powder metallurgy, welding.

13.1 Introduction

In PM materials processing techniques, all or some constituents of a part are employed in particulate form with certain characteristics of composition, morphology and size, compacted into a high precision product (Figs 13.1 and 13.2).

The ability of PM to produce high quality, complex parts with close tolerances and high productivity presents significant advantages, such as energy efficiency, with potentially low capital costs. PM is widely used for a range of applications, such as dental restorations, implants, bearings and automotive transmission parts, from biomedical to automotive industry sectors.

PM components are becoming increasingly attractive as substitutes for wrought and cast materials in various applications. However, it is possible to increase further the use of PM by exploiting the ability to manufacture complex geometrical configuration by joining PM parts to one another or to other cast/wrought products. The main issues restricting the welding of PM parts have been porosity, impurities and the fact that some PM parts have a high carbon content. Of these, the most important characteristic of a

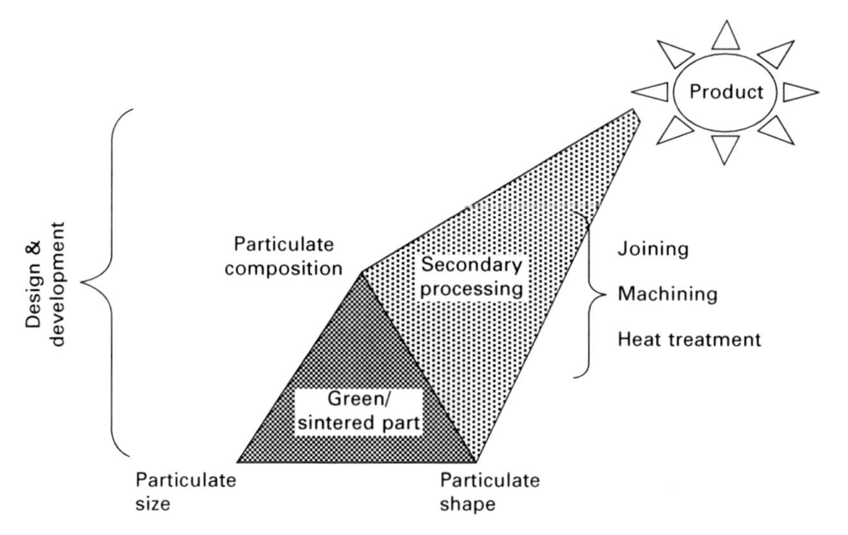

13.1 Important factors to consider in development of PM products. Joining is an enabling step.

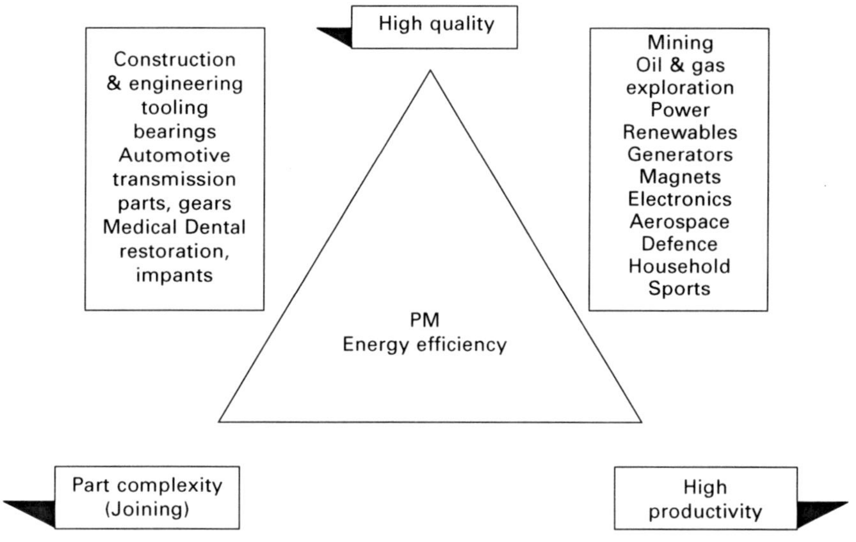

13.2 Ability of PM to create complex geometries for industry with the help of joining.

PM part for welding has been porosity, created either deliberately to make a porous part or incidentally due to insufficient densification. Powder particle characteristics (such as particle shape, size and surface area) determine the porosity or relative density of a powder compact, which in turn influences several important physical properties of the preform, such as thermal conductivity and hence hardenability, as well as thermal expansion. It has been noted that porosity can also act as a trap for impurities/inclusions that could potentially have an impact on secondary operations such as welding when entrapped impurities in the pores could be deleterious to the weldability of the PM part, for example by encouraging solidification cracking. PM parts have been reported to be susceptible to cracking in the heat affected zone when welded, owing to the porosity of the preform and limited area of interparticle bonding giving low ductility adjacent to the joint. Hence these locations may be unable to resist the thermal stresses generated as a result of contraction in a fusion weld. In addition, welding is commonly associated with resultant distortion, whereas PM parts are known to provide good dimensional and geometrical accuracy and thus, if machining operations are to be avoided, distortion must be minimised via selection of appropriate joining technology.

Therefore, widespread success in welding PM parts requires understanding of the influence of porosity, chemical composition, impurity level and overall cleanliness, upon weldment properties such as weld metal and heat affected zone (HAZ) cracking, ductility and toughness, residual stresses and distortion. These issues are addressed in this review.

13.2 Welding processes for powder metallurgy parts

13.2.1 Introduction

Joining processes applicable for PM parts can be categorised as solid state and liquid state. The solid state processes such as diffusion bonding and brazing have been predominantly used for lower density porous parts. In comparison, parts with higher densities or minimal porosity are typically treated as fully dense wrought materials, and these are typically welded using fusion-based joining processes, including arc welding, that is gas tungsten arc (GTA), gas metal arc (GMA), electron beam (EB) and laser welding. Further joining techniques such as adhesive bonding and shrink fitting may be used for some applications but are not considered here.

13.2.2 Arc welding

Arc welding of PM parts may give porosity with an associated detrimental influence on the weld integrity.[1] Gas metal arc (GMA) welding of powder compacts, which are not fully dense, can result in porous welds and weld toe cracking, the latter presumably resulting from low ductility of the original PM part. The density of a PM part and its composition are expected to affect the tendency for porosity during welding.[2,3]

Welding of ferrous PM parts can form a soft pearlitic microstructure as a result of slow cooling, caused by porosity reducing the thermal conductivity, which can allow strains to be accommodated that would otherwise give rise to cracking. An additional benefit of the porous structure is that hydrogen can diffuse out of the metal into and through the interconnected pores, hence reducing susceptibility to fabrication hydrogen cracking. Sintered steels are thus considered to be more resistant than wrought steels.

It is also worth mentioning that, when subjecting ferrous PM parts to steam treatment (heating the parts to a temperature in the region of 550°C and exposing them to water vapour) which is a common secondary process in production environment, a thin layer of Fe_3O_4 is formed both on the outer surface and on the surfaces of the interconnected porosity. This treatment is for improved corrosion resistance, increased surface hardness, compressive strength and wear resistance. This or a similar post sintering heat treatment of ferrous sintered parts will, however, prevent satisfactory welding. The problem is due to a resulting oxide film, which is slightly porous, probably containing moisture and has an insulating effect and hence is potentially a source of weld metal porosity and a potential cause of cracking in weldments.

Arc welding (gas tungsten arc welding, GTAW, and gas metal arc welding, GMAW) has also been attempted for non-ferrous metal matrix composite (MMC) systems, for example, SiC particle-reinforced aluminium. Gross porosity and delaminations in both weld metal and HAZ were reported to be frequent. It was claimed that despite large volumes of particulates in the Al alloy matrix, a wide range of MMCs can be fusion welded, with weldability similar to Al.[4] However, particulate characteristics (shape, size and distribution) of reinforcement, their proportion within the matrix, along with the homogeneity of the material and chemistry are critical for welding. Any interfacial reactions between the reinforcement and matrix that may result in the presence of secondary compounds or flux derivatives can influence the weldability and its success. Filler metal compositions, with respect to inclusions such as oxides and impurity contents (e.g. silicon and phosphorous levels in notably ferrous parts), can change the output of welding by influencing the microstructures attainable, especially in the heat affected regions. This can manifest itself in formation of low melting point eutectic phases at grain boundaries owing to segregation of impurities which can lead to solidification cracks in the heat affected zone (HAZ). Porosity distribution across the weld metal, HAZ and parent metal will affect the strength of the joint. PM components have the potential to replace their wrought counterparts in several applications owing to the advantages of net shape capability, low cost of production and high added value in terms of performance (e.g. strength, wear and fatigue). However, this can only be fully achieved if they can be confidently joined to other components especially with respect to design and structural considerations. Weldability of PM parts is therefore crucial to establish when it comes to widening the horizons for PM and future applications.

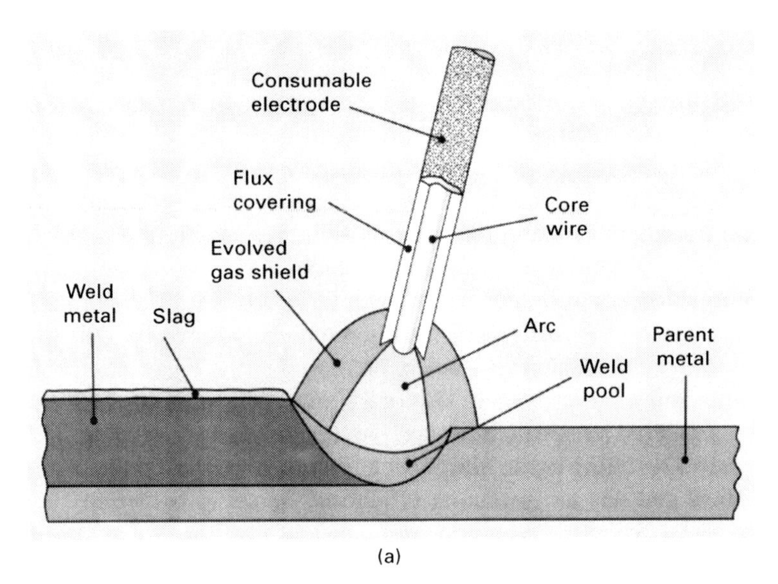

(a)

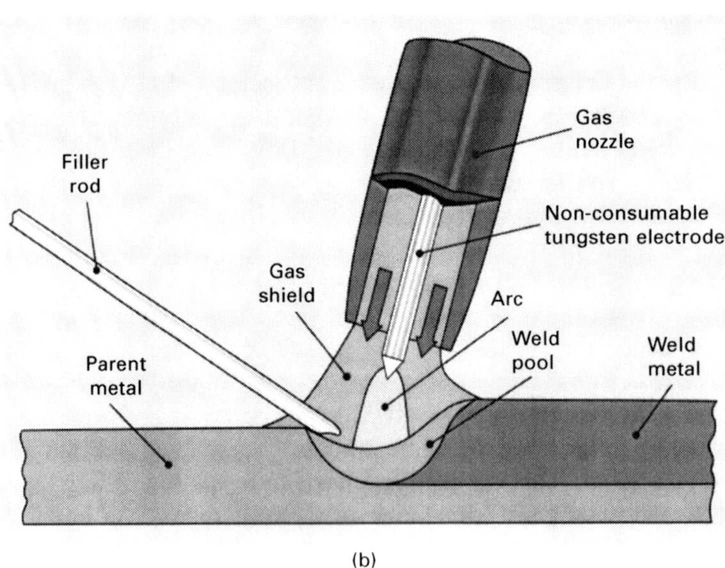

(b)

13.3 Schematic representations of manual metal arc: (a) gas–metal arc GMAW welding and (b) tungsten–inert gas gas–tungsten arc-GTAW welding (courtesy of TWI Ltd).

13.2.3 Laser welding

There are advantages of laser welding for PM parts, as it is a highly automated process, offering precision and control (Fig. 13.4). Welding speeds of several

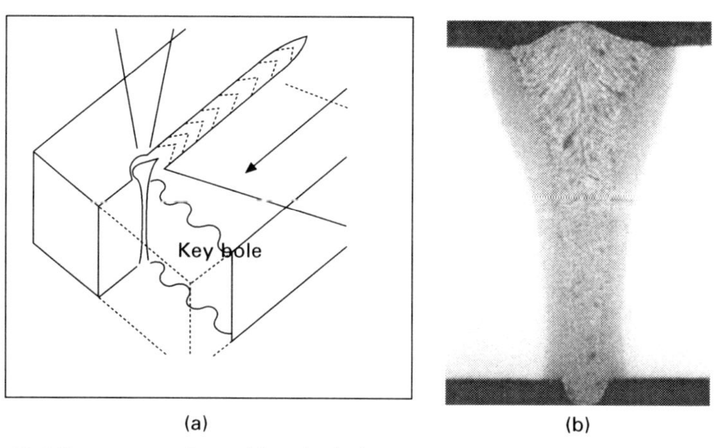

13.4 Representation of key hole laser welding and a typical weld (courtesy of TWI Ltd).

metres per minute are possible, with a low heat input, resulting in a small HAZ and limited thermal distortion and residual stress.[5]

However, various defects, such as blowholes that are probably due to entrapment of gases that cannot leave the melt during rapid solidification, were observed in laser welding of sintered steel parts in general, together with the occurrence of hydrogen cracking of medium carbon parts, cold, and hot cracking which have been observed in laser welding of sintered steel parts.[6] It was recommended by one author that a filler wire could be used to prevent all of these types of defect, although filler wire addition is an extra complication in laser welding. Low C-steels (typically up to 0.3% C) can be satisfactorily laser welded and it has been reported that a sintered steel part can be acceptably laser welded to a wrought counterpart, as long as both components have low carbon content.[7] However, laser welding of medium C-steels (typically 0.3–0.6% C) appears to be difficult, owing to the formation of a hard, brittle and hydrogen crack sensitive martensitic structure in the joint, caused by rapid cooling. Pre-heating may help by inducing a softer bainitic microstructure with some fine pearlite, which will increase toughness and hence improve the defect tolerance of the joint and reduce sensitivity to hydrogen cracking.

Gas carburised and oil quenched steels are considered to be difficult to weld, because of significant blowhole formation, probably due to entrapped oil and gas in the pores. Elsewhere, it was noted that smooth, discontinuity-free welds can be produced at slower travel speeds and lower beam powers.[8] It was also reported that a beam weaving laser welding technique would suppress porosity and that an increased width to depth ratio of the molten metal was beneficial for the escape of bubbles in the weld zone.[9]

Laser welding of sintered austenitic stainless steel (grade 316L) was reported to be very easy, resulting in good joints.[7] In trials with sintered Al alloys, it proved difficult to obtain a sound joint because of porosity formation, resulting in spongy welds.[7] It is thought that oxidation was the probable cause, possibly creating locally overheated spots when more laser energy is absorbed by the oxides than the metal; another possibility is that the oxide absorbs moisture and this is released during welding.

13.2.4 Electron beam (EB) welding

EB welding is normally carried out in high vacuum (e.g. 10^{-6} mbar). Therefore it is a batch process and may be expensive and thus restricted to high value parts only. It has a tendency to give high cooling rates and high hardness in C-steels, similar to laser welding, but is likely to have a greater tendency towards pore formation as the vacuum encourages trapped gas to try to escape during welding. It is reported that porosity increased with reduced travel speed. However, it has been demonstrated that weld metal porosity content in sintered ferrous compacts with a range of porosities can be controlled by beam parameters[10] and a non-vacuum EB welding process has been used for sintered parts.[11]

Any residual films, such as heat treatment quench oil trapped in the pores of a PM part, were found to have a detrimental effect on EB welding.[12] A fine grain size (< ASTM 8–9) in a sintered PM part was reported to be essential for good EB weldability, presumably due to improved ductility, in high temperature PM superalloys.[13] This is related to increased ductility and toughness of a fine grained material for absorbing strains upon solidification. Cracking has been observed in EB welding of PM superalloys designed for aerospace applications (engine components such as turbine discs) which required further investigation.[14] However, despite the problems described above, EB welding has the ability to give low distortion, again similar to laser welding, or at least uniform distortion effects, which is important for preserving the dimensional stability of near net shape PM components (Fig. 13.5).

13.2.5 Resistance projection welding

Projection welding is one of the most widely applied welding processes for sintered PM parts. A significant aspect of projection welding is the limited distortion associated with it, which is certainly advantageous in terms of geometrical stability. However, one potential difficulty for the success of projection welding is the cleanliness of the parts or presence of surface films that can inhibit bonding. This can manifest itself in the likely presence of an oxide layer on some parts, for example ones subjected to steam treatment,

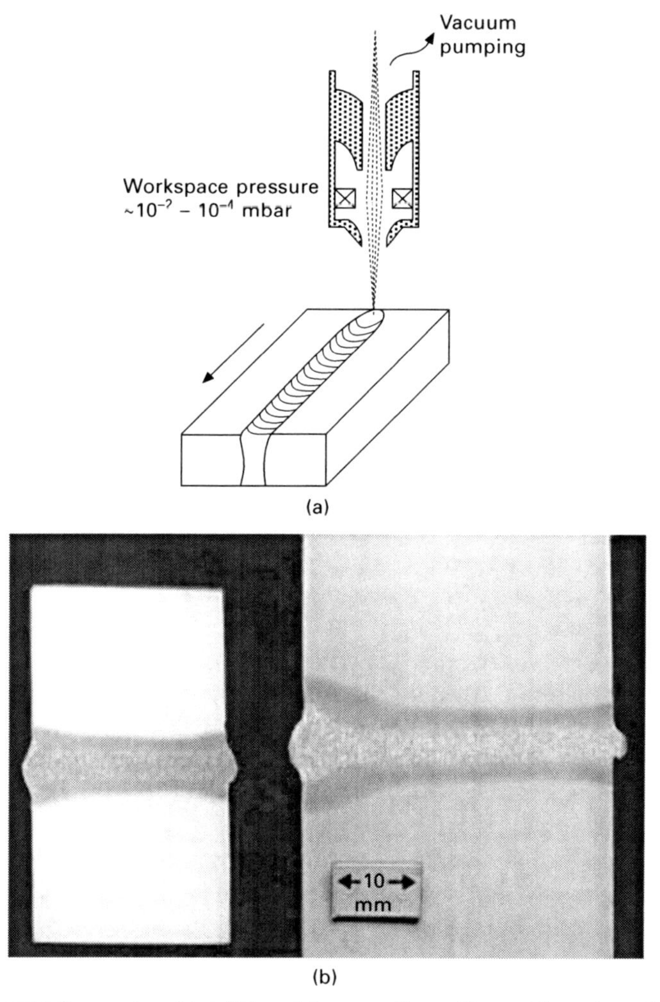

13.5 Example of (a) EB welding configuration with reduced pressure and (b) reduced pressure EB welds in C-Mn steel (courtesy of TWI Ltd).

typically on ferrous sintered parts. Such a layer could prevent satisfactory welding through an insulation effect at the interface and as a potential source of moisture and hence porosity.[15,16] It is therefore recommended that any steam treatment on PM parts, as described earlier, should be done after welding. It has been possible to weld high carbon PM steels and case hardened parts using resistance projection welding.[17] Light alloys have also been projection welded and the combined effect of pressure and temperature at the joint can, in fact, help to densify the region by closing the pores. This can help improve the strength of the weld region. The process enables PM parts to be joined to wrought materials and therefore allows a degree of flexibility in creating complex geometries and dissimilar joints (Fig. 13.6).

13.2.6 Friction welding

Friction welding is a solid phase joining process, mainly used for wrought products (e.g. light alloys of Al and Ti) in a range of geometries such as extrusions (Fig. 13.7). Particular advantages are the absence of flux, filler or need for a protective atmosphere, which makes the process extremely attractive.[15] The friction welding process is highly suitable for welding PM parts as it enables pore closure, which can potentially lead to a pore-free weld interface and a refined microstructure. This is particularly valid for Al-based PM parts and MMCs.[18] An additional advantage of friction welding for Al alloy PM components, is that it is useful in breaking down any oxide layer deposited on the particles by intense deformation within the weld region. This in turn results in exfoliation of oxide and its rejection from the joint. As a consequence, a clear bond line is achieved and often the bond homogeneity and strength is improved. One potential disadvantage of friction welding may be associated with the change in microstructure caused by reorientation and deformation of sintered metal grains, which may create a potential weak region in the joint, thereby reducing fatigue performance.[16] Since its invention in the early 1990s at TWI in the UK, friction stir welding (FSW), which is a variant of friction welding, has come a long way and it is widely applied in various industry sectors notably transport and aerospace. Al has been a good material to demonstrate the applicability of process in its fully industrialised form. The tooling and agility of the process have been critical for complex parts and relatively difficult to weld materials. FSW presents a good opportunity for PM materials, not only for joining but also for creating MMCs in bulk or on the surface. The process can be utilised to combine what would otherwise be difficult to co-exist alloy systems. This is partly due to the fact that it is a solid state process like PM.

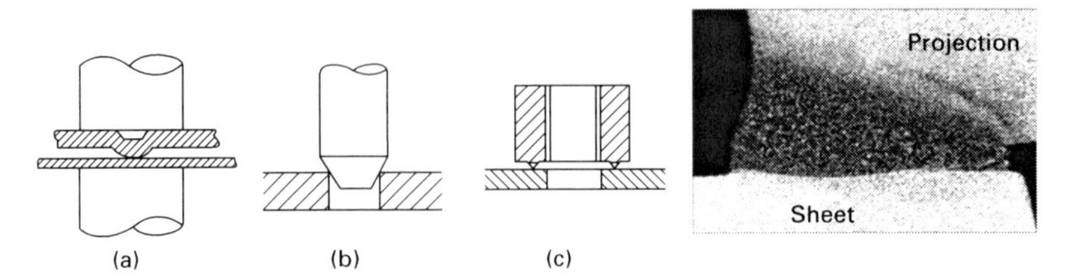

13.6 Projection welding configurations: (a) embossed projection section, (b) stud to plate, (c) annular projection, and an example microstructure from a typical weld (courtesy of TWI Ltd).

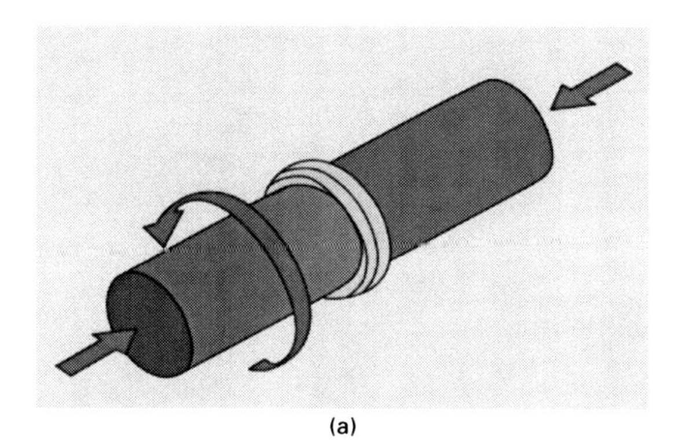

(a)

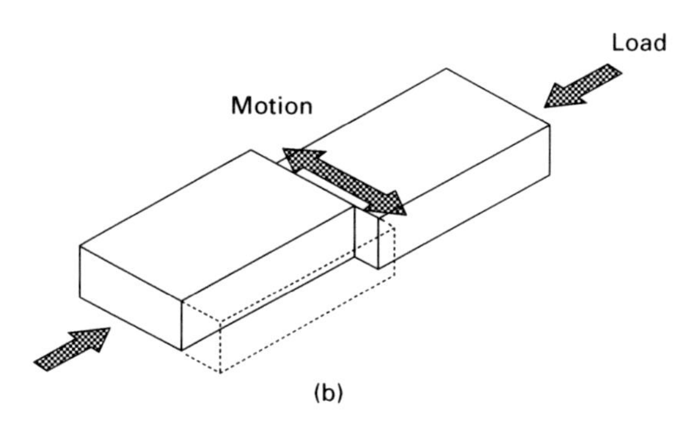

(b)

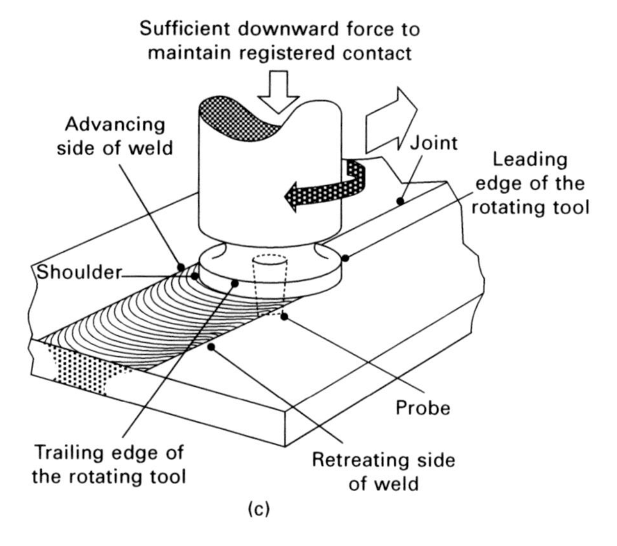

(c)

13.7 Examples of friction welding: (a) rotary, (b) linear, (c) friction stir configurations, and typical weld macrograph and microstructure in Ti alloy forging, shown in (d) and (e) respectively (images courtesy of TWI Ltd).

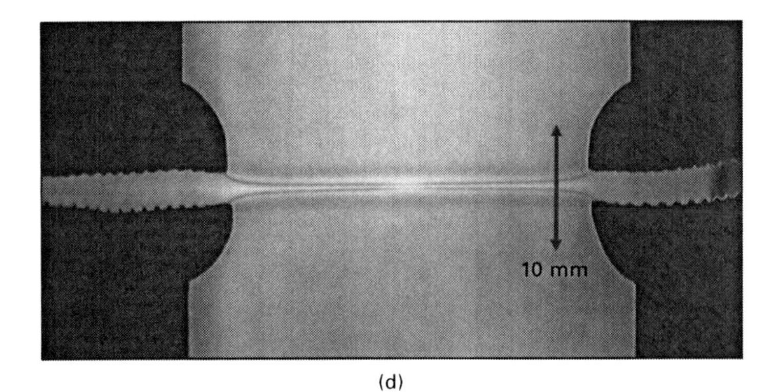

(d)

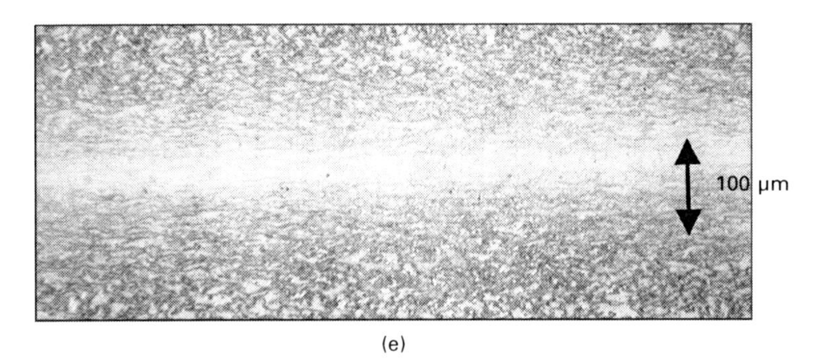

(e)

13.7 Continued

13.3 Other joining processes for powder metallurgy parts

13.3.1 Brazing

In brazing, the porosity of the PM part can, in fact, draw away the brazing alloy from the joint region. This can leave insufficient material for bonding. To overcome this problem, techniques have been developed where brazing is integrated in the sintering process.[19,20] Another potential issue which is encountered commonly in brazing is the presence of secondary products, for example due to flux reactions which would stop further infiltration by blocking the pores. Factors that can affect the brazed joint in terms of strength are surface condition of the particles (as is the case for other techniques) and the part's surface roughness;[21] their influence can be significant. There are new brazing techniques, such as laser brazing, and new brazing alloys being developed for joining sintered components in large volume.[22] Brazed assemblies can typically consist of PM to PM or PM to wrought or cast structures, which can end up as a functionally graded material. Brazing, however has been selected as a joining method for hot isostatically pressed (HIP) parts.

13.3.2 Diffusion bonding

Diffusion bonding is typically utilised for ferrous parts which are fairly small in size. Hence, part geometry can be a concern. To avoid this complication, sintering and diffusion bonding can potentially be carried out in the same furnace. A eutectic reaction is often employed to provide a transient liquid phase for bonding at the interface.[23] The presence of resulting reaction products such as oxide compounds may reduce the bond strength.[24] To induce high bond strength diffusion can be activated via the addition of suitable elements, such as Cu in, for example, ferrous PM parts.[25] When compared with other joining techniques, rather low strength joints can be expected from diffusion bonding, which may be limited to certain geometries and alloy compositions. Diffusion bonding has been a choice when handling more exotic compositions where the chemistry of the part plays a significant role in the joining process and its feasibility. Hence, applications have been focused on light materials such as Ti alloys, MMCs and special products. More conventional C–Mn steels have not been commonly diffusion bonded. Although an attractive option, the success of the bond strength gained by this joining technique critically depends on compositions and hence phase transformations and may require protective atmospheres (inert such as argon, or reducing, such as hydrogen) which bring additional considerations such as complexity and cost of production. Small volume high performance and added value components may therefore be more favourable for diffusion bonding and the technique may well be the only option for joining such materials. Similar to brazing, diffusion bonding has also found applications in HIP components.

13.3.3 Shrink fitting (press fitting)

In shrink fitting, dimensional changes between parts are employed to form the joint, which will not be necessarily gas-or liquid-tight. This may be a disadvantage in some applications.[15] It has been demonstrated that, for ferrous PM parts, dimensional change will be dependent on the powder selection and the nature of the constituents.[26] Shrink fitting is a cost effective option for many applications, reducing machining, especially for parts that could be manufactured by other methods such as metal injection moulding (MIM) for particular geometrical reasons.[19] Cylindrical and ring-shaped compacts are most often joined by press fitting. Joint strength is dependent on variations in densities between the parts, fitting pressure and degree of taper angle, if any.[27]

13.3.4 Adhesive bonding

The porosity of sintered parts is ideal for adhesive bonding, since surface porosity can facilitate mechanical keying of the adhesive. Most PM parts can simply be joined by gluing.[12] However, the surface has to be cleaned and free of contaminants. Potential advantages include uniform distribution of stress and the ability to join thin to thick parts, making complex geometries possible. The major disadvantage tends to be poor resistance to elevated temperatures where the adhesive degrades; as a result the strength of the joint can be significantly reduced. It is worth noting that the performance of any adhesive strongly depends on the loading mode and therefore any testing should be representative of the actual loading conditions in the application of interest.[19]

13.3.5 Joining metal injection moulded (MIM) parts

At this juncture it is worth noting that the above considerations for different processes are also valid for MIM parts even more so than PM parts owing to the higher binder content in MIM which can affect the weldability, porosity content and impurity levels in the parts. Therefore, careful handling of the component chemistry, including residual levels as well as geometrical considerations and strength/porosity of the parts to be welded, is paramount in defining a window for joining and will enable choice of a suitable process for MIM parts. To date both diffusion bonding and brazing have been favourable processes for joining. It is worth highlighting that powder injection moulding (PIM) of ceramics and metals can also be regarded as a potential route for creating tool materials for friction stir welding of metals and alloys.

13.3.6 Laser metal deposition

Laser metal deposition (LMD) is a unique technique, combining laser and powder processing, which enhances material utilisation by enabling manufacture of high precision near net shape components from powders. Owing to the cost of specialised powder feedstock and laser equipments with protective atmosphere in many cases, the technique is regarded as an expensive but versatile option. Therefore, LMD has been of most interest in high value added applications such as in aerospace and medicine which can afford this developing process as a whole, in spite of the criticality of performance and acceptance criteria in these industry sectors.

There are several benefits to using LMD for repair. Highly complex parts can be repaired in an automated fashion and, because the heat input is low, the heat affected zone is small. Therefore the strength of material is not affected. Distortion is also lower than for conventional welding techniques. Anisotropy in the mechanical properties occurs, however, due to the layered microstructure and residual stresses that are commonly present because of steep thermal gradients. These are usually detrimental to the mechanical properties of the parts produced. A current concern with LMD is that it is very hard to predict the properties of LMD fabricated components because there are so many process variables involved in the process. The parts have very complex thermal histories, which depend on process variables. The effects of process parameters on the microstructure are complex with a strong dependence on the material system.

Despite the major advantages of LMD, there is, however, a continuous need to develop a full understanding of the process–structure–property relationships, with a particular emphasis on the effect of powder characteristics, such as powder particle size, shape and distribution on process variables, and the metallurgy and resulting mechanical properties, which are crucial for many in service applications (Fig. 13.8, Table 13.1).[28]

13.4 Discussion

A broad range of powder metallurgy parts is available, in a wide range of alloys, and there is no single best way to join them. However, there are a number of welding characteristics of PM parts that are different from those associated with wrought or cast equivalents, either as a consequence of the PM production route or the typical applications of PM parts. For example, as PM parts are used in a variety of high precision applications, it is desirable to weld with a process that gives minimal distortion. This favours low heat input processes such as laser and EB welding but any low heat input process will also inevitably give rapid cooling and hence high hardness in steel parts, particularly for higher C contents. It is not clear whether sintered parts can be EB welded in a vacuum however, considering inherent porosity where gases and impurities can be retained and entrapped in the weld. Reduced pressure EB welding, which only requires a vacuum of the order of 10^{-3} mbar, may be more suitable for welding sintered PM parts, to overcome difficulties in achieving an adequate vacuum, but this is still under development.

As with any welding, the main requirements for welding PM parts is that the process should not introduce defects (Fig. 13.9). Powder metallurgy parts contain porosity, either deliberately, and hence with a fairly high volume fraction, or as a consequence of the inability to obtain complete densification, in which case it is typically at a low level. Any porosity in the PM part will tend to trap contaminants and gas, which can cause pores in the weld metal and introduce species that increase sensitivity to both hot and cold cracking mechanisms, for example sulphur and phosphorus

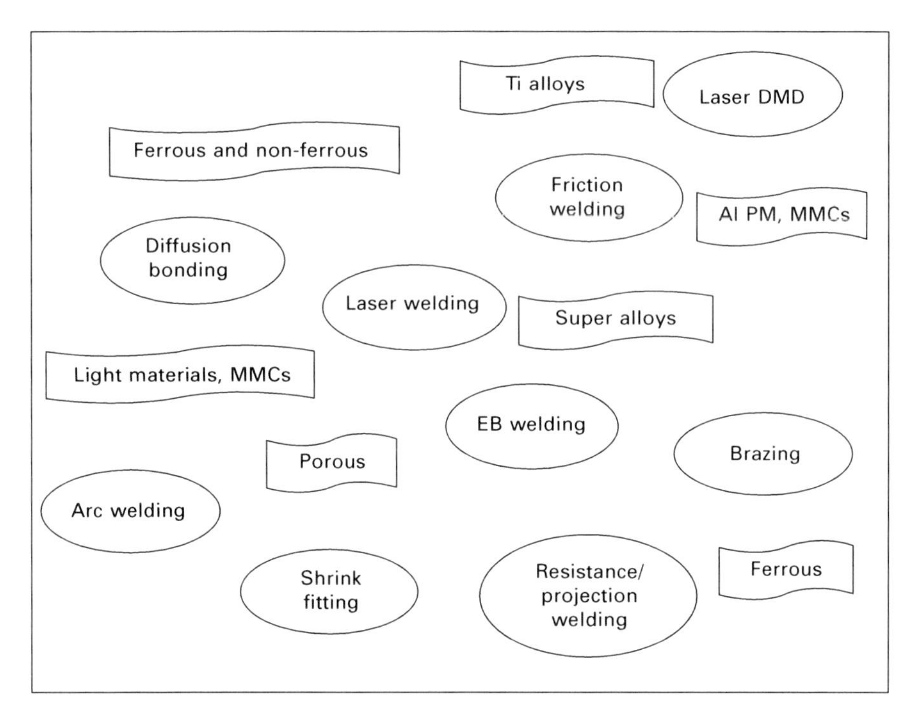

13.8 Layout showing relative disposition of joining processes with respect to applications in terms of the nature of materials.

contamination will encourage solidification cracking, whilst moisture and carbon contamination will encourage hydrogen cracking. In order to minimise these problems, cleanliness of parts is vital. In this respect, avoiding steam treatment will be beneficial and degreasing is important prior to welding. Where contamination exists, use of a filler metal that is more tolerant of contamination than the parent material, for example a nickel alloy, may be beneficial, in which case arc welding processes are preferred. One possible advantage of an interconnected porosity may be that hydrogen can diffuse out via the open porous structure, in welding PM steel parts, which may make them more resistant to hydrogen fabrication cracking.

Where present at significant levels, porosity of the parent material may lead to tearing of the material adjacent to the weld, simply due to the development of plastic strain beyond the capacity of the PM part, perhaps exacerbated by geometric effects at the joint. In such cases, use of low heat input is preferred, to reduce the amount of material strained, and friction welding may be advantageous, as the compression involved tends to close pores. Indeed friction welding may be generally useful for PM parts owing

Table 13.1 Qualitative ranking of joining processes for PM parts in terms of relative distortion, geometrical constraint, porosity and cost as high (H), medium (M) or low (L)

Joining process	Distortion	Geometrical constraint	Porosity	Cost	Typical applications
Arc welding	H/M	H/M	H/M	M/L	Ferrous and non-ferrous materials
Laser welding	M/L	M/L	H/M	H/M	Ferrous and non-ferrous materials
EB welding	M/L	M/L	H/M	H/M	PM superalloys
Resistance projection welding	M/L	M/L	M/L	M/L	Ferrous materials
Friction welding	M/L	M/L	M/L	H/M	Al based PM parts, MMCs
Brazing	M/L	M/L	M/L	M/L	Ferrous: PM to PM or PM to wrought materials
Diffusion bonding	M/L	H/M	M/L	M/L	Ferrous, light materials, Ti, MMCs
Shrink fitting	M/L	M/L	M/L	M/L	Porous materials

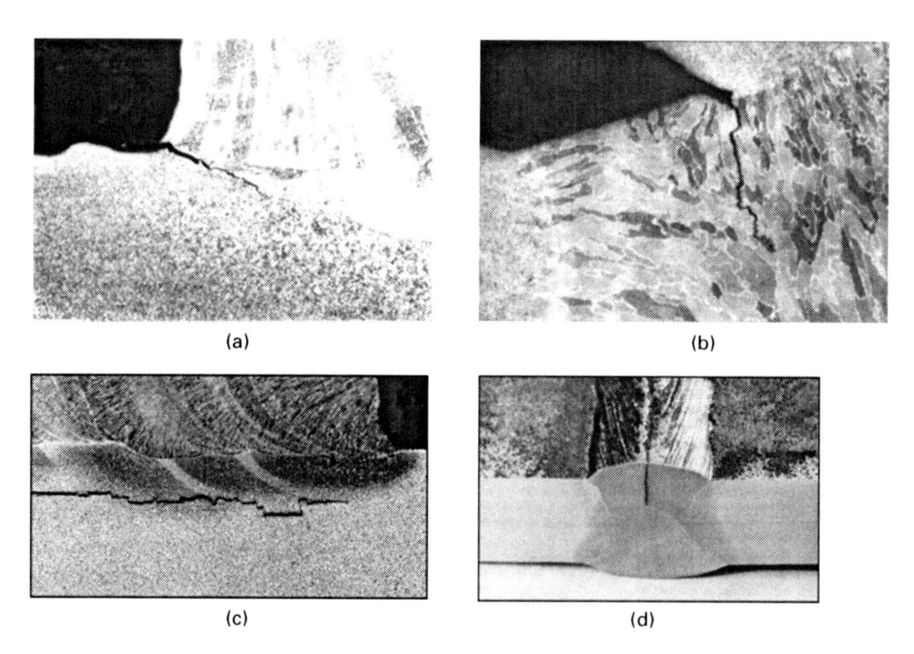

13.9 Representation of typical welding flaws: (a) HAZ hydrogen cracking and (b) weld metal hydrogen cracking, (c) lamellar tearing and (d) solidification (hot) cracking (courtesy of TWI Ltd).

to the compressive force involved and the fact that friction welding squeezes the original surface layer, which may be contaminated, out of the joint.

It is apparent that powder particle characteristics, which influence densification of a PM part and therefore its final porosity, have not received much emphasis in relation to welding studies. In order to achieve better control of porosity and minimise its detrimental effects in welding, consideration should be given to the influence of powder particle characteristics such as particle shape, size and surface area on the density and porosity of a powder compact, as well as any interfacial reactions and subsequent formations such as secondary phases for improved weldability and joint strength.

13.5 Conclusions

The following conclusions are drawn:

- Welding is widely used for a range of PM components for diverse applications across several industry sectors but limitations exist owing to the inherent porosity, contamination within the pores and the effect of porosity on the ductility of the material and hence its ability to withstand strain in a weld heat affected zone.
- For porous materials, use of low heat input is recommended to reduce strains that develop in the heat affected zone and minimise the risk of tearing adjacent to the weld. Where applicable, friction welding and projection welding may be advantageous as they involve compression, which tends to close pores in the joint area. Porosity in the weld metal is likely to result when any fusion-based process is used and rapid cooling rates may increase porosity owing to the limited opportunity for bubbles to escape from the molten metal.
- Where contamination causes weld metal cracking, the use of a welding process that allows introduction of a consumable filler material, such as the various common arc welding processes, is beneficial. In extreme cases, use of a tolerant nickel-based filler metal may be necessary to avoid cracking. Any process that introduces contamination or oxide to porous parts, such as steam treatment is likely to be detrimental to the ability to make sound welded joints and should be avoided. Similarly, cleaning the joint surfaces prior to welding is beneficial.
- Laser and EB welding have also found application in welding PM components when the inherent low distortion of these processes is an advantage, that is for high precision parts, but these processes have rapid thermal cycles, which encourage hardening and cracking of the weld metal, with limited opportunity for filler addition, and are not always applicable. EB welding in particular is likely to suffer from porosity and the necessary high vacuum might not be achieved when gas contamination

is extensive. Reduced pressure EB welding, which operates with higher gas pressures, is attractive for sintered porous parts.

13.6 References

1. J. C. Thornley, *Welding Design and Metal Fabrication*, 1973, **12**, 399–402.
2. K. Couchman, M. Kesterholt and R. White, 'Seminar on secondary operations', *Proceedings International Conference PM 1988*, Orlando, 33–9.
3. M. A. Greenfield, R.F. Geisendorfer, D. K. Haggend and L.P. Clark, *Welding Research Supplement*, 1977, May, 43–148.
4. J. H. Devletian, *Welding Journal*, 1987, June, 33–9.
5. A. Rocca and G. Capra, *SPEI GCL-7th International Symposium on Gas Flow and Chemical Lasers*, Austria 1988, Vol. 1031, 635–45.
6. A. Joskin, J. Wildermuth and D. F. Stein, *International Journal Powder Metallurgy and Powder Technology*, 1975, **11**(2), 137–142.
7. E. Mosca, A. Marchetti and U. Lampugnani, *Proceedings International Conference PM Powder Metallurgy*, Florence, 1982, 193–200.
8. S. Chiang and C. E. Albright, *Journal of Laser Applications*, 1988, **1**(1), 18–24.
9. X. Zhang, W. Chen, G. Bao and C. Zhao, *Science and Technology of Welding and Joining*, 2004, **9**(4), 379–76.
10. G. M. Alexander-Morrison, A. G. Dobbins, R. K. Holbert and M. W. Doughty, *Journal Materials for Energy Systems*, 1986, **8**(2), 79.
11. J. A. Hamill, Jr, *Welding Journal*, 1993, February 37–45.
12. G. W. Halldin, S. N. Patel and G. A. Duchon, *Progress in Powder Metallurgy*, **39**, 267–280.
13. J. H. Davidson and C. Aubin, *Proceedings: High Temperature Alloys for Gas Turbines*, 1982, Liege, Belgium, 853–86.
14. P. Adam and H. Wilhelm, *Proceedings: High Temperature Alloys for Gas Turbines*, 4–6 October 1982, Liege, Belgium, 909–30.
15. W. V. Knopp, *Automobile Engineering Meeting Toronto*, Canada, Oct 21–25, Society of Automotive Engineers, 1974, 740984.
16. J. E. Middle, *Chartered Mechanical Engineer*, 1980, **27**(7), 55–60.
17. L. J. Johnson, G. J. Holstand, M. J. O'Hanlon, *Fall Powder Metallurgy* Conference, 19–20, Detroit, Michigan, MPIF/APMI, 1971, 193–203.
18. W. A. Baeslack III and K. S. Hagey, *Welding Research Supplement*, 1988, July 1395–495.
19. P. Beiss, *Powder Metallurgy*, 1989, **32**(4), 277–84.
20. W. V. Knopp, *Materials Engineering*, 1975, 12–75, 34.
21. K. Okimoto and T. Satoh, International Journal Powder Metallurgy, 1987, **23**(3), 163–69.
22. N. Janissek, DVS Berichte, *Proceedings Brazing, High Temperature Brazing and Diffusion Welding Conference*, no. 243, Auchen, 19–21 June 2007, 1–5.
23. H. Duan, M. Kocak, K. -H. Bohm and V. Ventzke, 2004, *Science and Technology of Welding and Joining*, 2004, **9**(6), 513–17.
24. A. Akutso and M. Iijima, 1985, *Modern Developments in Powder Metallurgy*, **16**, 195–208.
25. T. Tabata, Nasaki, H. Susuki and B.G Zhu, *International Journal Powder Metallurgy*, 1989, **25**(1), 37–41.

26. J. C. Thornley, *Welding and Metal Fabrication*, 1972, 390–5.
27. T. Tabata and S. Masaki, *International Journal Powder Metallurgy Powder Technology*, 1979, **15**(3), 239–44.
28. C. Selcuk, *Powder Metallurgy*, 2011, **54**(2), 94–9.

14 Process optimization in component manufacturing

G. M. LEE, Pusan National University, South Korea and
S. J. PARK, Pohang University of Science and Technology, South Korea

DOI: 10.1533/9780857098900.3.399

Abstract: This chapter introduces the concept of optimization in the area of component manufacturing. A short introduction explains the associated concepts, applications, formats and approaches and familiarizes the reader with the terminology. The main body of the chapter examines approaches to optimization in four different component manufacturing applications: die compaction process design, powder injection moulding process design, sintering process design and steady-state conduction design. The methodologies used in the applications include both mathematical iterative methods and experimental optimization methods.

Key words: algorithms, design of experiments, design problem, iterative methods, optimization, optimization algorithms.

14.1 Introduction

Optimization is defined as an act, process or methodology of making something (as a design, system, or decision) fully perfect, functional or effective as possible. In a narrower meaning, it is a mathematical procedure to maximize or minimize a function by finding the best available alternatives, satisfying a set of constraints. The theory of optimization can be extended to a wide spectrum of engineering and science fields, so that its applications comprise a huge body of applied mathematics. In general, an optimization problem can consist of three components: (1) objective function(s) to be achieved, maximized or minimized, (2) decision variables or alternatives to be chosen and finally (3) constraints to be satisfied. In the areas of optimization, various techniques have been proposed and developed for unconstrained or constrained optimization problems. In other fields, different terminologies areas also used to characterize their fields of interest. In the area of mechanical system designs, the design variables and state equations may be used to refer to decision variables and constraints, respectively.

However, not all problems in practice are able to be described or modelled mathematically. It is not unusual to come across problems in modelling that are so complex that it is difficult even to identify the decision variables which determine the system performances or outputs. The level of detail, abstraction and significance are other dimensions that can add to the complexity of a problem. Hence, the task of modelling a process and system as a functional optimization model is not easy to complete or achieve and it does not always produce a well-defined mathematical model.

In broader terms, optimization is also used to refer to procedures used to improve processes or systems, even if the optimality cannot be reached at the end. It is common that a system or a process is too complex to be mathematically modelled or described. It is not unusual for all the factors influencing the performances or results to be unknown or uncontrollable at the time of optimization. In this case, a progressive trial-and-error approach is the only possible option to improve the system or the process. A typical example of this situation exists when a system needs to be optimized for the best results while the underlying mechanism or theory has not been completely understood. Then, in general, a series of experiments can be conducted to understand the effects of decision (or design) variables under consideration and to choose the optimal combination of those variables.

This chapter introduces some optimization approaches which have been successfully applied to several powder metallurgy methods for component manufacturing. Sections 14.3, 14.4, 14.5 and 14.6 introduce different powder metallurgy processes and their related optimization approaches.

14.2 Formal optimization

An optimization problem can be described as follows: given a function or a system $f: D \to R$ that is a mapping from domain D to a set of real numbers in n dimensions, an element x^* in D needs to be identified so that $f(x^*) \leq f(x)$ for all x's in D for the minimization or so that $f(x^*) \geq f(x)$ for all x's in D for the maximization. This kind of formulation is called a mathematical programming problem. The term 'programming' is not directly related to computer programming but is widely in use, for example in linear programming. Many engineering or science problems exist in the aforementioned formulation. For example, consider automotive engines there are many car makers and they make various kinds of engines for different car models. Common goals for engine designers are better fuel efficiency and higher performance. These goals need to be converted or modelled into quantitative measures by a thorough study of the problems appropriate to the context of the optimization. Then the controllable design parameters for various feasible engine designs need to be identified or defined. Therefore, the effects of design parameters on the fuel efficiency and performance can be studied so that the mechanical design can be optimized. The effects of design parameters on the objective functions are nonlinear in general and are not easy tasks to describe or quantify. Typically, D is a subset of the Euclidean space R^n, which is bounded by a set of constraints. These constraints are an equality or inequality that the elements of D have to satisfy. This domain D is called the search space and all elements in D are called feasible solutions. The function f is called an objective function. A feasible solution that minimizes or maximizes the objective function is called an optimal solution.

The maximization problem can be easily converted to the minimization problem, and *vice versa*. Well-established mathematical approaches and algorithms have been developed over the years for convex programming problems, which have a convex objective function and the convex feasible region in minimization problems. In a convex programming problem, there is a single optimal solution that is a global minimum. However, many problems in the component manufacturing are non-convex problems by nature of physics. In non-convex problems, there may be several local minima; a local minimum x^* is defined as a point for which there exists a small number $\varepsilon > 0$ so that for all x such that $|x - x^*| \leq \varepsilon$, $f(x^*) \leq f(x)$ is true. That is to say, for some neighbouring region around x^*, the objective function values are greater than or equal to the objective function value at the local optimal point. Many design problems in component manufacturing can be modelled as optimization problems. To solve problems, there are various algorithms or iterative methods that can converge toward optimal solutions. However, the solution methods sometimes do not converge or take too much time to increase the quality of solutions. In that case, it is common to settle on the reasonably good solutions even if they are not optimal, in exchange for computational times.

14.3 Optimization in the die compaction process

Nowadays, sintered products have been broadly adopted by the automotive industry for the purpose of reducing cost and weight (German, 1994). There are more versatile forming processes, such as powder injection moulding and extrusion, which have been developed for shaping the desired geometry (German and Bose, 1997), but uniaxial die compaction is still one of the most popular forming approaches. In the die compaction process, after metal powders are compacted into the desired shape, the green compact needs to go through the sintering process. The sintered or brown compact shrinks and often distorts due to the sintering dimensional change caused by the green density. The non-uniform density distribution caused during the compaction process even leads to the distortion and an inability to hold tight tolerances. There are two main causes for the non-uniform green density: (1) the friction between the particles and tool surface and (2) the non-uniform stress on the compaction body caused by the pressure decay along with the depth in the green body. Therefore, additional expensive finishing steps, such as machining and grinding, are often needed to achieve the desired tolerances.

To reduce the variability of the green density in the compact, field engineers have mainly focused on reducing the friction between powders and tooling using functional lubricants. However, the non-uniform green density cannot be eliminated even though the powder-tooling friction is controlled carefully. A systematic methodology for optimizing the compaction process parameters, including upper and lower punch speeds as well as the die shape is greatly needed.

Diverse optimization techniques have been used for forming processes such as extrusion, rolling, forging and powder forging. They include backward tracing schemes (Hwang and Kobayashi, 1984; Zhao *et al.*, 1995), evolutionary algorithms such as genetic algorithms (Roy *et al.*, 1997; Chung and Hwang, 1998a) and derivative-based searches (Kusiak and Thompson, 1989; Joun and Hwang, 1993; Byon and Hwang, 1997; Michaleris *et al.*, 1994; Barinarayanan and Zabaras, 1995; Fourment and Chenot, 1996; Fourment *et al.*, 1998; Chung and Hwang, 1998b; Chung *et al.*, 2000, 2001). Among these techniques, derivative-based searches have shown good performance in terms of the quality of the solutions and the computation times. In derivative-based searches, the calculation of the design sensitivity is very important. The design sensitivity is the derivative of the objective function with respect to design variables. To calculate the design sensitivity, analytical methods such as the adjoint variable method (AVM) and direct differentiation method (DDM) and numerical methods such as finite difference method (FDM) are being used in the field. Since the FDM requires additional finite element calculations which depend on the number of design variables, it is not in general used in the optimization iterations due to their complexities. Rather it is used for verifying results about the design sensitivity which is obtained by other analytical methods.

In considering ways of treating the derivatives of the finite element solutions, such as velocity fields with respect to design parameters, analytical methods can be classified into DDM and AVM, as mentioned above. In DDM, the derivatives can be calculated directly while in the AVM, the calculation of derivative requires introduction of Lagrangean multipliers on the finite element equations. The Lagrangean multipliers technique provides a strategy for finding the local maxima and minima of a mathematical system which consists of equality constraints (Bazaraa *et al.*, 2006). In the DDM, in order to calculate the design sensitivity, it is necessary to calculate the derivative of the finite element solutions such as the velocity field with respect to the corresponding design variables. In general, the calculation of this derivative costs much in terms of time and is proportional to the number of design variables. In contrast to DDM, AVM does not directly calculate the derivative by treating adjoint variables, which leads to shorter computation times. Therefore, AVM is practically more useful and efficient in calculating the design sensitivity in terms of calculation time when the number of design variables is large (Haug *et al.*, 1986). However, the treatment of adjoint variables is very complicated and difficult to derive in non-steady forming. Therefore, AVM has generally been used in steady-state forming, while DDM has been used in non-steady-state forming. However, Chung *et al.* (2003) derived the equations in AVM to find the optimal intermediate die shape in a non-steady-state forging process. In the following paragraphs, AVM is adapted to reduce the computation times through the derivation of the equations for AVM in non-steady-state forming of porous materials. See van Keulen *et al.* (2005) for the detail of design sensitivity analysis.

In general, the die compaction is conducted at a room temperature and the material properties depend only on the relative density and effective strain. Therefore, the derivatives related to the relative density and effective strain need to be included to derive the AVM. The deformation behaviour of the powder caused during the die compaction process is based on a yield criterion. Unlike bulk solids, the volume change in the die compaction process requires the yield criterion for a powder to include the hydrostatic pressure as follows:

$$AJ_2' + BJ_1^2 = \bar{\sigma}^2 \qquad [14.1]$$

where A and B are material parameters that are functions of the relative density, J_1 is the first invariant of the stress tensor σ_{ij}, J_2' is the second invariant of the deviatoric part of stress tensor σ_{ij}, and $\bar{\sigma}$ is the effective stress of powder continuum. The effective stress of powder continuum can be expressed by the function of the effective stress on the base material and the relative density as follows:

$$\bar{\sigma} = \sqrt{\beta(\rho)} \cdot \bar{\sigma}_m(\bar{\varepsilon}_m, \dot{\bar{\varepsilon}}_m, T) \qquad [14.2]$$

where $\beta(\rho)$ is the material parameter which is a function of relative density ρ, $\bar{\varepsilon}_m$ the effective strain of the base material, $\dot{\bar{\varepsilon}}_m$ the effective strain rate of base material, T the temperature, and $\bar{\sigma}_m(\bar{\varepsilon}_m, \dot{\bar{\varepsilon}}_m, T)$ is the effective stress of base material.

The uniaxial tension or compression test usually shows that the relation of A and B is $A = 3(1-B)$. According to the definition of A and $\beta(\rho)$, many criteria have been suggested (Chung *et al.*, 2003; Gurson, 1977; Shima and Oyane, 1976; Doraivelu *et al.*, 1984; Kim, 1988). In this section, Shima and Oyane's criterion is used as follows:

$$A = \frac{3}{1 + 0.694(1-\rho)} \quad \text{and} \quad \beta(\rho) = \frac{\rho^5}{1 + 0.694(1-\rho)} \qquad [14.3]$$

Assume an example that represents the tool–workpiece interface, as shown in Fig. 14.1. Consider a deforming body Ω with velocity $u_i = \bar{u}_i$ prescribed on a part Γ_u. Let Γ_c be the remainder of the part surface. The boundary value

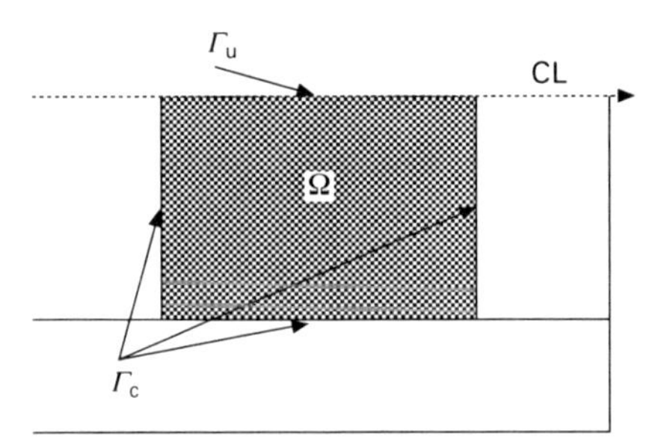

14.1 Diagram of domain and boundary conditions.

problem associated with the current moment in the non-steady-state plastic deformation process is to find a velocity field u_i satisfying the mass balance equations, equilibrium equation, constitutive equation, yield criterion and boundary conditions (see Chung *et al.* (2009) for the detailed introduction of these conditions).

During the finite element simulation, the position of a die-contacted node and die surface needs to be calculated by the contact algorithm. The size of projection method is generally used so that the position of the die-contacted node is projected on the die surface at each incremental time step. The size of incremental time step can be changed to confine the amount of projection, which leads to an additional computation time in the finite element simulation. In addition, the projection scheme cannot be differentiated analytically for the AVM. Therefore, a new contact algorithm suggested by Fourment *et al.* (1999) is required, where the time step size is kept constant by allowing the penetration or separation distance of a die-contacted node to a certain degree. The derivative function g of the tangential component σ_t of stress vector in friction boundary condition is based on the suggestion of Chen and Kobayashi (1978) so as to deal with both sticking and sliding friction as follows:

$$g(\Delta u_i) = -\frac{2}{\pi}\tan^{-1}\left(\frac{|\Delta u_i|}{a}\right) \qquad [14.4]$$

where a is very small positive value.

The finite element approximation on the coordinate x_i and velocity u_i results in the following finite element equation:

$$R = R(q, \mathbf{X}, \rho, \bar{\varepsilon}_m, \mathbf{V}) \qquad [14.5]$$

where R is the residual of the finite element equation, which is equivalent to the state equation in an optimization problem, q is the design variable, and **X** and **V** are the coordinate and velocity at a nodal point, respectively. Variables affecting the finite element equations can be classified into three categories as defined by Chung and Hwang (1998b). The first category is the design variables q, which needs to be controlled and selected by the process or part designers, the second category is the state variables S which are the solutions of the finite element analysis, and the third category is governing parameters L that are the function of the design variables and determine the state variables. In the die compaction process, the design variables include the upper and lower punch speed, die shape, friction and reduction ratio. The heating cycle and friction in the sintering process can be design variables as well.

Since the only solution of the finite element analysis of porous material is the velocity field, the velocity field at each incremental time step is the state variable. The last governing parameters include the coordinates of nodal points, relative density and effective strain of base material. The design variables affect both the state variables and governing parameters. They can directly affect the initial values of governing parameters like the initial shape of powder body and in addition, indirectly affect the state variables and governing parameters through the finite element equation for punch speed, die shape and friction.

The governing parameters at each incremental time step are updated by the following equations using the velocity field obtained from the finite element analysis:

$$X_i = X_{i-1} + V_{i-1} \cdot \Delta t_{i-1} \quad [14.6]$$

$$\rho_i = \rho_{i-1} \exp[-(\dot{\varepsilon}_v)_{i-1} - \Delta t_{i-1}] \quad [14.7]$$

$$(\bar{\varepsilon}_m)_i = (\bar{\varepsilon}_m)_{i-1} + (\dot{\bar{\varepsilon}}_m)_{i-1} - \Delta t_{i-1} \quad [14.8]$$

where X_i and V_i are coordinates and velocity of nodal point at i-th time step, ρ_i and $(\dot{\varepsilon}_v)_{i-1}$ are the relative density and the first invariant of strain tensor at i-th time step, $(\bar{\varepsilon}_m)_i$ and $(\dot{\bar{\varepsilon}}_m)_{i-1}$ are the effective strain and the effective strain rate at i-th time step and finally Δt_i is the incremental time step.

The optimization problem pivots on finding design variables which minimize the objective function of concern. For example, two objective functions, the relative density distribution and the forming energy, can be considered. Since there is no die in sintering process, no confined deformation takes place. As mentioned above, the distortion of the powder compact during sintering can occur due to the non-uniform relative density distribution in powder compact. This defective invokes some problems in designing the press-sinter process. In general, the minimization of the forming energy is often considered to be the main objective function in the metal forming process. By minimizing the forming energy, the manufacturing cost can be reduced and furthermore the tool life can be increased by reduced tool wear. Many other objective functions can be considered but the objective function of concern can be expressed as follows:

$$\Phi(q) = \Phi(q, \mathbf{X}, \rho, \bar{\varepsilon}_{\mathrm{m}}, \mathbf{V}) \qquad [14.9]$$

Then, assume that there are N state equations which the process or part design must satisfy. By introducing the Lagrangean multipliers λ_i for each state equation, the Lagrangean function can be derived as follows:

$$\Lambda(q, \lambda) = \Phi(q) + \sum_{i=0}^{N-1} \lambda_i R_i(q, \mathbf{X}, \rho, \bar{\varepsilon}_{\mathrm{m}}, \mathbf{V}) \qquad [14.10]$$

Differentiating this Lagrangean function with the corresponding design variable q produces the design sensitivity with respect to q as follows:

$$\frac{d\Lambda}{dq} = \frac{\partial \Phi}{\partial q} + \sum_{i=0}^{N-1} \lambda_i \frac{\partial R_i}{\partial q} + \Sigma_{-1}^{xX} \cdot \frac{dX_0}{dq} + \frac{\Sigma_{-1}^{\rho} + \Sigma_{-1}^{\bar{\varepsilon}_{\mathrm{m}}\rho}}{\rho_0} \frac{d\rho_0}{dq} + \sum_{-1}^{\bar{\varepsilon}_{\mathrm{m}}} \frac{d(\bar{\varepsilon}_{\mathrm{m}})_0}{dq} \qquad [14.11]$$

The objective functions in the optimization of the metal forming process may be classified into two categories: the first is a function of only the final state variables and the second is the sum of incremental objective functions. The first category includes the uniform relative density in the final state of net-shape forming and the second category includes the minimization of the forming energy.

The design sensitivities of two design variables, the die shape and the punch speed, have been obtained by this method based on the AVM in Chung *et al.* (2009). The tests show that the design sensitivities can be accurately calculated by the AVM over diverse objective functions and design variables in a non-steady-state forming process. The upper and lower punch speeds are optimized in cylindrical compaction processes to obtain the uniform relative density distribution. The objective function is the deviation of the relative density distribution.

In direction-based searches, the search direction and step size are very important parameters. One of the most popular methods is the conjugate gradient method (CGM). The search direction can be determined by the design sensitivity obtained by the AVM and the step size can be determined by second-order polynomial curve fitting.

14.4 Powder injection moulding optimization

This section introduces an optimization technique for process parameters using the design of experiments. Many researchers have investigated opportunities for powder injection moulding (PIM) in the fabrication of microsystem components because of its flexibility in using a wide range of materials combined with its complex shape-forming ability. PIM combines the shape-forming capability of plastic injection moulding, the precision of die-casting and the material selection flexibility of powder metallurgy (German, 1993). The basic steps of the PIM process include powder/binder selection and characterization, mixing of the powder/binder (feedstock preparation), pelletization, part and mould design, injection moulding, debinding, sintering and finishing operations. It is very difficult to anticipate accurately the interactions between the powder/binder feedstock, thermal and mass transfer phenomena combined with the formation of the frozen layer at the melt and the mould wall interface.

For the economic and practical success of multi-scale technology, the primary challenge is the availability of suitable mass-fabrication techniques. Plastic injection moulding has already successfully proved its ability in fabricating multi-scale components (Piotter *et al.*, 2003). But most typical, thermal and chemical applications of multi-scale components, in general, demand much better material properties than the polymers can offer. Multi-scale devices typically have high aspect ratios of 20 or more (Paul and Peterson, 1999). Aspect ratio is defined as the ratio of width to the height of a particular feature on a component or the whole component itself. Typical PIM applications have aspect ratios of about 8 but as high as 70 have been achieved (German, 1990). The challenge of fabricating micro components relates to the high aspect ratios combined with micro features, which have not been previously attempted to reproduce using PIM. To injection-mould a micro component successfully, an appropriate process condition needs to be maintained.

The design methodology currently employed in the PIM industry is based on a trial-and-error approach. With advancements in computer technology, CAE (computer-aided engineering)-based design has been widely accepted as the most competitive technique for product/process design. The numerical/computer simulation of the PIM process provides a platform to shed some light on the interactions between the material properties, the processing conditions and geometric attributes. The main advantage of using a simulation tool is in the reduction of cost and lead time for new products, as there is little or no need to manufacture and rework moulds before the optimal design is found. The simulations give predictions of pressure, velocity and temperature profiles throughout the flow region. These data provide information about the velocity distribution, shear rates and possibility of defect formation which are valuable in determining the mould design parameters and moulding conditions (Najmi and Lee, 1991).

The process conditions studied in PIM processes can be classified into pressure-related, temperature-related and velocity-related measures (Atre *et al.*, 2007). The pressure-related measures include injection pressure, clamping force and the maximum wall shear stress. The injection pressure and the clamping force are limited by maximum capacity of the injection moulding machine.

Insufficient injection pressure may lead to defects like short shots and too high a value may result in jetting. The amount of clamping force needed to mould a part sets the limit to the number of parts that can be simultaneously moulded using multi-cavity moulding. Insufficient clamping force may result in defects such as flashing. Controlling the maximum wall shear stress helps to reduce the amount of residual stress that is induced owing to improper mould-filling (Atre *et al.*, 2007). The locations of high wall shear stresses, if over a certain threshold value, may result in powder–binder separation.

The temperature-related measure of concern is melt front temperature difference (ΔMFT) and cooling time. Controlling ΔMFT results in maintaining uniform temperature during mould-filling. The parameter, ΔMFT is the temperature difference between the highest and smallest value of the melt front temperatures (Atre *et al.*, 2007). A large melt front temperature difference may lead to variations in part shrinkage and lead to warpage. In plastic injection moulding, a difference of less than 10°C is considered acceptable (Turng, 1998). Controlling the cooling time results in shorter processing times and in turn higher productivity. Longer cooling times result in a longer cycle time and a short cooling time may lead to the creation of residual stresses and uneven cooling of the part (Najmi and Lee, 1991).

The velocity-related measures under consideration are the maximum shear rate and the standard deviations for both melt front velocity (MFV) and melt-front-area (MFA). Controlling this condition leads to reduction and uniform deformation of the moulded part. High shear rates may lead to powder–binder separation. Urval *et al.* (2008, 2010) studied the maximum shear rate and the MFA depends on the geometry of melt flow during its passage through the mould cavity. In turn, the MFV changes as the injection flow rate and the MFA change (Fig. 14.2). MFV depends on MFA and determines the injection speed. Variations in MFA result in undesirable packing and powder concentration gradients in the moulded part. Variations in MFV increase surface stress, differential molecular orientation of binder and non-uniform physical properties, resulting in warpage. Hence, it becomes necessary to have a uniform MFV (Tsugawa and Yokouchi, 1992; Urval *et al.*, 2008).

In order to design the PIM process, quantitative models for predicting temperature and pressure fields for varying process conditions are required. The coupling of the filling/packing and the cooling stages complicates the design of the injection moulding process. The aforementioned process conditions and measures can be monitored and observed using commercially available software such as Moldflow® (Autodesk, US) or PIMSolver® (Cetatech, South Korea), which is a 2.5D finite element method (FEM) software package developed to simulate the powder injection moulding process. The simulation package PIMSolver® closely predicts the mould-filling behaviour observed in empirical results (Ahn *et al.*, 2008; Urval *et al.*, 2008, 2010).

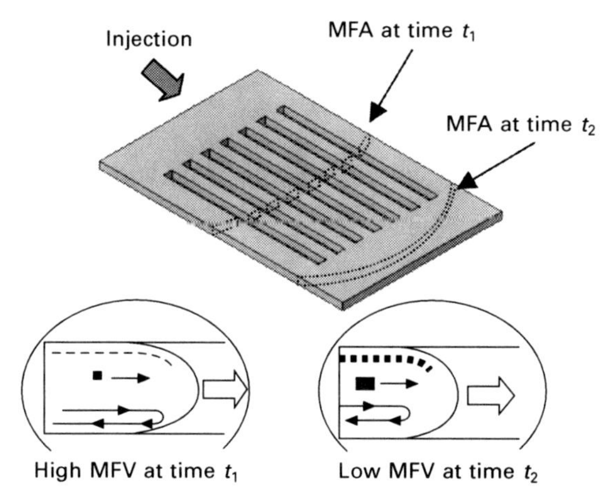

14.2 Depiction of the variation of the melt front velocity (MFV) with varying melt front area (MFA).

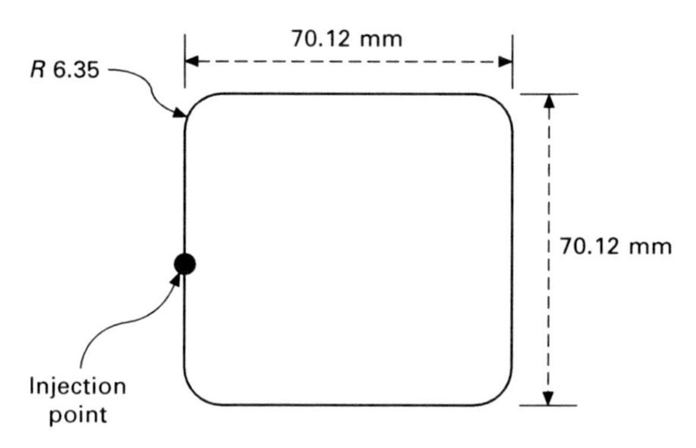

14.3 Plain plate geometry for 1 mm thickness.

The focus of this section is to propose a robust design method to optimize the process conditions in the PIM process for a thin, high aspect ratio part, the features of which can be frequently found in micro components.
For example, a component of interest is a simple plate shown in Fig. 14.3. The rationale for choosing this geometry is to reproduce the high aspect ratios and thin cross-sections, which are typical features of multi-scale components (Paul and Peterson, 1999).

The material selected for the components was gas-atomized 316L stainless steel with a mean particle size of 10 μm. Its high corrosion and high temperature resistance were some of the favourable attributes preferred in materials for fabricating multi-scale components operating in harsh environments (Paul and Peterson, 1999). The binder system used for preparing the feedstock was a mixture (in weight percentage) of 60% paraffin wax (PA), 35% polyethylene wax (PE) and 5% stearic acid (SA), with a solids volume fraction of 53%. The material properties of the binder system for the simulation are measured and given in Table 14.1. The viscosity model and the slip lay model for the binder system as an example are given by Atre *et al.* (2007).

In Table 14.1, B_b and $T_{b,b}$ are amplitude coefficient and Arrhenius type coefficient in binder–viscosity model, respectively. Notations α, β, and m are amplitude coefficient, Arrhenius type coefficient and power law exponent in slip layer model, respectively. The powder and binder were mixed to prepare a feedstock using twin extruder. The material properties of the feedstock for the simulation are given in Table 14.2. The viscosity model for the feedstock as an example is given by Atre *et al.* (2007).

In Table 14.2, n, B_f, C, T_b and τ_y are power law exponent, amplitude coefficient, coefficient for transient region, Arrhenius type coefficient and yield stress in feedstock viscosity model, respectively.

14.4.1 Design parameters of the injection moulding process

The four process parameters, whose values need to be manually fed into the injection moulding machine control, were chosen as the input variables;

Table 14.1 Input data: binder material properties

Density	ρ_b	739 kg m^{-3}
Specific heat capacity	$c_{p,b}$	2286 J kg^{-1} K^{-1}
Thermal conductivity	k_b	0.178 W m^{-1} K^{-1}
Viscosity	B_b	5.72 × 10^{-3} Pa·s
	$T_{b,b}$	3652 K
Slip layer	α	2.37 × 10^{-9} m
	β	4.232 × 10^{-3} K^{-1}
	m	0.5126

Table 14.2 Input data: feedstock material properties

Density	ρ	3975 kg m^{-3}
Specific heat capacity	$c_{p,f}$	685 J kg^{-1} K^{-1}
Thermal conductivity	k	1.84 W m^{-1} K^{-1}
Transition temperature	T_g	52.4°C
Eject temperature	T_e	50°C
Viscosity	n	0.18
	B_f	5.19 × 10^{-3} Pa·s
	C	1.15 × 10^{-4} Pa$^{(n-1)}$
	T_b	5370 K
	τ_y	100 Pa

fill time (t_f in s), feedstock injection temperature (T_m in °C), mould-wall temperature (T_w in °C) and switch-over position (SO in % of total cavity volume). To assess the process capabilities, the following performance metrics have been adopted as process condition measures.

1. Injection pressure (P_i): The P_i rises as the mould cavity begins to fill and continues to do so until switch-over occurs. Improper specification of the P_i can result in several types of defects in the moulded component. A mould cavity can be over-packed, creating a flash or it can also be under-packed, resulting in a short shot. Other mould filling problems include jetting and formation of weld lines. The unit of measurement of P_i is MPa.
2. Clamping force (f_c): The clamping force is the force required to hold the mould plate together to facilitate defect-free mould filling. In common practice, the f_c is usually specified as a fraction of the P_i. Clamping force is proportionally related to the projected area of a part, the larger the projected area the more clamping force that is needed and *vice versa*. This is an important factor when selecting a moulding machine for a given mould specification. Too low a clamping force can lead to flashing or non-filled parts, while its over-specification results in an increase in power consumption and possibly a reduction in mould life. Thus, operating in the optimal range is of foremost importance. The unit of measurement of f_c is the ton.
3. Maximum wall shear stress (τ_{max}): Shear stress is the force between layers of material that causes the layers to initiate motion. Maximum wall shear stress, τ_{max}, occurs at the gates or at the mould walls when the molten feedstock flows into the mould cavity. A sufficiently high shear stress can break the polymer chains of the binder system. This phenomenon reduces the strength of the injection moulded part called the green part. The high shearing also results in the frictional heating of the feedstock, which causes binder degradation. The unit of measurement of τ_{max} is also MPa (Urval *et al.*, 2008).
4. Melt front temperature difference (ΔMFT): Uniform temperature throughout the part is important to produce a quality part. At the end of filling, all areas of the filled cavity should have similar temperatures. A large ΔMFT leads to uneven cooling of the part, leading to the formation of residual stresses. These residual stresses may deform the part on ejection or while at the sintering stage. The unit of measurement of ΔMFT is °C.
5. Cooling time (t_c): Mould cooling accounts for a major part in the injection molding cycle (more than two-thirds of total cycle time). Cooling time, t_c, is a function of mould wall temperature, melt temperature, material properties and part wall thickness. It is essential to have uniform and even

cooling to reduce residual stresses and maintain dimensional accuracy and stability to improve part quality. For production reasons it is best to keep the shortest t_c, but if it is too short the parts could have defects. Some of the common defects seen caused by premature part ejection are warpage and sink marks. In addition, parts sticking to the mould may occur, leading to problems in ejection. Cooling time is measured in seconds (s) (Urval *et al.*, 2008).

6. Maximum shear rate ($\dot{\gamma}_{max}$): The shear rate is represented by the velocity gradient perpendicular to the shear stress. Shear rates affect the suspension viscosity and slip phenomena. The unit of measurement of ($\dot{\gamma}_{max}$) is s^{-1}.
7. Standard deviations of melt front area and melt front velocity (σMFA and σMFV, respectively): During injection moulding, the melt spreads out from the gate into the mould cavity in different directions by splitting into streams. For uniform freezing of the melt, all melt fronts should spread inside the cavity with constant velocity. The melt streams may divide into a number of streams if and when they come across obstructions in their path which also results in variation in the MFV. The melt velocity may also vary owing to variations of the wall thickness and the mould surface temperature. As shown in Fig. 14.2, the cross section of a melt stream or melt front keeps changing depending on the geometry of melt flow during its passage through the mould. Therefore, the velocity of melt front changes as the MFA (cross section area of the moving melt front) changes during the passage of melt through the mould. MFV depends on MFA and the set injection speed depends on machine control (melt flow rate = MFV × MFA). The MFV increases when the melt divides around any obstruction, as MFA is reduced while passing around the slot. This increase in MFV, in turn, increases surface stress and orientation resulting in warpage. Hence, it becomes necessary to have a uniform MFV and MFA in the melt front which can be implemented by using a varying injection speed called the ram-speed profile. The MFA is measured in mm^2 and the MFV is measured in mm s^{-1}. The standard deviations were calculated by performing time integration of the instantaneous melt front areas and melt front velocities from the flow analyses results and followed by taking the square root of integrated value (Urval *et al.*, 2008).

To define the baseline parameters, mould-filling simulations were performed on the part. Experimental runs were conducted by systematically varying one variable at a time and these were progressively conducted till a complete mould-fill was achieved. Complete mould fill was achieved with t_f = 1 s, T_m = 140°C, T_w = 45°C and SO = 97% with the 1 mm plate and represent the worst-case-scenario or the lower bound for this simulation study (Urval *et al.*, 2008).

The processing conditions were varied over three different levels each and these levels were based on the above limiting conditions, material properties and typical values used in practice. For example the levels of T_m, were selected based on the available operating range for the binder system used, which was between 130°C and 180°C. Below this range, the binder is a solid and above this range, degradation/burn-out of the binder occurs. Based on these levels, 81 runs were required to be conducted for each individual plate thickness to evaluate the effects of each processing condition. The Taguchi robust method helped to perform the analysis systematically with the smallest number of experiments and also with economy of time, cost, and materials.

The Taguchi method's orthogonal array-based design of experiments was used to optimize the processing conditions. The Taguchi method, also known as the 'robust design method', focuses on improving the fundamental function and designing a product or process that is robust in the event of variations in the process conditions. This method is used to reduce product cost, improve quality and simultaneously reduce interval time during the product development process (Taguchi, 1987).

The experiments were conducted based on 'orthogonal array' which is balanced with respect to all control factors and yet requires a minimum number of experiments to be performed. This is a derivative of the fractional factorial design. For the problem in this section, L_9 orthogonal array designs can be used for the four factors each at three levels. Table 14.3 shows experimental parameters for the L_9 orthogonal array design. The four input processing conditions were the control factors and the four output parameters were taken as the optimizing characteristics. In the Taguchi method, the optimization procedure involves the determination of 'best' levels of control factors. The 'best' levels of control factors are those that maximize the 'signal-to-noise

Table 14.3 The L_9-Taguchi array, with nine runs and variable parameters

L_9 (3^4) Orthogonal array					Actual values			
Run no.	*A*	*B*	*C*	*D*	t_f (s)	T_m (°C)	T_w (°C)	SO (%)
1	1	1	1	1	1.0	140	45	97
2	1	2	2	2	1.0	150	50	98
3	1	3	3	3	1.0	160	55	99
4	2	1	2	3	0.9	140	50	99
5	2	2	3	1	0.9	150	55	97
6	2	3	1	2	0.9	160	45	98
7	3	1	3	2	0.8	140	55	98
8	3	2	1	3	0.8	150	45	99
9	3	3	2	1	0.8	160	50	97

A = fill time (s), *B* = feedstock injection temperature (°C), *C* = mold wall temperature (°C) and *D* = switch-over position (%).

(S/N)' ratios. The signal-to-noise ratios are log functions of desired output characteristics (Taguchi, 1987).

The S/N ratio is a statistical quantity representing the power of a response signal to the power of the variation in the signal due to noise (Ross, 1988). The S/N is derived from the loss function and depending on the optimization objectives, it takes different forms. Maximizing the S/N ratio results in minimizing any property which is sensitive to noise. In this example, the larger-the-better characteristic of the S/N ratio was chosen because a high value of S/N implies that the signal is much higher than the uncontrollable noise factors. The S/N ratios were calculated using Equation [14.12] which is measured in unit decibels (db):

$$S/N = -\log_{10} \text{ (mean sum of squares of the measured data]} \quad [14.12]$$

The data was then analysed using the statistical methods of ANOM (analysis of means) and ANOVA (analysis of variance). The ANOM was used to optimize the process for the performance metric. The ANOVA was used to assess the significant parameters and their relative contributions to the variations in the observed parameter. These relative contributions helped identify the factors influencing the relative sensitivity to noise (Urval *et al.*, 2008).

Urval *et al.* (2008, 2010) conducted various simulation runs, based on the orthogonal array, and the corresponding output parameter values were tabulated. These simulations predicted the flow paths, pressure, temperature profiles and the other process factors that were of interest to the mould-filling process. Table 14.4 gives nine simulation runs for injection pressure P_i. For other process parameters discussed earlier, see Urval *et al.* (2008, 2010).

Figure 14.4 shows the pressure distribution in a cavity at the end of mould-filling for 3 mm thick plates. The process conditions were T_m = 140°C, T_w = 45°C, t_f = 1.0 s, SO = 97%. The highest pressure is at the gates and there is a gradual decay as the melt moves farther and farther away from the gates. The pressure completely decays as the feedstock-melt reaches the farthest

Table 14.4 Injection pressure as an output parameter from the simulation runs

Thickness	Run no.	P_i (MPa)
3 mm	1	7.18
	2	6.63
	3	6.14
	4	7.18
	5	6.25
	6	6.63
	7	6.72
	8	7.35
	9	6.38

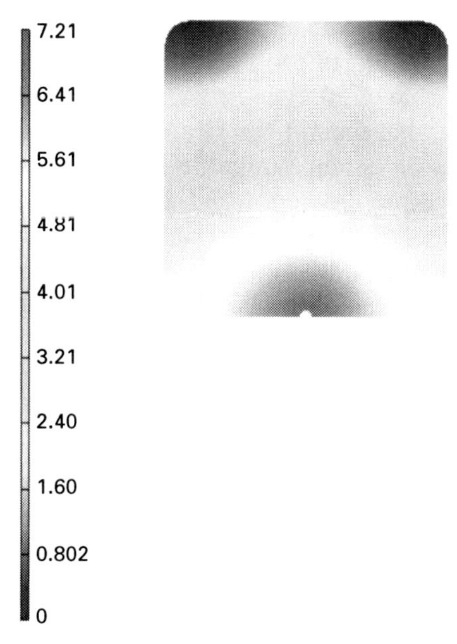

14.4 Injection pressure distribution across the mould cavity for the 3 mm-thick plate with a set of process conditions (T_m = 140°C, T_w = 45°C, t_f = 1.0 s, SO = 97%).

Table 14.5 ANOM and ANOVA calculations of the injection pressure for the three plate thicknesses

P_i	Levels	t_f	T_m	T_w	SO
S/N for 3 mm	1	**−16.44**	−16.93	−16.96	**−16.38**
	2	−16.49	−16.56	**−16.08**	−16.74
	3	−16.66	**−16.10**	−16.55	−16.47
Relative contribution		0.10	0.36	0.39	0.15

corners (opposite the gate, shown in Fig. 14.4) of the mould cavity. One key aspect to be pointed out here is the amount of pressure drop from the gate to the extreme points. For the 3 mm plate, the pressure drop is from 7.2 MPa to 0 MPa. For defect-free moulding this pressure drop should be low.

From the ANOVA, relative contributions of each process parameter were calculated. Table 14.5 gives the relative contribution of each process parameter to the variation seen in the injection pressure. For the 3 mm plate, the significant factors that affected the injection pressure were mould wall temperature and feedstock–melt temperature at 36% and 38%, respectively.

Based on the Taguchi method, the optimization of P_i was done for the 3 mm thick plates using ANOM. The process parameter level corresponding to the highest *S/N* ratio in bold font in Table 14.5 was chosen as the optimum level. In the present processing window, the optimum levels, corresponding to the injection pressure, were a fill time of 1 s, a feedstock-melt temperature of 160°C, a mould wall temperature of 50°C and a switchover point of 97% of the ram stroke. In summary, this section has laid down the framework for optimizing the PIM process to produce microfluidic devices, in order concurrently to solve an engineering problem which is based on machine, material and geometry interactions. The current mould-filling study will be extended to study the packing and cooling stages during the PIM process in future. The debinding and the sintering of the thin-walled high-aspect ratio micro-fluidic devices are other areas that require optimization for successful operations.

14.5 Sintering optimization

This section shows the optimization approach when more than a single objective needs to be achieved simultaneously. To demonstrate the methodology, two kinds of powders were prepared. A standard WC-10Co powder (powder S) produced by carburizing W powder, milling it with Co powder, and spray drying it with 2 wt% paraffin was acquired from Kennametal Inc. (Latrobe, PA). The second powder is a nanocrystalline WC-12Co, which was produced by Inframat Corp. (Farmington, CT) by a continuous wet chemical process based on the spray conversion process in which W, C, and Co precursors are mixed and reacted together. Then, powder N was prepared from this powder by ball milling it for 24 h with grain growth inhibitors, 1 wt% TiC and 3 wt% TaC. The powders were characterized by their tap and pycnometer densities and BET surface area. Owing to the presence of the wax in powder S, no BET surface area was possible, but its pycnometer density was measured after binder removal from the as-received powder. The BET equivalent spherical particle diameter (D_{BET}) of powder N was calculated from the equation:

$$D_{BET} = 6/\rho_p A_{BET} \quad [14.13]$$

where A_{BET} is the specific BET surface area of the powder in m^2 g^{-1}, ρ_p is the pycnometer density of the powder in g cm^{-3} and D_{BET} is in µm.

A die compaction process has been used as a fabrication process. Right cylindrical compacts, approximately 3.20 mm in diameter and 30 g mass, were made at three different compaction pressures of 110, 400 and 565 MPa from powder S with 2 wt% paraffin wax as lubricant and powder N with 3 wt% paraffin wax. The die compaction was carried out using a hand press. Twelve samples were compacted for grain size measurement and an additional three samples for dilatometry experiments at each pressure level. A typical variation of 2% was observed in the green density at each pressure level.

The grain growth data are collected from the quenching experiment and image data are obtained from microscopic analysis (Park *et al.*, 2008). For the wax removal and pre-sintering, the samples were heated at 2°C min^{-1} up to 250°C with a 1 h hold, heated again at 3°C min^{-1} to 1000°C with a 1 h hold, and cooled at 10°C min^{-1} to room temperature in a H_2 atmosphere. Sintering was performed in a H_2 atmosphere at four different sintering temperatures of 1100, 1200, 1300 and 1400°C. The samples were heated at the rate of 5°C min^{-1} up to the designated temperature with a 1 h hold and cooled at 10°C min^{-1} to room temperature.

The polished and etched microstructures were imaged using scanning electron microscopy (SEM; TopCon Model ABT-32 SEM, Tokyo, Japan). The WC grain size was measured using image analysis software using the average grain area and perimeter. The image analysis software was developed using Clemex (Longueuil, Canada) Version PE 3.5 software. Grain boundaries were visually identified by considering the penetration by the Co matrix. The diameter of an equivalent circle was calculated to represent the average WC grain size $G_{equivalent} = 4A/P$, where A is the average grain area and P is the average grain perimeter.

A dilatometer was employed for *in situ* measurement of shrinkage, shrinkage rate and the temperature at which phase changes take place. Sample specimens are compacted using both powders S and N, at pressures of 110, 400 and 565 MPa. The dilatometry experiment has been conducted at heating rates of 4, 7 and 10°C min^{-1} using a vertical push rod dilatometer in hydrogen (ANTER Unitherm™ Model 1161). The sintering cycle consisted of a ramp of 4, 7 or 10°C min up to 1400°C with a 1.5 h hold in a H_2 atmosphere before slow cooling to room temperature in a H_2 atmosphere. The dilatometer was also used to measure the thermal expansion of each composition during a controlled cool down at a rate of 20°C min^{-1} from the sintering temperature.

The modelling procedure for grain growth is explained by Park *et al.* (2008). The mean grain size increases over sintering time. To sinter hard materials, a differential form of the classical model for interface-controlled grain growth with isothermal condition is:

$$\frac{dG}{dt} = \frac{K_0}{2GT} \exp\left(-\frac{Q_G}{RT}\right) \qquad [14.14]$$

where G is the mean grain size, t is the time, T is the absolute temperature, K_0 is the associated pre-exponential factor, Q_G is the apparent activation energy, and R is the universal gas constant. The integration of the above equation leads to:

$$G = \sqrt{G_0^2 + \Theta_G} \qquad [14.15]$$

where G_0 is the initial mean grain size and

$$\Theta_G = \int_0^t \frac{K_0}{T} \exp\left(-\frac{Q_G}{RT}\right) dt \qquad [14.16]$$

The three parameters (G_0, K_0 and Q_G) are determined by grain size measurement at the given die compaction pressures for both powders S and N, as shown in Table 14.6 with average errors. Figure 14.5 shows the grain growth model for powders S and N.

The master sintering curve (MSC) theory enables a systematic analysis of solid state densification of both powders. The apparent activation energy for sintering (Q_ρ) can be calculated from the experimental data as a constant in the temperature range of interest and can be used to simulate the densification process (Park *et al.*, 2006). It leads to a parameter that is equivalent to the thermal work performed in reaching the density. This parameter Θ_ρ with slight modification is termed the work of sintering,

$$\Theta_\rho(t, T) \equiv \int_{t_0}^t \frac{r_{\text{ref}}}{T} \exp\left[-\frac{Q_\rho}{R}\left(\frac{1}{T} - \frac{1}{T_{\text{ref}}}\right)\right] dt \qquad [14.17]$$

Table 14.6 Material parameters used in the grain growth model

Powder	Die compaction pressure (MPa)	G_0 (nm)	Q_G (kJ mol^{-1})	K_0 (m^2·K s^{-1})	Average error (%)
S	110	480	171	1.47×10^{-8}	2.48
	400		209	4.39×10^{-7}	2.54
	565		224	1.81×10^{-6}	3.05
N1	110	188	64	9.96×10^{-13}	3.27
	400		88	5.85×10^{-12}	3.80
	565		65	9.56×10^{-13}	2.59

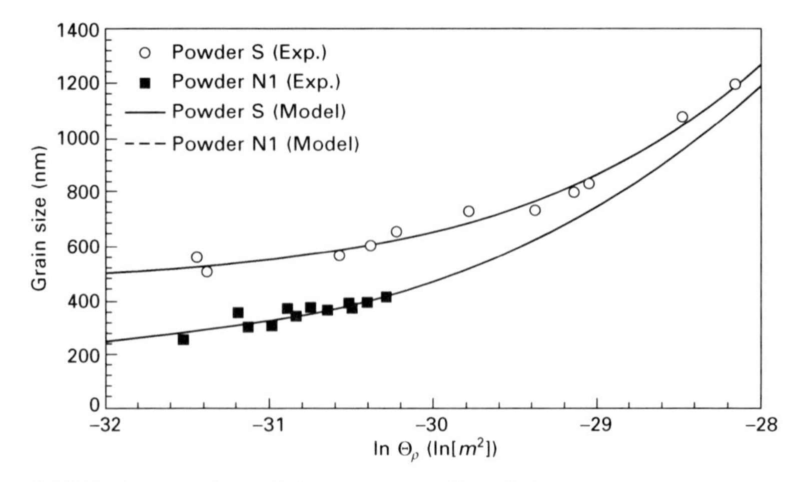

14.5 Grain growth model for powders S and N.

where r_{ref} is a reference heating rate for the dimensionless Θ_ρ parameter and T_{ref} is a reference temperature. In this case, r_{ref} is 10°C min^{-1} and T_{ref} is 1400°C for both powders.

A sigmoid function provides a good curve fit between the densification parameter Ψ over the range of $\rho_0 < \rho < 1$ and the natural logarithm of the work of sintering, ln Θ_ρ. The sigmoid equation used to define the MSC is:

$$\Psi = \frac{\rho - \rho_0}{1 - \rho_0} = \frac{1}{1 + \exp\left[-\frac{\ln\Theta_\rho - a}{b}\right]} \qquad [14.18]$$

where ρ_0 is the initial relative density, and a and b are constant parameters for the sigmoid function. An alternative form of the above equation is:

$$\Phi \equiv \frac{\rho - \rho_0}{1 - \rho} = \frac{\rho - \rho_0}{\theta} = \left(\frac{\theta_\rho}{\theta_{ref}}\right)^n \qquad [14.19]$$

or

$$\ln\Phi = \ln\left[\frac{\rho - \rho_0}{1 - \rho}\right] = n \cdot \ln\left(\frac{\Theta_\rho}{\Theta_{ref}}\right) = n(\ln\Theta_\rho - \ln\Theta_{ref}) \qquad [14.20]$$

with ln $\Theta_{ref} = a$ and $n = 1/b$. Φ is called the densification ratio over the range of $\rho_0 < \rho < 1$ and is defined as the ratio of the density difference between the current density and the initial density to the current porosity θ, n is a slope, power law exponent or densification function, which defines the rate of change of ln Φ during the sintering process, and ln Φ_{ref} is the natural logarithm of the work of sintering ln Θ at $\rho = (\rho_0 + 1)/2$, that is the mid-point of densification or densification to a parameter Ψ of 0.5.

Figure 14.6 shows the MSC plot for both powders. Linearization of Fig. 14.6 based on Equation [14.20] is very useful for understanding the densification mechanism, as shown in Fig. 14.7. In the example under consideration, only solid state sintering (T < 1310°C) is considered for minimizing the grain growth. Based on Fig. 14.7(a), these two regions for densification behaviour of powder S can be roughly identified as: (1) region I, solid state sintering region (T < 1310°C) and (2) region II, liquid phase sintering region ($T \geq$ 1310°C). Similarly, based on Fig. 14.7(b), these three regions for densification behaviour of powder N can be roughly identified as: (1) region I, the first solid state sintering region (T < 1100°C), (2) region II, the second solid state sintering region (1100°C $\leq T <$ 1310°C), and (3) region III, liquid phase sintering region ($T \geq$ 1310°C). The second solid state sintering region for powder S shows a faster densification rate owing to the effect of additives (1 wt% TiC and 3 wt% TaC).

The three parameters (Q_ρ, Θ_{ref}, and n) were determined by dilatometry

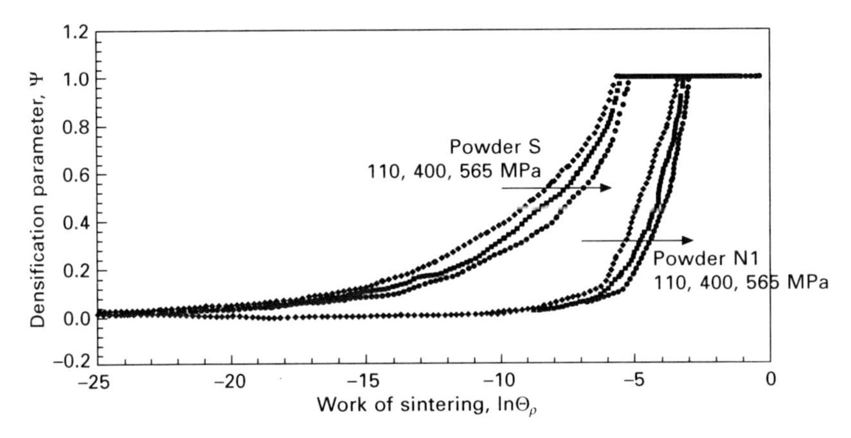

14.6 MSC of densification parameter Ψ for powders S and N.

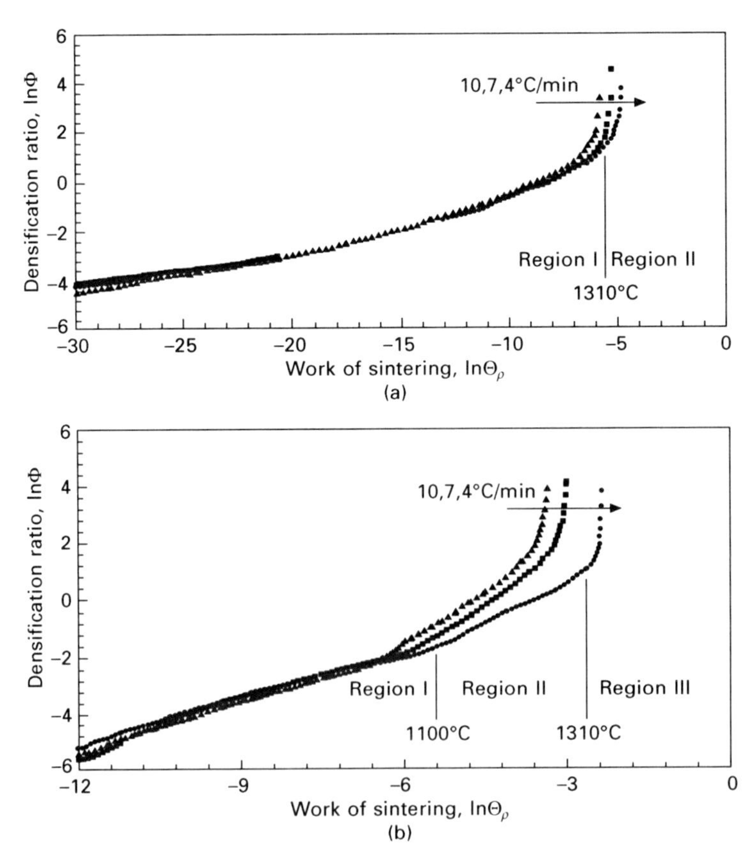

14.7 MSC plot of densification parameter Φ for powders S and N: (a) powder S at die compaction pressure of 565 MPa and (b) powder N at die compaction pressure of 565 MPa.

data at the given die compaction pressures for both powders S and N, in regions of solid state sintering in Table 14.7. The activation energy shows a big difference between two powders, which shows that powder S is much more sensitive to temperature in densification than powder N. This result is confirmed by the sensitivity analysis below.

For the sensitivity analysis, input parameters were varied over a fixed range (± 5%) and the response of the output parameters was monitored. The sensitivity value S was calculated as the gradient or the slope of the dimensionless dependent variable with respect to the dimensionless independent variable according to the following equation:

$$S \equiv \frac{\text{percentage change in output}}{\text{percentage change in input}} = \frac{\dfrac{\text{increment in output}}{\text{initial value of output}}}{\dfrac{\text{increment in input}}{\text{initial value of input}}} \qquad [14.21]$$

This definition of sensitivity is made to compare input and output parameters that have different dimensions. For example, a sensitivity value of –1.5 means that the percentage decrement of output is 1.5% if the percentage increment of input is 1.0% and is independent of the units of the input and output parameters.

Tables 14.8 and 14.9 show all normalized sensitivity values for the grain growth and the densification in both powders S and N. For the grain growth, activation energy Q_G and sintering temperature are the most critical parameter for both powders and die compaction pressure is significant for powder N (see the values in bold in Table 14.8). For the densification, sintering temperature is the most critical parameter for both powders and initial green density is significant for powder S (see the values in bold in Table 14.9).

The integration of the grain growth and the densification models can lead to the minimum grain size for a given specification of sinter density. To obtain maximum density and minimum grain size, the following objective function F is proposed. It is the linear combination of two objective functions:

Table 14.7 Material parameters for solid state sintering used in densification model

Powder	Compaction pressure (MPa)	Q_p (kJ mol^{-1})	Region I		Region II
			Θ_{ref}	n_1	n_2
	110	688	–5.30	0.237	
S	400	780	–6.33	0.218	–
	565	806	–6.77	0.211	
	110	163	–2.19	0.622	1.36
N1	400	185	–2.61	0.600	1.32
	565	210	–2.83	0.589	1.29

Table 14.8 Sensitivity analysis for grain growth

Parameter	Powder S		Powder N	
	Reference value	Sensitivity value *S*	Reference value	Sensitivity value *S*
Final grain size (nm)	997	–	421	–
Initial grain size (nm)	480	0.1116	188	0.0887
Activation energy (kJ mol^{-1})	209	**–2.8201**	88	**–2.0554**
Pre-exponent ($m^2 \cdot K\ s^{-1}$)	4.39×10^{-7}	0.2961	5.85×10^{-12}	0.3038
Heating rate (°C min^{-1})	10	–0.0563	10	–0.0724
Sintering temperature (°C)	1300	**2.3681**	1300	**1.7581**
Holding time (h)	1	0.2396	1	0.2313
Die compaction pressure (MPa)	400	0.1192	400	0.9888

Table 14.9 Sensitivity analysis for densification

Parameter	Powder S		Powder N	
	Reference value	Sensitivity value *S*	Reference value	Sensitivity value *S*
Final sinter density (%)	83.40	–	98.60	–
Die compaction pressure (MPa)	400	0.0322	400	0.0088
Initial green density (%)	58.58	**0.2815**	50.81	0.0147
Activation energy (kJ mol^{-1})	780	–0.0937	185	–0.0209
$\ln \Theta_{ref}$	–6.33	0.1646	–2.61	0.0217
n_1	0.218	0.0480	0.600	0.0260
n_2	–	–	1.32	0.0749
Heating rate (°C min^{-1})	10	–0.0011	10	–0.0073
Sintering temperature (°C)	1300	**1.2611**	1300	**0.2035**
Holding time (h)	1	0.0249	1	0.0162

$$F = \alpha\left[\frac{\Delta\rho}{\rho}\right] + (1-\alpha)\left[-\frac{\Delta G}{G}\right] \qquad [14.22]$$

where α is adjustable parameter.

The first optimization is to obtain the minimum grain size and corresponding holding time with the following constraints: (1) sintering cycle consists of heating at 10°C min^{-1} to 800°C for 1 h holding, the second heating at

10°C min^{-1} to the target sintering temperature, and cooling at 10°C min^{-1} to 20°C, (2) target sintering temperature range is 1100–1300°C, and (3) target density is 97%. Figure 14.8 shows the optimization results. The higher sintering temperature and higher die compaction pressure can reduce the final grain size. The minimum grain size is 332 nm to obtain 97% density at a die compaction pressure of 565 MPa, sintering temperature of 1300°C for 20 min, and $\chi = 0.68$, where χ is an adjusting parameter between two objective functions. Note that powder S cannot reach to 97% density with reasonable holding time.

The second optimization is to obtain the minimum grain size and corresponding sinter density with a varying target density of 80–100%, varying the second heating rate of 0–20°C, and no target sintering temperature.

Figure 14.9 shows the optimization results. The higher die compaction pressure can reduce the final grain size. The minimum grain sizes in the case of die compaction pressure of 565 MPa are 289 nm at 90% density, 295 nm at 92% density, 304 nm at 94% density, 319 nm at 96% density and 353 nm at 98% density. Note that the minimum obtainable density is 82% and the minimum is 98% for minimizing grain size.

14.6 Design optimization of steady-state conduction

Many engineering problems involve heat transfer, stress analysis, fluid mechanics, electronics, acoustics, and so on in complex three-dimensional geometries with a closely spaced surface and design features (for example, circular cooling channels). Typical examples of such geometries in the heat transfer problems are moulds of the injection moulding and compression moulding processes. This section is concerned about an optimization problem of steady-state conduction heat transfer in these kinds of geometries. In such

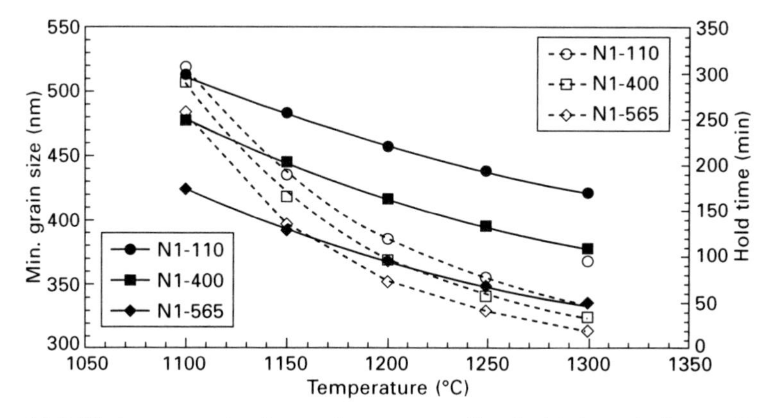

14.8 Minimum grain size and corresponding hold time during solid state sintering (1100–1300°C) to obtain 97% density.

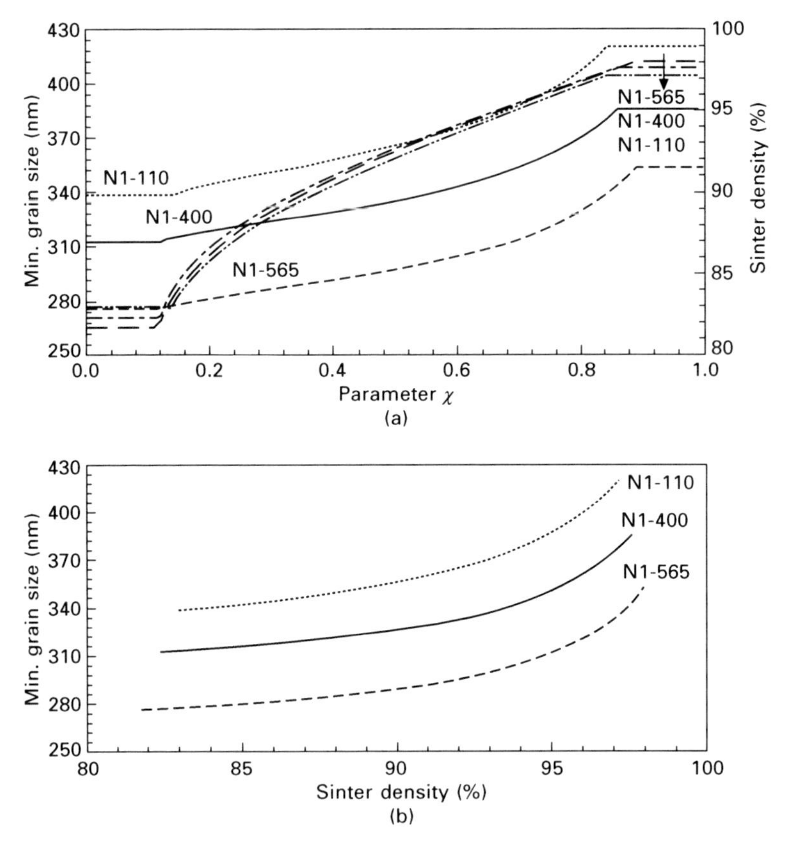

14.9 Minimum grain size and corresponding sinter density during solid state sintering (1100–1300°C): (a) G-ρ-χ plot and (b) *G*-ρ plot.

a thermal system, the cooling stage would be of utmost importance because it significantly affects the productivity of the manufacturing process and the quality of the final part. It is well known that more than three-quarters of the cycle time in the injection moulding process is spent in cooling the hot polymer melt sufficiently so that the part can be ejected without any significant deformation. An efficient cooling channel design *can* considerably reduce the cooling time and, in turn, increase the productivity of the injection moulding process. On the other hand, severe warpage and thermal residual stress in the final product may result from non-uniform cooling. Warpage and sink marks can significantly affect part quality, especially in terms of appearance and precision. Accordingly, there are at least two important objectives to consider when designing cooling systems and determining processing conditions during the cooling stage of the injection moulding process: (a) to minimize the cooling time and (b) to achieve uniform cooling. Toward these objectives, the designer may need to optimize the system design of concern.

A typical optimization process starts with a preliminary design (or initial) and searches for a better design with the help of the numerical analysis and the design sensitivity analysis. Considering design sensitivity coefficients obtained in the current iteration, a new improved design can be suggested in the next iteration. A non-linear programming search algorithm such as the steepest descent method and the CONMIN conjugate gradient method proposed by Haarhoff and Buys (1970) can be used during the iterative processes. There can be various terminal conditions for the iterative process. The iterative process can be terminated after a pre-determined number of iterations or if no improvement over a pre-determined number of iterations is observed. Otherwise, the iterative process in moving toward optimal design continues.

In order to use any first-order method for this optimization process, the thermal analysis and the corresponding design sensitivity analysis need to be conducted for optimal design. The design sensitivity analysis in a heat transfer problem generates design sensitivity coefficients (DSCs) that are rates of change in response variables such as temperatures or heat fluxes with respect to the corresponding design variables. In the literature, there are some research works on the subject of optimization problems using finite element method and boundary element method. Saigal and Chandra (1991) studied two-dimensional and axisymmetric heat diffusion problems and Awa *et al.* (1992) investigated an optimal pultrusion die design. Tang *et al.* (1997) and Matsumoto *et al.* (1993a) independently optimized the two-dimensional cooling system design of injection moulds. Matsumoto *et al.* (1993b) studied the three-dimensional cooling/heating design of compression moulds. As far as the thermal analysis is concerned, Rezayat and Burton (1990) proposed a special boundary integral formulation for the steady heat conduction problem in these complex geometries. There are some research works on the design sensitivity analysis for several design problems in mechanics such as elasticity or heat transfer: Choi *et al.* (1990) for two-dimensional elastic structures, Saigal and Chandra (1991) for two-dimensional and axisymmetric heat diffusion problems, and so on.

There are three different approaches in design sensitivity analysis; the finite-difference approach (FDA), the adjoint structure approach (ASA) and the direct differentiation approach (DDA). These three approaches are used in conjunction with the finite element method (FEM) or the boundary element method (BEM) to obtain the calculated design sensitivity coefficients. Park and Kwon (1996) proposed a design sensitivity analysis formulation for the steady heat conduction problem using DDA based on the special boundary integral formulation proposed by Rezayat and Burton (1990).

This section introduces an efficient optimization procedure to achieve the goal of uniform temperature distribution over a closely spaced surface for the steady heat conduction problem. Radii and locations of circular holes are considered as design variables. The CONMIN conjugate gradient algorithm has been adopted to obtain the optimal configuration of the design variables in conjunction with the special boundary integral formulation by Rezayat and Burton (1990) and the design sensitivity analysis formulation by Park and Kwon (1996).

The optimal design problem of steady-state conduction heat transfer in an infinite domain with a closely spaced surface (see Fig. 14.10) is defined as follows: given a constant heat flux boundary condition on the closely spaced part surface, find optimal radius and location (design variables) of each circular hole with a constant heat transfer coefficient and a constant bulk temperature to minimize the non-uniformity of the temperature distribution (or maximize the uniformity of the temperature distribution) on the closely spaced part surface with side constraints (i.e. realistic intervals for each design variable). In this problem, the closely spaced surface and circular holes correspond to the part surface and circular cooling channels, respectively.

The geometry of this simple problem consists of the closely spaced surfaces and the circular holes in an infinite domain as shown in Figs 14.10 and 14.11. In these figures, Ω, S_P, Γ, S_C, S_E and $\hat{n}$ denote the infinite domain, the part surfaces (closely spaced surfaces), the mid-surfaces of the part, the circular hole surfaces, the exterior surfaces of the domain (at infinite point), and the outward unit normal vector in the part, respectively. The part surface can be described by only the mid-surface of the part instead of the whole surface (positive and negative surface as indicated by S_P^+ and S_P^-, respectively in Fig. 14.11) because the part surfaces are closely spaced. The governing differential equation for the steady-state heat conduction equation may be written as:

$$\nabla^2 T = 0 \quad \text{in } \Omega \qquad [14.23]$$

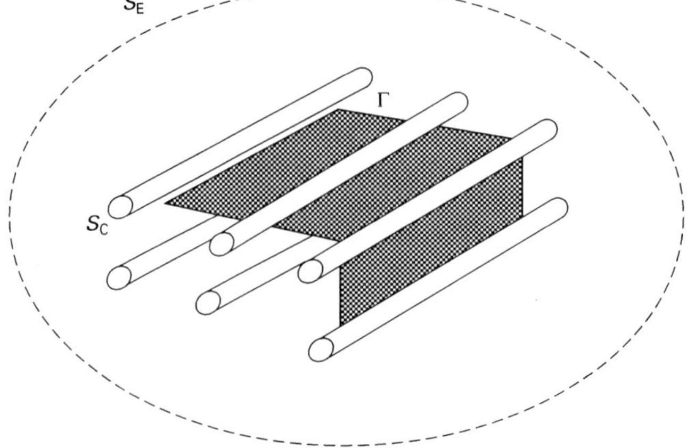

14.10 Geometry of a simple model problem bounded by Γ, S_c and S_E.

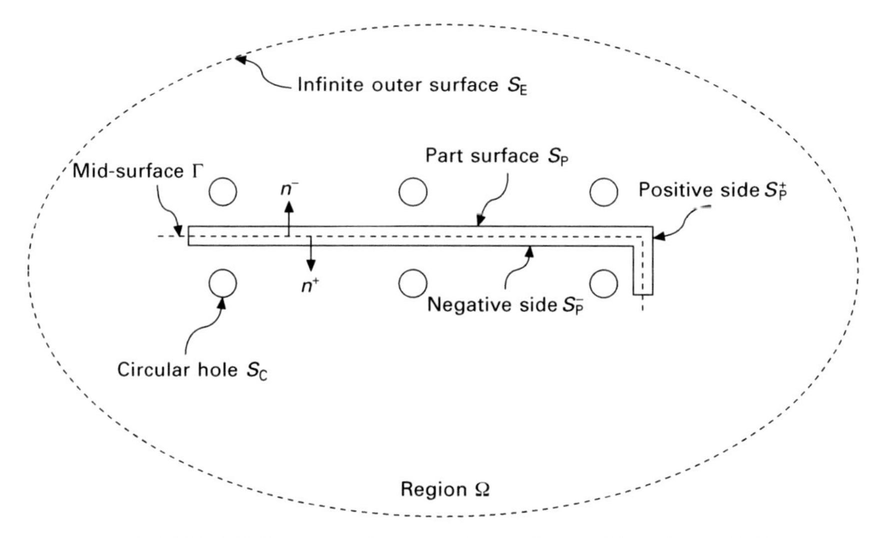

14.11 Detail diagram of a sample model problem for notations.

where T is the temperature. In this problem a constant heat flux is imposed over the part surfaces as a boundary condition and the boundary condition on the circular hole surfaces is treated as a mixed boundary condition with a specified heat transfer coefficient and a specified internal mean temperature (for example, a coolant mean temperature when the circular hole is a cooling/heating channel). It is also assumed that the exterior surface of domain (at infinite point) may be treated as an infinite adiabatic sphere. Thus, the boundary conditions on the boundary $S = S_P + S_C + S_E$ are given as:

$$\frac{\partial T}{\partial n} = \left(\frac{\partial T}{\partial n}\right)_0 \quad \text{on } S_p \qquad [14.24]$$

$$-k\frac{\partial T}{\partial n} = h(T - T_m) \quad \text{on } S_c \qquad [14.25]$$

$$\frac{\partial T}{\partial n} = 0 \quad \text{on } S_E \qquad [14.26]$$

where k, h and T_m denote the thermal conductivity of the material in domain, the specified heat transfer coefficient and the specified internal mean temperature, respectively, and $(\partial T/\partial n)_0$ is the specified temperature gradient.

A standard boundary element formulation for the three-dimensional Laplace's equation given by Equation [14.23] governing the steady conduction leads to:

$$\alpha T(\mathbf{x}) = \int_S \left[\frac{1}{r}\left(\frac{\partial T(\xi)}{\partial n}\right) - T(\xi)\frac{\partial}{\partial n}\left(\frac{1}{r}\right)\right] \mathrm{d}S(\xi) \qquad [14.27]$$

Here $\mathbf{x}$ and $\boldsymbol{\xi}$ are points in space, $r = |\boldsymbol{\xi} - \mathbf{x}|$, and α denotes a solid angle formed by the boundary surface. Note that $\alpha = 2\pi$ at the smooth boundary surface, and $\alpha = 4\pi$ or 0 at internal points or external points of the domain, respectively. Kwon (1989) and Forcucci and Kwon (1989) have used Equation [14.27] for mould cooling system analyses.

For any two closely spaced surfaces, such as the part surfaces in the example of this section, because Equation [14.26] leads to redundancy in the final system of linear algebraic equations, a modified procedure as described by Rezayat and Burton (1990) needs to be considered. According to this modification, the mid-surface, Γ, is considered rather than two closely spaced surfaces, S_P^+ and S_P^- as schematically depicted in Fig. 14.11. For each mid-surface element, a derivative of Equation [14.27] with respect to the normal direction vector, $\hat{v}$ ($\hat{n}^+$ in Fig. 14.11), at the mid-surface element, is taken to derive the extra equation corresponding to the additional degree of freedom. For circular hole surfaces, S_C, a special formulation based on the line-sink approximation is derived to avoid the discretization of the circular channels along the circumference and thus save a large amount of computer memory and time. See Rezayat and Burton (1990), Himasekhar *et al.* (1992) and Park and Kwon (1996) for further details of this modified approach for part surfaces, circular holes and exterior surface.

The final BEM formulae for these modifications are given in the following:

For a point x on the mid-surface of the part, Γ, the following pair of integral equations holds:

$$\alpha^- T^+(x) + \alpha^+ T^-(x) = \int_\Gamma \left[\frac{1}{r}\left(\frac{\partial T^+}{\partial n^+} + \frac{\partial T^-}{\partial n^-}\right) - \frac{\partial}{\partial n}\left(\frac{1}{r}\right)(T^+ - T^-)\right] \mathrm{d}S(\xi)$$

$$+ \Sigma_{k=1}^{N} \int_{l_k} \left[\left(\frac{\partial T}{\partial n}\right)\int_0^{2\pi}\left(\frac{1}{r}\right)\mathrm{d}\theta - T\int_0^{2\pi}\frac{\partial}{\partial x}\left(\frac{1}{r}\right)\mathrm{d}\theta\right] a_k dl(\xi) + 4\pi T_\infty \quad [14.28]$$

$$\alpha^- \frac{\partial T^+(x)}{\partial n^+} - \alpha^+ \frac{\partial T^-(x)}{\partial n^-}$$

$$= \int_\Gamma \left[\frac{\partial}{\partial v}\left(\frac{1}{r}\right)\left(\frac{\partial T^+}{\partial n^+} + \frac{\partial T^-}{\partial n^-}\right) - \frac{\partial}{\partial v}\left(\frac{\partial}{\partial n}\left(\frac{1}{r}\right)\right)(T^+ - T^-)\right] dS(\xi)$$

$$+ \Sigma_{k=1}^{N} \int_{l_k} \left[\left(\frac{\partial T}{\partial n}\right)\int_0^{2\pi}\frac{\partial}{\partial v}\left(\frac{1}{r}\right)\mathrm{d}\theta - T\int_0^{2\pi}\frac{\partial}{\partial v}\left(\frac{\partial}{\partial x}\left(\frac{1}{r}\right)\right)\mathrm{d}\theta\right] a_k dl(\xi) \quad [14.29]$$

For a point on the axis of the cylindrical segment of the cooling channels, the following equation can hold:

$$0 = \int_{\Gamma} \left[\frac{1}{r} \left(\frac{\partial T^{+}}{\partial n^{+}} + \frac{\partial T^{-}}{\partial n^{-}} \right) - \frac{\partial}{\partial n} \left(\frac{1}{r} \right) (T^{+} - T^{-}) \right] \mathrm{d}S(\xi)$$

$$+ \sum_{k=1}^{N} \int_{l_k} \left[\left(\frac{\partial T}{\partial n} \right) \int_{0}^{2\pi} \left(\frac{1}{r} \right) \mathrm{d}\theta - T \int_{0}^{2\pi} \frac{\partial}{\partial x} \left(\frac{1}{r} \right) \mathrm{d}\theta \right] a_k \mathrm{d}l(\xi) + 4\pi T_{\infty} \qquad [14.30]$$

In Equations [14.28]–[14.30], T_{∞} denotes the temperature on the exterior surface at infinite point, N is the total number of cooling channels, a_k is the radius of kth cooling channel, l is the local coordinate in the axial direction of each cooling channel, and θ is the local circumferential coordinate of each cooling channel as illustrated in Fig. 14.12.

Once the integrals on each of the elements are calculated, the discretized boundary element formulae for the analysis can be manipulated to the following form:

$$[H_{ij}]\{T_j\} = [G_{ij}] \left[\left. \frac{\partial T}{\partial n} \right|_j \right] \qquad [14.31]$$

where $[H_{ij}]$ and $[G_{ij}]$ are functions of boundary surface geometries. Some of $[H_{ij}]$ and $[G_{ij}]$ are singular integrals, which can be evaluated analytically in a simple manner as suggested by Rezayat and Burton (1990). The integrals over θ in Equations [14.28]–[14.30] can be evaluated as a closed form when the complete elliptic integrals are performed using the method proposed by Park and Kwon (1996). Other integrals can be evaluated by the Gaussian quadrature rule. Next, boundary conditions can be introduced to obtain a system of linear algebraic equations;

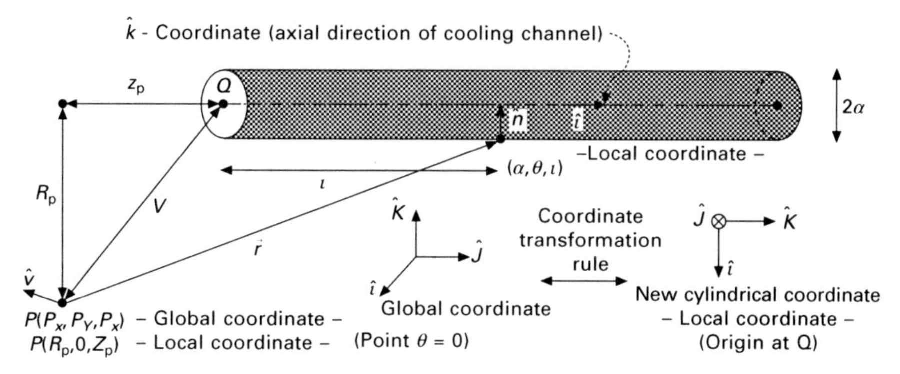

14.12 Coordinate transformation rule in one cooling channel element.

$$[A_{ij}]\ \{T_i\} = \{f_i\} \qquad [14.32]$$

where $[A_{ij}]$ and $\{f_i\}$ reflect boundary conditions. In this kind of problem, temperature, T, is taken to be an unknown on each element with $\partial T/\partial n$ being eliminated with the help of a boundary condition. This system of equations can be solved through LU-decomposition by Gaussian elimination or an under-relaxation iterative method. It may also be mentioned that T_∞ can be easily determined during the iteration procedure to satisfy the heat balance.

The boundary element method is also an efficient numerical simulation in design sensitivity analyses when the problem can be treated without the domain discretization and the design variables can be defined on the boundary. In the example in this section, the part surface is fixed, but the size and the arrangement of the circular holes (only defined on boundaries) can be changed to achieve other design objectives. Therefore, the boundary element method can be effectively applied to the design sensitivity analysis. In this example, the radii and the locations of the circular holes are considered as design variables.

The design sensitivity analysis is essentially to determine the variation of an objective function with respect to a variation in the design variable. In this model problem, the design sensitivity coefficients of the temperature on the mould surface with respect to all design variables provide valuable information for the optimal design using the first-order optimization techniques. By the implicit differentiation of Equations [14.28]–[14.30] with respect to each design variable (the direct differentiation approach), various boundary integral formulae are derived for each design variable: in design sensitivity equations, all T and $\partial T/\partial n$ in Equations [14.28]–[14.30] are replaced by $\partial T/\partial n$ and $\partial(\partial T/\partial n)/\partial x$, respectively, X being a design variable; in additions to this replacement, design sensitivity equations should include extra boundary integral terms because the mould geometry itself also depends on all the design variables. Refer to Park and Kwon for the final forms of design sensitivity equations and further details of this sensitivity approach for each design variable.

Once the integrals on each element are calculated, the discretized boundary element formulae for the sensitivity analysis can be converted to the following form:

$$[H_{ij,\mathrm{X}}]\{T_i\} + [H_{ij}]\ \{T_{i,\mathrm{X}}\} = [G_{ij,\mathrm{X}}]\left\{\frac{\partial T}{\partial n}\bigg|_j\right\} + [G_{ij}]\left\{\frac{\partial T}{\partial n}\bigg|_{j,\mathrm{X}}\right\} \qquad [14.33]$$

where x represents a derivative with respect to a design variable X. Note that $\{T_j\}$, $\{(\partial T/\partial n)_j\}$, $[H_{ij}]$ and $[G_{ij}]$ have already been obtained in the thermal analysis. The integrals for coefficients of the matrices of $[H_{ij,x}]$ and $[G_{ij,x}]$ can be evaluated in a manner similar to the thermal analysis (see Park and Kwon (1996) for details of evaluating these integrals). The sensitivities of boundary conditions can be introduced into Equation [14.33] to obtain a system of linear algebraic equations as follows:

$$[A_{ij}]\{T_{j,\mathbf{X}}\} = \{f_{i,\mathbf{X}}\} - [A_{ij,\mathbf{X}}][T_j] \equiv \{f'_{i,\mathbf{X}}\} \qquad [14.34]$$

where $T_{j,\mathbf{X}}$ is taken to be an unknown on each element. Note that the only forcing terms $\{f'_{i,\mathbf{X}}\}$ vary with the design variable X. Therefore, the sensitivity analysis can be completed simultaneously for all design variables with multiple forcing terms, which leads to a significant saving in the computation time. The major advantage of using LU-decomposition lies in the fact that the LU-decomposition of the matrix $[A_{ij}]$, is formed during the analysis. Therefore, not only is the overall computation time reduced but it can also be reused through matrix multiplication whenever necessary during the sensitivity analysis. Nonetheless, an iterative method takes less computer memory and computation time than a direct solver, especially when a large number of elements are to be used for complicated part geometries. The following example introduces an iterative method.

In the example of this section, an optimal configuration of circular holes is sought to make the temperature distribution over the part surface as uniform as possible. It may be mentioned that the injection moulded part quality increases with the temperature uniformity. Towards this design goal of the uniform part surface temperature distribution, the objective function is chosen as:

$$\mathbf{F}(\mathbf{X}) = \frac{\int_{S_p} (T - \bar{T})^2 \mathrm{d}A}{\bar{T}^2 \int_{S_p} \mathrm{d}A} \qquad [14.35]$$

where

$$\bar{T} = \frac{\int_{S_p} T \mathrm{d}A}{\int_{S_p} \mathrm{d}A} \qquad [14.36]$$

In Equation [14.35], $\mathbf{X}$ is a design variable vector and $\bar{T}$ is the average temperature over the part surface. Proper constraints have to be imposed on the design variables to keep the design feasible for this optimization problem. Upper and lower bounds are placed on the radius and position of each circular hole to keep the design practical from a manufacturing viewpoint. Inequality constraints associated with a design variable X_i that has the upper and lower inequality bounds are:

$$A_i \le X_i \le B_i \quad \text{where} \quad i = 1, \ldots, n$$

They can be modified to an equality constraint by introducing a slack design variable Y_i as follows:

$$G_i(X_i, Y_i) = X_i - A_i - (B_i - A_i)\sin^2 Y_i = 0 \quad i = 1, \ldots, n$$

where n is the number of design variables (Fox, 1971).

In order to solve the above constrained minimization problem, the CONMIN conjugate gradient algorithm can be applied. The algorithm has been successfully used in an optimal design of heating systems in compression moulds by Barone and Caulk (1990) and Forcucci and Kwon (1989). The CONMIN conjugate gradient algorithm employs the augmented Lagrangian multiplier (ALM) method to deal with the equality constraints, and Davidon–Fletcher–Powell method (Fletcher and Powell, 1963) for the unconstrained minimization iterative processes.

Figure 14.13 shows the initial design (dotted lines) and the final optimal configuration (solid lines). The result indicates that the optimization method changes the configuration in the direction of increasing the cooling effect on the plus plane and decreasing it on the minus plane. This optimization method quite significantly improves the temperature uniformity.

14.7 Conclusions

The chapter has focused on the concept of optimization in component manufacturing, looking at the optimization of design and processes in different

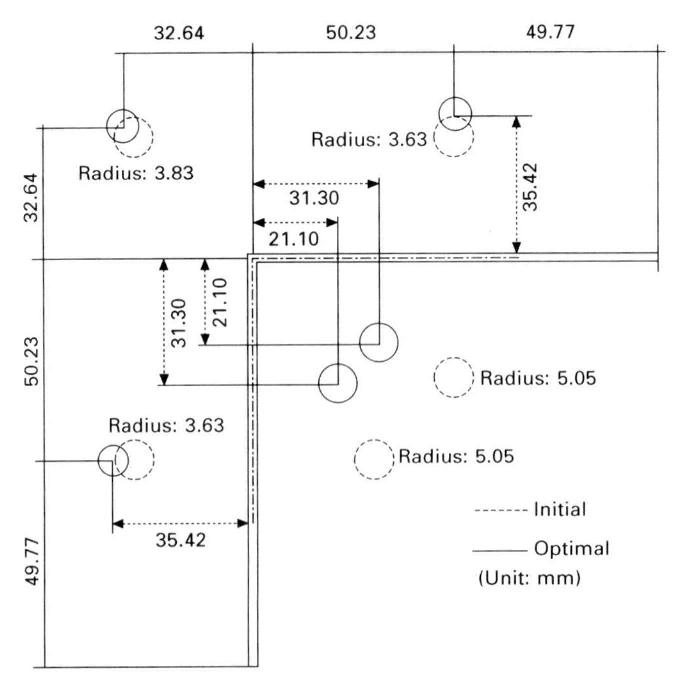

14.13 Initial design and optimal configuration of cooling channels.

manufacturing applications. Optimization in this context is a mathematical procedure to maximize or minimize different functions by finding the best available alternatives within certain constraints, in order to ensure that a component works as well as possible.

Techniques examined here include die compaction, powder injection moulding, sintering and steady-state conduction. Issues in the die compaction process include the need to optimize the compaction process parameters to reduce variability in a component's green density. In powder injection moulding, it is the process conditions that need to be optimized, which involves altering the process parameters. Sintering has been used as an example of a process where more than one objective needs to be achieved to optimize the process, in this case density and grain size. Finally, optimization of the cooling process was focused on in terms of steady-state conduction.

14.8 References

Ahn, S., Park, S.J., Lee, S., Atre, S.V. and German, R.M. (2009). 'Effect of powders and binders on material properties and molding parameters in iron and stainless steel powder injection molding process', *Powder Technol.*, **193**(2), 162–9.

Atre, S. V., Park, S. J., Zauner R. and German, R. M. (2007). 'Process simulation of powder injection molding. I. Identification of significant parameters during the mold filling phase,' *Powder Metall.*, **50**(1), 76–85.

Awa, T.W. West and Price (1992). 'Heater configuration design of a pultrusion die using design optimization techniques', in Proceedings 1992 *ASME International Computers in Engineering Conference Vol. 1, Computers in Engineering 1992*, Gabriele GA (ed.), San Francisco.

Barinarayanan, S. and Zabaras, N. (1995). 'Preform design in metal forming', in *Proceedings of the Fifth International Conference on Numerical Methods in Industrial Forming Processes*, 533–8.

Barone, M.R. and Caulk, C.A. (1990). 'Optimal arrangement of holes in a two-dimensional heat conductor by a special boundary integral method', *Int. J. Numer. Methods Eng.*, **18**, 675–85.

Bazaraa, M.S. Sherali H.D. and Shetty, C.M. *Nonlinear Programming*: theory and algorithms, Wiley, New York, 2006.

Brebbia, C.A., Telles, J.C.F. and Wrobel, L.C. (1984). *Boundary Element Techniques Theory and Application in Engineering*, Springer, Berlin.

Byon, S.M. and Hwang, S.M. (1997). Die shape optimal design in bimetal extrusion by the finite element method, *Trans. ASME J. Manuf. Sci. Eng.* **119** 143–150.

Chen, C.C. and Kobayashi, S. (1978). 'Rigid plastic finite element analysis of ring compression', *Applications of Numerical Methods to Forming Processes*, **28**, 163–74.

Choi, J.H. Kwak, B.M. and Lee, D.H. (1990). 'A unified approach in shape design sensitivity analysis of elastic structures in the boundary integral equation formulation', in M. Tanaka, C.B. Brebbia and T. Honma (eds), *Boundary Element XII Proceedings 12th International Conference*, Vol. 2, Computational Mechanics Publications, Boston, 225–36.

Chung, J.S. and Hwang, S.M. (1998a). 'Application of genetic algorithm to process optimal design in non-isothermal metal forming', *J. Mater. Proc. Tech.* 80–81, 136–143.

Chung, S.H. and Hwang, S.M. (1998b). 'Optimal process design in non-isothermal, non-steady metal forming by finite element method', *Int. J. Numer. Methods Eng.* **42**, 1343–1390.

Chung, S.H., Lee, J.H., Chung, H.S. and Hwang, S.M. (2000). 'Process optimal design in nonsteady forming of porous metals by the finite element method', *Int. J. Mech. Sci.* **42**, 965–990.

Chung, S.H., Park, H., Jeon, K.D., Kim, K.T. and Hwang, S.M. (2001). 'An optimal container design for metal powder under hot isostatic pressing', ASME *J. Eng. Mater. Technol.* **123**, 234–239.

Chung, S.H., Fourment, L., Chenot, J.L. and Hwang, S.M. (2003). 'Adjoint state method for shape sensitivity analysis in non-steady forming applications', *Int. J. Numer. Methods Eng.*, **57**, 1431–44.

Chung, S.H., Kwon, Y.S., Park, S.J. and German, R.M. (2009). 'Sensitivity analysis by the adjoin variable method for optimization of the die compaction process in particulate material processing, *Finite Element Anal. Design*, **45** 836–44.

Doraivelu, S.M., Gegel, H.L., Gunasekera, J.S. and Morgan, J.T. (1984). 'A new yield function for compressible P/M materials', *Int. J. Mech. Sci.* **26**, 527.

Fletcher, R. and Powell, M.J. (1963). 'A rapidly convergent descent method for minimization', *Comput. J.*, **6**, 163–8.

Forcucci, S.J. and Kwon, T.H. (1989). 'A Computer Aided Design System for Three-Dimensional Compression Mold Heating', *ASME J. Eng. Ind.*, **111**, 361–368.

Fourment, L. and Chenot, J.L. (1996). 'Optimal design for non-steady state metal forming processes–I. Shape optimization method', *Int. J. Numer. Methods Eng.* **39**, 33–50.

Fourment, L., Vieilledent, D. and Chenot, J.L. (1998). 'Shape optimization of the axisymmetric preform tools in forging using a direct differentiation method, *Int. J. Form. Proc.* **1**, 399–423.

Fourment, L., Chenot, J.L. and Mocellin, K. (1999). 'Numerical formulations and algorithms for solving contact problems in metal forming simulation', *Int. J. Numer. Methods Eng.* **46**, 1435–1462.

Fox, R. L. (1971). *Optimization Methods for Engineering Design*, Addison-Wesley, Menlo Park.

German, R.M. (1990). *Powder Injection Molding*, Metal Powder Industries Federation, Princeton, NJ.

German, R.M. (1993). 'Technological barriers and opportunities in powder injection molding', *Powder Metall. Int.*, **25**, 165–9.

German, R.M. (1994). *Powder Metallurgy Science*, 2nd edition, Metal Powder Industries Federation, Princeton, NJ.

German, R.M. and Bose, A. (1997). *Injection Molding of Metals and Ceramics*, Metal Powder Industries Federation, Princeton, NJ.

Gurson, A.L. (1977). Continuum theory of ductile rupture by void nucleation and growth-Part i. Yield criteria and flow rules for porous ductile media, *ASME J. Eng. Mater. Technol.* 99 (2).

Haarhoff, P.C. and Buys, J.D. (1970). 'A new method for the optimization of a non-linear function subject to non-linear constraint', *Comput. J.*, **13**, 178–184.

Haug, E.J., Choi, K.K. and Komkov, V. (1986). *Design Sensitivity Analysis of Structural System*, Series in Mathematical Science Engineering, Academic Press, Orlando, FL.

Himasekhar, K., Lottey, J. and Wang, K.K. (1992). 'CAD of mold cooling in injection molding using a three-dimensional numerical simulation', *ASME J. Eng. Ind.*, **144**, 213–221.

Hwang, S.M. and Kobayashi, S. (1984). 'Preform design in plane-strain rolling by the finite element method', *Int. J. Machine Tool Design Res.* **24**, 253–266.

Joun, M.S. and Hwang, S.M. (1993). 'Optimal process design in steady-state metal forming by finite element method—I. Theoretical considerations—II. Application to die profile design in extrusion', *Int. J. Machine Tools Manuf.* **33**, 51–70.

Van Keulen, F., Haftka, R.T. and Kim, N.H. (2005). Review of options for structural design sensitivity analysis. Part 1: linear systems, *Comput. Methods Appl. Mech. Eng.* **194**, 3213–3243.

Kim, K.T. (1988). 'Elastic-plastic response of porous metals under triaxial loading', *Int. J. Solids Struct.* **24**, 937.

Kusiak, J. and Thompson, E.G. (1989). 'Optimization techniques for extrusion die shape design', in *Proceedings of the Third International Conference on Numerical Methods in Industrial Forming Processes*, A. Samuelsson, E.G. Thompson, R.D. Wood and O.C. Zienkiewicz (eds), NUMIFORM 89: Numerical Methods in Industrial Forming Processes, Fort Collins, June 1989, 569–74.

Kwon, T.H. (1989). 'Mold cooling system design using boundary element method', *ASME J. Eng. Ind.*, **110**, 384–394.

Matsumoto, T., Tanaka, M. and Miyagawa, M. (1993a). 'Boundary element system for mold cooling heating design', in *Boundary Element XV, Vol. 2; Stress Analysis*, C. A. Brebbia and J. J. Rencis (eds), Computational Mechanics Publications, Boston, 461–75.

Matsumoto, T., Tanaka, M., Muyagama, M. and Ishii, N. (1993b). 'Optimal design of cooling lines in injection moulds by using boundary element design sensitivity analysis', *Finite Elements Anal. Design*, **14**, 177–185.

Michaleris, P., Tortorelli, D. and Vidal, C. (1994). 'Tangent operators and design sensitivity formulations for transient non-linear coupled problems with applications to elastoplasticity', *Int. J. Numer. Methods Eng.* **37**, 2471–2499.

Najmi, L.A. and Lee, D. (1991). 'Analysis of mold filling for powder injection molding processes', *Proceedings of ANTEC '91: In Search of Excellence: Conference Proceedings*, ANTEC'91, Montreal, 1991, Volume **37**, 508–11.

Park, S.J. and Kwon, T.H. (1996). 'Sensitivity analysis formulation for three-dimensional conduction heat transfer with complex geometries using a boundary element method', *Int. J. Numer. Methods Eng.*, **39**, 2837–2862.

Park, S.J., Martin, J.M., Guo, J.F., Johnson, J.L. and German, R.M. (2006). 'Densification behavior of tungsten heavy alloy based on master sintering curve concept,' *Metall. Mater. Trans.* A, **37A**(9), 2837–48.

Park, S.J., Cowan, K., Johnson, J.L. and German, R.M. (2008). 'Grain size measurement methods and models for nanograined WC–Co,' *Int. J. Refract. Met. Hard. Mater.*, **26**(3), 152–163.

Paul, B.K. and Peterson, R.B. (1999). 'Microlamination for microtechnology-based energy, chemical and biological systems', *Am. Soc. Mech. Eng., Adv. Energy System (AES)*, **39**, 45–52.

Piotter, V., Merz, L., Ruprecht, R. and Hausselt, J. (2003). 'Current status of micro powder injection molding', *Mater. Sci. Forum*, 426–432, 4233–4238.

Rezayat, M. and Burton, T. (1990). 'A boundary-integral formulation for complex three-dimensional geometries', *Int. J. Numer. Methods Eng.*, **29**, 263–273.

Ross, P. J., *Taguchi Techniques for Quality Engineering*, McGraw-Hill Book Company, New York, 1988.

Roy, S., Ghoshi, S. and Shivpuri, R. (1997). 'A new approach to optimal design of multi-stage metal forming processes with micro genetic algorithms', *Int. J. Machine Tools Manuf.* **37**, 29–44.

Saigal, S. and Chandra, A. (1991). 'Shape sensitivities and optimal configurations for heat diffusion problems: a Pow approach', *ASME J. Heat Transfer*, **113**, 287–295.

Shima, S. and Oyane, M. (1976). 'Plasticity theory for porous metals', *Int. J. Mech. Sci.* **18**, 285.

Taguchi, G. (1987). *System of Experimental Design*, ASI Press and UNIPUB-Kraus International Publications, Dearborn, Michigan, and White Plains, New York.

Tang, L.Q., Chassapis, C. and Manoochehri, S. (1997). 'Optimal cooling system design for multi-cavity injection molding', *Finite Elements Anal. Design*, **26**, 229–251.

Tsugawa, Daisuke; Yokouchi, Hirotaka, (1992). 'Analysis of flow of polymer melt in molds considering temperature distribution on injection molding', *J. Jpn. Soc. Precision Eng.*, **58**, 2031–2036.

Turng, L.S., (1998). *C-mold Design Guide – A resource for plastics engineers*, 3rd edition, Advanced CAE Technology, Troy, NY.

Urval, R. (2004). *CAE-Based Process Design of Powder Injection Molding for Thin-walled Micro-fluidic Device Components*, MSc Thesis, Oregon State University, Corvallis, OR.

Urval, R., Lee, S., Atre, S.V., Park, S. and German, R.M. (2008). 'Optimization of process conditions in powder injection molding of microsystem components using a robust design method: Part I primary press parameters', *Powder Metall.*, **51**(2), 133–42.

Urval, R., Lee, S., Atre, S.V., Park, S. and German, R.M. (2008). 'Optimization of process conditions in powder injection molding of microsystem components using a robust design method: Part I. Primary process parameters', *Powder Metall.*, **51**(2), 133–42.

Urval, R., Lee, S., Atre, S.V., Park, S. and German, R.M. (2010). 'Optimization of process conditions in powder injection molding of microsystem components using a robust design method: Part II. Secondary process parameters', *Powder Metall.*, **53**(1), 71–81.

Zhao, G., Wright, E. and Grandhi, R.V. (1995). 'Forging preform design with shape complexity control in simulating backward deformation', *Int. J. Machine Tools Manuf.* **35**, 1225–1239.

15
Non-destructive evaluation of powder metallurgy parts

C. SELCUK, Brunel Innovation Centre, UK

DOI: 10.1533/9780857098900.3.437

Abstract: Powder metallurgy (PM) manufacture of parts is one of the most energy and material efficient forms of net-shape production, particularly, for automotive industry. PM allows repeatable mass production which makes it unique. However, it is well known that the quality of sintered parts can be variable. There can be typically around a 5% scrap rate in existing PM manufacturing lines. Current efforts are being made to develop non-destructive testing (NDT) techniques that will allow inspection of PM parts, notably sintered ones ideally in line with production to increase the quality of output batch and reduce scrap, as much as possible. This chapter presents an overview of non-destructive evaluation methods for PM components. It also seeks to capture latest NDT strategies such as digital radiography (DR) and identify apparent technology gaps in NDT of PM parts, in terms of applicability issues, with an emphasis on offering solutions to detection problems. It also seeks to highlight future work.

Key words: digital radiography, non-destructive testing, powder metallurgy, production, quality control.

15.1 Introduction

Sintered parts obtained by the powder metallurgy process are employed in several industry sectors. They are typically intricate, complex-shaped parts produced in near net shape by compaction of powders into a geometry (referred to as the green state) followed by sintering of the compacts for consolidation, where powder particles are bonded upon heating. The PM market has had an estimated size of 6 billion Euros and an anticipated growth rate of 5–10% per annum. Annual worldwide metal powder production exceeds one million tonnes with most of the output comprising ferrous parts for the automotive industry (80%), with aerospace, defence and general engineering having their share of the remaining 20%.

The PM process, by its nature, is suited to high volume production and therefore any flaws/defects in the parts can have a significant impact on the production output, for example loss of material and efficiency, as well as potential failures in use later. Therefore, there is a need for automated inspection by non-destructive means, for determining and separating the good and bad batches during production, preferably as early as possible, without having to seek destructive examination carried out manually, which can have a negative impact on the production flow and output. More critically, any faulty part that is overlooked may cause more problems later on, such as unexpected premature failures in application. Depending upon the component and the criticality of the application (automotive, construction, oil and gas or aerospace) this can have drastic consequences, such as accidents.

In this chapter, several aspects of a non-invasive inspection for quality control of PM parts are analysed. An NDT strategy for online inspection of PM parts is presented. Particularly digital radiography (using X-rays) is introduced as a modern approach for advanced NDT which should allow fast and in-depth inspection in combination with application of advanced image processing algorithms that can lead to software for the detection of small cracks, flaws and density variations *in situ*. There are several benefits of developing a reliable NDT strategy for PM parts and an advanced inspection system tailored to the manufacturing environment, preferably in line with production and adaptable to handle large scales for full coverage (100%). These are indicated in Table 15.1.

15.2 Need and incentive for NDT

NDT is instrumental in achieving a holistic quality control regime not only for an efficient production environment but also for more reliable performance of PM parts in various applications. Inspection and structural health monitoring should be two important elements of what would be an ideal life management scenario for end users and part suppliers, which would no doubt increase

Table 15.1. Benefits of NDT of PM parts; overall opportunities in the light of PM market and end user needs

Wider benefits of NDT of PM parts, from an expanding European perspective	
Political	Efficient use of materials & processes for maintenance of improved quality. Reduced waste, reject, failures.
Economical	Competition with overseas. Processes and products need to have the edge in terms of notable quality over that offered elsewhere.
Societal	Improved integrity, warranty. Reduced risk of failure. Increased quality.
Technological	A leading technology adaptable to high production environment for *in situ* real time quality monitoring in a continuous manner.
Legislative	Materials and processes that can be modified for consistent product quality. A reliable automated non-destructive control method tailored to production. Less dependence on manual NDT.
Environmental	Reduced reject parts and risk of part failures, hence less waste, higher efficiency in production resulting in efficient use of resources.

confidence in PM and also open up new applications for the manufacturing industry. Figure 15.1 presents some important considerations in developing an inspection and monitoring strategy for PM products.

There is a growing need for efficient use of materials and processes while delivering improved quality for the whole of manufacturing across the globe in order to remain competitive. PM is currently one of the most material effective forms of manufacturing and with its high material utilisation and low energy requirement it has the potential to give significant competitive advantage to manufacturers. However this manufacturing process does have a major drawback and weakness. There is currently no true 'in line' NDT technique that can inspect parts immediately after pressing or sintering. This is an issue because it has been reported that even with a 100% end-of-line inspection (which is a time consuming process) typically 6–8% of production parts are scrapped and there are still field failures returned by customers

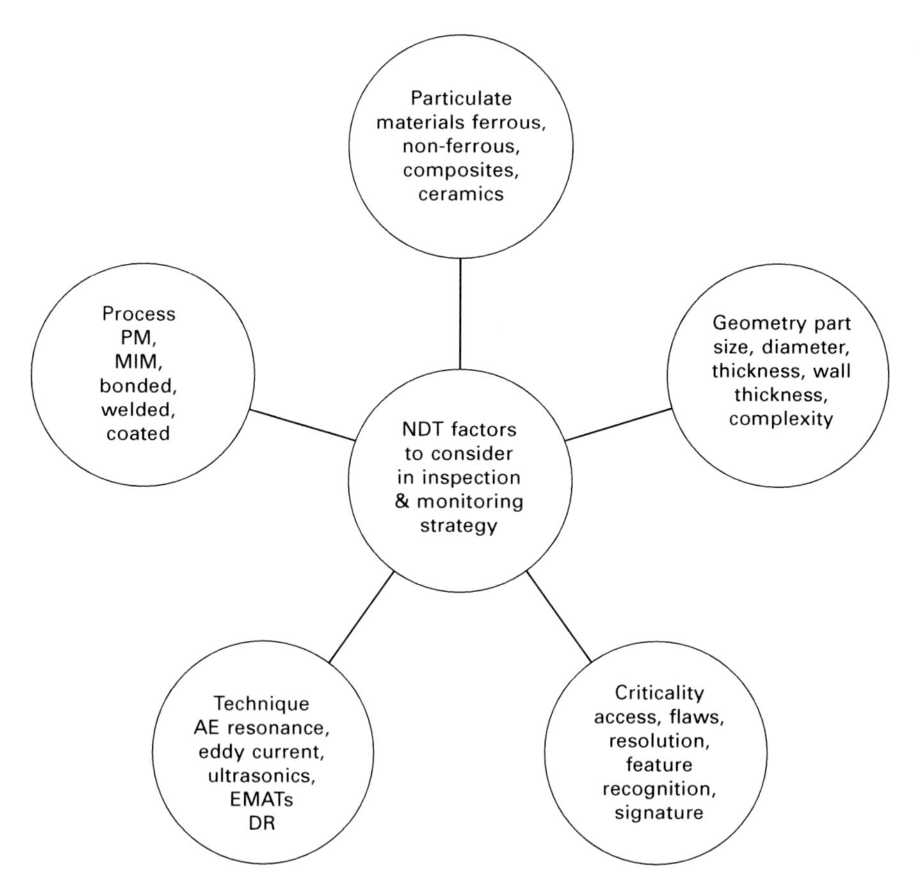

15.1 Important factors to consider in development of NDT regime for PM products.

(private communication). To reduce the inspection time manufacturers of PM parts usually have no choice but to perform spot testing on batches of parts from a given production run. Nevertheless, this means that defective parts are even more likely to reach the customer and because of the high expectations of both primary manufacturers and end consumers, defects cannot be tolerated even in million piece quantities. Increasingly, PM parts are used in the aerospace and defence industries where 100% inspection is required. This is apparent from the recent approval of titanium PM parts for use by the aerospace industry, which is a significant stepping stone for PM in general. The current end-of-line inspection techniques used are not suitable for high volume 100% manufactured part inspection because they involve and rely upon strongly manual or part manual processes to a degree, are therefore slow and require subjective interpretation by an operator. Therefore, there is a need for new inspection equipment that is automated, can inspect with a high throughput and can reliably identify good and bad components. This will allow greater use of sintering for a variety of batch sizes suitable for both small and medium-sized industries or larger manufacturing plants and will result in more efficient production, reduced waste, fewer rejects, and minimal failures in a range of products. This will also help create more efficient and higher quality production processes with improved utilisation of materials and resources available allowing better and more expensive materials and alloys to be used with confidence in their end quality.

As the processes and products need to be more competitive and hence efficient, the requirements from the material performance become increasingly demanding. Therefore, improved integrity and warranty are crucial. With a typical failure rate of 6–8% parts produced in existing PM manufacturing processes, there is also a need for reduced risk of failure. This is even more important considering there are increasingly demanding and more aggressive operating environments (higher speed, higher pressure, higher temperature (e.g. hot isostatic pressing, HIP) and the associated higher performing materials, etc). Increased quality and the consistency of the quality are paramount to the success of these applications. An automated reliable inspection technique for PM production (compaction or sintering dependent upon choice) that can be operated on-line in the production environment will help meet such requirements. This will also necessitate improvement in knowledge and skills enhancement in the NDT environment. Although detection of flaws and or faulty parts in the green state can be an advantage, for some it is the final check after sintering that matters and hence the approach is versatile.

It is worth capturing at this juncture that there is a need for an inspection technology that is also adaptable to a lower production volume environment (e.g. more selective metal injection moulded (MIM)/power injection moulding (PIM) parts) for *in situ* real time quality monitoring in order to reduce production downtime, which costs energy and money and replaces the need for destructive testing, especially for high value added parts and bespoke processing that may be tailored for special components in terms of alloy, function and geometry.

15.3 Problem/approach concept

The PM process, by its nature, is suited to high volume production (in hundreds of thousands) and therefore any flaws/defects in the parts can have a significant impact on the production output, for example loss of material (of the order of millions of parts in a medium to large size PM plant) and efficiency, as well as potential failures in use later. Figure 15.2 presents how closely linked the manufacturing technology, quality control/NDT and end user applications should be for a production environment.

Currently there is no effective *in situ* automated NDT inspection system available for PM parts before or after sintering. In such a mass production environment or even a specialised small batch production, any defects that may be present in the parts may present further problems for the end users (Fig. 15.3). If any flaws are critical and this is not known, the manufacturer may, for example, end up with a poor batch of parts in their thousands returned due to poor quality or simply as a faulty batch. More critically, any faulty part that is overlooked may cause more problems later on such as unexpected premature failures in application. Depending on the component

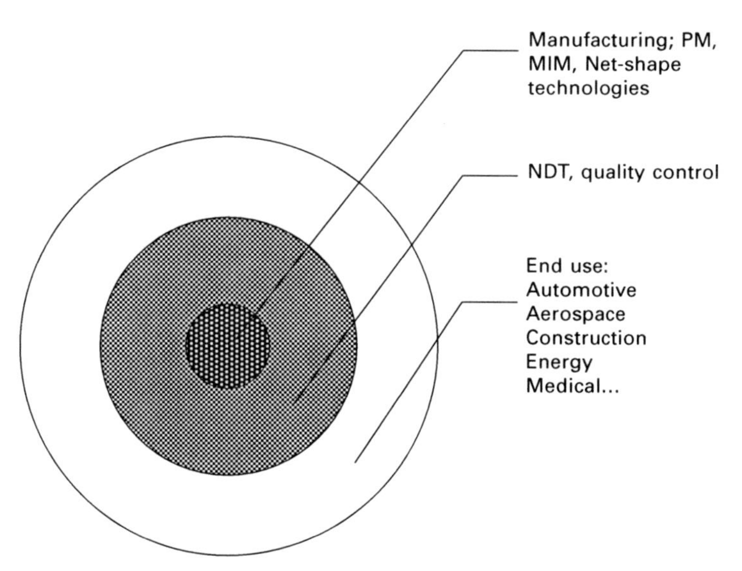

15.2 Schematic diagram highlighting the link between manufacturing technologies, NDT/quality control and end user sectors. A fully integrated approach in production is paramount for success.

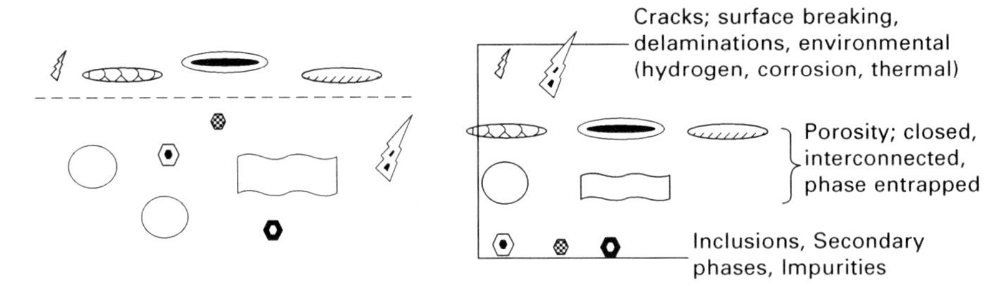

15.3 Schematic from a cross-sectional view based on simple part geometry, showing potential forms of flaws as closed, deformed and interconnected porosity, cracks near edges, delamination (tearing, thermal) or surface breaking in a finished component as well as impurities or secondary phases that NDT can be useful in capturing in terms of nature and extent.

and the criticality of the application this can have drastic consequences, such as accidents. Failure of moving parts of engine components, for example valve seats and gears in automobiles, although repairable, can be complicated in huge numbers and some rotating components (turbines) in aeroplanes, where service conditions and criteria are even more stringent.

The recall of millions of components from mass production, especially as has happened in the last couple of years with its impact on the automotive industry, has raised significant awareness through wider media (newspapers and TV, etc) highlighting the importance of quality control and the need for automated inspection in mass production environments. Products need to be inspected before they are released to the public, to avoid faults being discovered later which can cause failures and accidents that can be disastrous. An automated inspection regime for NDT, not only for PM parts, is therefore a very timely strategy in the pursuit of an improved quality control in an increasingly mass productive industrial environment.

It is not viable to examine destructively each and every part to check the quality in such a production environment. Therefore there is a need for automated inspection by non-destructive means, to determine and separate the good and bad batches during production, preferably as early as possible, without having to seek destructive examination carried out manually, which can have a negative impact on the production flow and output while also making small volumes of high value parts uneconomic.

The solution is an automated non-destructive inspection system that can be applied in real time *in situ* in production opening up the use of sintering on smaller batch sizes as well as better quality in the larger volume areas. A digital radiographic system for online inspection of PM parts can provide a technique that allows fast inspection and application of image processing for the detection of small cracks/flaws/density variations *in situ*. This will enable manufacturers to separate good and bad batches instantly and will help them to address any issues with the production process or raw material on the spot and to correct any faults before resuming their production with the quality required and even improved.

However, the proposed technology can only be developed through industrial and academic collaboration between PM part manufacturers, NDT technique and equipment providers, advanced materials characterisation expertise and software development. The approach should be able to cover the conventional 'press and sinter' products that are most commonly available, but also the powder MIM components which are increasingly employed in the PM industry. Currently major research work is underway across wider Europe based on digital radiograph (DR) of PM parts. This work (e.g. AutoInspect and DiraGreen projects with a focus on sintered and green parts, respectively) seeks to increase the competitiveness of the European industry sector by developing a fast, automated inspection system that can be offered in high production environments globally while opening up the sintering/PM processes to small and medium enterprises (SMEs) for smaller batch sizes of high value added net-shape products.

15.4 Quality control by digital radiographic (DR) inspection in production

The primary function of a DR-based inspection system for a typical PM production line is to receive PM sintered parts after sintering, for example, inspect them at high throughput and designate them as good or bad components accordingly. To ensure maximum throughput, the PM components are fed continuously via a conveyor. A moveable X-ray radiation source allows multi-planar inspection of the component without having to manipulate the component mechanically. Figure 15.4 shows the general or common process cycle with an integrated digital radiography inspection system. The steps in the process include (1) powder/raw material, (2) compaction/moulding, (3) sintering, (4) in-line automated digital radiography inspection, (5) image/data processing, (6) quality check.

Any flaws detected after sintering can be identified by digital radiographic image acquisition and processed/analysed for quality check. Any problems with the process can be identified on-line before the goods are dispatched. This will allow improved control over the consistency of the product quality. The DR inspection system is expected to aid profitability by achieving near zero defects and lowering overall production costs by reducing variances in the production process. Because it is a fast enough technique (a few seconds of bulk inspection time per part) to test 100% of the parts, the system can be designed to function as a process monitoring tool in addition to part quality inspection and insures against quality-related expenses while functioning as a process control monitor. DR will open up a completely new generation

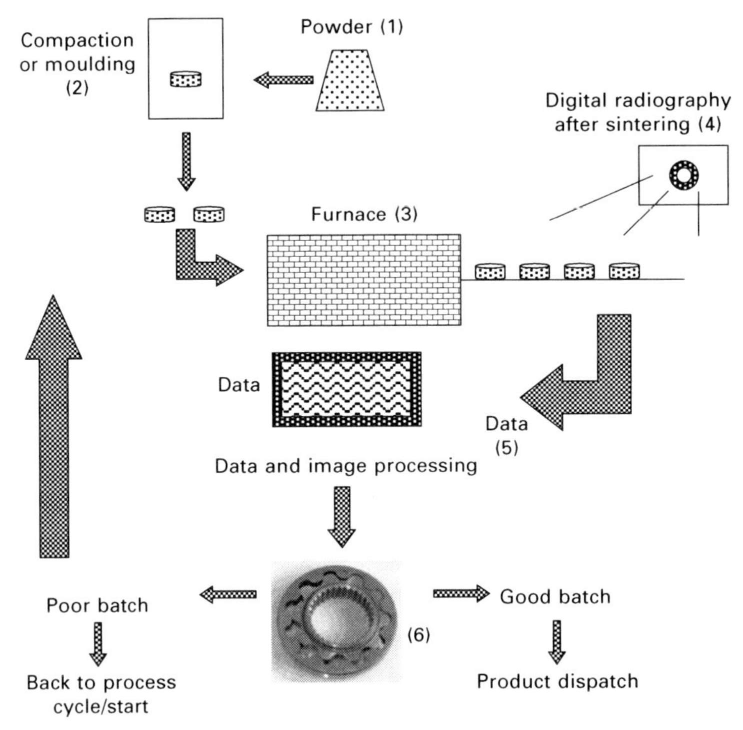

15.4 Schematic of PM process cycle with integrated DR inspection system.

inspection capability specifically for the PM industry.

The proposed concept is a new approach to the problem of performing 100% inspection of PM components and will require significant research effort beyond current industrial knowledge. Hence, the research and development output will be a significant improvement beyond the state-of-the-art. This will entail: (1) an advanced X-ray source generator and collimator including excitation driving electronics and control software, preferably to allow multi-planar component inspection without having to move the PM component; (2) high resolution miniature X-ray detection technology for fast capture of images and acquisition software; (3) software-driven automated defects detection in PM component radiographs and image processing algorithms; and (4) component feeding and manipulator mechanics.

15.5 Challenges in relation to the state-of-the-art

In recent years the complexity of PM parts has increased. Mass production components have placed greater demands on examination for quality assurance. For assemblies, components and connections need technologies that allow inspection in areas that are visually inaccessible. Testing the integrity of these joints is a problem because of lack of access and increasingly complex shapes. Currently, NDT techniques are applied off line, mostly relying on visual inspections and operator interpretation. In some cases if a serious defect is found, it is possible for the whole batch to be scrapped because the manufacturing conditions that lead to defects will have been applied to the whole batch.

Often for the inspection of defects in MIM or PM parts, optical microscopy and scanning electron microscopy (SEM) are used, and most often these require cross sectioning of specimens to facilitate observation of inner defects.[1] However, it is a destructive and time consuming technique, mostly applied on an individual part by part basis in a laboratory environment off line. Therefore there is a need to inspect non-destructively. Once a fault is found by one NDT technique, it is typically necessary to confirm this with another technique. This whole process will still be time consuming and typically takes at least one hour; the production line cannot be restarted before the investigation is complete because corrective actions on the manufacturing conditions may have to be made and these actions may depend on the nature of the defects found.

Sintered components typically suffer from porosity (hence density variations), cracks and impurities that may be present. Such defects negatively affect the mechanical properties of the part and hence their performance.[2] As they are produced in their thousands in a production environment, if the defected parts and the cause(s) are not determined at an early stage, the whole production output can be rejected, if considered unsuitable for the intended application. There is an increasing push in industry (e.g. automotive) for fuel economy that will create weight savings targets which require higher performance requirements and better materials/alloys for sintered parts.[3] PM parts will need to be as defect free as possible for improved integrity to meet such requirements, which advanced NDT will enable. The need for defect-free manufacturing across the industry has motivated manufacturers to seek reliable cost effective inspection methods for eliminating the defect output in production.[4] On-line automated non-destructive inspection can offer the solution required. However the inspection method should have ideally:

- high inspection speed (a few seconds)
- high accuracy (micrometre–mm range)
- high throughput (100 mm min^{-1}).

Traditional NDT techniques focus on detecting and diagnosing defects. They are based on visual techniques or imaging to scan for any indication of defects. Scanning methods include magnetic particle testing (MT), ultrasonic testing (UT), eddy current/electromagnetic testing (ET), dye penetrant testing (PT) and X-ray/radiographic testing (RT). These methods often are manual and require subjective interpretation by an operator. Although diagnosing and/or imaging specific defects are applicable when evaluating an individual part or system, they are not appropriate for high-volume, 100% manufactured part inspection. In these cases, it is of primary importance to detect whether a part is non-conforming, rather than why, which can be addressed separately off line. Identifying the type of defect itself is secondary to identifying the nonconforming parts. Therefore, an end-of-line 'go/no go' objective inspection is preferred over that of a slower subjective diagnosis.

The possible techniques that can be automated for inspection of PM parts include:

- automated ultrasonic testing (UT)
- on-line resonance inspection
- automated optical inspection
- current off-line X-ray inspection

However, each of the techniques presented have characteristics and limitations as outlined below.

15.5.1 Ultrasonic inspection technique

Ultrasonic testing (UT) is an NDT test technique that interrogates components and structures to detect internal and surface breaking defects and measures wall thickness on hard (typically metallic or ceramic) components and structures. In UT, the maximum flaw size that can be detected is typically 1 mm but this will require a high frequency greater than 10 MHz. This will require full contact with the specimen and in some cases full immersion in liquid medium. In immersion testing, real-time inspection of PM parts is not possible as the inspected items must be water resistant, when using water-based couplants that do not contain rust inhibitors. It is also difficult to use the immersion technique for real-time inspection of PM parts as residue left behind from the water will degrade the final quality of the sintered part. Additionally, the acquisition time is in minutes or hours (depending on resolution) as the UT probe must be scanned over the component surface.

An alternative method for inspecting these parts is to use ultrasonic electromagnetic acoustic transducers (EMATs). However this technique relies upon point contact with a part to induce and receive Rayleigh surface waves which are sensitive to surface breaking cracks and density variations. The transducers can be shaped to inspect both flat and curved geometries effectively such as boreholes. Feasibility studies performed on both green and sintered parts have been successful in penetrating ultrasonic energy into the parts for defect detection. Owing to the low aspect ratio of many powder metal parts, ultrasonic guided waves can provide full part characterisation. Shear horizontal waves were used to inspect the flange of a transmission part and torsional waves have been used to inspect the welded region of a sintered porous filter. It would also be of interest to explore HIP parts.

The surface wave technique is an efficient way to detect defects in PM parts with different geometries including multilevel parts. The EMATs can be shaped to fit specific part geometries, such as the rounded profiles which fit boreholes. EMATs also can provide global inspection of sintered porous media and green parts. Weld uniformity and weld quality of the sintered porous media can be obtained by signal analysis. Although there are lots of benefits of using the UT technique, it is still difficult to deploy the technology in the production line. The parts can be rough, inhomogeneous and irregular in shape, very small or thin, owing to the versatile nature of net-shape processing and so can be complex to inspect. In these cases, different shapes of probes are required to compensate for any variation in PM part types. For post production this can be costly and tedious when it comes to maintenance and repair. In addition, the detectability and sensitivity of flaws also depend on the orientation of the cracks, because the beam has directional properties. Therefore the response will be sensitive to orientation of the suspect features (e.g. cracks, secondary phases, inhomogeneities).

15.5.2 On-line resonant technique

The resonant acoustic method (RAM) is a 'full part' inspection technique for detecting flaws on a component level. The inspection spectrum includes materials ranging from ductile iron to powder metal to ceramics, and part sizes ranging from less than an ounce up to 50 lb (23 kg). The technique is regarded as easily automated to eliminate human error, with fast throughput providing cost-effective inspection with minimal disruption to production. The non-destructive testing on-line resonant method (ORM) is considered to be the only possible in-line inspection on production lines for PM and cast parts. The measurements are based upon the resonant frequency of a structure which is a function of part geometry and material properties (such as Young's modulus), and are therefore defined by the nature and characteristics of the materials (such as stiffness and microstructure). The resonant frequency of material (single or dual phase) can be obtained by impedance measurements and visually using laser scanning vibrometry. The ORM systems detect frequency shifts that can be caused by imperfections such as cracks, porosity and voids, as well as variances in the morphology of reinforcements or phases in a matrix, for example nodular graphite (in ductile cast irons), dimensions, geometry, weight, density and manufacturing processes.

After defective parts have been sorted by ORM, complementary visual or imaging NDT techniques may be required to diagnose and identify the defect on the smaller subset of parts. This is useful in determining the root cause of a defect and ultimately improving the product design or production process. This can, however, be very time consuming (usually several hours, e.g. from a minimum of 1–2 h up to 8 h and more depending on the criticality of the application) for analysis. Neither ORM nor UT can provide a visual three dimensional (3D) picture of the part in order to simplify the defect detection process. For ORM, the data is recorded as frequency values which have to be correlated to a particular feature. Hence it requires significant interpretation, which can be different, depending on the individual, time-consuming and monotonous. Therefore the consistency of the output and its meaning can be variable and lead to more reject errors.

15.5.3 Automated optical inspection

The NDT method most commonly used for in-line (production and assembly) applications is automated optical inspection (AOI) but this relies upon sight vision, is only effective for the detection of surface flaws and is challenged by more complex designs. The technique is also influenced by the surface finishes and profiles found on lead-free joints. In summary it can be said that optical inspection can only inspect the outer surfaces of optical objects.

15.5.4 Use of X-rays

DR systems are well developed in the medical and dentistry markets and have been for several years. The smaller industrial inspection radiographic market is dominated by the use of conventional radiographic film, currently estimated to be over 90% of the market place. However, digital X-ray systems are more efficient than X-ray film and, as a result, exposure time is reduced (resulting in a faster inspection). Also digital radiography can be conducted in enclosed lead shield cabinets and is therefore safer for operators. Although industrial DR systems are available, their take up has been limited to a few specialist inspection applications. Indeed, DR systems are commercially available for quality analysis of components and are commonly used for manual inspection of single components. However, currently almost all X-ray technology is conducted off line.

The main reason for performing off-line measurements as opposed to during production are: (1) traditional radiography inspection is too slow and relies on manual manipulation for appropriate interaction between X-rays and different planes of the part in terms of orientation to capture any flaws that may be present; (2) common DR equipment is typically unsuitable for use for in-line inspection as only inspection of single placed components is possible and requires an operator to perform the loading and component sorting; (3) additionally, the equipment is often designed to be general purpose allowing different sized samples and materials to be inspected. This requires higher energy rated sources and energy rated detectors typically in the range 160–450kV. One issue with high kV is that it requires extra X-ray shielding.

These factors naturally lead to large equipment size and increased cost which prohibits the adoption of X-ray inspection by industry. The total off-line measurement process, including the image processing time, mechanical manipulation of the component and image interpretation and sentencing can be carried out in a few minutes at best. However this can still equate to significant efficiency losses to the industry of the order of €100 000 per inspection in lost production time.

15.6 Real-time on-line powder metallurgy parts inspection

Generally, X-ray imaging or radiography is considered to be among the most promising *in situ* automated inspection methods for non-destructive evaluation.[5] Indeed on-line digital radiography inspection is already finding application in other industries including food and electronics inspection.[6–8] However, radiography inspection of PM components is significantly more challenging than inspection of food or electronics components where inspection is merely looking for contamination of foodstuffs or differences inside electronic components and their joints, which are more easily revealed using radiography because of differing distinct density materials present.

DR investigations have already been widely reported for applications in PM component inspection.[9] However, all applications shown have been for inspection of PM parts off line, for example X-ray inspection has been researched for metal injection moulding parts.[10] The technique was used to demonstrate detection capability for identifying porosity in the form of bubbles and cracks that may be present in PM parts, in order to establish their cause(s) and eliminate these for structural improvement of the parts. The images are formed on a fluorescent screen from X-rays. However the described technique is off line and therefore it does not provide images in real time. Moreover the output is not digital. In contrast to the use of a fluorescent screen, a direct digital radiography (DDR) detector-based inspection system can provide an *in situ* flaw detection capability in real time by developing DR for high definition image processing and defect characterisation. Instead of a stop/start conveyor and manipulator to pick and place the component to ensure that it is presented to the detector at the correct angle and ensure defects are revealed in all planes as in existing systems, a continuous conveyor will allow continuity in inspection. In addition, a moveable X-ray source can enable multiple planar inspections of the PM component whilst it moves along the conveyor.

The high cost of X-ray detectors has prevented the uptake of automated radiography in the past. However, by modifying and developing existing detector technology from the medical and dentistry market it is now feasible to produce an affordable in-line automated radiography inspection system for PM components in a factory production line. Recent work at TWI, UK has resulted in technology for inspection that uses a highly sensitive digital detector, capable of rivalling fine-grain film in sensitivity, contrast and resolution. The samples that have been used range from 1 mm thick magnesium castings to 10 mm thick steel welds.[11] A DR system has been proved in a laboratory environment by TWI, UK in a project called MICROSCAN.[12] Accordingly, the system is capable of inspecting small voids in ball grid array electronics components, as small as several mm of ball volume with a 99.97% detection probability. The system, however, does not have a high resolution detector and does not make use of linear array detector and a continuous feed conveyor. In addition component throughput was restricted because each component was mechanically lifted from the conveyor and mechanically oriented for different planar radiography images. There was no image processing and automated defect recognition in the system. Data analysis has to be carried out off line by human interpretation.

Existing radiography technology has previously been developed for examination of manufactured component parts. Although the radiography techniques described are already in existence and routinely used in specific market applications, the development work required to enable these techniques to be applied into a prototype system suitable for automated 100% inspection of PM components is considerable. In an advanced DR-based inspection system, flaws and defects, if at critical size (such as cracks and pores down to sub millimetre scale), can be detected through scanning as soon as the PM parts come out of the furnace on to the conveyor belt. The digital data from the scan will then feed into a microprocessor for data processing and analysis thus allowing the PM parts to be accepted or rejected immediately without halting the production process. In addition, defect detection in the green state can allow parts to be healed in subsequent sintering. This will instantaneously allow a zero defect manufacturing process, which will avoid the production of large numbers of faulty PM parts that would otherwise need to be recycled. By avoiding such waste, the cost of manufacturing will be reduced and better process control achieved, making especially high value lower volume parts an economic possibility and high volume production more reliable. This will further develop existing radiography technologies used in industrial NDT, medical and dentistry markets and extend their capability into new PM parts inspection applications where there is a genuine requirement based on safety, environmental and economic issues as components are employed in ever more aggressive environments (high temperature, pressure, corrosion).

15.7 Prior art in relation to radiography of particulate matter and near net-shape parts

There have been studies on inspection of PM parts by X-ray techniques for prediction of density in porous materials at laboratory scale, however these have not been implemented in mass production.[13,14] There is hardly any patent that is directly related to DR-based on-line inspection technology. The most related application found is a method for detecting the presence and locations of occlusions in hollow passages in investment castings by radiographic analysis.[15] However this technique is dependent upon the use of a radiographically detectable powder, which is injected into the casting which is then vibrated to help distribute the powder throughout the passages. Firstly, powder characteristics such as particle shape, size, distribution and surface roughness may affect the powder flow properties and may result in insufficient powder in the specific regions to be detected and hence flaws can be left undetected. Secondly, mechanical forces from vibration could potentially be destructive for fragile components such as porous materials or parts that are not fully dense, as the vibration may cause parts to weaken and crack, if they do not have sufficient integrity to sustain such forces. Also the method relies upon a radiation sensitive film which is not digital and therefore does not have the advanced pattern recognition function. The film has to be scanned whereby resolution of the image will degrade and definition will be reduced. DR technology however offers digital output and results in high definition images than can be processed electronically.

Another described technique requires a chemical, radiographic agent to be coated on the part that is to be inspected. In addition, a leaching agent is used prior to the radiographic inspection which can introduce undissolved chemicals and these can interfere with the inspection results because of foreign additives in the original material.[16] Such interferences will potentially be misleading and make the interpretations and analyses difficult. The DR system does not require any chemicals and therefore it will not introduce additives or foreign species into the material during inspection. It is possible to determine the presence of foreign materials, such as inclusions in hollow passages, by gravimetric means but such gravimetric analyses are extremely limited in accuracy, particularly where the residual core inclusions are small and the passages have narrow openings. It is also possible to detect the presence of such inclusions by using selective neutron absorption radiography where the materials are doped with neutron absorbers but this procedure is very expensive and necessitates chemicals to be absorbed in the material that is the subject of inspection.

In a production environment, where hundreds and thousands of parts are produced on line, it is highly disruptive and costly to impregnate each and every part with a chemical. This could also induce more unwanted species into the parts while trying to inspect them for defects and therefore it becomes very unattractive and impractical to employ such a technique in a production environment, without stopping the process and thereby introducing a separate step into the production cycle. DR technology does not require dopants and offers a fully automated non-invasive and non-destructive inspection system in line with production.

Other studies include determination of density variation in powder compacts using ultrasonic tomography[17] and X-ray tomography on tablets.[18] UT inspection has been studied for sintered parts but detection was limited by the surface condition of the parts.[19] DR technology for inspection of PM parts will not rely on surface contact or any chemical interface between the inspection system and the part, as it is solely based on DR and produces images resulting from the interaction between the X-rays and the material being inspected. This technological progress will permit a substantial increase in the reliability of PM part production and the safety of assemblies owing to increased measurement accuracy, flaw detection sensitivity and reduced dependence on operator subjectivity during inspection. Therefore the use of X-rays is obvious because these can penetrate the opaque structure of a PM part and access the defect region regardless of the complexity of the structure.[20] It is worth noting that image processing algorithms are critical for the success of clear inspection and flaw detection (Figs 15.5 and 15.6).[21]

15.8 Summary

Several NDT techniques have been deployed over the years for inspection of PM parts. Some have relied on basic methods common in industry such as dye penetrant or magnetic particle inspection, which necessitated a sufficiently surface breaking flaw to be present so that defect-like feature(s) can manifest themselves. Other techniques which could be regarded as more elaborate have

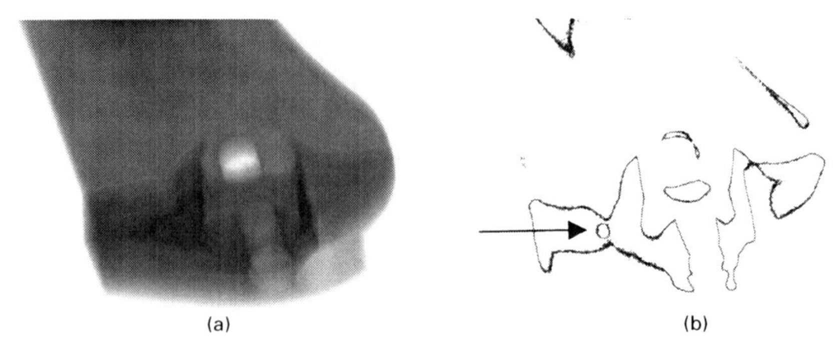

15.5 Digital radiographic image of a PM–MIM part (safety lever) in green state before and after image processing: (a) initial radiography, (b) after image processing. The arrow points to a volumetric defect-like feature in the part that would not have been revealed otherwise.

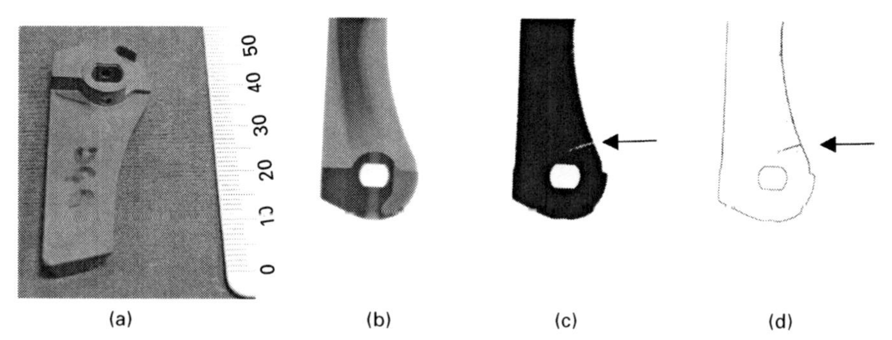

15.6 Images of a safety lever (a) before (scale in mm/cm) and (b) after X-ray plus subsequent image processing shown in (c) and (d) pointing to a defect-like feature, which appear to be a crack finely breaking the surface.

employed ultrasonics: surface/guided waves, using EMAT type transducers or similar transduction. Acoustic emission has also been explored to relate defect-like features to signals and is open to interpretation in classifying a signal as a defect. None of these have provided information on volumetric features or provided an image that can correspond to the actual part. DR on the other hand has proved to be useful in providing an exact image that can match the part and point towards, after careful image processing, flaws and defects that may be present in the part. This digital technique is tailored to the production environment to provide a fully comprehensive on-line inspection technique for improved quality control. This novel approach can provide classification possibilities for a range of flaws preferably by a referenced-based approach, where a library of defects and features can be built in to use as a signature for flaw identification and classification.

A comprehensive DR system will enable full quality control where each part can be inspected rather than representative samples and that will be open to statistical analyses in making a judgment on the quality output in production. As such, the system is able to provide a high throughput, inspecting each and every individual PM part on-line, thereby offering traceability per part as each part will have a corresponding image on the system or database, allowing 100% quality check. This will help manufacturers or part suppliers to be ready for full compliance and investigation, if need be, further down the supply chain, identifying any weakness in production and addressing whether it is linked to the stock or the manufacture (compaction or sintering, for example). This ability to detect flaws/defects in production can enable possible healing opportunities, such as fine tuning the compaction stage or heating cycle in sintering or the environment, or establishing if there are any deficiencies that must either be removed or replaced by alternative methods. In some cases, it may be possible to recycle the parts, if flaw detection is made after pressing, for example, or if it is found after sintering, it will enable a fully checked output for delivery with 100% confidence in part quality control.

In pursuing effective quality control by novel NDT methods, it is paramount that industry and academia collaborate closely. In this respect, the complete supply chain can be engaged, from part manufacturers to equipment supplier to designers, software developers and system integrators from across the engineering sector and, of course, with end user involvement that will be critical to the usefulness and future deployment of any solution that will be created. The key to a successful NDT method for PM parts does not solely lie with the PM part manufacturers or the PM supply chain as we know it, but also with interdisciplinary research and development in mathematics, statistics, electronics and mechanical engineering and design. A combined approach will result in a range of possibilities for a more versatile and comprehensive solution to quality control issues in production and naturally help to develop and deploy advanced non-invasive methods of inspection for industry.

15.9 References

1. S. Libb *et al.*, *Progress in Powder Metallurgy*, 1986, **42**, 81.
2. C. Selcuk *et al.*, *Proceedings EuroPM2008, Mannheim, Germany*, Volume 3, 255–60.
3. J. A. Sheilds, Jr., *IJPM*, 2009, **45**(5), 12.
4. Schiefer *et al.*, *Advances in Powder Metallurgy and Particulate Materials*, 2006.
5. R. E. Green Jr, *MRS Bulletin* 1988, **13**, 44.
6. M. Milner *et al.*, Food Technology, February 1952, **6**(2), 44–45.
7. T. Brosnan and Da-Wen Sun, *Journal of Food Engineering*, January 2006, **61**(1), 3–16.
8. Chu, Y., Vision '87; Detroit, Michigan; USA; June 1987, 8–11 4/41-4/56.
9. Gail R. Stultz and Richard W. Bono, http://www.pcb.com/techsupport/
10. Y. Q. Fu *et al.*, *Journal of Materials Science Letters*, 1997, **16**, 1873–1875.
11. Insight, February 2006, **48**(2).
12. P. I. Nicholson and P. Wallace, Circuit World, 2007, **33**(4) 31–42.
13. A. Bateni *et al.*, 17th World Conference on NDT (25–28 Oct 2008, China).
14. J. U. Ejiofor, R. G. Reddy and G. F. Fernando, *Journal of Materials Science*, **33**(16), 4029–4033.
15. US Patent no 3,624,397.
16. US Patent no 2,812,562.
17. Zhao *et al.*, IEEE Transactions on Ultrasonics, Ferroelectrics and Frequency control, 2006.
18. Sinka *et al.*, J. Pharmaceutics, 2003.
19. Partridge, P. G., AGARD Lecture series no 154, 1987.
20. M. Iovea *et al.*, Proceedings EuroPM2012 (Basel, Switzerland), **1**, 145–150.
21. M. Ponomarev *et al.*, Proceedings EuroPM2012 (Basel, Switzerland), **1**, 489–494.

16
Fatigue and fracture of powder metallurgy steels

N. CHAWLA and J. J. WILLIAMS,
Arizona State University, USA

DOI: 10.1533/9780857098900.3.455

Abstract: Sintered ferrous powder metallurgy (PM) components are replacing wrought alloys in many applications, owing to their low cost, high performance and ability to be processed to near-net shape. The fracture and fatigue behavior of these materials is very important. In this chapter the microstructure, tensile and fatigue behavior of sintered steels is reviewed.

Key words: fatigue, fracture, microstructure, porosity, powder metallurgy, residual stress, steel.

16.1 Introduction

Sintered ferrous powder metallurgy (PM) components are replacing wrought alloys in many applications, owing to their low cost, high performance, and ability to be processed to near-net shape. The fracture and fatigue behavior of these materials is very important. In this chapter the microstructure, tensile and fatigue behavior of sintered steels is reviewed.

16.1.1 Microstructure of powder metallurgy steels

Sintered materials are typically characterized by residual porosity after sintering, which is quite detrimental to the mechanical properties of these materials (Salak, 1997; Hadrboletz and Weiss, 1997; Chawla *et al.*, 2001a, 2001b, 2002, 2003; Polasik *et al.*, 2002). The nature of the porosity is controlled by several processing variables such as green density, sintering temperature and time, alloying addition and particle size of the initial powders (Salak, 1997). In particular, the fraction, size, distribution and morphology of the porosity have a profound impact on the mechanical behavior of materials (Hadrboletz and Weiss, 1997; Chawla *et al.*, 2001a, 2001b, 2002, 2003, Polasik *et al.*, 2002). In general, the porosity in sintered ferrous alloys is bimodal in nature and can be divided into primary or secondary porosity. Primary porosity consists of larger pores, owing primarily to the powder packing characteristics, which result in less than complete densification during sintering. Secondary porosity, on the other hand, consists of much smaller pores, often caused by the transient liquid phase of one or more alloying additions that form during sintering. An example of an alloying element that forms a transient liquid phase in sintered steels is Cu. Upon melting, the Cu particles leave behind small, rounded 'secondary' pores (on the order of the original Cu particle size). In addition to the residual porosity, ferrous PM alloys also exhibit a heterogeneous microstructure that develops owing to inhomogeneous distribution of powder particles, as well as incomplete diffusion of alloying additions during sintering (Salak, 1997). Graphite, for example, is typically used as a carbon source in PM steels so a microstructure consisting of coarse and fine pearlite, and bainite is not uncommon caused by local variations in carbon content and/or cooling rate. Nickel, on the other hand, often remains in the solid state during sintering and diffuses only partially into iron, contributing to a heterogeneous microstructure where nickel is present predominantly at the periphery of the pores (Chawla *et al.*, 2001a, 2001b, 2002; Polasik *et al.*, 2002). Thus, the heterogeneous nature of microstructure in PM steels certainly plays a role in the onset and evolution of damage under applied stress.

Typical sintered steel microstructures, in terms of pore size, morphology and distribution, are shown in Fig. 16.1. The pores at the lowest density appear

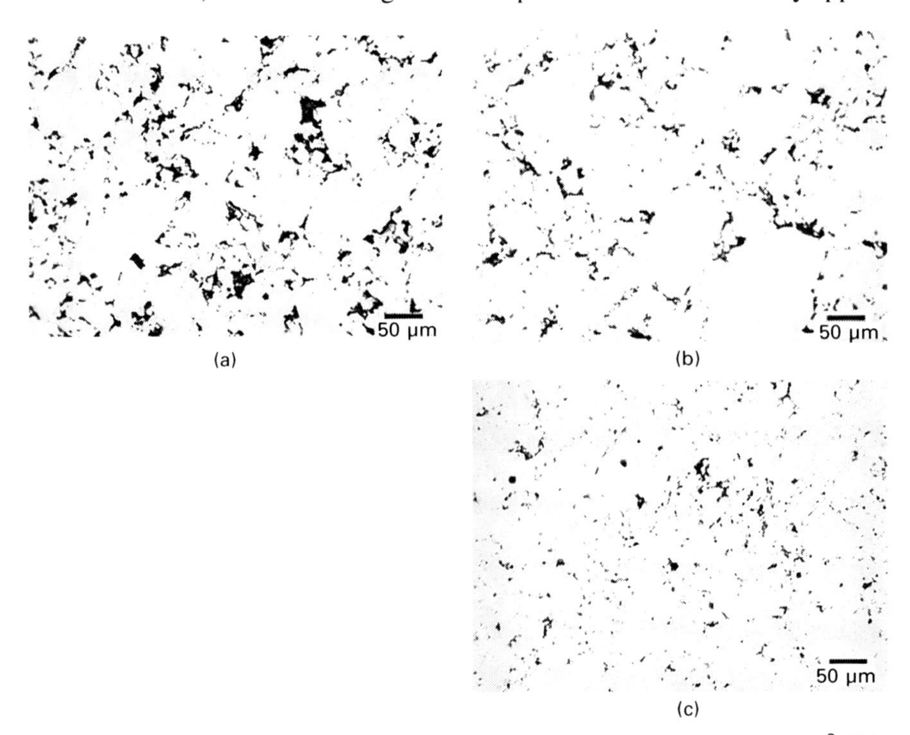

16.1 Microstructure of PM steels at three densities: (a) 7.0 g cm^{-3}, (b) 7.4 g cm^{-3} and (c) 7.5 g cm^{-3}. Note the higher fraction of porosity, as well as larger, more irregular pores at the lowest density.

to be much larger and more irregular than pores in the other two alloys. Pores at the lowest density also appear to be more clustered and segregated along the interstices between particles. Scanning electron microscopy (SEM) revealed the phases in the steel, which include pearlite (coarse and fine), bainite and Ni-rich austenite around the periphery of the pores, Fig. 16.2. An example of a microstructure in alloy steel, for example, Fe–Cr–Ni–Mo steel, is shown in Fig. 16.3. Here the microstructure consisted primarily of tempered martensite and bainite. Some isolated, Ni-rich areas were also

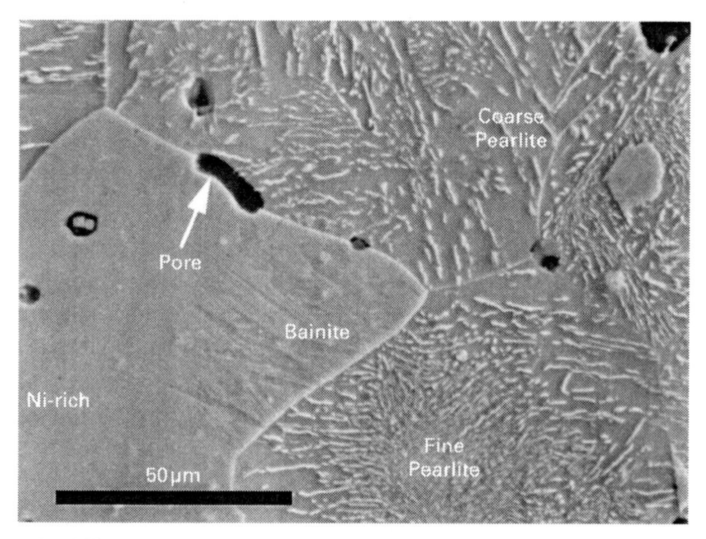

16.2 Microstructural phases in the PM steel including pearlite, bainite and Ni-rich austenite.

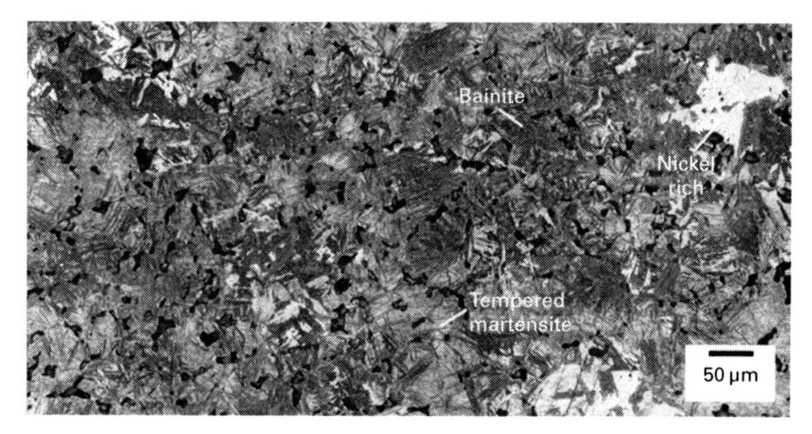

16.3 Optical micrograph of a sample tempered at 205°C, polished and then etched with Nital. Phases, from darkest to lightest, are bainite, tempered martensite and retained austenite and/or nickel rich ferrite.

observed, probably due to agglomeration of Ni powders during mixing. The sintering temperatures used are not high enough to cause Ni to form a liquid, so only solid state diffusion of Ni takes place.

Three dimensional (3D) X-ray tomography can also be used to visualize and quantify the degree and interconnectivity of porosity. Figure 16.4(a) and (b) show the 3D reconstructed volume of the porosity. The total pore volume is about 12 v/o, which correlates well based on a calculated sintered density of 7.0 g cm^{-3}. The pore network in yellow consists of one single interconnected network and makes up about 93% of the total pore volume. The green pores correspond to isolated pores. Two interconnected pores, shown in light blue and magenta, are highlighted in Fig. 16.4(b). Note the highly tortuous and irregular nature of the porosity, which is clearly highlighted by the X-ray tomography technique, Fig. 16.4(c).

16.2 Fracture behavior

Under monotonic tensile loading, porosity reduces the effective load bearing cross-sectional area and acts as a stress concentration site for strain localization and damage, decreasing both strength and ductility (Hadrboletz and Weiss, 1997). With an increase in porosity fraction (> 5%), the porosity tends to be interconnected in nature, as opposed to the situation where pores are relatively isolated (< 5%). Interconnected porosity causes an increase in the localization of strain at relatively smaller sintered regions between particles,

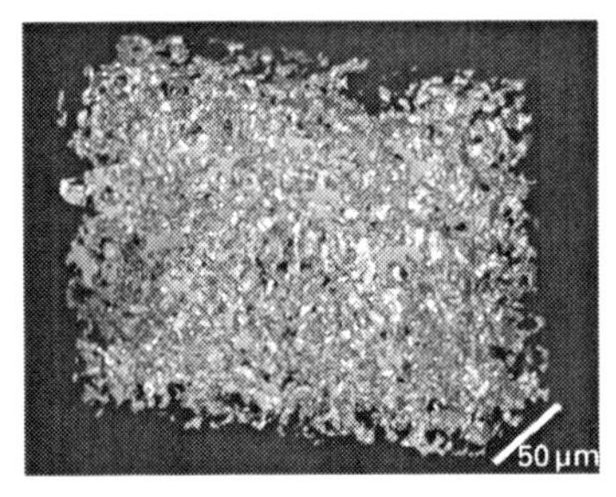

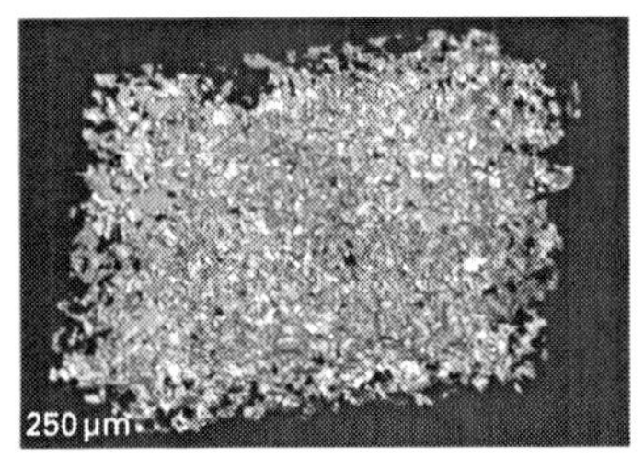

16.4 3D reconstruction of porosity in PM steel. Porosity in yellow shows one large interconnected network: (a) front view, (b) back view, (c) high magnification image of two pore networks.

while isolated porosity results in more homogeneous deformation. It is also not uncommon for the porosity distribution in the material to be inhomogeneous. In this case, strain localization will take place at 'pore clusters' (Danninger *et al.*, 1993). Thus, for a given amount of porosity, interconnected porosity is more detrimental and reduces macroscopic ductility to a greater extent than isolated porosity.

The monotonic behavior of the alloys is clearly influenced by the fraction of porosity and the pore morphology. Increasing density results in an increase in Young's modulus, proportional limit stress or 2% proof stress (taken as the onset of yielding in the material, as measured by a 2% deviation from the linear elastic portion of the stress–strain curve), tensile strength and strain-to-failure, Table 16.1. Figure 16.5 shows a comparison of the effect of density on monotonic tension. Note that even a moderate fraction of porosity (~10–12%) is sufficient to induce a significant decrease in strength and strain-to-failure.

Table 16.1 Tensile properties of Fe–Mo–Ni steels

Density (g cm^{-3})	Young's modulus (GPa)	Proportional limit stress (MPa)	Ultimate tensile strength (MPa)	Strain-to-failure (%)
7.0	138.6 ± 1.2	160.0 ± 19.4	570.6 ± 8.1	2.1 ± 0.1
7.4	171.6 ± 0.8	189.0 ± 8.1	745.2 ± 24.0	4.6 ± 0.4
7.5	182.6 ± 1.9	196.4 ± 9.3	784.4 ± 16.3	6.5 ± 0.7

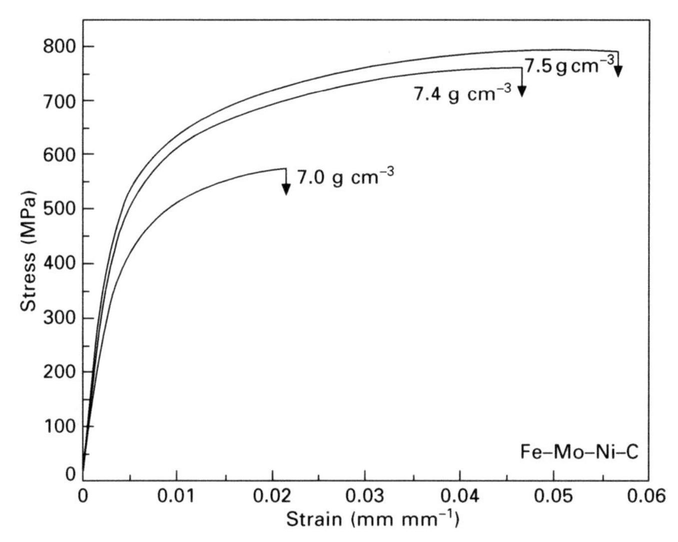

16.5 Effect of density on tensile behavior of the PM steels. A marked decrease in strength and ductility in the 7.0 g cm^{-3} alloy is observed, relative to the close to fully dense material, 7.5 g cm^{-3}.

Fractography of tensile and fatigue fracture surfaces provided further insight into the role of porosity in fracture of these materials. At the lowest density, tensile fracture takes place primarily by localized void nucleation and growth in sintered necks of the material, Fig. 16.6(a). At higher densities, however, the fracture morphology is more characteristic of that of fully dense materials, with a combination of ductile rupture as well as brittle fracture from fully dense pearlitic grains, Fig. 16.6(b).

It is well known that porosity decreases the Young's modulus of a material (Salak, 1997). Analytical models, such as that of Ramakrishnan and Arunachalam (R-A) (1993), can be used to model the effect of porosity on Young's modulus. In the R-A model a single spherical pore is surrounded by a spherical matrix shell, by considering the intensification of pressure on the pore surface caused by interaction of pores in the material. The Young's modulus of a material, E, with a given fraction of porosity, p, is given by (Ramakrishnan and Arunachalam, 1993):

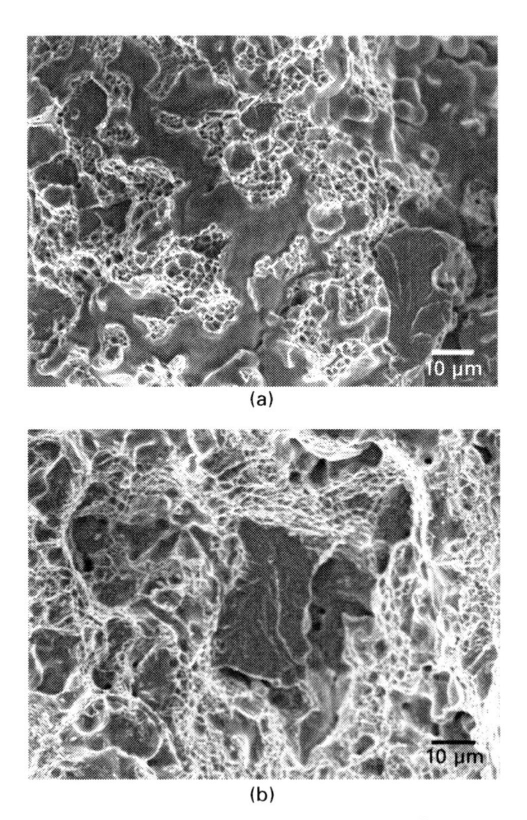

16.6 Tensile fracture in (a) 7.0 g cm^{-3} alloy and (b) 7.5 g cm^{-3} alloy. Localized void growth is observed at the lower density, while a combination of void growth and brittle cleavage fracture is observed at higher density.

$$E = E_0 \left[\frac{(1-p)^2}{1 + \kappa_E p} \right] \quad [16.1]$$

where E_0 is the Young's modulus of the fully dense steel (taken by extrapolating the experimental data to zero porosity, which yielded a value of approximately 201 GPa), and κ_E is a constant in terms of the Poisson's ratio of the fully dense material, ν_0:

$$\kappa_E = 2 - 3\nu_0 \quad [16.2]$$

For a fully dense steel, the Poisson's ratio is approximately 0.3. A comparison of the experimental data and the R-A prediction is shown in Fig. 16.7. The R-A predicts the experimental data very well, for the range of porosity examined here. The relatively good agreement between experiment and theory, based on a simple spherical pore geometry, indicates that the elastic properties of these materials do not appear to be significantly influenced by the shape and morphology of the porosity.

To simulate the effect of microstructure on local stress and strain distributions, the microstructures in Fig. 16.1 can be used as a basis for finite element method (FEM) analysis of uniaxial loading. Figure 16.8 shows the finite element mesh and boundary conditions. The analysis is two dimensional (2D) under plane strain condition. The left edge is fixed in the horizontal direction ($U_1 = 0$) while load is applied to the right edge horizontally under displacement rate control. A 1% applied strain is applied

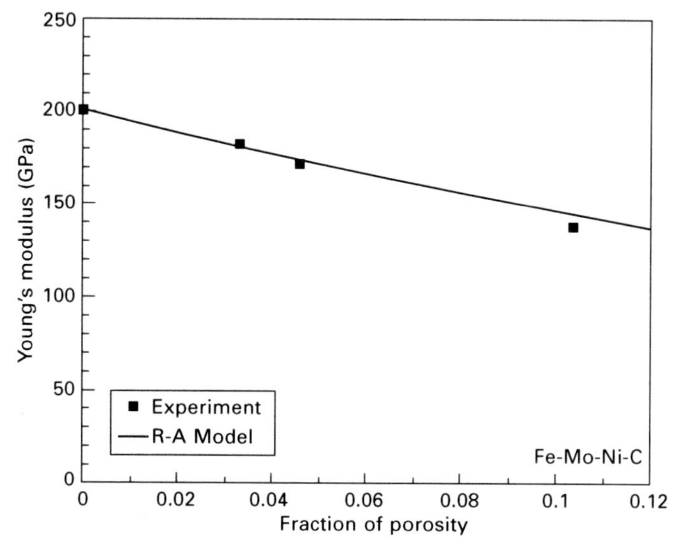

16.7 Young's modulus of PM steels versus porosity. The R-A model predicts the experimental data very well.

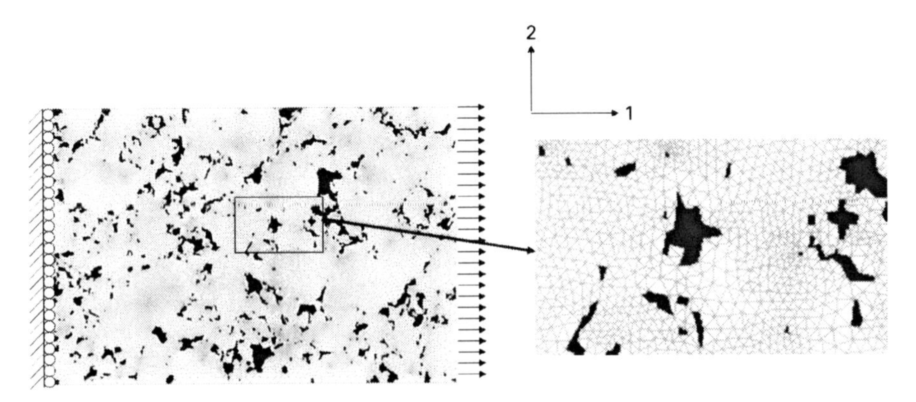

16.8 Finite element mesh and boundary conditions used for finite element analysis based on the real microstructure of PM steel.

to the model. A quadratic triangular mesh is employed in this simulation to conform to the irregular nature of the microstructure. A finer mesh is used in regions of pore clusters and a coarser mesh is employed in matrix-rich areas. The constitutive behavior of the steel matrix is extrapolated from tensile test experiments. In order to yield accurate simulation results, the microstructure used in the model should be as large as possible to give a large enough statistical representation. A balance must also exist between the size of the model and computational resources.

The macroscopic stress–strain behavior predicted by the model is shown in Fig. 16.9. The stress–strain input for the fully dense steel (extrapolated from the experimental stress–strain behavior at the three different densities) is also shown. Note that even a slight decrease in porosity (between 4–10%) results in a significant decrease in strength of the steel, as is observed experimentally.

The reason for this can be gleaned from the equivalent plastic strain evolution in each of the three microstructures, shown in Fig. 16.10. A large amount of strain localization takes place in the sintered regions between pores. In particular, networks of pores are quite effective in localizing the strains in the steel ligaments between the pores. Thus, a very small section of the microstructure is actually being plastically deformed, so that a large portion of the materials is largely undeformed. The modeling results are confirmed by experimental observations that porosity causes deformation to be localized and inhomogeneous (Vedula and Heckel, 1981; Spitzig *et al.*, 1988; Straffelini and Molinari, 2002). The strain intensification in the sintered ligaments between pores, probably serves as areas for crack initiation. Once the onset of crack initiation takes place, the large pores will be linked and the effective load-bearing area of the materials locally will decrease very quickly, resulting in fracture of the material. An increase

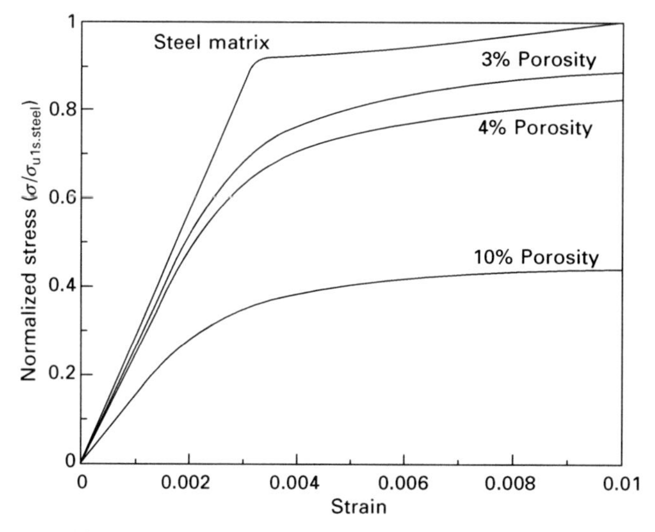

16.9 Modeled stress–strain behavior by 2D FEM analysis. A sharp decrease in strength is observed at 10% porosity, commensurate with the experimental data.

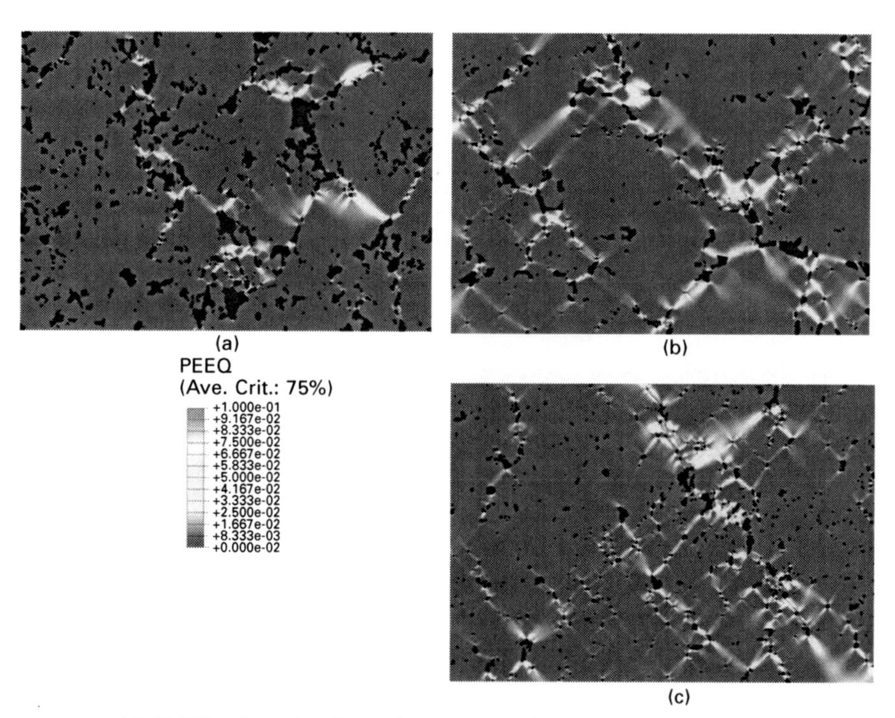

16.10 Effective plastic strain contours in modeled microstructures: (a) 7.0 g cm^{-3}, (b) 7.4 g cm^{-3}, and (c) 7.5 g cm^{-3}. Larger and interconnected pores cause strain intensification, while smaller, more homogeneously distributed pores contribute to more homogeneous deformation.

in porosity decreases the overall sintered ligament fraction and spacing between pores, thus accelerating the intensification of strain in the matrix material. This modeling also shows that plastic strain intensification begins at the tips of the irregular pores in the microstructure. Vedula and Heckel (1981) compared the damage mechanisms between round and angular pores in materials with identical pore fractions and observed that highly localized slip bands formed at the sharp tips of angular pores, producing an uneven distribution of strain around angular pores. This resulted in highly localized and inhomogeneous plastic deformation compared to the deformation around round pores which was much more homogeneous.

The distribution of the pores is also important, since it has been shown that plasticity may initiate at pore clusters because of the higher localized stress intensity associated with these defects (Hadrboletz and Weiss, 1997; Polasik *et al.*, 2002; Bourcier *et al*, 1986). The plastic strain distribution in the modeled microstructure for the densest steel, 7.5 g cm^{-3}, Fig. 16.10(c), shows that when the pores are much smaller and more homogeneously distributed, the plastic strain distribution is more uniform and the deformation is more uniformly distributed throughout the material. This explains why only a slight increase in density from 7.43 g cm^{-3} to 7.53 g cm^{-3} resulted in a significant increase in strain-to-failure, although the strength of the material increased only slightly. This may be attributed to a narrower and more homogeneous distribution of pores in the 7.5 g cm^{-3} alloy compared with the 7.4 g cm^{-3} alloy, although the total amount of porosity in the latter alloy was not significantly higher. Most of the strain localization takes place at the shortest distance between pores or pore clusters. In particular, most of the plastic deformation bands tend to be at an angle to the tensile direction, so the orientation of pores with respect to the loading axis may also play a significant role in plastic deformation. Thus, while the strength of the material is controlled by the fraction of pores, macroscopic ductility is also influenced by the size distribution, orientation and degree of clustering of the pores, since the sintered ligaments of the steel control fracture of the material. An equally important result of the model is that, even in the highest density material, a large amount of strain intensification takes place at a single pore cluster in the microstructure, Fig. 16.10(c). Thus, even when the overall amount of porosity is relatively low (4–5%), strain intensification may take place around pore clusters. It follows that the homogeneity and distribution of the porosity is as important as the fraction of porosity in controlling the evolution of plastic strain and thus the onset of crack initiation.

16.3 Fatigue behavior

Porosity also significantly affects fatigue behavior, although the role of porosity in fatigue is quite different from that in monotonic tension. In many investigations (Hadrboletz and Weiss, 1997; Chawla *et al.*, 2001a, 2001b, 2002; Polasik *et al.*, 2002; Holmes and Queeney, 1985; Christian and German, 1995; Lindstedt *et al.*, 1997), crack initiation was reported at pores or pore clusters located at or near the specimen surface. Holmes and Queeney (1985) proposed that the relatively high stress concentration at pores, particularly surface pores, is responsible for localized slip leading to crack initiation. Christian and German (1995) showed that fraction of porosity, pore size, pore shape and pore spacing are all important factors that control the fatigue behavior of PM materials. In general, more irregular pores have a higher stress than perfectly round pores (Salak, 1997). Pores have also been shown to act as linkage sites for crack propagation through interpore ligaments (Hadrboletz and Weiss, 1997; Polasik *et al.*, 2002). Polasik *et al.* (2002) showed that small cracks nucleate from the pores during fatigue and coalesce to form a larger crack leading to fatigue fracture. Here the heterogeneous nature of the microstructure played an important role by contributing to crack tortuosity. Crack arrest and crack deflection were observed, owing to microstructural barriers such as particle boundaries, fine pearlite and nickel-rich regions.

An example of the significant influence of porosity on fatigue behavior is shown in Fig. 16.11. Stress versus cycle curves revealed that the 7.0 g cm^{-3} alloy had a significantly lower fatigue endurance limit than the other two alloys. It is well known that single large pores or clusters of pores act as stress concentration sites for fatigue crack initiation (Hadrboletz and Weiss,

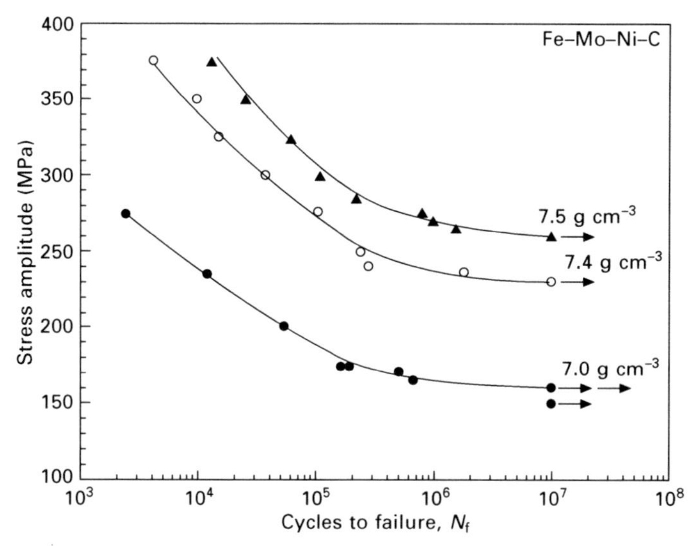

16.11 Stress versus cycles (*S–N*) fatigue behavior of PM steels as a function of density. Increasing density significantly increases fatigue life.

1997; Chawla *et al.*, 2001a, 2001b, 2002; Polasik *et al.*, 2002; Danninger *et al.*, 1993; Holmes and Queeney, 1985; Christian and German, 1995). Fatigue crack propagation, on the other hand, is largely influenced by the interconnectivity of the porosity, for example, interconnected porosity is more conducive to fatigue crack propagation because the crack grows through the path of least resistance. As in the case of tensile loading, pores with a more irregular shape will have a higher stress concentration and are more likely sites for crack initiation. Thus, the higher observed fatigue strength with increasing density can be attributed significantly to a lower overall porosity, pore clustering and more rounded pores.

The heterogeneous microstructure of PM steels also has a significant effect on fatigue crack growth in these materials. Chawla *et al.* (2005) showed that fatigue cracking appears to be highly dependent on the phase(s) at the crack tip, Fig. 16.12. For the Ni-rich regions, cracks tend to propagate in a linear fashion, suggesting that the Ni-rich regions offer little to no resistance to crack propagation. This is further supported by the Vickers hardness data, which showed the Ni-rich phase to be very soft, indicating that it might be Ni-rich austenite phases (Prasad *et al.*, 2003). For the pearlite regions, cracks tend to be highly deflected, with some evidence of the Fe_3C particles in the ferrite matrix bridging the crack, Fig. 16.13. Pores inside the Ni-rich regions can also act as nucleation sites for secondary cracks. It has been demonstrated that these cracks, which originate at pores ahead of the crack tip, often join the main crack (Polasik *et al.*, 2002; Prasad *et al.*, 2003), Fig. 16.14. Cracks propagating through the coarse pearlite, fine pearlite and bainite all show large increases in the degree of fatigue resistance owing to crack deflection, Fig. 16.14. Crack arrest is often present and further induces crack deflection through branching. Quantitative measurements of the crack growth rate of the different microstructural constitutents (in the Paris law

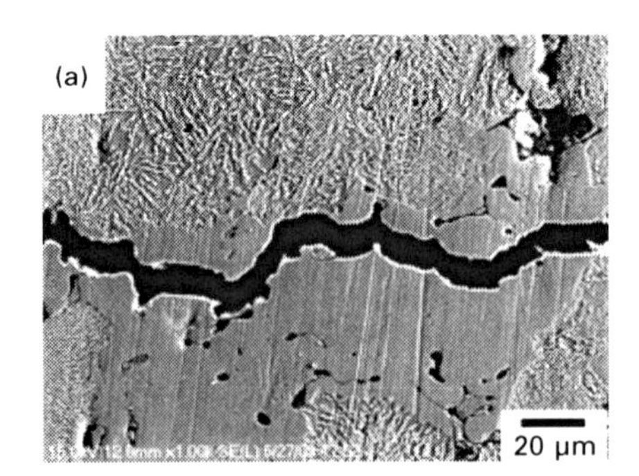

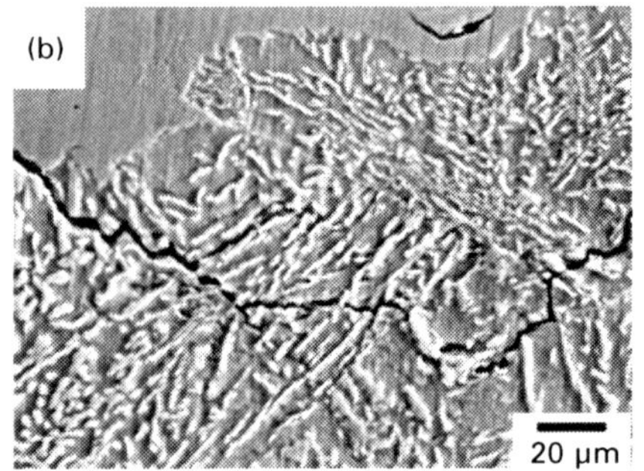

16.12 Fatigue crack behavior through heterogeneous microstructure: (a) Ni rich and (b) pearlite. The crack propagates through the Ni-rich areas, but it is tortuous and deflected through pearlite owing to the Fe_3C needles.

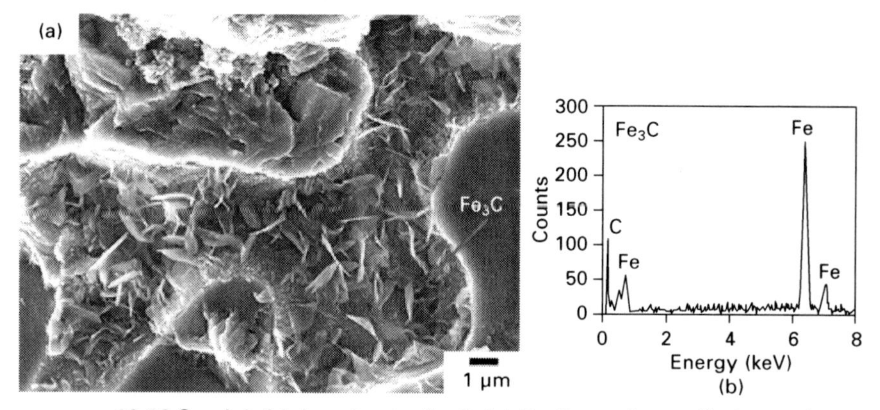

16.13 Crack bridging due to Fe_3C: (a) Fe_3C needles pulled out of the ferrite matrix during fatigue, and (b) energy dispersive X-ray spectroscopy (EDS) analysis showing the composition corresponding to Fe_3C.

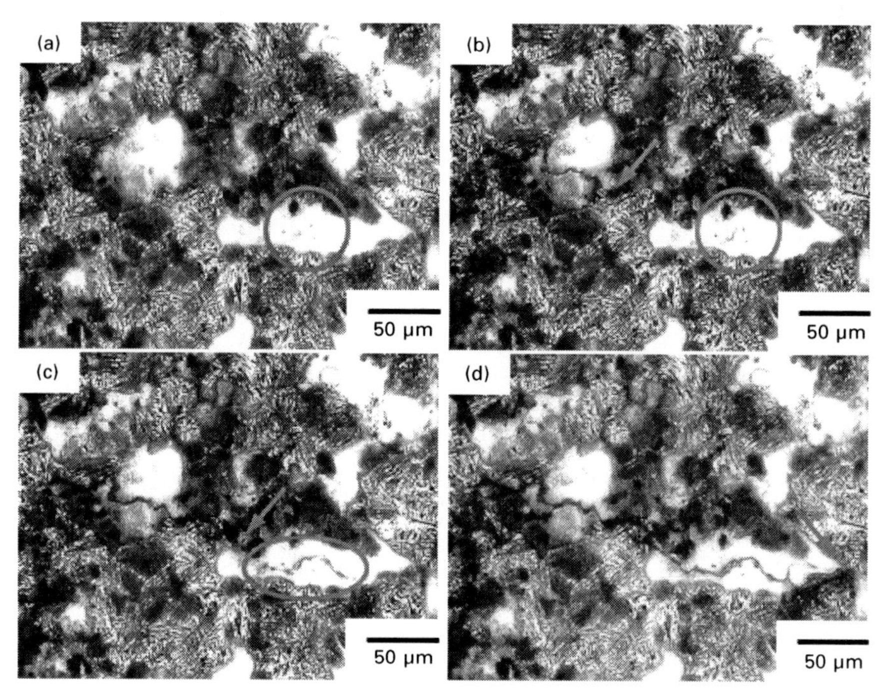

16.14 In situ observation of cracking indicated by arrows and circles during fatigue: (a) initial, (b) 10 000 cycles, (c) 16 000 cycles, and (d) 22 000 cycles.

regime) were conducted and these are presented in Table 16.2. As expected, the crack growth rate was highest in the Ni-rich regions, followed by coarse pearlite, fine pearlite and bainite. The crack growth rate in the Ni-rich regime

Table 16.2 Crack growth rate through microconstituents in the microstructure (nominal crack growth rate of 4×10^{-9} m/cycle)

Phase	Fatigue crack growth rate (10^{-8}m/cycle)	Fraction of total crack path (%)
Ni rich	1.74 ± 0.89	35.9
Coarse pearlite	0.55 ± 0.04	41.7
Fine pearlite	0.18 ± 0.04	19.0
Bainite	0.08 ± 0.01	3.5

was three times faster than that of coarse pearlite. Since the Ni-rich areas surround the periphery of the pores, the fraction of the total crack path was highest in this phase.

In general, the fatigue strength of PM materials is correlated to the ultimate tensile strength, by computing the fatigue ratio ($\sigma_{fat}/\sigma_{uts}$), in order to obtain an estimate of the fatigue strength relative to the monotonic strength (Salak, 1997). In PM materials the fatigue ratio is typically between 0.3 and 0.4, which is much lower than that reported for conventional wrought steels, 0.4–0.5 (Salak, 1997). The lower fatigue in PM materials can be attributed to the presence of porosity. Nevertheless, correlating fatigue strength to the ultimate tensile strength may not be the best approach, since the ultimate tensile strength is really a measure of the large-scale macroscopic damage that takes place at relatively large applied stress or plastic strain. Fatigue damage, on the other hand, is more complex, and typically takes place at much lower applied stress by localized plasticity at defects in the material. Thus, a more appropriate measure of fatigue strength may be the proportional limit stress or the Young's modulus of the porous material. The proportional limit stress is a measure of the onset of plasticity during monotonic loading. Thus, while the onset of plasticity under monotonic conditions is certainly not equivalent to cyclic plasticity, it may be a better measure of the fatigue strength than the ultimate tensile strength, since the onset of plasticity is related to the onset of damage in the material. Young's modulus may also be a good measure of the fatigue strength in heterogeneous materials, such as porous materials, since for a given stress, an increase in Young's modulus will result in a lower applied strain. A lower macroscopic applied strain will result in a lower degree of strain localization, which should decrease fatigue damage and increase fatigue life.

Figure 16.15 shows a plot of proportional limit stress and Young's modulus versus fatigue strength. While there is no exact correlation between σ_{pl} and E with σ_{fat}, there is certainly a good correlation between these parameters. It should also be noted that the lower proportional limit stress relative to fatigue strength may be attributed to our definition of fatigue strength as fatigue runout at 10^7 cycles. It has been shown that fatigue failures can take

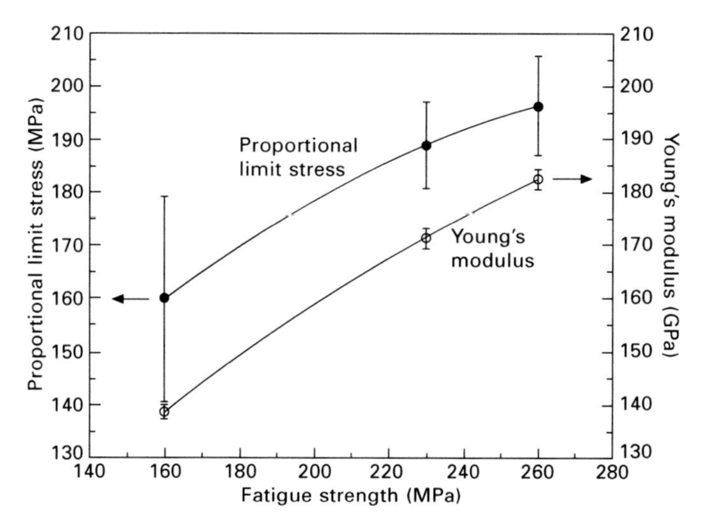

16.15 Correlation between fatigue strength and: (a) proportional limit stress and (b) Young's modulus.

place at longer fatigue lives than 10^7 cycles (Spoljaric *et al.*, 1996). Thus, at very long fatigue lives it is possible that the fatigue strength may approach the proportional limit stress.

In order to quantify further the damage evolution during fatigue of the sintered steels, stress–strain hysteresis experiments can be measured as a function of fatigue cycles. Figure 16.16 shows the evolution of the hysteresis loops, at a stress amplitude of 300 MPa, for all three densities. The changes with increasing cycles are most predominant at 7.0 g cm^{-3}. With increasing cycles, the width of the loop increases and the slope of the loops decrease significantly.

The evolution of plasticity can be quantified by plotting the plastic strain amplitude with increasing cycles, Fig. 16.17 (a). At lower densities, the plastic strain amplitude increases sharply until failure. The evolution of damage can also be quantified by a damage parameter, D_E, which is given by (Lemaitre, 1992):

$$D_E = \left(1 - \frac{E}{E_0}\right) \quad [16.3]$$

where E_0 is the modulus of the material in the undamaged state, and E is the modulus of the material at a given number of fatigue cycles. This parameter has been used to quantify damage in sintered steels (Polasik *et al.*, 2002) as well as other materials (Xu *et al.*, 1995). Figure 16.17(b) demonstrates that in addition to plastic strain evolution, there is a significant increase in D_E, particularly at the lower density. This shows that in addition to extensive

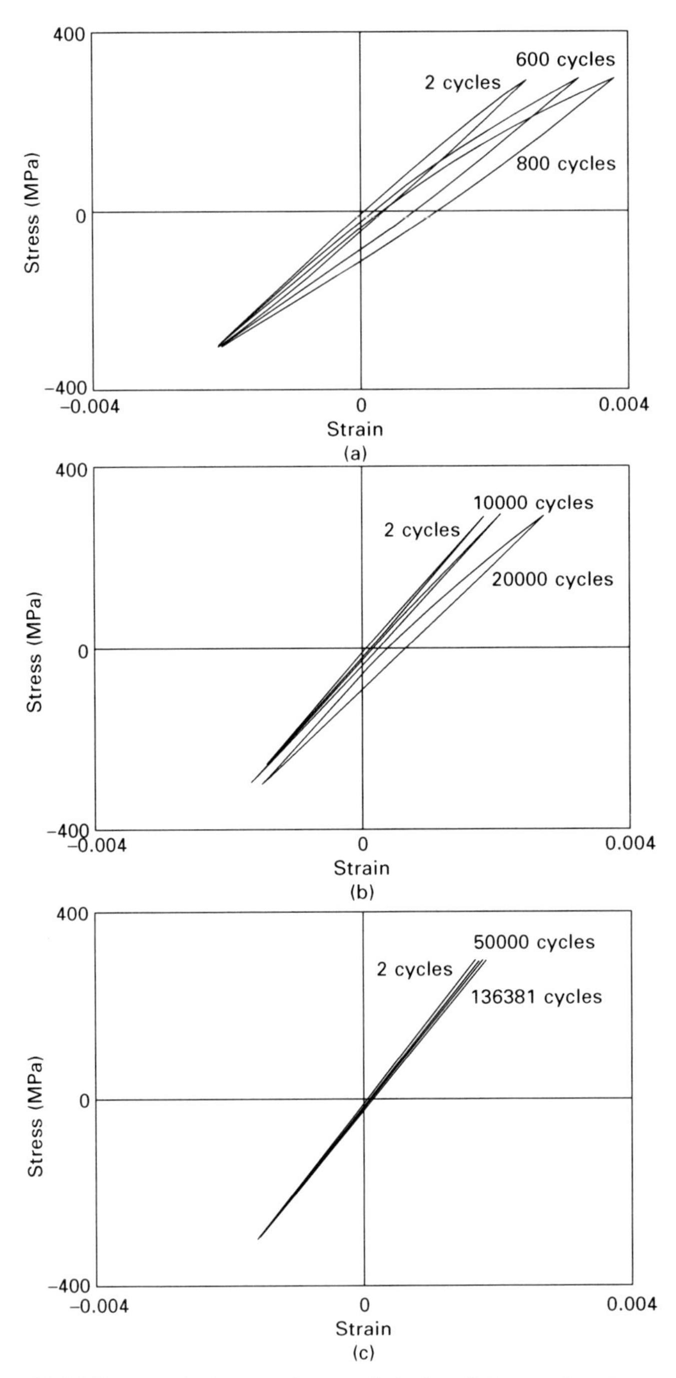

16.16 Hysteresis loops observed during fatigue at a stress amplitude of 300 MPa for the different densities: (a) 7.0 g cm^{-3}, (b) 7.4 g cm^{-3}, and (c) 7.5 g cm^{-3}.

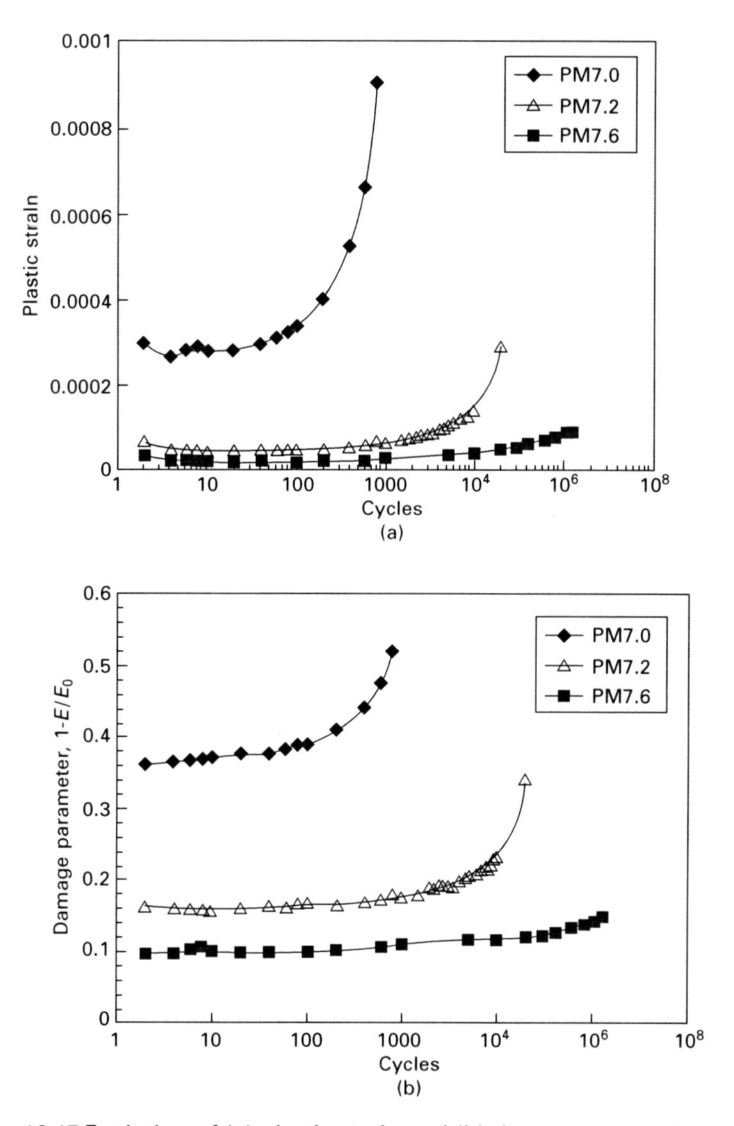

16.17 Evolution of (a) plastic strain and (b) damage parameter as the function of fatigue cycles. For a given stress (shown here at 300 MPa) the rate of plastic strain and damage is highest at the lowest density.

plastic deformation in the sintered ligaments, microcrack formation and propagation are also taking place. In the highest density materials, only plastic deformation appears to be taking place, at this stress level. It is also interesting to note that both plastic strain amplitude and damage parameter increase sharply in the last few fatigue cycles, immediately prior to failure.

At lower density, the rate of increase of plastic strain and damage parameter is highest.

Fatigue fractography shows that the microscopic damage mechanisms in fatigue are quite different from those observed in tension. Figure 16.18 shows the fatigue fracture surface of a 7.0 g cm^{-3} alloy. The fracture surface characteristics are somewhat similar to those observed under tensile loading, showing localized dimple rupture and evidence of void nucleation and coalescence in the sintered necks bonding the particles. A large number of spherical submicrometer MnS particles are also observed, although the influence of MnS inclusions on fatigue crack nucleation and/or propagation is unclear. It would appear that the effect of the inclusions would be negligible when the pore size is much larger than the inclusion size, since the pores act as the weakest links for crack initiation in the material. Localized fatigue striations are also apparent, although the propensity of these increased with density. Figure 16.19 shows a region of localized fatigue striations in an alloy with density of 7.4 g cm^{-3}. Localized striations have also been observed by Polasik *et al.* (2002) and have been hypothesized to form in areas of the microstructure that are favorably orientated to the loading axis.

Williams *et al.* (2007) mapped the type of fracture on the fatigue fracture surface of a PM alloy steel. The map of the mode of fracture, as a function of position in the specimen, is shown in Fig. 16.20. Fatigue striations are observed emanating from the crack initiation sites. The regions of fatigue striations merge into regions of mixed fracture mode. These include combinations of tearing and tearing/ductile rupture. Finally, the latter region evolved into the fast fracture regime, which is characterized, almost exclusively, by ductile rupture and cleavage.

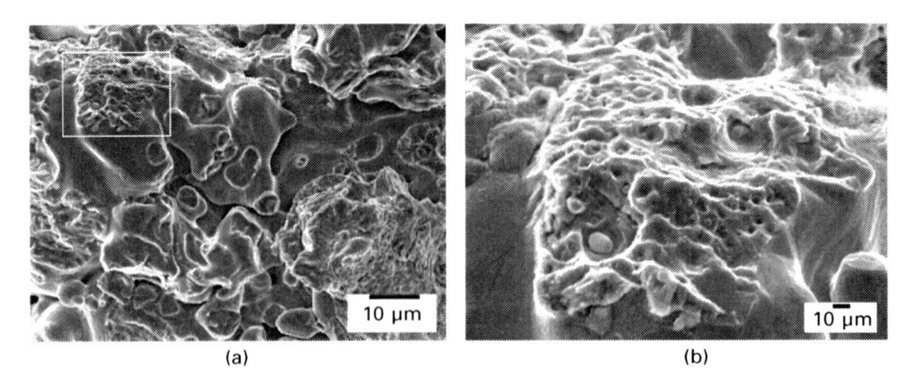

16.18 Fatigue fracture surface of 7.0 g cm^{-3} alloy: (a) low magnification and (b) high magnification inset of (a) showing localized void growth at MnS particles.

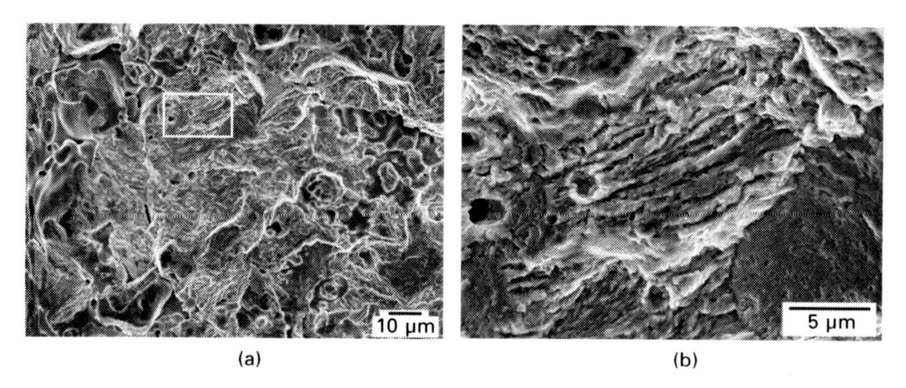

16.19 Fatigue fracture surface of 7.5 g cm^{-3} alloy: (a) low magnification and (b) high magnification inset of (a) showing localized fatigue striations.

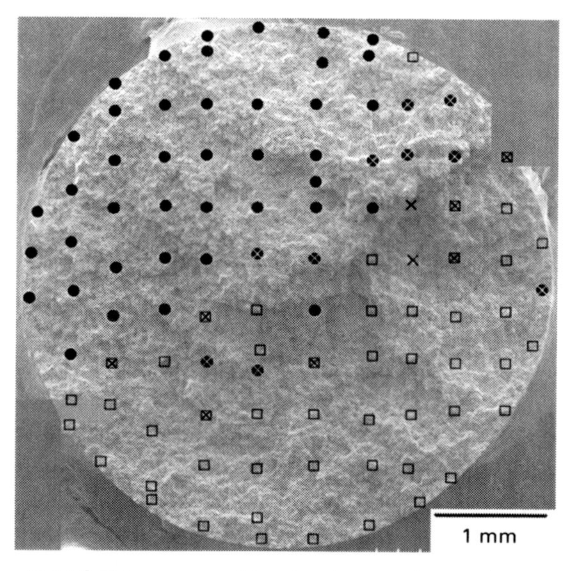

16.20 SEM survey of the fracture surface of a sample fatigued at 290 MPa. ●, ductile rupture and cleavage; □, fatigue crack striations; × tearing (overlapping symbols mean two different modes were present).

16.4 Residual stress effects on fatigue

Surface quality and residual stresses can affect the fatigue behavior of engineering material (Dieter, 1986). In wrought materials, the surface quality is often determined by final machining, which alters the surface roughness, can cause phase transformations and usually generates significant residual stresses (Kaczmarek, 1976). In electro-discharge machining (EDM), cutting occurs by melting the surface of the specimen. Upon solidification, the contraction of the hotter surface, as it cools, leads to compressive stresses in the underlying material and tensile stresses at the surface. These residual tensile stresses have been shown to reduce the fatigue strength significantly (Ghanem *et al.*, 2002). In contrast, grinding usually produces residual compressive stresses at the surface (Yahata, 1987; Vohringer, 1987; Matsumoto *et al.*, 1991). These compressive surface stresses may improve the fatigue resistance, although the increase in surface roughness may counteract this effect. Additionally, a poor choice of grinding parameters can cause the surface temperature of the work piece to increase significantly, such that on cooling, residual tensile stresses form on the surface (Kaczmarek, 1976). Finally, hard-turned parts show a large range of residual surface stress (Matsumoto *et al.*, 1986; Meurling *et al.*, 2001). Several studies on hard-turning have shown that residual surface stresses can vary from tensile to compressive, depending on complex relationships between the cutting tool's geometry and orientation, cutting rates and the mechanical properties of the work piece (Matsumoto *et al.*, 1986, 1991; Wu and Matsumoto, 1990; Hua *et al.*, 2006).

Sonsino *et al.* (1988, 1992) reported that surface rolling and shot peening improved fatigue strengths of sintered steels by roughly 20%. This was attributed to surface densification and the introduction of residual compressive stresses at the surface. In porous sintered steels, local densification at the outer surface has been observed after machining (Salek *et al.*, 2005; Agapiou and DeVries, 1988).

Williams *et al.* (2007) showed that machining resulted in a surface layer about 2 μm thick that was nearly fully dense (< 0.2% porosity). In contrast, the average area fraction of porosity throughout the specimen was measured to be 10.4%. The porosity continuously increased from the surface to the interior, until the mean value of 10.4% was reached about 7 μm below the machined surface. Fatigue specimens were not tested in the as-machined condition. Instead, the gauge sections were polished to a 1 μm finish to remove surface roughness. Approximately 15 μm was removed from the surface during polishing, more than the thickness of the dense layer measured in the as-machined sample. Examination of the polished surfaces, however, revealed only about 1% surface porosity (compared to 10.4% seen in the interior of the sample, Figs 16.21 and 16.22). Also, the amount of porosity remained below the mean level for a depth of about 7 μm below the surface. This suggests that some plastic deformation of the surface also occurred during polishing, although the gradient in density was not as high as that in the as-machined samples.

The residual surface stress that developed along the axial direction after each processing step is illustrated in Fig. 16.23. After sintering, the cooled specimen has a small residual tensile stress on the surface. The exact cause of this tensile stress is unknown, but is likely to be a result of volume changes

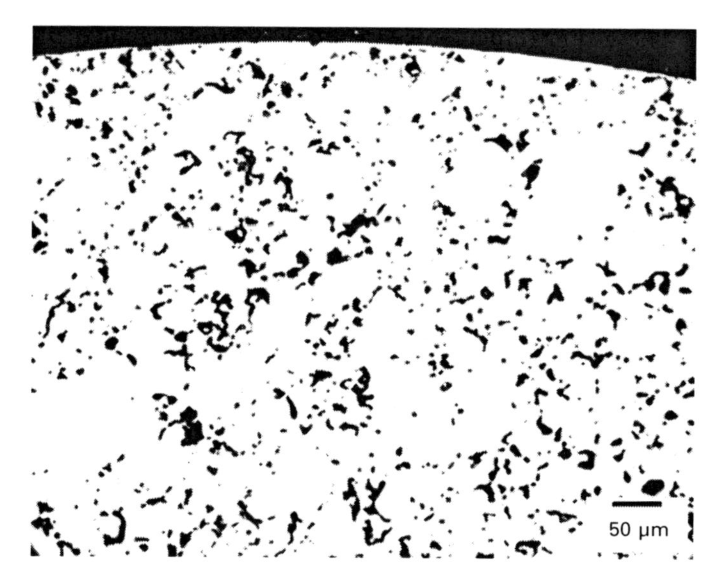

16.21 Optical micrograph of the cross section of the gauge section of an as-machined fatigue specimen. In cross section, pores are irregularly shaped and clustered. There is a relatively dense 7 μm layer at the surface caused by machining.

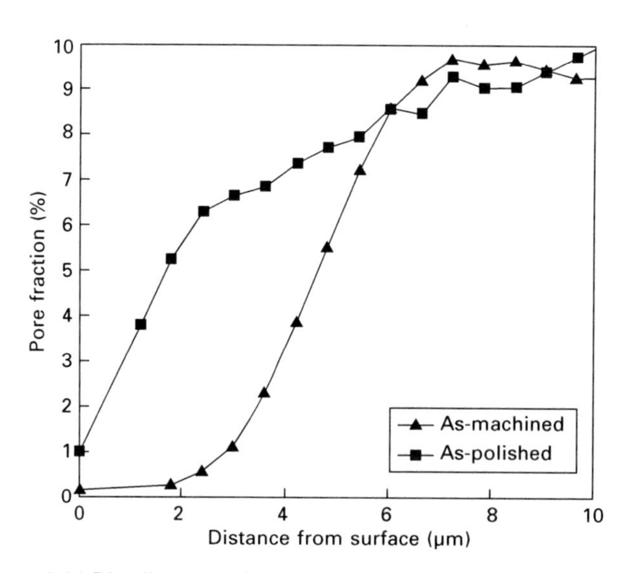

16.22 Distribution of porosity near the surface of as-machined and as-polished fatigue specimens. Machining generated a nearly fully dense layer 2 μm thick at the surface of the specimens. Although the entire machined layer was removed during polishing, polished surfaces were still denser than the interior, suggesting that some plastic deformation occurred during the polishing procedure.

caused by phase transformations on cooling or to differential coefficients of thermal expansion between phases. During tempering, some relief of the residual surface stress takes place and the stresses become slightly compressive, with no significant difference between samples tempered at 205°C and 315°C.

As seen in Fig. 16.23, the largest change in residual surface stress occurs due to machining. The magnitude of this compressive surface stress is well within the expected range of grinding stresses. Based on other studies (Yahata, 1987; Matsumoto *et al.*, 1991), this compressive stress is expected to be at a maximum at the specimen's surface and rapidly decay to near zero over the plastically deformed surface layer. This is in contrast to cutting (Matsumoto *et al.*, 1991) and turning (Matsumoto *et al.*, 1986), where the magnitude of the compressive stress reaches a maximum well below the surface. No significant or systematic change in residual surface stress was measured as a result of the 'stress-relieving' heat-treatment at 175°C of the machined specimens. Polishing, which was the final processing step before fatigue testing, reduces the compressive surface stresses by about 20%. This reduction in stress is due to the removal of the plastically deformed machined layer. Nevertheless, as shown above, some plastic deformation of the surface was observed after polishing as well. Thus, complete removal of residual surface stresses may require less aggressive polishing techniques, such as electropolishing or finer polishing compounds.

The effect of tempering, stress relief treatment after machining and residual stress on fatigue behavior is shown in Fig. 16.24. A comparison of the fatigue behavior between samples tempered at 205°C and 315°C is shown in Fig.

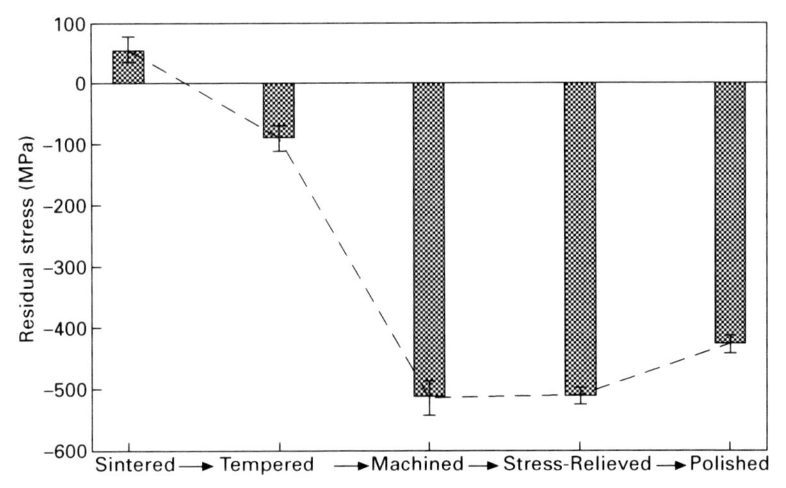

16.23 Measured residual axial surface stresses in five different specimens after each of the processing steps. The most significant residual stresses were generated during machining.

16.24. The samples tempered at 315°C have a slightly higher endurance limit (225 MPa) than those tempered at 205°C (220 MPa). This can be attributed to the slightly higher hardness in the 315°C tempered samples. Wrought steels commonly show an increase in fatigue strength as the hardness is increased (Dieter, 1986). The effect of the stress-relief anneal at 175°C, for specimens tempered at 205°C, is shown in Fig. 16.25. As discussed earlier, the stress-relieving heat treatment has no effect on residual surface stresses. It also had no effect on fatigue behavior; samples from either condition had identical fatigue endurance limits of 220 MPa.

No obvious correlation exists between fatigue strength and residual surface stress, Fig. 16.26. This behavior can be explained by the fact that residual compressive surface stresses extended roughly 8 μm below the surface of these samples (through the plastically deformed zone). The bulk of the material, however, is still filled with porosity. In cases where the residual stress extends into a larger volume of the material, for example, hundreds of micrometers by shot peening, a significant improvement in fatigue strength of the PM steels has been observed (Sonsino *et al.*, 1988, 1992).

16.5 Constitutive behavior of microstructural constituents

One of the challenges in quantifying the behavior of sintered steels is the difficulty, to date, of investigating the properties of the individual

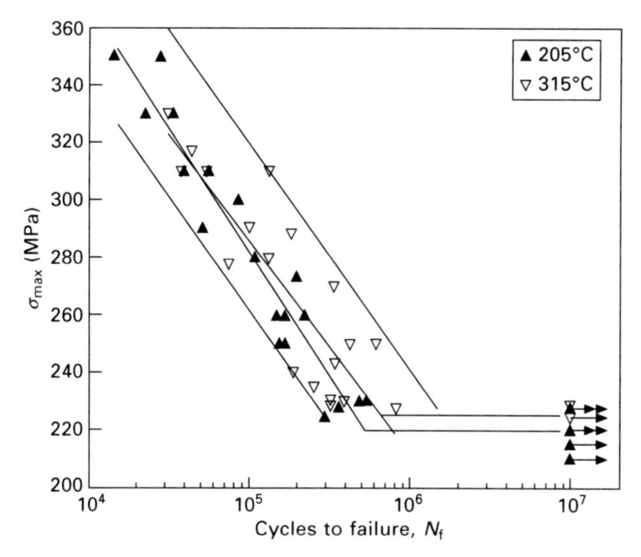

16.24 Comparison of fatigue behavior between samples tempered at 205°C and 315°C. Samples tempered at 315°C showed more scatter and slightly higher fatigue strengths.

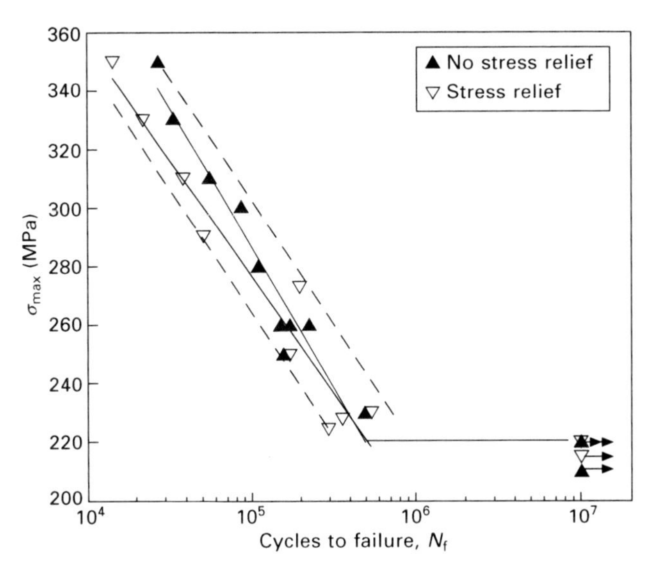

16.25 *S–N* curves showing a fatigue endurance limit of 220 MPa, irrespective of the stress-relieving heat treatment.

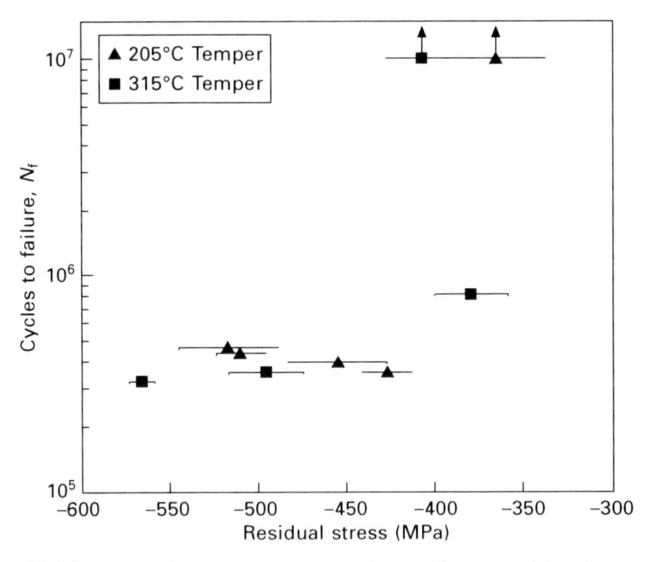

16.26 Residual stress versus cycles to failure at the two tempering conditions for specimens tested at a constant stress amplitude (228 MPa). No clear correlation between residual axial surface stress and cycles to failure was observed.

microconstituents (e.g. pearlite, bainite, martensite, etc). Nanoindentation and micropillar compression can be used to probe the local mechanical properties of the steels' microconstituents to understand better the composite-like behavior. This technique is particularly valuable in dual phase steels owing to the capability of indenting individual phases.

An example of nanoindentation data, in a dual phase stainless steel, on ferrite and martensite is illustrated in Fig. 16.27 (Walters *et al.*, 2012a). Two sets of samples are shown: low martensite (LM) and high martensite (HM) content. The LM specimens had ferrite and martensite phase fractions of 29% and 81% of the fully dense material, respectively. In the HM samples, significantly lower ferrite and higher martensite fractions of 9% and 92%, respectively, were observed. Attention was taken to probe the centers of the

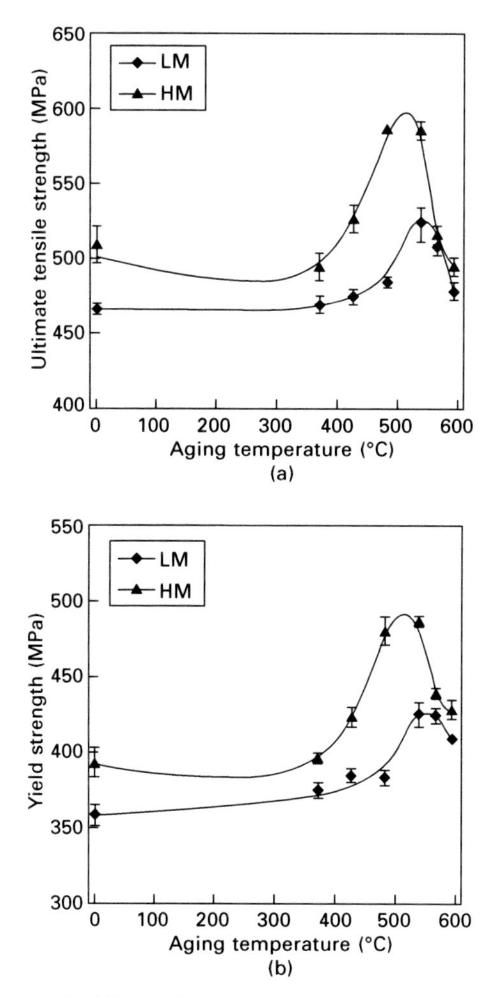

16.27 Effect of thermal aging on (a) ultimate tensile strength, (b) yield strength, (c) elongation to failure, and (d) Young's modulus on low martensite (LM) and high martensite (HM) specimens. Note concurrent increases in strength and ductility with aging.

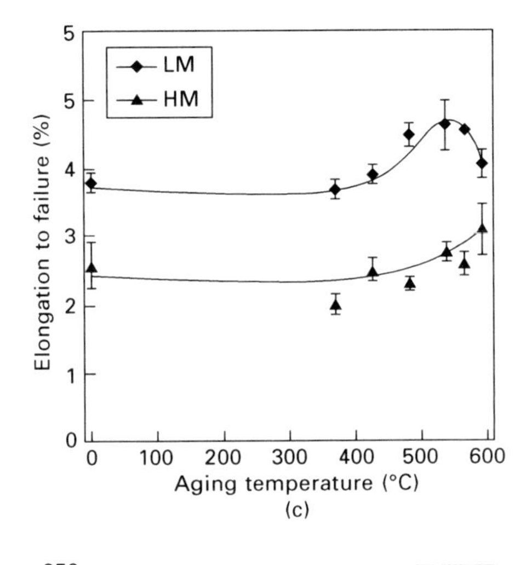

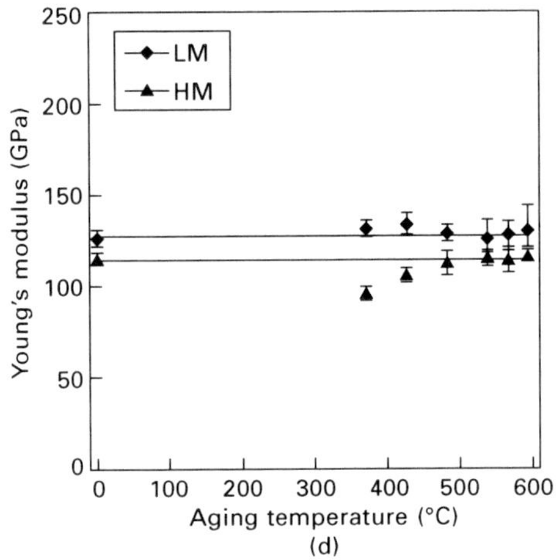

16.27 Continued

grains to reduce effects from the surrounding phases, Fig. 16.28. As expected, Fig. 16.29 shows that the ferrite had lower hardness than the martensite in all samples. The HM specimens also had higher ferrite and martensite hardness values than those of the LM specimens in accordance to aging temperature which may be attributed to different cooling rates.

In the LM specimens, the hardness of the ferrite and martensite increased with aging temperature showing maxima at 538°C. The tendency is consistent with trends observed for the yield and ultimate tensile strengths of the bulk material during tensile testing and suggests that thermal aging influences the mechanical properties of both the ferrite and martensite. Furthermore, these results indicate both ferrite and martensite contribute to overall strengthening

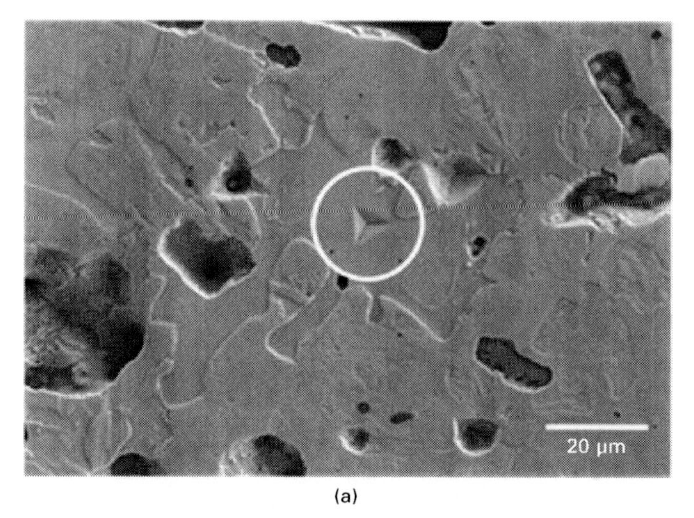

(a)

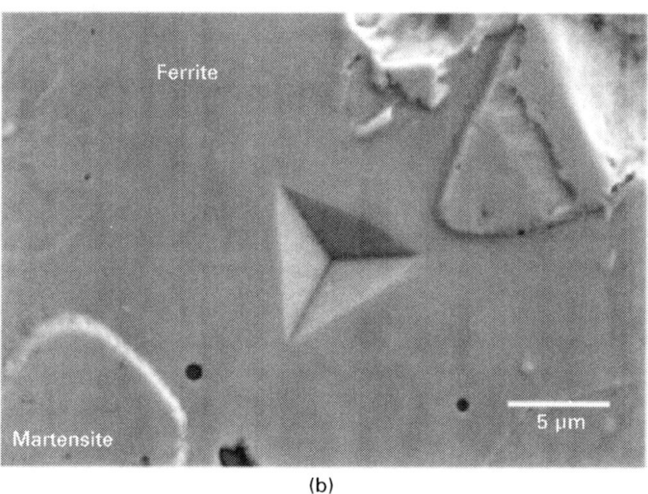

(b)

16.28 Example of nanoindentation targeting individual microconstituents on etched steel surface (a) and (b) high magnification of circled area. Ferrite and martensite are denoted by labels.

in the material for the LM material with a relatively high ferrite phase fraction of 29%.

Nanoindentation results from HM specimens exhibiting the same trend in which the maximum hardness was observed at 538°C for martensite, but no significant maximum was observed in the hardness of the ferrite. However, aging did increase the ferrite hardness when compared to the as-sintered condition.

While tensile testing showed strengthening behavior of the bulk LM and

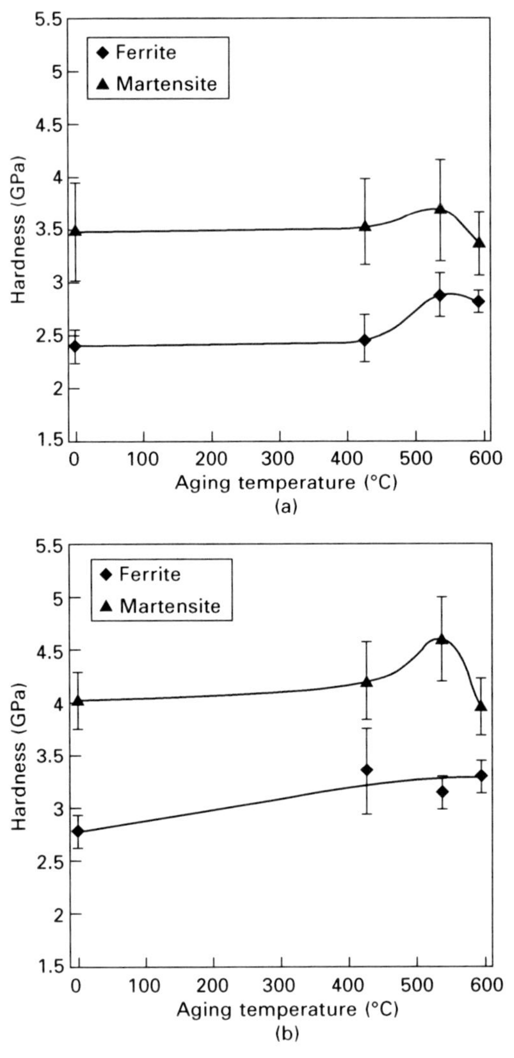

16.29 Effect of thermal aging on nanohardness of ferrite and martensite in (a) low martensite (LM) and (b) high martensite (HM) specimens. Note increased hardness of both microconstituents with aging.

HM materials, nanoindentation of the ferrite and martensite constituents shed light on the microstructural mechanism. It is shown that both ferrite and martensite are strengthened with aging and therefore both contribute to strengthening of the bulk LM and HM materials, although strengthening from the martensite may be dominant owing to its much higher phase fraction. Several mechanisms are at work here. As previously discussed, aging causes the supersaturated solution to precipitate into small intermetallic particles.

The particles strengthen the metal matrix through mechanisms such as Orowan bowing that make dislocation motion difficult (Moon *et al.*, 2008). Since precipitates in the matrix are much smaller than the indentation volume currently used, the nanoindentation results include contributions from both the matrix and its precipitates. Increases in hardness with aging suggest that precipitation hardening occurs in both the ferrite and martensite microconstituents and the most effective precipitation response is found in specimens aged at 538°C.

At higher temperatures, overaging occurs in which precipitates have grown large enough to allow dislocations to bend and pass between adjacent particles, corresponding to a decrease in hardness. Precipitate growth also results in increased interparticle spacing which contributes to this softening (Moon *et al.*, 2008). Strengthening of the ferrite phase has also previously been attributed to the grain size and solid solution hardening from the alloying elements (Speich and Miller, 1979; Furnemont *et al.*, 2002). The latter is more plausible in this case because there are no apparent grain size differences between aging temperatures in the LM and HM specimens. Lastly, tempering of the martensite occurs upon aging which relieves residual stresses, results in short-range diffusion and contributes to enhancement in strength and ductility. At higher temperatures, temper softening may be observed owing to rearrangement of carbon atoms and recovery of dislocation structures (Ohmura and Tsuzaki, 2008).

As expected, the Young's moduli for the ferrite and martensite were similar over the various aging temperatures (Fig. 16.27(d)), and the LM and HM specimens exhibited similar moduli. As previously explained, the bulk material's modulus from tensile testing is dependent upon porosity, that is, as the porosity increases the modulus decreases. This is not a factor in the nanoindentation modulus experiments because the small areas which were probed by the indenter were free from voids and thus considered fully dense.

Understanding the contributions of the individual microconstituents to the mechanical behavior of dual phase steels has proven difficult owing to the inability to obtain accurate constitutive relationships for each individual constituent. The properties of martensite or ferrite in bulk form are not representative of their behavior at the microscale. Conventional nanoindentation using a sharp tip Berkovich indenter has been used to probe the local mechanical properties of dual phase steels (Delince *et al.*, 2006; Hernandez *et al.*, 2010a, 2010b; Stewart *et al.*, 2012) but is limited to the determination of the Young's modulus and hardness of the microconstituents. Owing to confinement of plastic deformation to a very small volume, non-uniform strain and stress distributions result during indentation.

Novel and creative techniques must be used to quantify the constitutive behavior of individual microconstituents. Micropillar compression of microsized pillars is a promising technique for obtaining the stress–strain behavior at small-length scales. The technique consists of fabrication of free-standing pillars in the micrometer to nanometer scale, which can be isolated to individual phases, followed by compression using a nanoindenter with a flat punch. For the most part, this technique has been used to study size effects on mechanical properties (Uchic *et al.*, 2004; Uchic and Dimiduc, 2005; Greer *et al.*, 2005; Greer and Nix, 2005, Motz *et al.*, 2005; Nix *et al.*, 2007; Frick *et al.*, 2008; Schneider *et al.*, 2009) of single crystal materials. In addition, Jiang and Chawla (2010) have used the technique to obtain the constitutive behavior of intermetallic phases formed in Sn-based alloys.

Stress–strain plots for ferrite and martensite, in a dual phase steel, were obtained by micropillar compression (Walters *et al.*, 2012b). The top diameter of the pillar was used to calculate the nominal cross-sectional area of the pillars as in previous studies (Schneider *et al.*, 2009; Singh *et al.*, 2010). The strain was calculated as the ratio of the measured displacement to the original pillar height less its plastic compressive displacement (expressed as a percentage). The method of Greer et al. (2005) was used to correct the stress–strain curve, whereby the pillars are assumed to be perfectly cylindrical and the volume during plastic deformation is assumed to be conserved during compression. The resulting stress–strain curves are shown in Fig. 16.30. Two tests for each microconstituent per aging condition were completed. A summary of the compressive properties is shown in Table 16.3.

As expected, micropillar compression tests show higher strengths for martensite when compared to ferrite. The effect of aging is also examined. The yield strength of ferrite is observed to increase with aging temperature, while the fracture strength remains relatively constant. It is arguable that both the yield and fracture strengths of ferrite are within test scatter and therefore may be considered to remain constant with aging. Martensite, however, exhibits both increased yield and fracture strengths with aging which is consistent with results from tensile tests of the dual phase bulk specimens. This increased strength with aging is attributed to the growth of copper-containing precipitates which hinder dislocation motion by mechanisms such as Orowan bowing (Stewart *et al.*, 2011). It is therefore concluded that martensite is the primary driver of increased strength with aging in the bulk, especially with its predominate martensite phase fraction of 92%. Micrographs of representative pillars in ferrite and martensite (pre- and post-deformation) are shown in Fig. 16.31. Nearly constant initial volumes and dimensions were maintained for all fabricated pillars. Deformation is observed to occur by crystallographic slip.

In the work of Walters *et al.* (2012b), the objective was to obtain the constitutive stress–strain behavior of the individual microconstituents, and not to investigate size effects. As such, the volumes tested here are likely to be large enough to be in the regime where size effects do not play a

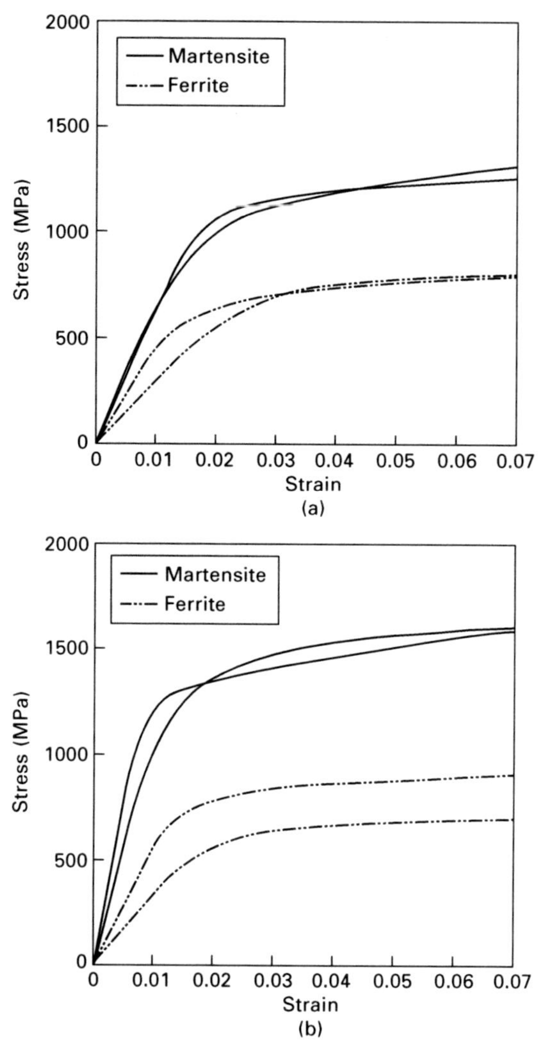

16.30 Stress–strain curves from ferrite and martensite micropillar compression of (a) as-sintered and (b) aged at 538°C specimens. Note increased martensite strength with aging.

Table 16.3 Yield and fracture strengths of ferrite and martensite from micropillar compression

Aging temperature	Yield strength (MPa)		Fracture strength (MPa)	
	Ferrite	Martensite	Ferrite	Martensite
As-sintered	569 ± 8	887 ± 53	823 ± 40	1475 ± 143
538°C	610 ± 151	1177 ± 54	808 ± 146	1858 ± 348

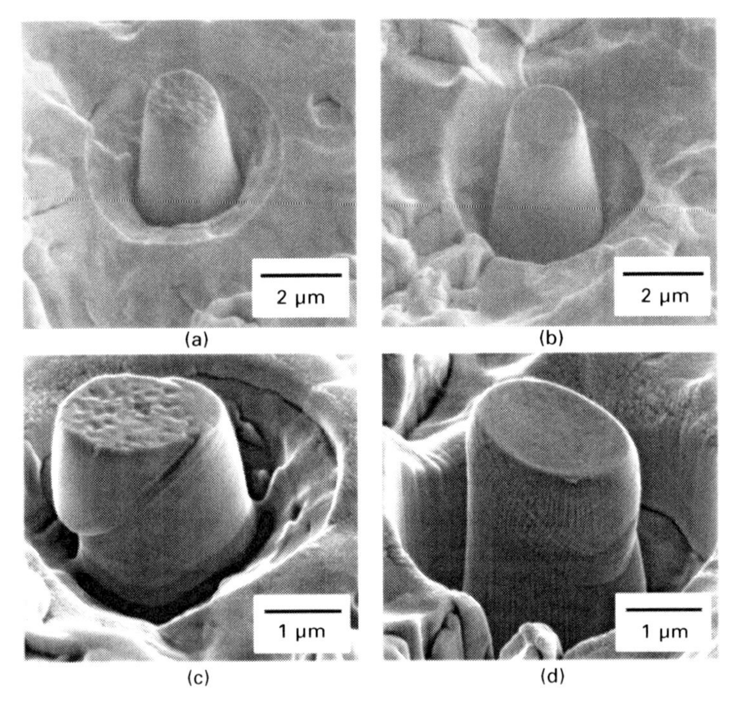

16.31 Scanning electron microscope images of (a) ferrite and (b) martensite pillar pre-(a)&(b) and post-(c)&(d) deformation. Note pillar deformation occurs by crystallographic slip.

role since the current pillar diameter is greater than the critical diameter of approximately 1 μm previously noted (Greer and Nix, 2005). In addition, the steel studied here is not expected to suffer from size effects owing to its inherent defect and dislocation density. Pouchon *et al.* (2010), in fact, observed good agreement in the yield stresses measured by tensile testing and micropillar compression tests of irradiated ferritic steel and therefore concluded that size effects were not present in the material. This absence of the size effect was attributed to the presence of defects in the material. It should also be noted that Pouchon's pillars were of similar diameter to the pillars investigated here.

These strength–porosity models were applied to the fracture strengths from the aforementioned rule of mixture analyses of the micropillar compression of ferrite and martensite. The resulting strengths which are, thus, inclusive of porosity, were then compared to the ultimate tensile strengths from tensile tests of the bulk steel and reasonable agreement was found. This agreement validates the approach of micropillar compression to determine the strength of individual microconstituents as a basis of predicting the overall material behavior.

16.6 Summary

The mechanical behavior of sintered steel is a complex function of porosity, microstructure and residual stress. In this chapter we have critically reviewed and summarized the tensile, fatigue and fracture behavior of powder metallurgy steels.

16.7 Acknowledgments

The authors are grateful for long-term research support from Hoeganaes Corporation (Dr Sim Narasimhan). NC acknowledges research contributions and wonderful discussions with many former students and postdocs, including, Dr Xin Deng, George Piotrowski, Steven Polasik, and Jennifer Walters, as well as Tom Murphy at Hoeganaes. We thank Huxiao Xie for his help with formatting the manuscript.

16.8 References

Agapiou, J.S. and DeVries, M.F. (1988). 'Machinability of powder metallurgy materials', *Int J Powder Metall*, **24**(1), 47–57.

Bourcier, R.J., Koss, D.A., Smelser, R.E. and Richmond, O. (1986). 'The influence of porosity on the deformation and fracture of alloys', *Acta Metall*, **34**, 2443–53.

Chawla, N., Polasik, S., Narasimhan, K.S., Murphy, T., Koopman, M. and Chawla, K.K. (2001a). 'Fatigue behavior of binder-treated powder metallurgy steels', *Int J Powder Metall*, **37**, 49–57.

Chawla, N., Murphy, T.F., Narasimhan, K.S., Koopman, M. and Chawla, K.K. (2001b). 'Axial fatigue behavior of binder-treated versus diffusion alloyed powder metallurgy steels', *Mater Sci Eng A*, **308**, 180–8.

Chawla, N., Babic, D., Williams, J.J., Polasik, S.J., Marucci, M. and Narasimhan, K.S. (2002). 'Effect of Ni and Cu alloying additions on the tensile and fatigue behavior of sintered steels', *Adv Powder Metall Part Mater*, **5**, 104.

Chawla, N., Piotrowski, G.B., Deng, X., Narasimhan, K.S. and Marucci, M.L. (2005). 'Fatigue crack growth of Fe-0.85Mo-2Ni-0.6C steels with a heterogeneous microstructure', *Int J Powder Metall*, **41**, 31–41.

Christian, K.D. and German, R.M. (1995). 'Relation between pore structure and fatigue behavior in sintered iron–copper–carbon', *Int J Powder Metall*, **31**, 51–61.

Danninger, H., Tang, G., Weiss, B. and Stickler, R. (1993). 'Microstructure and mechanical properties of sintered iron. Part II-Experimental study', *Powder Metall Int*, **25**, 170–5.

Delince, M., Jacques, P.J. and Pardoen, T. (2006). 'Separation of size-dependent strengthening contributions in fine-grained dual phase steels by nanoindentation', *Acta Mater*, **54**, 3395–3404.

Dieter, G.E. (1986). *Mechanical Metallurgy*, 3rd edition, McGraw-Hill, New York.

Furnemont, Q., Kempf, M., Jacques, P., Goken, M. and Delannay, F. (2002). 'On the measurement of nanohardness of the constitutive phases of TRIP-assisted multiphase steels,' *Mater Sci and Eng A*, **328**, 26–32.

Frick, C.P., Clark, B.G., Orso, S., Schneider, A.S. and Arzt, E. (2008). 'Size effect on strength and strain hardening of [111] nickel sub-micron compression pillars', *Mater Sci Eng A*, **489**, 319–29.

Gerard, D.A. and Koss, D.A. (1991). 'The Influence of Porosity on Short Fatigue Crack Growth at Large Strain Amplitudes', *Int J Fatigue*, **13**(4), 345–52.

Ghanem. F., Braham, C., Fitzpatrick, M.E. and Sidhom, H. (2002). 'Effect of near-surface residual stress and microstructure modification from machining on the fatigue endurance of a tool steel', *J Mater Eng Perf*, **11**, 631–9.

Greer, J.R. and Nix, W.D. (2005). 'Size dependence of mechanical properties of gold at the sub-micron scale', *Appl Phys A*, **80**, 1625–9.

Greer, J.R., Oliver, W.C. and Nix, W.D. (2005). 'Size dependence of mechanical properties of gold at the micron scale in the absence of strain gradients', *Acta Mater*, **53**, 1821–1830.

Hadrboletz, A. and Weiss, B. (1997). 'Fatigue behavior of iron based sintered material: A review', *Int Mater Rev*, **41**, 1–44.

Hernandez, V.H.B., Panda, S.K., Okita, Y. and Zhou, N.Y. (2010a). 'A study on heat affected zone softening in resistance spot welded dual phase steel by nanoindentation', *J Mater Sci*, **45**, 1638–47.

Hernandez, V.H.B., Panda, S.K., Kuntz, M.L. and Zhou, Y. (2010b). 'Nanoindentation and microstructure analysis of resistance spot welded dual phase steel', *Mater Lett*, **64**, 207–10.

Hua, J., Umbrello, D. and Shivpuri, R. (2006). 'Investigation of cutting conditions and cutting edge preparations for enhanced compressive subsurface residual stress in the hard turning of bearing steel', *J Mater Process Tech*, **171**(2), 180–187.

Jiang, L. and Chawla, N., 'Mechanical Properties of Cu6Sn5 Intermetallic by Micropillar Compression Testing', *Scrip Mater*, 2010, **63**, 480–3.

Kaczmarek, J. (1976). *Principles of Machining by Cutting, Abrasion and Erosion*, Voellnagel, A. and Lepa, E. (translators). Peter Peregrinus Stevenager.

Lindstedt, U., Karlsson, B. and Masini, R. (1997). 'Influence of porosity on deformation and fatigue behavior of PM austenitic stainless steels', *Int J Powder Metall*, **33**, 49–61.

Matsumoto, Y., Barash, M.M. and Liu, C.R. (1986). 'Effect of hardness on the surface integrity of AISI 4340 steel', *J Eng Ind*, **108**, 169–75.

Matsumoto, Y., Magda, D., Hoeppner D.W. and Kim, T.Y. (1991). 'Effect of machining processes on the fatigue strength of hardened AISI 430 steel', *Trans ASME*, **113**, 154–9.

Meurling, F., Melander, A., Tidesten, M. and Westin, L. (2001). 'Influence of carbide and inclusion contents on the fatigue properties of high speed steels and tool steels', *Int J Fatigue*, **23**(3), 215–24.

Moon, J., Kim, S., Jang, J., Lee, J. and Lee, C. (2008). 'Orowan strengthening effect on the nanoindentation hardness of the ferrite matrix in microalloyed steels', *Mater Sci Eng A*, **487**, 552–7.

Motz, C., Schoberl, T. and Pippan, R. (2005). 'Mechanical properties of micro-sized copper bending beams machined by the focused ion beam technique', *Acta Mater*, **53**, 4269–79.

Nix, W.D., Greer, J.R., Feng, G. and Lilleodden, E.T. (2007). 'Deformation at the nanometer and micrometer length effects of strain gradients and dislocation starvation', *Thin Solid Films*, **515**, 3152–57.

Ohmura, T. and Tsuzaki, K. (2008). 'Analysis of grain boundary effect of bulk polycrystalline materials through nanomechanical characterization', *J Phys D: Appl Phys*, **41**, 1–6.

Polasik, S.J., Williams, J.J. and Chawla, N. (2002). 'Fatigue crack initiation and propagation in binder-treated powder metallurgy steels', *Metall Mater Trans*, 2002, **33A**, 73–81.

Pouchon, M.A., Chen, J., Ghisleni, R., Michler, J. and Hoffelner, W. (2010). 'Characterization of irradiation damage of ferritic ODS alloys with advanced microsample method', *Exp Mech*, **50**, 79–84.

Prasad, S.N., Mediratta, S.R., Sarma, D.S. (2003). 'Influence of austenitisation on the structure and properties of weather resistant steels', *Mat Sci Eng A*, **358**, 288–297.

Ramakrishnan, N. and Arunachalam V.S. (1993). 'Effective elastic moduli of ceramic materials', *J Am Ceram Soc*, **76**, 2745–52.

Salak, A. (1997). *Ferrous Powder Metallurgy*, Cambridge International Science Publishing, Cambridge.

Salak, A., Miskovic, V., Dudrova, E. and Rudnavova, E. (1974). *Powder Metall Int*, **6**(3), 128–32.

Salak, A., Selecka, M. and Danniger, H. (2005). *Machinability of Powder Metallurgy Steels*, Cambridge International Science Publishing, Cambridge.

Schneider, A.S., Clark, B.G., Frick, C.P., Gruber, P.A. and Arzt, E. (2009). 'Effect of orientation and loading rate on compression behavior of small-scale Mo pillars', *Mater Sci Eng A*, **508**, 241–6.

Singh, D.R.P., Chawla, N., Tang, G. and Shen, Y.-L. (2010). 'Micropillar compression of Al/SiC nanolaminates', *Acta Mater*, **58**, 6628-36.

Sonsino, C.M., Müller, F., Arnhold, V. and Schlieper, G. (1988). 'Influence of mechanical surface treatments on the fatigue properties of sintered steels under constant and variable amplitude loading', *Mod Dev Pow Metall*, **21**, 55–66.

Sonsino, C. M., Müller, F. and Mueller, R. (1992). 'The improvement of fatigue behaviour of sintered steels by surface rolling', *Int J Fatigue*, **14**(1), 3–13.

Speich, G.R. and Miller, R.L. (1979). 'Mechanical properties of ferrite-martensite steels', in *Structure and Properties of Dual-Phase Steels*, American Institute of Mining Metallurgical and Petroleum Engineers, 145–182.

Spitzig, W.A., Smelser, R.E. and Richmond, O. (1998). 'The evolution of damage and fracture in iron compacts with various initial porosities', *Acta Metall*, **36**, 1201.

Spoljaric, D., Danninger, H., Weiss, B., Chen, D.L. and Ratzi, R. (1996). *Proceedings International Conference Deformation and Fracture in Structural PM Materials*, L. Parilak, H. Danninger, J. Dusza, and B. Weiss (eds), IMR-SAS Kosice, Volume 1, 147.

Stewart, J.L., Williams, J.J. and Chawla, N. (2012). 'Influence of thermal aging on the microstructure and mechanical behaviour of dual-phase, precipitation-hardened, powder metallurgy stainless steels', *Metall Mater Trans A: Phys Metall Mater Sci*, **43**(1), 124–35.

Straffelini, G. and Molinari, A. (2002). 'Evolution of tensile damage in porous iron', *Mater Sci Eng*, **334**, 96–103.

Uchic, M.D. and Dimiduk, D.M. (2005). 'A methodology to investigate size scale effects in crystalline plasticity using uniaxial compression testing', *Mater Sci Eng*, **400–1**, 268–278.

Uchic, M.D., Dimiduk, D.M., Florando, J.N. and Nix, W.D. (2004). 'Sample dimensions influence strength and crystal plasticity', *Science*, **305**, 986–9.

Vedula, K.M. and Heckel, R.W. (1981). *Modern Developments in Powder Metallurgy*', Metal Powder Industries Federation, Princeton, NJ.

Vohringer, O. (1987). 'Relaxation of residual stresses by annealing or mechanical treatment', *Advances in Surface Treatments. Technology–Applications–Effects*, Volume 4. Pergamon, Oxford, 367–96.

Walters, J.L., Jiang, L., Williams, J.J. and Chawla, N. (2012a). 'Prediction of bulk tensile behavior of dual phase stainless steels using constituent behavior from micropillar compression experiments,' *Mater Sci Eng A*, **A534**, 220–7.

Walters, J.L., Williams, J.J. and Chawla, N. (2012b). 'Influence of thermal aging on the microstructure and mechanical behavior of sintered dual phase stainless steels', *Metall Mater Trans*, **43**, 124–135.

Williams, J.J., Deng, X. and Chawla, N. (2007). 'Effect of residual surface stress on the fatigue behavior of a low-alloy powder metallurgy steel', *Int J Fatigue*, **29**, 1978–84.

Wu, D. W. and Matsumoto, Y. (1990). 'The effect of hardness on residual stresses in orthogonal machining of AISI 4340 steel', *J Eng Ind*, **112**, 245–51.

Yahata, N. (1987). 'Effect of lapping on the fatigue strength of a hardened 13Cr-0.34C stainless steel', *Wear*, **115**, 337–48.

Part IV
Applications

17
Automotive applications of powder metallurgy

P. RAMAKRISHNAN, Indian Institute of Technology Bombay, India

DOI: 10.1533/9780857098900.4.493

Abstract: Powder metallurgy (PM) is an established green manufacturing technology for the production of net-shape components. The ability to use PM to mass produce reliable precision parts consistently at a cheap rate is very attractive to the automotive industry. This chapter discusses the use of a wide variety of components produced from different metallic materials in traditional and eco cars. Key factors in the increase in use of PM parts in automotive applications and the challenges related to this increase are examined. Finally, the chapter looks at emerging trends and prospects for innovative PM technology in the automotive industry.

Key words: automotive PM parts, ferrous and non-ferrous, functional parts, nanotechnology, structural parts, traditional and eco vehicles.

17.1 Introduction

The automotive industry is one of the key sectors of the economy noted for its global value chain and is a major customer of the powder metallurgy (PM) industry. Automobile manufacturers are facing the challenge of procuring components with light weight, superior performance and enhanced durability and reliability in order to produce comfortable, safe vehicles at low cost. PM is an energy saving, green technology offering improved performance and greater design flexibility compared with traditional manufacturing methods such as casting, extrusion, forging, stamping and machining. The automotive industry has accepted PM technology, with the result that over 70% of its production worldwide is aimed at the automobile manufacturing sector. There is tremendous scope for PM to penetrate further, considering the wide variety of vehicles such as cars, buses, trucks, vans and three-wheeled vehicles, as well as two-wheeled vehicles such as scooters, motorcycles and bicycles.

Many powder metallurgy processing techniques are available for manufacturing different components, ranging from pressing and sintering to hot isostatic pressing, powder forging, metal injection molding and rapid prototyping; these are discussed in various publications (MPIF, 1998, 2008; EPMA, 2008; ASM International, 1998; Ramakrishnan, 2002).

The use of PM in the automobile sector began with the use of sintered self-lubricating bearings and low density structural parts in the 1920s and 1930s, followed by small parts (as a substitute for cast iron) and shock absorbing components in the 1950s. Larger components and more complicated shapes such as timing gear sprockets, synchronizer hubs, valve seats, belt pulleys and power steering parts were developed in the 1960s and 1970s. During the 1980s, efforts were concentrated on the substitution of PM for forged steel in components such as connecting rods, composite cam shafts and emission control system parts. The number of parts in use and their weight per vehicle has increased substantially since the 1920s and 1930s to the current level of over 1000 parts with more than 300 applications, weighing (in the USA) in the range of 13 to 45 kg depending upon the type of vehicle. The use of finite element analysis in the design of components with complex geometry, process control and vendor certification has reduced the development and acceptance time of PM components. Continuous improvements in raw materials, powders, equipment such as compacting presses (including isostatic and sintering furnaces), and processes such as warm and hot pressing, sinter bonding, hot isostatic pressing, powder forging, rolling and injection molding, have aided in the efficient adoption of PM technology.

Ferrous alloys were developed from diffusion alloyed, fully alloyed, bonded mix and admixed elemental powders, along with lubricants and additives, to improve their microstructure and mechanical properties. Efforts were made to reduce the cost by using cheaper alloying elements, modifying the sintering atmospheres, using higher sintering temperatures and developing processes such as sinter hardening to replace conventional sintering and heat treatment. To increase the size, range of shapes and complexity of components, as well as improve the economy of the process, a wide variety of presses have been developed. Some examples include computer numerical control (CNC) presses with closed loop controlled movements, multi platen adopters, high speed presses, servo-driven presses and high velocity compaction (Hinzmann and Sterkenburg, 2007). In order to reduce the energy consumption and improve product quality, a variety of batch and continuous sintering furnaces with controlled atmospheres, higher temperatures and more sophisticated spark plasma sintering (field assisted sintering technology) have been introduced. The wider use of computation and simulation has resulted in reductions in the cost and processing time, and has also improved product quality. The overall result is the acceptance of PM as a low-cost alternative by the automotive industry.

17.2 Powder metallurgy parts

The use of PM in automotive vehicles such as cars, buses, trucks, vans, and so on can be broadly classified into transmissions, engines, chases and other components. The potential areas of application for PM in a car are

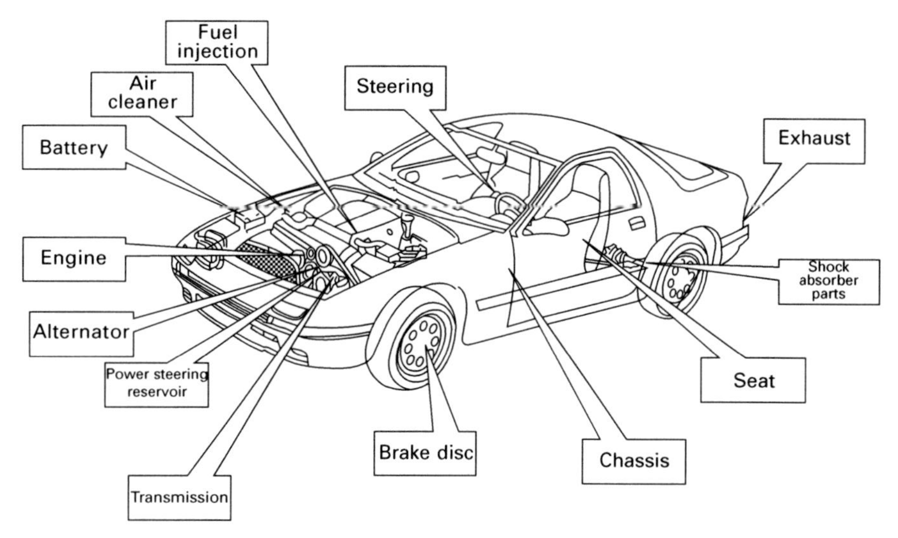

17.1 Powder metallurgy application areas in a car.

17.2 Oil pump rotors and gears.

shown in Fig. 17.1. Some PM parts such as bearings and gears (shown in Fig. 17.2) are common to all vehicles, although their size, strength, geometry and function may vary depending upon their specific application. The mass production of scooters and motorcycles also uses PM parts; some of these are shown in Fig. 17.3. Detailed coverage of the automotive applications of PM has been reported by Hall and Mocarski (1985), Mocarski *et al.* (1989)

17.3 Scooter/motorcycle parts.

17.4 Gear box/transmission (clutch hub, gear shifting yokes etc).

and Fujiki (2001). However, continuous improvements and innovation are taking place in automobile production and PM is rising to the occasion to meet these challenges.

17.2.1 Transmissions

Automotive transmission can be manual, semiautomatic, automatic, or use variable valve timing. The parts used in manual transmissions are as follows: synchronizer hub, blocking ring, interlock sleeve, gear shift lever, synchronizer plate, clutch hub, clutch release hub, detent lever, guide plate, sleeve interlock and lever control. Some transmission parts are shown in Fig. 17.4. As automatic transmissions become more widely used, the demand for

PM parts will increase. In an automatic transmission, the following parts are used: vane-type and gear-type pumps, the converter turbine hub, clutch plates, the clutch hub, converter clutches, outer and inner races, lock-up converter parts, parking gears, kick-down levers and chain sprockets.

In a manual transmission used in passenger cars and trucks, synchronizer units make gear shifts smoother and easier for the driver. Sigl and Rasch (2009) have covered advances in components for the synchronization module in manual transmission gear boxes. The performance of the synchronizer unit is critical for the correct operation of the transmission as well as the comfort of the driver when changing gear. The unit contains the following components: shaft, bearings, a hub mounted on the input shaft and a sliding sleeve which transfers the torque from the shaft-hub assembly to the clutch gear. This sleeve is firmly attached to the clutch gear, which is firmly attached to the synchronization cone. A two-shaft manual transmission is shown in Fig. 17.5, and the key components of the synchronization system are shown in Fig. 17.6.

The clutch-gear-cone unit is welded to a freely spinning transmission gear. The sliding sleeve connects to the clutch gear; the free gear becomes locked to the input shaft and transfers the torque to the commuter transmission gear which is rigidly fixed to the output shaft. The synchronizing ring, in combination with the clutch gear cone, ensures that the synchronization of the rotational speeds between the input and output shafts is achieved before the sliding sleeve engages with the clutch gear. The synchronization process during a gear change requires the gears to accelerate or decelerate against the inertia of transmission; this is achieved using a synchronizer ring and a cone-type friction clutch. When the driver moves the gear lever, a shift fork slides the sliding sleeve with the clutch gear-transmission gear unit. The torque flow through the system begins when the sleeve engages with the

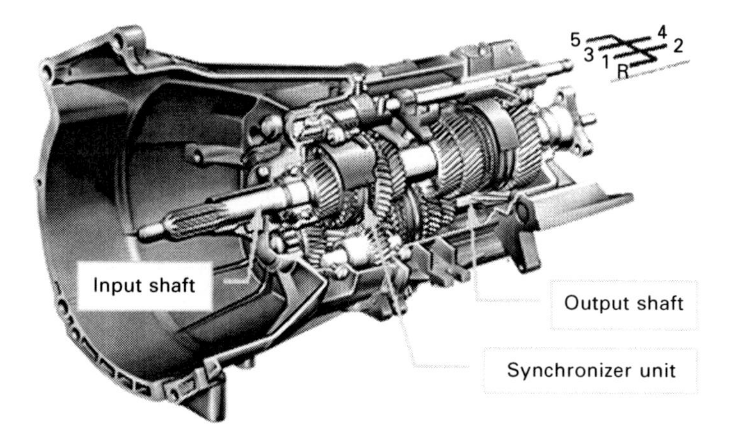

17.5 Two shaft manual transmission of a synchronization system.

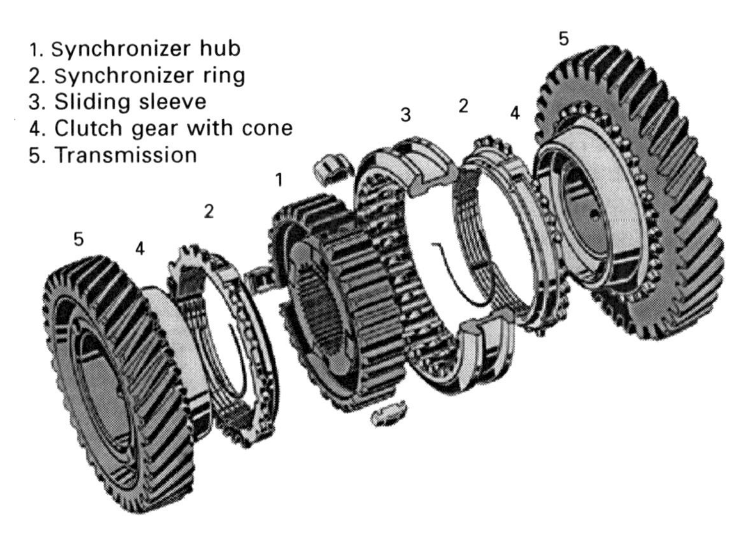

17.6 Key components of a synchronizer system.

clutch gear. In transmissions, the main object of a gear is the reliable and silent transmission of torque; hence gears in gear box applications require a high degree of geometric precision. There is considerable scope to increase the number of PM components in the synchronization module in manual transmission gear boxes. Arnold (2009) gives a survey of the complex design, material combinations, as well as unique solutions for manufacturing advanced products. Hubs, races, clutch plates and clutch plate carriers are all important components in automatic transmissions that are manufactured using PM. Individual PM parts are joined by brazing or welding with tight tolerances. The assembled parts and their carriers are shown in Fig. 17.7.

17.2.2 Engines

Automobile engines have undergone substantial changes, which have resulted in increased performance and reduced fuel consumption and environmental impact. Engines can be classified into different categories depending on the type of fuel used: gasoline, diesel, hybrid or fuel cell. A schematic view of a piston engine with PM parts is shown in Fig. 17.8. PM has played an important role in the development of new engines, either by improving existing components or by improving the design possibilities of components in newly developed systems, as discussed by Molins (2012). The use of PM components has improved engine performance, as well as reducing costs.

Cam shafts are traditionally forged from solid bars and are heavily machined. PM technology has allowed for the use of cams and intermediate rings on a hollow tube, as shown in Fig. 17.9, resulting in dimensional

17.7 Auto transmission carriers: (a) assembled components, (b) individual PM parts, (c) 2 piece carrier brazed, (d) 3 piece carrier welded.

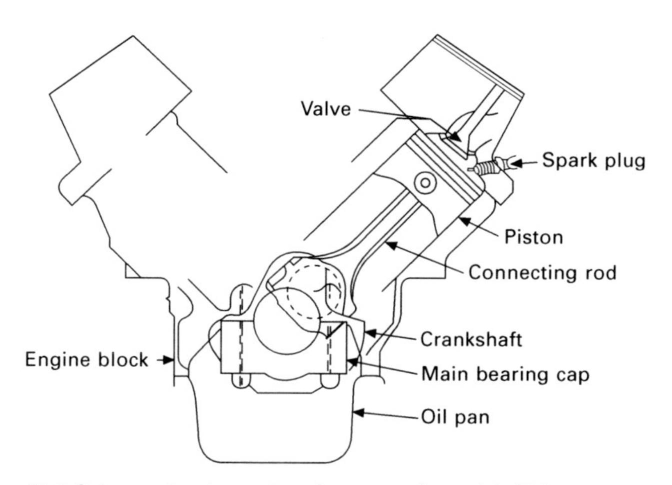

17.8 Schematic view of a piston engine with PM parts.

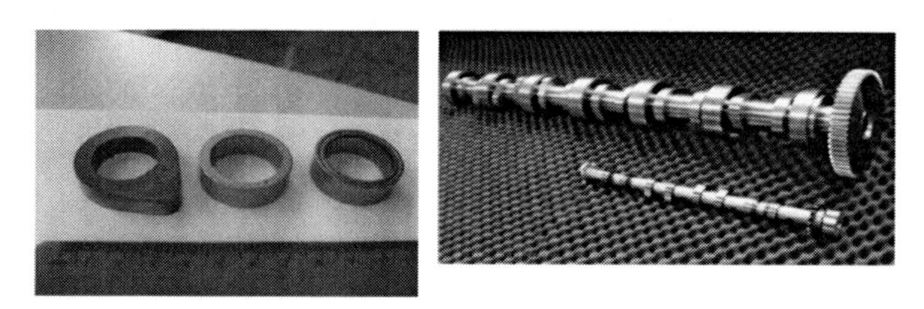

17.9 Composite cam shaft (courtesy of Thyssen).

precision, stringent wear resistance and a lighter assembly. Powder forged connecting rods, shown in Fig. 17.10, have much tighter tolerances and the introduction of fracture splitting of the rod's main bearing head has resulted in improved balance in the engines. Other parts include camshaft and crank shaft sprockets, timing gears, G rotor and gear type oil pumps, water pump impellers, pulleys, flange distributer gears, rocker arm fulcrums and inserts, double chain sprockets, governor weights, valve seat inserts, valve lifter guides, stainless steel throttle cam inserts, turbochargers and fuel injector parts.

The use of PM technology has also helped to reduce the weight of distribution pulleys, as shown in Fig. 17.11, and the transition from carburetor to gasoline

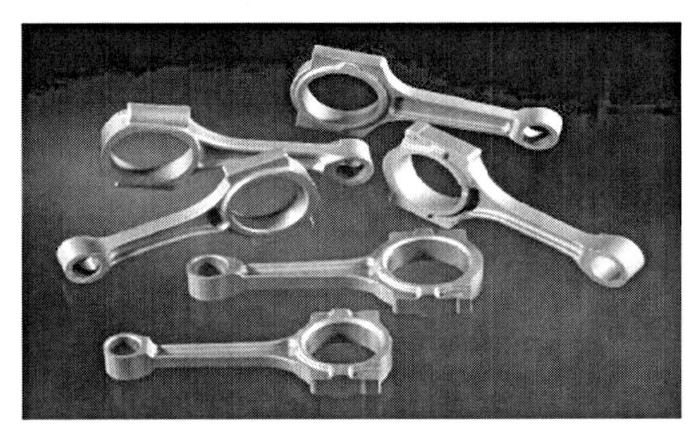

17.10 Powder forged connecting rods (courtesy of GKN).

17.11 Distribution pulleys.

fuel injection systems has resulted in a reduction in fuel consumption and emissions. Fuel injection pump parts are shown in Fig. 17.12. Soft PM magnets are used in the injectors, as shown in Fig. 17.13.

In exhaust gas recirculation systems, some of the exhaust gas is fed back into the combustion chamber to reduce the amount of NO*x* in the exhaust. This requires valves and actuators that can work at high temperatures in an environment containing aggressive combustion gases, for which PM technology is required. Some of these PM parts are shown in Fig. 17.14.

PM stainless steels and special bronzes are used as sliding bushes and valve bodies. Changes in combustion conditions and fuel composition have

17.12 Fuel injection pump parts.

17.13 Soft magnetic materials for gasoline injection system.

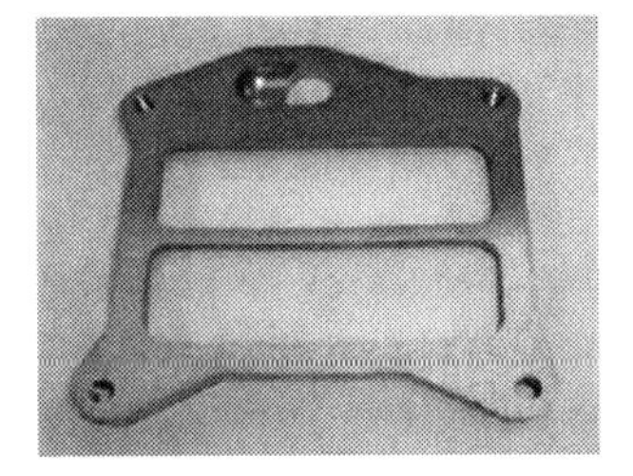

17.14 Parts for an exhaust gas circulation system.

17.15 Manifold flange and exhaust converter outlet flange.

caused an increase in the temperature of exhaust gases and the use of PM stainless steel flanges, shown in Fig. 17.15, has increased durability. In the engine, gasoline is converted to kinetic energy through combustion and a relatively large amount of heat is exhausted through emission gases. Efforts are being made to generate electricity from the exhaust heat of engines using thermoelectric converters (Fujiki, 2001), in order to increase the efficiency of engines. Many such thermoelectric devices are manufactured using PM.

The applications of metal injection molding (MIM) in the automotive industry are increasing, owing to the volume requirements of the industry. MIM parts are finding applications in the engine, transmission, body and other areas (Kato, 1998). Parts being manufactured using MIM include the engine valve seat, engine filter and shock absorber parts, as shown in Fig. 17.16.

17.2.3 Chases and others

These areas include steering, suspension, shock absorber parts, door lock parts, brake parts, and so on. Passenger cars with front-wheel drive and rack-

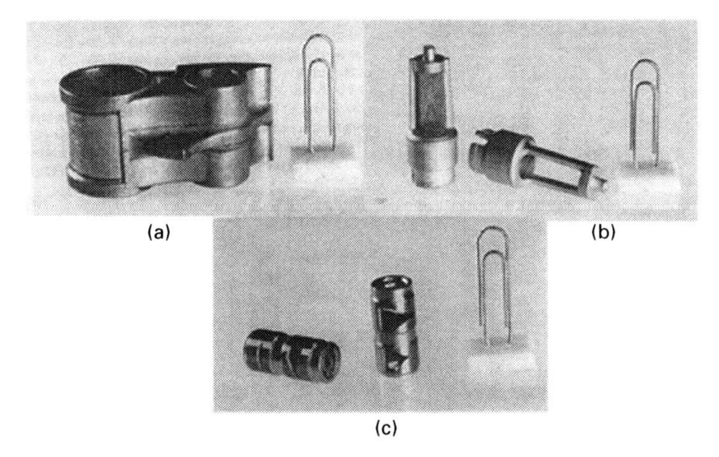

17.16 Engine and synchronizer stainless steel MIM parts: (a) engine valve seat, (b) engine filter (c) shock absorber parts.

17.17 Power steering parts.

and-pinion steering mechanisms have a PM steel bulkhead assembly, end plate and steering rack guide. Steering columns use a variety of ratchets to lock the tilting steering wheel in position. The steering column ignition key systems also use a PM ignition-key lock gear. In power steering mechanisms, PM technology is used for the power pump, which consists of a rotor, upper and lower end plates, an actuator and a cam insert. Some power steering parts are shown in Fig. 17.17. Shock absorber parts such as piston cylinder ends, strut rod guides and compression valves and seat reclining parts (see Figs 17.18 and 17.19) are also made using PM. An anti-lock braking system (ABS) using sensor rings, a sensor pad holder and friction materials such as brake pads is shown in Fig. 17.20 (Babu and Dheeraj Golla, 2008). Further developments in these areas are discussed in Sections 17.5 and 17.5.3.

17.18 Shock absorber parts.

17.19 Seat reclining parts.

17.20 Metallic brake pad (credit PMAI).

17.3 Materials

A variety of iron, steel, alloy steels, stainless steel, copper, bronze, aluminium and titanium alloys and composites are used for automotive components, in addition to electrical, magnetic and electronic materials. The general trend is to reduce weight, increase strength, improve the net shape, utilize heat treatment and special surface treatments and improve precision in transportation.

17.3.1 Iron and steel

In the ferrous category Fe, Fe–C, Fe–Cu, Fe–Cu–C and Fe–Mo–Cu–Ni–C (low Mo and Ni and high Mo and Ni) are used for structural applications such as camshaft and crankshaft sprockets, connecting rods, synchronizer rings, bearing caps, oil pump gears, and so on. The expensive Mo, Ni and Cu elements are being substituted by cheaper Cr, Mn and Si with high temperature sintering. Thus Fe–Mn–C, Fe–Mn–Si–C, Fe–Si–Cr–Mo–C (low Mo) and Fe–Si–Cr–Mo–Ni–C (low Ni) compositions have been developed to reduce the cost of the components. For wear-resistant applications such as valve seats, valve guides, cam shafts, Fe–Cr–Mo–W–V–C, Fe–Cu–Sn–P–C and Fe–Cr–Ni–Mo–P–C compositions are used. Potential applications of stainless steels such as 304 L, 430 L, 17-4PH include engine valve seats, Fe–Cr–Mn–Si materials, shock absorber parts, filters for hydraulic systems in engines, manifold flanges and exhaust converter outlet flanges, and exhaust gas recirculation systems.

17.3.2 Aluminium, titanium and others

Aluminium alloy parts are used to reduce the weight and increase the fuel efficiency of the vehicle. Examples include: aluminium camshaft-bearing caps, pulleys, oil pump rotors, cylinder liners and shock absorber parts. Al–Si, Al–Cu–Mg, Al–Zn–Mg–Cu, Al–Cu–Si–Mg and Al–Fe–Ce alloys are used for these applications. In the press–sinter–sizing process, elemental powders are mixed, or aluminium powders are mixed with master alloy powders of the desired composition, and the liquid phase sintered to produce components with the desired properties. Further improvements in properties can be achieved through rapid solidification of the powders and then processing by hot pressing, extrusion, forging or hot isostatic pressing to form components such as cylinder liners, rotors and vanes for automotive air conditioners and oil pump rotors (Delarbre and Krehl, 2000; Hunt, 2000). Selig and Doman (2011) studied finite element simulations of compaction of aluminium-based PM gears. Aluminium-based metal matrix composites utilizing SiC, Al_2O_3 or B_4C particles will provide improved wear resistance and high cycle fatigue resistance at moderately elevated temperatures. Another method used to improve properties is through reinforcement with whiskers or fibres made from C, SiC or other materials. Some of the cast aluminium and ferrous materials in valve train components can be replaced with aluminium PM parts. Aluminium metallic foams, which are produced by blending metal powders with foaming agents and heat treating them, will result in products with a porosity of 40–80%. The cellular structures produced will provide a high stiffness-to-weight ratio and absorb high amounts of energy during compressive deformation. They are therefore attractive for use in lightweight and energy absorption applications.

Titanium is another metal with low density, high strength and good corrosion resistance. The feasibility of producing titanium parts for automotive applications such as inlet valves, timing pulleys, fuel pump filters, screws and nuts has been established for Ti-6Al-4V using elemental and alloy powders (Ivasishin *et al.*, 2002; Pattanayak *et al.*, 2007). Using PM titanium in power train components allows the manufacturer to reduce the weight of components, particularly in reciprocating applications if produced at a lower cost than the conventional titanium route. This could contribute to the production of increasingly energy-efficient vehicles.

Copper, alloys such as bronze and brass, and the bearing alloys, Cu–Sn, Cu–Zn, Cu–Pb–Sn, and Fe–Cu–Pb–C are used in large numbers for self-lubricating brush and sleeve bearings, filters, brake bands, linings, clutch segments, balance shaft mechanisms, locks and keys. Cu–C materials are used in motors, starters, governors, voltage regulators, circuit protectors, relays and contact brushes. They are also used in electrical control systems such as headlights, fog lights, window lifters, wipers and electromagnetic vibrators such as buzzers and horns. Electrical contact materials include: copper-, silver- and tungsten-based alloys, multi-component systems combining precious metals such as platinum, palladium, gold and silver with other metals such as Cu, Ni, Mo or W, and silver-based composites such as Ag–CdO, Ag–SnO, Ag–C (Joshi *et al.*, 1998). Figure 17.21 shows a schematic view of a car indicating typical contact applications. Other electrical, magnetic and electronic materials are covered in the section on functional materials.

17.3.3. Nanomaterials

Materials dealing with atomic and molecular levels, with lengths ranging from about 1–100 nm, can provide new properties and functions. These materials have shown improved mechanical, electrical, optical, chemical, magnetic and electronic properties. There are several applications in the automotive industry for nanotechnology, including lightweight construction, powertrains, energy conversion, pollution reduction, sensors and the reduction of wear in driving dynamics (Presting and Konig, 2003). Nano-based light alloys and composites have improved mechanical properties, as demonstrated by Melnyk *et al.* (2011), so they can be used for higher performance applications.

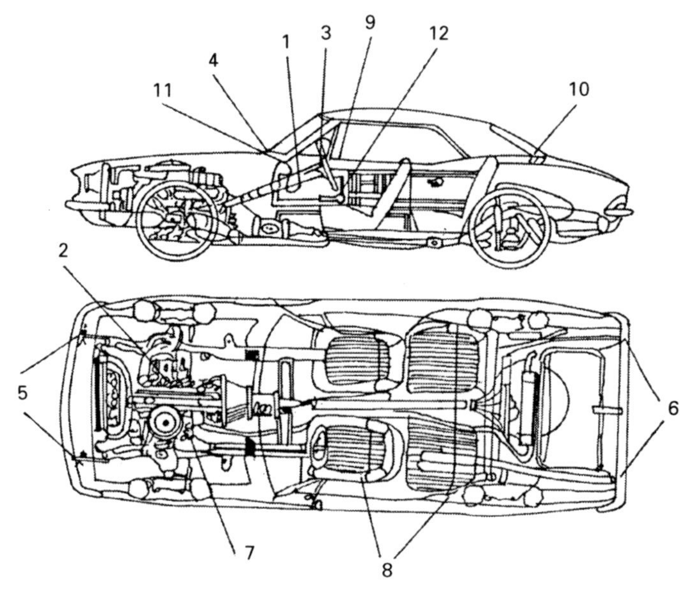

1. Motor starter
2. Voltage regulator
3. Horn
4. Wiper
5. Headlight
6. Tail lamps
7. Buzzer
8. Safety belts
9. Window lifters
10. Rear window defogger
11. Heating and airconditioning system
12. Door locking

17.21 Schematic view of a car indicating typical contact applications.

While the tensile strength of a conventional aluminium alloy, Al 7075, is of the order of 517 MPa, that of the Al 7075–Al_2O_3 nanocomposite is 710 MPa and that of the Al–Mg–SiC nitrogen cryomilled nanocomposite is 758 MPa. In the automotive industry, uses include: nanocomposites for structural reinforcement and safety; nanoparticle catalysts for fuel economy and nano-additives for applications such as lubricants and self-cleaning, anti-fogging, anti-abrasion, anti-corrosion and self-repairing coatings.

17.4 Innovative powder metallurgy products

There are several outstanding applications demonstrating precision, complexity, performance, economy and innovation that can be obtained from PM materials and components, some of these products are described in this section. Figure 17.22 shows an oil-impregnated bronze pivot bushing used in the manual transmission of a light truck, which has resulted in considerable cost savings.

The aluminium camshaft-bearing parts with complex designs, shown in Fig. 17.23, represent an evolution in the manufacturing of aluminium PM

17.22 Bronze oil impregnated pivot bushing (credit MPIF).

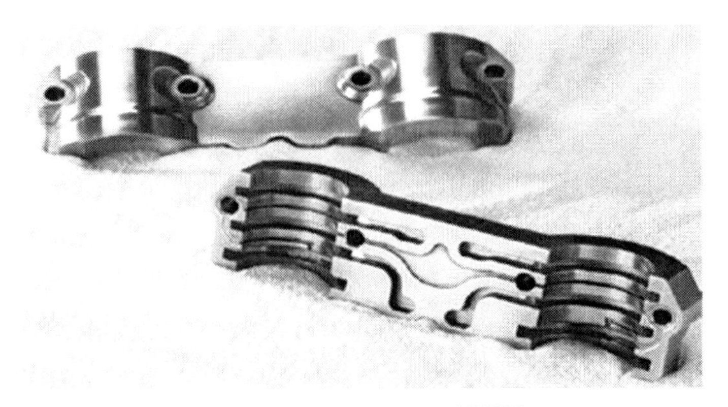

17.23 Aluminium cam caps (courtesy of MPP).

parts and have resulted in considerable weight savings compared to the earlier wrought steel parts used in engines. Another example is the aluminium sprocket and rotor weighing 450 g (Fig. 17.24) used in automotive cam phaser applications; this can be compared to the earlier use of sintered iron weighing 900 g.

In addition to weight savings, the rotating inertias are also reduced, which improves the dynamic response; this results in improved engine performance and efficiency. The variable valve timing system (Fig. 17.25) contains three complex parts: a vane rotor, sprocket and thrust plate. The use of PM in

17.24 Aluminium sprocket and rotors (credit EPMA).

17.25 Variable valve timing (credit MPIF).

this system offers substantial cost savings and results in improved engine efficiency and power over a wide range of engine revolutions.

Figure 17.26 is a forward clutch hub, a torque transmission part for forward and backward movement in a chain-type, continuously variable transmission. The component has complicated features such as a large thin rim with a long outer spline, and a long boss with inner and outer splines. Using the PM process resulted in both a lower production cost and improved strength properties. A complex PM steel crankshaft sprocket used in engines is shown in Fig. 17.27. The multilevel part has resulted in significant cost savings. The planetary carrier and one-way clutch assembly used in automotive transmissions (Fig.

17.26 Forward clutch hub for CVT systems (credit JPMA).

17.27 Complex PM steel crankshaft sprocket for auto engine (credit MPIF).

17.28) consists of three-piece, sinter-brazed planetary carriers and is made using both PM copper steel and sinter-hardened materials assembled with a single pressed cam plate, which is also sinter hardened. The PM assembly is replaced by a two-piece riveted planetary carrier, attached by a spline interface to a full one-way clutch assembly; this resulted in cost savings of more than 25%.

Another automotive transmission (Fig. 17.29) is a carrier and one-way rocker clutch assembly containing five PM steel parts; its application provided cost savings of about 20% when compared to other competitive processes. The differential bearing adjuster (Fig. 17.30) used in automobile chassis is a steel part, which preloads the bearing and is locked in place through the side holes. Using a PM part to replace the cast component (which required extensive machining) resulted in considerable cost savings. Figure 17.31 shows the intake sprocket and exhaust gear used in a coupling assembly operating in diesel engines. PM steel gears replaced the machined wrought steel gears, resulting in highly efficient material utilization and energy conservation.

17.5 Emerging trends

The automotive industry is undergoing several transformations to improve sustainability, which are resulting in changing technology. The technologies explored in eco-cars, such as battery hybrid, plug-in-hybrid, fuel cell hybrid and fully functioning electric vehicles, will meet the demand for both improved

17.28 Planetary carrier one-way clutch assembly for transmission (credit MPIF).

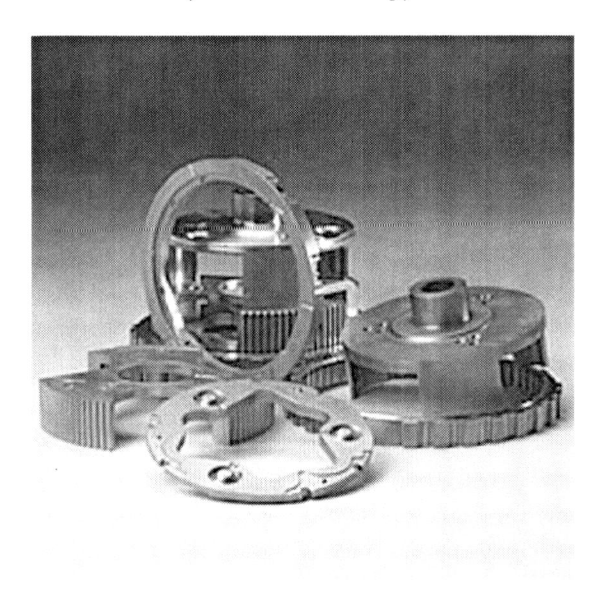

17.29 Carrier and one-way rocker clutch assembly for automatic transmission (credit MPIF).

17.30 Differential bearing adjuster for auto chassis (credit MPIF).

energy efficiency and substantial reductions in emissions. The fuels approved for use in eco-cars are ethanol, biodiesel, compressed hydrogen gas and energy carrier electricity. To address climate change and rising oil prices, major automotive industries across the world are engaged in research and development of eco-cars. The global car fleet is expected to grow from 800

17.31 Intake sprocket gear and exhaust gear for diesel engine (credit MPIF).

million vehicles in 2012 to 1.6 billion by 2030. It is expected that by 2025, sales of new hybrids will reach about 40% and sales of electric vehicles will rise to 10% of total car sales. There will also be a shift towards the Asian market with the demand for smaller cars. In the USA, car manufacturers will be required to achieve a new corporate average fuel economy (CAFE) of 87 km per gallon for all cars by 2025. Current CAFE standards call for 57 km per gallon by 2016. Other factors include the European Green Cars Initiative, the 2020 target of 95 km per gallon for passenger cars, and the long-term 2030 perspective. These targets will force car manufacturers to overhaul their approach to vehicle design and engine and transmission design, using lighter materials. Some of the design features include variable valve timing (VVT) and high-speed transmissions such as six- and eight-speed gear boxes, which provide more kilometers per gallon over a wide range of speeds. The traditional hydraulic VVT system will not be able to operate with the desired precision immediately after the engine starts, owing to viscous oil flow. The electric VVT system can operate with precision due to the use of a motor, which directly adjusts the valve timing even at low engine revolutions and low engine temperatures.

In principle, the electric car will be a threat to the PM industry, as it uses fewer moving parts than internal combustion engines and drive lines, which contain hundreds of parts. However, automotive original equipment manufacturers (OEMs) of electric and hybrid vehicles have announced that there is no plan to switch transmissions from mechanical to electrical systems (see Capus, 2011). Hybrid vehicles could use conventional transmissions such as standard manual or dual clutch transmissions. Furthermore, the electric engine may not replace the heavy lifting carried out by trucks. Again, there are several opportunities for PM in electric and hybrid vehicles. Electromagnetic parts will increasingly be used in such drives and the availability of extra power will allow the use of electromagnetic actuators, sensors and controls instead of mechanical and hydraulic alternatives.

17.5.1 Denser, stronger and lighter parts

There is continuous demand from the automotive industry for denser and stronger components, such as clutch race assemblies, engine transmission systems, connecting rods, gears and sprockets. These requirements are met using new alloy designs, higher compacting pressures and sintering temperatures and techniques such as liquid phase sintering, double pressing and double sintering, warm compaction and powder forging. The higher densities of 7.3–7.5 g cm^{-3} (95% theoretical densities), with spherical porosities and martensitic microstructures, could provide substantial improvements in mechanical properties. In selective densification, components are fully densified only where necessary, while the core remains at the sintered density.

Endurance limits in both the bending fatigue and surface fatigue of high-density 1.5 Cr–0.2 Mo surface-rolled sintered steel gears completely match those of case-carburized wrought-steel gears made from typical Cr–Mo steel (Takemasu *et al.*, 2010). Sprockets made of Fe–O.5 Cr–0.5Mn–0.5Mo steel powder, using compaction and high-temperature sintering at 1270°C to a density of 7.0 g cm^{-3}, have been selectively densified to 7.8 g cm^{-3} through working by rolling. A wide variety of gears, helical gears and sprockets with exceptional properties are manufactured cost effectively by selective densification. For warm compaction, special powder mixes have been developed using iron, steel, and low-alloy steel powders with optimized lubricant content. The powders are heated to 130°C and pressed in the tooling, which is heated to 150°C (Engstrom *et al.*, 2002). Since a higher green strength can be achieved using warm compaction, these products can be machined if necessary, thereby reducing the manufacturing cost. Considerable improvements in tensile strength, yield strength, impact strength and fatigue strength are possible through the use of warm compaction, in addition to reduced density variations within a compact, which are important for gears and synchronizing steel parts. Typical examples of parts that can be manufactured using warm compaction are gears, synchronizing rings, synchronizing latch cones and complex multilevel torque converter hubs. Further developments in this area allow the manufacture of larger parts (up to 40 mm high) and complex shapes, by heating the tools to 83°C without powder heating (warm die compaction). This results in densities of the order of 7.5 g cm^{-3} after sintering at 1120°C (Narasimhan, 2012).

Reducing the weight of automobile parts by using aluminium and titanium will result in improved fuel economy and lower emissions. The conventional press-and-sinter routes have been used for Al–2.5Cu–0.5Mg–14Si pre-alloyed powders, followed by shot-blasting to eliminate surface porosity. These aluminium chain sprockets are used by BMW, Germany, in camphaser systems (Pohl, 2006).

Another approach is to use cold isostatic pressing followed by pressure sintering. These new approaches open up additional opportunities for PM aluminium in the automotive industry. Although titanium PM components are expensive, Toyota's Altezza car uses a PM Ti–6Al–4V inlet valve and a TiB/Ti–Al–Zr–Sn–Mo–Si exhaust valve. PM offers an alternative to wrought titanium fabrication by reducing the number of processing steps.

17.5.2 More complex shapes

A complex shape like a synchronizer hub for a passenger car is mass produced in the net shape by a proper selection of raw materials, compacting press, tooling and stages of operation. A high-alloy powder containing Fe, Ni, Cu and Mo, together with carbon, is either diffusion bonded or glued and pressed in a sophisticated press and tooling system where it is possible to transfer the powder before compaction for uniform density. The powder is then compacted in three levels with the use of six punches together with die movements. All these operations should be synchronized so that the final density distribution is uniform and the green product is free from cracks. These parts are manufactured in CNC automatic presses, where the machines are operated with closed-loop controlled multi-platen tools systems, with optimum programming of their compacting sequence. This ensures the reproducible fabrication of the parts. Another capability of the PM process is the assembly of the different parts using sinter bonding to produce multifunctional structural parts (Eckart *et al.*, 2002).

Complex clutch plate carriers for automatic transmissions are manufactured using brazing or welding, as reported earlier and shown in Fig. 17.7. Other methods of manufacturing complex components include sinter bonding of a complex valve plate channel for a compressor, sintering the lobes of an assembled camshaft (see Fig. 17.9) and powder-forging a one-way clutch racer for caged rollers. A variety of gear shift cams for use in gearboxes for two-wheeled vehicles like scooters and motorcycles are produced economically by co-sintering. Complex shapes can also be produced using cold and hot isostatic pressing in flexible molds and hot isostatic pressing can also be used for further densification of sintered and MIM parts.

MIM will enable the production of complex shapes, which it is not possible to produce using rigid die pressing. The process involves hot mixing of powders of the required characteristics with a binder and granulates to form the feedstock, which is molded in a similar manner to plastic injection moulding. The molded parts are then subjected to debinding and sintering to form the finished parts. Some stainless steel MIM parts are shown in Fig. 17.16. Other applications include complex rocker arm components, fuel injectors, transmission components for transverse gear synchronization and parts for lock systems (Whittaker, 2007). Automotive engine manufacturers, faced with more stringent emission regulations and demands for increased efficiency, are hoping to reduce engine weight in order to improve efficiency. Turbo-charging technology has given diesel engines an enormous boost and manufacturers are turning their attention to gasoline engines. Diesel turbochargers operate at 850°C; gasoline-powered vehicles have to withstand temperatures of 1000°C. Turbocharger parts such as vanes, adjustment rings, roller, and other parts in the exhaust gas recirculation, have been made from nickel-based super alloys using MIM. Some components used in exhaust gas recirculation systems are shown in Fig. 17.14. Compared to investment-cast, nickel-based superalloy turbocharger components, MIM parts provide superior mechanical properties with regard to yield strength at elevated temperatures (Kern *et al.*, 2011).

17.5.3 Functional parts

Functional parts such as filters, sensors, actuators and electrical and magnetic components are made using PM. Electric motors are used in the following systems: seat positioning, window lifts, door mirror positioning, engine cooling fans, air conditioning, anti-lock braking systems, anti-skid systems, speakers, radios, CD players, electrical steering systems and braking systems. A few hundred magnets cover all these applications. The use of a 42 volt system, higher than the 12 volt system currently in use, will allow full electric steering to replace the current heavy hydraulic-assisted rams and engine-driven hydraulic pumps. The extra performance will provide opportunities to improve combustion and fuel efficiency and reduce emissions.

Traditionally Alnico (Al–Ni–Co) alloys are used in sensors and instrument applications. The increasing cost of cobalt has resulted in the use of other materials. Sintered ferrites in DC motors are more widely used. For higher strength magnetic applications, samarium–cobalt alloys are used. The development of cheaper sintered neodymium–iron–boron magnets has resulted in the miniaturization of components, which has in turn resulted in improvements in efficiency.

Nd–Fe–B magnets are also processed by bonding and injection molding, thereby avoiding the machining necessary to produce complex parts, and are replacing sintered ferrite segments in motors. The development of sintered soft magnetic materials from insulated iron powder is a relatively recent advance and is finding applications in automotive sensors and actuators (Hultman,

2010). The development of soft magnetic composite (SMC) material has led to materials with improved performance, including lower losses, higher permeability and increased mechanical strength. The combination of high magnetic saturation, fast responses and cost-effective production of compact three-dimensional shapes makes SMC technology very attractive for fast-switching actuator applications.

17.6 Conclusions

Powder metallurgy is a sustainable manufacturing technology for the economical mass production of precision engineered components using a wide variety of materials. It is increasingly attracting the attention of the automotive industry. Many of the developments in PM applications have emerged from collaborations of automotive designers with materials, manufacturing and value engineers.

The automotive industry has undergone substantial changes in terms of increased performance, reduced fuel consumption and environmental impact and improved safety. PM has played an important role in these developments, both by improving existing components and by allowing the introduction of innovative components to new systems. The introduction of eco-cars could be seen as a threat, but is providing opportunities for the use of PM in the development of new products in the areas of functional electromagnetic controls and drives.

17.7 References

Arnold V (2009), 'Future potential of powder metallurgy applications in automotive industry', *Powder metallurgy for automotive and engineering industries*, Ramakrishnan P (ed), Narosa Publishing House Pvt. Ltd., New Delhi, 33–44.

ASM International (1998), *Powder Metal Technologies and Applications, Handbook*, Volume 7, Materials Park, Ohio.

Babu G and Dheeraj G (2008), 'Sintered metallic disc brake for automobiles', *Trans PMAI*, **34**, 43–44.

Capus J (2011), 'Powder metallurgy, progress and the eco friendly car', *Metal Powder Rep*, **2**, 16–18.

Delarbre P and Krehl M (2000), *Powder Metallurgy Aluminium and Light Alloys for Automotive Applications*, W F Jondeska, Jr and R A Chernenkoff (eds), Metal Powder Industries Federation, New Jersey, 33–39.

Eckart S, Dollmeier K, Byrd K and Neubert H (2002), 'An industry overview on applied powder metal joining techniques', *Powder Metallurgy in Automotive Applications-11*, Rama Mohan T R R and Ramakrishnan P (eds), Oxford and IBH Publishing, New Delhi, 63–72.

Engstrom U, Johansson B and Senad D (2002), 'Experience with warm compaction of densmix powders in the production of complex parts', *Powder Metallurgy in Automotive Applications-11*, Rama Mohan T R and Ramakrishnan P (eds), Oxford and IBH Publications, New Delhi, 135–142.

European Powder Metallurgy Association (EPMA) (2008), *Powder Metallurgy. The Processes and its Products*, Shewsbury, UK.
Fujiki A (2001), 'Present state and future prospects of powder metallurgy parts for automotive applications', *Mater Chem Phys*, **67**, 298–306.
Hall H W and Mocarski S (1985), 'PM automotive applications', *Int J Powder Metall*, **21**(2), 79–109.
Hinzmann G and Sterkenburg D (2007), 'High density multilevel PM components by high velocity comp action', *Powder Metallurgy: Processing for Automotive, Electrical, Electronic and Engineering Industry*, Ramakrishnan P (ed), New Age International Publishers, New Delhi, 63–67.
Hultman (2010), 'Advanced magnetic materials and their applications. Advances in SMC technology–materials and applications', *Trans PMAI*, **36**, 50–58.
Hunt Jr. (2000), 'New directions in aluminium based PM materials for automotive applications, *Int J Powder Metall*, **36**(6), 51–60.
Ivasishin O M, Sarvakin D G, Moxson V S, Bondareval K A and Froes F H (2002), 'Titanium powder metallurgy for automotive components', *Mater Technol*, **17**(1), 20–25.
Joshi P B, Prabha S Krishnan, Patel R H, Gadgeel and Ramakrishnan P, (1998), 'New silver-zinc oxide DC contacts for automotive applications', *Powder Metallurgy in Automotive Applications*, Ramakrishnan P (ed), Oxford and IBH Publishers, New Delhi, 231–240.
Kato Y (1998), 'Ultrafine powder for metal injection molding of automotive parts', Ramakrishnan P (ed), *Powder Metallurgy for Automotive Applications*, Oxford and IBH Publishers, New Delhi, 295–303.
Kern A, Blomacher M, Johan T and Artnd T (2011), 'Manufacturers mull prospects of turbo boost for gasoline', *Metal Powder Rep*, **2**, 22–26.
Melnyk C, Weinstein B, Lujan D, Grant D and Gansert R (2011), 'Production of nano-based high alloys and composites for aerospace applications', *Adv Mat Process*, **169**(5), 42–44.
Metal Powder Industries Federation (MPIF) (1998), *Powder Metallurgy Design Manual*, MPIF, New Jersey.
Metal Powder Industries Federation (MPIF) (2008), *Powder Metallurgy The preferred metal forming solutions*, MPIF, New Jersey.
Mocarski S, William Hall D, Joginder K and Sang-Kee Suh (1989) 'Parts for automotive applications–Part III', *Int J Powder Metall*, **25**(2), 103–124.
Molins C (2012), 'Developments in PM engine components', *Powder Metallurgy for Automotive and High Performance Engineering Industries*, Ramakrishnan P (ed), New Age International Publishers, New Delhi, 23–35.
Narasimhan K S (2012), 'New products and processes and the global growth of powder metallurgy', *Powder Metallurgy for Automotive and High Performance Materials in Engineering Industries*, Ramakrishnan P (ed), New Age International Publishers, New Delhi, 1–13.
Pattanayak D K, Panigrahi B B, Rama Mohan T R, Godkhindi M M, Dabhade V V and Ramakrishnan P (2007), 'Titanium and nano crystalline titanium for the automotive industry', *Powder Metallurgy Processing for Automotive, Electrical, Electronic and Engineering Industry*, Ramakrishnan P (ed), New Age International Publishers, New Delhi, 111–125.
Pohl A (2006), 'PM aluminium drives forward in new BMW break through', *Metal Powder Rep*, **2**, 13–15.

Presting H and Konig U (2003), 'Future nano technology developments for automotive applications', *Mater Sci Eng C*, **23**, 737–741.

Ramakrishnan P (2002), 'Residual stresses in powder metal processing', *Handbook of Residual Stresses and Deformation of Steel*, Totten G, Howes M and Inoue T (eds), ASM International, Materials Park, Ohio, 397–423.

Selig S and Doman D A (2011), 'Finite element simulation of the compaction of aluminium based PM gears', *Int J Powder Metall*, **47**(4), 9–10.

Sigl L S and Rasch C (2009), 'PM components for synchronizer systems', *Powder Metallurgy for Automotive and High Performance Engineering Industries*, Ramakrishnan P (ed), Narosa Publishing House Pvt. Ltd, New Delhi, 45–56.

Takemasu T, Takao K, Yoshinoba T, Toshinaka S and Akihiro S (2010) 'Performance of surface rolled PM steels for automotive gears', *Trans PMAI*, **36**, 66–69.

Whittaker (2007), 'Powder injection moulding looks to automove applications for growth and stability', *Powder Injection Moulding Int*, **1**(2), 14–22.

18
Applications of powder metallurgy in biomaterials

M. BRAM, Institute of Energy and Climate Research, Germany, T. EBEL and M. WOLFF, Institute of Materials Research, Germany, A.P. CYSNE BARBOSA, Universidade Federal do Rio Grande do Norte, Brazil and N. TUNCER, Anadolu University, Turkey

DOI: 10.1533/9780857098900.4.520

Abstract: Powder metallurgy (PM) of biomaterials is still a niche market, but considerable progress in related manufacturing technologies opens up the possibility of participating in the emerging market for medical devices and surgical implants within the next decade. PM technologies like metal injection moulding (MIM) are promising manufacturing routes if large quantities of complex-shaped parts are required. In addition, porous implants or coatings that improve implant fixation by bone ingrowth are preferentially made by PM technologies. In this chapter, the most promising PM routes for biomedical applications are introduced. Challenges and specific properties of implant materials are discussed, which were made starting from titanium, magnesium or Nitinol powders. The potential of PM is demonstrated in four case studies.

Key words: case studies, functional porosity, green machining, magnesium, metal injection moulding (MIM), Nitinol, space holder materials, titanium.

18.1 Introduction

Metallic materials like 316L steel, CoCr alloys, titanium and titanium alloys are established for biomedical applications owing to their biocompatibility and outstanding combination of strength and ductility. Therefore, they are preferentially applied for load-bearing implants provided for joint replacement, trauma or dentistry. In most cases, metallic implants are manufactured by ingot metallurgy, which includes casting semi-finished parts with subsequent cold or hot working to improve the microstructure and mechanical properties and finally net-shaping by mechanical machining. Nevertheless, powder metallurgy (PM) is an attractive alternative under economic considerations if a large number of complex-shaped parts is required or to avoid secondary operations like extensive joining processes. Furthermore, PM might become advantageous for metallic biomaterials like titanium, magnesium or Nitinol, which are difficult to machine by conventional methods. Last but not least, it offers the possibility of manufacturing bone implants with well-defined surface roughness or functional porosity, combining improved implant fixation due to bone ingrowth and adaptation of the elastic properties to the human bone reducing the risk of stress shielding.

Nevertheless, PM has rarely been used for the net-shape manufacturing of implants and biomedical devices so far. One of the main reasons for this is the fact that in PM manufacturing considerable effort is required to achieve parts with mechanical properties that are competitive with parts made by established ingot metallurgy. Ductility and fatigue resistance are reduced by the residual microporosity resulting from the sintering process as well as the uptake of impurities like oxygen and carbon, which are introduced during processing by the contact of metal powders with organic binders or a sintering atmosphere. These impurities may lead to embrittlement of the materials owing to solid–solution hardening or precipitation of secondary phases.

In this chapter, state of the art of PM manufacturing of net-shaped biomedical implants is summarised. We focus on titanium, titanium alloys, magnesium and Nitinol, which are exceptionally susceptible to impurity uptake during PM processing. Solutions are provided for confronting the challenges of closing the gap between PM and ingot metallurgy. Secondly, PM processes are presented, which are used to achieve well-defined surface roughness or functional porosities. Four case studies of biomedical implants are given, which demonstrate the potential of PM processing for biomedical applications.

18.2 Challenges of powder metallurgy biomaterials

18.2.1 Titanium and titanium alloys

Titanium and titanium alloys are relatively young and expensive structural materials, used in off-shore industry, chemical industry, aerospace and automotive industries as well as for luxury goods and medical devices and implants. For medical purposes, advantage is taken of all the specific properties of titanium: high specific strength, high corrosion resistance, low stiffness and excellent biocompatibility. The lower elastic modulus compared to surgical steel is beneficial for reduction of the so-called stress-shielding effect in bone implants. This effect is related to the mismatch between the elastic modulus of bone (about 1–20 GPa) and that of steel (around 200 GPa) or titanium (around 100 GPa), which increases the risk of loosening the implant. Bone tissue is degraded because most of the mechanical load is concentrated on the stiff implant.

Obviously, in all application areas excellent corrosion resistance is an important issue for usage. It is mainly caused by the high affinity of titanium for oxygen which effects spontaneous formation of a dense oxide layer at the surface. This ceramic-like surface is the reason for the high corrosion resistance and outstanding biocompatibility (Panigrahi *et al.* 2005). On the other hand, the great affinity for oxygen and other elements, like carbon and nitrogen, is mainly responsible for the high costs of raw material and the challenges of further processing. The Kroll process is a multi-step, energy-intensive technique to produce metallic titanium from the stable oxide ore. During all subsequent processes and regardless of whether a melt or powder is involved, avoidance of contamination by oxygen, carbon and nitrogen is essential. Titanium features high solubility for these elements connected with a strong change in the mechanical properties even if small amounts are involved. From the point of view of solution hardening oxygen, nitrogen and carbon have the same effect, but with different potency. Therefore it is sensible to combine these elements in the form of an oxygen equivalent O_{eq}. A commonly used equation is (Conrad, 1966):

$$O_{eq} = c_O + 2\,c_N + 0.75\,c_C \qquad [18.1]$$

where c_O, c_N and c_C represent the concentration of oxygen, nitrogen and carbon, respectively. Because of the strong dependence of the mechanical properties on the O, N and C content, pure titanium is classified according to the specific amount of interstitials. Besides these elements only iron and hydrogen are listed and limited explicitly in the standards. As an example, Table 18.1 shows the values for the maximum content of oxygen according to ASTM standards B348-02 and F1295 for wrought bars and billets and its influence on tensile properties. In relation to powder metallurgy it is obvious that the protection against atmosphere during processing of fine powder with a comparably huge surface is an essential issue in order to avoid embrittlement. In Section 18.3.1 it will be shown that even carbon could be picked up during PM processing, when additional substances like polymeric binders are involved.

Table 18.1 Limit of oxygen content and tensile properties of different titanium-based materials according to ASTM standards B348-02 (Ti, Ti-6Al-4V) and F1295 (Ti-6Al-7Nb)

	Max oxygen content (wt%)	Min yield strength (MPa)	Min ultimate tensile strength (MPa)	Min ε_f (%)
Ti Grade 1	0.18	170	240	24
Ti Grade 2	0.25	275	345	20
Ti Grade 3	0.35	380	450	18
Ti Grade 4	0.40	483	550	15
Ti-6Al-4V Grade 5	0.20	828	895	10
Ti-6Al-4V Grade 23	0.13	759	828	10
Ti-6Al-7Nb	0.20	800	900	10

Furthermore, Table 18.1 shows the alloys, which are mostly used for medical devices. These are commercially pure (CP) titanium, denoted as Grades 1 to 4 and the alloy Ti-6Al-4V, which is available in two grades that essentially differ in the oxygen limit. For medical purposes Grade 23 is preferred, also specified in the ASTM F167 standard for surgical alloys. Ti-6Al-7Nb has nearly the same mechanical properties as Ti-6Al-4V Grade 5, but the toxic element vanadium is replaced by the biocompatible niobium. Release of Al and V ions from biomedical implants is suspected to be associated with long-term health problems, such as Alzheimer disease, neuropathy and osteomalacia (Singh *et al.*, 2010; Bahrami Nasab and Hassan, 2010). However, Ti-6Al-4V is still the most frequently used alloy for medical implants. A lot of research is done on so-called metastable beta-titanium alloys, which provide a lower elastic modulus down to 50–60 GPa. Titanium features two possible crystal structures: the hexagonal-closed-packaged alpha phase which is stable at temperatures below the beta transus at 885°C and the body-centred-cubic beta phase at higher temperatures. Alpha-beta alloys like Ti-6Al-4V show both phases even at room temperature, while in metastable beta-alloys mainly the beta phase is contained at room temperature (Leyens and Peters, 2003). This phase can be stabilised by biocompatible elements like molybdenum, zirconium or niobium, while aluminium can be omitted. However, they are not in practical clinical use yet. Another attempt to reduce the elastic modulus is by the introduction of functional porosity (Baril *et al.*, 2011; Ebel *et al.*, 2010) as introduced in more depth in Section 18.4.

Very roughly, the available powders can be divided into pure and expensive powders and comparably unclean and low cost powders. In this context, the purity relates especially to the oxygen and carbon content, but also to residuals like chlorides, calcium, sodium or magnesium from the specific production process. Furthermore, these powders also differ in geometry: the expensive ones are more spherical, the lower cost powders are typically irregular shaped and often agglomerated. Current techniques for the production of more expensive powders are inert gas atomisation (Hohmann and Jönsson, 1990) and plasma rotating electrode (REP) processing (Aller and Losada, 1990). Because alloys can be used as starting materials, pre-alloyed powder can be made by these techniques too. Lower cost powders are made by mechanical milling of sponge, scrap or ingots, for example as done in the HDH process (hydride–dehydride) (McCracken *et al.*, 2010). Much research effort has been devoted to the direct reduction of titanium oxide without the Kroll process mainly aiming at lowering the costs. However, at the moment HDH powders are mostly used as an economic alternative, if sufficient properties can be achieved after sintering. Generally, HDH powders show an enhanced content of oxygen. Nevertheless, even with HDH powders, low oxygen content might be achieved after sintering if hydrided powder is still used for PM processing. The hydrogen escapes during sintering and reacts with possible oxygen in the binder, powder and furnace atmosphere.

Summing up, PM of titanium requires careful consideration of contamination by oxygen, nitrogen and carbon owing to their strong influence on the mechanical properties. A variety of titanium materials are used in medicine and all show excellent biocompatibility. Titanium powders of these materials are commercially available.

18.2.2 Magnesium and magnesium alloys

In addition to their use as lightweight material, magnesium alloys are promising candidates for prospective surgical implants considering their excellent biodegradable properties (Staiger *et al.*, 2006; Witte *et al.*, 2007). The introduction of functional porosity would support the primary fixation and the degradation of magnesium implants by enabling the ingrowth of bone cells (osseointegration) into the degrading implant (Witte *et al.*, 2006) and, *vice versa*, corrosion products of magnesium, generated during biodegradation, support osteoconductivity of the bone (Janning *et al.*, 2010). Generally, magnesium alloys provide elastic modules and strengths matching those of bone tissue, reducing the risk of stress-shielding to a minimum (Poumarat and Squire, 1993).

The corrosion behaviour of magnesium in the body fluid permits well-controlled degradation of the material. Nevertheless, the corrosion rate of most magnesium-based alloys is still too high for biodegradable implants. In order to develop magnesium-based alloys with adapted corrosion rates, stability and biocompatibility, tailoring the base material with non-toxic alloying elements has to be performed. Recent studies have identified calcium and some rare earth elements as suitable candidates. On the other hand tailoring the material by applying functional surface coatings is another promising approach for adjusting the degradation rate.

PM technologies enable net-shape manufacturing of magnesium implants. Prospective MIM processing starting from magnesium-based powders is currently under development in order to manufacture small-sized and sophisticated biodegradable implants of a specific design with high reproducibility (Wolff *et al.*, 2010). Optionally, functional porosity can be introduced by adapting the particle size of the starting powders and sintering conditions. Furthermore, powder metallurgy allows flexibility of alloy design by blending pure magnesium powder with different elemental or master alloy powders. In this case, commercially available gas atomised pure magnesium powder with a particle size below 45 μm can be used as starting material. Because of the high affinity of magnesium for oxygen, all handling of powders and specimens, as well as subsequent sintering, has to be performed under a protective argon atmosphere. Shielding magnesium parts against interstitial impurities like oxygen during sintering is essential. This goal can be achieved by using getter material and a labyrinth-like crucible configuration (Wolff *et al.*, 2010). Under these conditions, mechanical properties matching those of cast material or cortical bone can be achieved. Depending on the sintering conditions, residual porosities can be varied between 2% and 45%. When approaching porosities near to 45%, interconnection of pores arises.

18.2.3 Nitinol alloys

The importance of shape memory alloys (SMA) for biomedical applications has increased significantly in the past few years. These alloys can return to their original shape even after they have been deformed well beyond the elastic deformation limit. Pseudoelastic shape recovery is the basic principle of self-expanding stents for treatment of embolisms. If shape recovery is inhibited, a well-defined, permanent load remains. The function of orthodontic wires is based on this effect. The shape memory effect is most pronounced in Nitinol alloys, which are based on the intermetallic phase NiTi (Massalski, 1996). These alloys have the highest technical relevance today. The production of complex-shaped components from these alloys has so far been limited by the fact that machining them is very difficult and expensive (Wu *et al.*, 1999). Because of the shape memory effect, the tools used for machining wear quickly and must be replaced at frequent intervals. In addition, the low workability makes producing semi-finished products from NiTi so costly even by means of ingot metallurgy that a loss of material in the subsequent shaping process is hardly acceptable.

PM is an alternative production method which enables near-net-shaped components with complex geometries to be produced from Nitinol powders with little loss of material. Furthermore, the well-known problems of ingot metallurgy like segregation and grain growth can be easily avoided. Blends of elemental nickel and titanium powders, as well as pre-alloyed Nitinol powders, can be used as starting materials (Wen *et al.*, 2010). In the first case, an exothermic reaction between nickel and titanium can be initiated by thermal treatment of powder compacts (self-propagating high temperature synthesis SHS), which leads to the formation of an open porous structure caused by the formation of a partial liquid phase in combination with the Kirkendall effect (Igharo and Wood, 1985; Hey and Jardine, 1994; Li *et al.*, 1998, 2000; Assad *et al.*, 2003). Generally, the exothermic reaction is difficult to control, leaving secondary phases which do not contribute to the phase transformation. Nevertheless, this method is used for the industrial production of a spinal implant, as will be discussed later. Pre-alloyed Nitinol powders are preferred as starting materials to achieve an essentially one-phase homogeneous structure. These powders are produced by inert gas atomisation of Nitinol melts resulting in spherical powders. Up to now, such powders are not available on a commercial scale.

Again, MIM processing is the most promising method of producing near-net-shape components with complex geometries from these powders (Krone *et al.*, 2005). Optionally, application of space holder particles like NaCl enables the manufacture of highly porous Nitinol components even preferentially starting from pre-alloyed powders (Köhl *et al.*, 2009; 2011). Powder metallurgy of Nitinol alloys is strongly influenced by the uptake of oxygen and carbon during each processing step. In contrast to titanium alloys, the intermetallic NiTi phase has almost no solubility of oxygen and carbon. Instead, a brittle Ti_2Ni phase with oxygen on the interstitial sites and TiC precipitates are formed (Frenzel *et al.*, 2010). Because both precipitates are known to initiate cracks during mechanical loading (Krone *et al.*, 2005; Mentz *et al.*, 2006), the oxygen and carbon uptake during processing must be reduced to a minimum. Furthermore, formation of titanium-rich precipitates shifts the nickel content of the NiTi-matrix to lower values, which is coupled with a strong reduction of martensitic phase transformation temperature by approximately 10°C/0.1 at% nickel if the nickel content exceeds 50.5 at% (Frenzel *et al.*, 2010). In the case of pre-alloyed Nitinol powders, the sintering process is influenced by the low sintering activity of the intermetallic, highly ordered NiTi phase. Sintering temperatures up to 95% of the NiTi melting temperature of 1310°C (Massalski *et al.*, 1996) are required to achieve residual porosities below 10% (Krone *et al.*, 2005, Köhl *et al.*, 2009). Under these conditions, the application of suitable sintering aids becomes a challenging issue. Another more general challenge of Nitinol as an implant material is its high nickel content. The risk of allergic reactions in the case of ongoing nickel release from the implant surface is quite controversially discussed in literature (Shabalovskaya *et al.*, 2008).

18.3 Production of powder metallurgy biomaterials

18.3.1 Metal injection moulding

As introduced in Chapter 6, metal injection moulding, (MIM), represents an excellent technology for the economic production of components with complex geometries. The great freedom with regard to the shape of the part makes it ideally suitable for the production of anatomically formed implants and functionally optimised medical devices. Therefore tiny and complicated parts like brackets for teeth correction were among the first medical products produced by MIM. However, today stainless steel is mostly processed by MIM, although titanium has some advantages regarding weight, biocompatibility and MRI-compatibility (magnetic resonance imaging). It could even be used for permanent implants without restrictions.

MIM of titanium, magnesium or Nitinol is still not standard, even if most technical problems of processing have been solved during the last few years. The strong affinity of these materials for oxygen and carbon requires special efforts regarding raw materials, equipment and handling. However, if starting from adequate powders, these materials can be processed to components with excellent mechanical properties matching those of machined parts and fulfilling standard requirements. This is at least true with respect to the tensile properties. Regarding fatigue behaviour, the properties are still promising and more than sufficient for most applications, but constraints compared to commonly processed materials have to be taken into account. Apart from residual porosity the main constraint is the change in microstructure caused by the sintering process. The coarser microstructure compared to standard wrought material particularly decreases the endurance limit. Later on, an example will be given of the excellent fatigue properties that can be achieved in the case of titanium alloys by changing the composition slightly and applying simple PM methods.

A detailed introduction to the MIM process can be found by German (1997). The first step in the process is the homogenisation of the starting materials, where a binder material is added to the powder. Normally at least two binder components are employed. After the homogenisation in a heatable kneader, the feedstock material is obtained. The feedstock is afterwards injected into the cavity of a mould using an injection moulding machine. The facilities and tooling for MIM are similar to those used for traditional plastics injection moulding. The green part obtained is subjected to a partial debinding step, where the first component of the binder system is removed. Solvent debinding is mainly employed for biomedical MIM parts if made from titanium-based powders. Related debinding facilities are commonly available on the market and no special needs for titanium, magnesium or Nitinol MIM parts exist. Increases in solvent temperatures (Wu *et al.*, 2006), application of low molecular weight binders (Bakan 1998) as well as lower volumes to surface area ratios of the moulded part (Shivashankar and German, 1999) lead to a faster debinding rate. The second binder component is usually removed by thermal decomposition at elevated temperatures in the range 400–600°C.

The choice of the suitable binder system is the key for successful production of MIM parts from titanium, magnesium or Nitinol powders. Although not present in the final part, the binder is in tight contact with the particle surfaces during processing and can therefore act as a source of contamination. Especially during thermal debinding, contamination-sensitive powders are susceptible to oxygen and carbon uptake. Owing to the usually organic nature of the binder system, the presence of carbon in the chemical composition of the binder system cannot be avoided. The presence of oxygen, on the other hand, may be minimised by using oxygen-free binder components with reduced oxygen content. A mixture of paraffin, polyethylene (either alone or as copolymer with vinyl acetate) and stearic acid has been successfully applied to titanium and titanium alloy powders and leaves an acceptable low amount of contamination in the metallic matrix (Aust *et al.*, 2005; Bideaux *et al.*, 2007; Cysne Barbosa *et al.*, 2012).

In addition to thermal debinding, sintering is another critical step related to contamination by oxygen and carbon. This is due to the high temperatures involved, forcing the pick-up of interstitials. Sintering of titanium-based materials takes place at temperatures between 1200°C and 1500°C. Thus, a furnace providing high-vacuum or a protective gas atmosphere of high purity is an essential requirement. Titanium is also often sintered under hydrogen atmosphere in order to keep the oxygen level low. In addition, hydrogen can improve the microstructure of the sintered part due to changing the beta-transus temperature.

The MIM technique can be easily combined with the space holder method (SHM) if production of porous titanium and Nitinol parts is required (Köhl *et al.*, 2009). NaCl particles were proved to be suitable space holders for MIM, as they possess sufficient mechanical and thermal stability during the injection process, while leaving a minimum amount of contamination after its removal. In the feedstock composition, the bimodal particle size distribution of coarse spacer particles (for example 300–600 μm) and fine metal powders (for example < 25 μm) leads to a need for increased solid loadings compared to feedstock without spacer materials. Subsequent processing steps are the same as for traditional MIM with the exception of an additional desalination step after partial debinding. After removal of the space holder material, the parts are sintered. Depending on the space holder particles used, microstructures with well-defined pore sizes and porosities in the range of 30–70 vol% are achieved. Recently, prospective two-component MIM was successfully applied to the production of titanium spine implants with a stepwise gradient in porosity (Cysne Barbosa *et al.*, 2009, 2012).

18.3.2 Pressing, green machining and sintering

Conventional press-and-sinter method is a relatively easy and cost-effective PM processing route, with which the high material price can be justified. In the case of biomedical applications, this method is preferentially used to manufacture highly porous materials. Pores can be created either by loose powder sintering (Thieme *et al.*, 2001; Oh *et al.*, 2003) or by the use of suitable space holder materials (Bram *et al.*, 2000; Wen *et al.*, 2001; Esen and Bor, 2007; Hong *et al.*, 2008). While the adjustment of porosity, pore shape and pore sizes is limited in the case of loose powder sintering, these parameters can be simply and accurately controlled within a wide range by the use of temporary space holder materials.

For net-shape manufacturing of porous titanium implants, a new powder metallurgical production route was developed. First, semi-finished compacts are made by pressing space holder/titanium powder mixtures. Shaping these compacts is preferentially done by machining them in the unsintered state, also called 'green machining' (Laptev *et al.*, 2004). In the case of small and medium implant dimensions, green machining can be done without any addition of organic binders to the powder compact, minimising the risk of oxygen and carbon uptake during further processing. After the removal of the space holder, a sintering step at temperatures in the range 1200–1400°C leads to sufficient strength to allow the application of porous titanium for load-bearing implants. In principle, machining of highly porous parts is also possible in the sintered state, but plastic deformation and smearing of the sintered struts may lead to partial closure of surface pores (Bram *et al.*, 2003).

Space holders can be removed before or after the sintering step. Removal can be either thermally or by leaching out in a suitable solvent. An ideal space holder should leave the system at relatively low temperatures to prevent any reaction with the host material. This temperature should be generally below 400°C for titanium, for example. The desired properties for a space holder material are, thus, low boiling/sublimation point, or a high solubility in specific solvents. Most commonly used space holders include carbamide (urea), ammonium hydrogen carbonate and sodium chloride (NaCl), although spacers such as magnesium, polypropylene carbonate (PPC), polymethylmethacrylate (PMMA) were also reported to have been used successfully as space holder material. Besides the macropores created by temporary space holder particles, PM biomaterials also contain micropores inside of the sintered struts, which are related to residual porosity from the sintering process. Both types of porosity are useful in implant applications since microporosity (pore diameter less than 10 μm) allows body fluid circulation and improves cell adherence, whereas macroporosity (pore diameters in the range between 100 μm and several mm, space holder contents up to 80 vol%) provides a scaffold for bone–cell colonisation in combination with vascularisation of blood vessels (Wintermantel and Ha, 2002).

PM provides a high degree of freedom and renders it possible to control internal architecture of the produced foams. This high degree of freedom brings a very high number of processing parameters that affect structural and chemical properties simultaneously (Tuncer *et al.*, 2011a, 2011b). PM titanium implants have been shown to be promising for biomedical applications, such as intervertebral discs, dental implants, or acetabular hip prostheses (Laptev *et al.*, 2004; Imwinkelried, 2007; Schiefer *et al.*, 2009). A related case study is given later.

18.3.3 Replica method

The replication of reticulated polymer foams by coating with Ti or Ti-6Al-4V powder slurries is another approach to achieve uniform and isotropic structures with porosities >80 vol%, fully interconnected pores and mean open cell diameters in the range of 0.5–5 mm (Li *et al.*, 2002; Quadbeck *et al.*, 2011). Polyurethane (PU) foam is used as a starting material, which is infiltrated by a water-based titanium or Ti-6Al-4V powder slurry. The rheological properties of the slurry play an important role in the impregnation process. Viscosity must be adjusted in a way that the slurry might penetrate the foam easily and be sustained homogeneously on the PU struts afterwards. The shear thinning behaviour of the slurry was found to be advantageous in achieving this aim. Removal of excess slurry is supported by the application of rubber rolls. Applying pressure supports the formation of a well distributed coating on the struts. After drying, the PU template is removed by a thermal treatment below 600°C leaving triangular holes inside the struts. Additional powder coatings are optional to remove the triangular holes and to enhance the thickness of the struts for improved mechanical properties. The remaining network of titanium or Ti-6Al-4V particles is sintered in a vacuum at temperatures in the range 1200–1400°C for several hours. After shaping the foam by mechanical machining, it can be directly used as a scaffold for bone tissue engineering.

18.3.4 Selective laser melting/electron beam melting

Foaming and conventional PM technologies have limitations regarding production of highly ordered porous structures as well as accurate control of the internal pore shapes, diminishing the reproducibility of morphology and physical properties. Therefore, computer aided design (CAD)/computer aided manufacturing (CAM)-based layered manufacturing techniques were recently adopted for near-net-shape fabrication of porous parts with controlled porosity and pore structure. Selective laser melting (SLM) or electron beam melting (EBM) are two examples of additive manufacturing processes that use a high powered laser or electron beam to create three-dimensional (3D) metal parts based on a digital information source by fusing metallic powders together. The principle of the method is to melt each layer of metal powders to the exact geometry defined by a 3D CAD model. The process begins with coating a layer of metallic powders by a wiper on a metal plate. The laser or electron beam melts the powder on the focusing level of the powder in welding beads. The metal melt merges with the metal plate. After treatment on the focusing level, the metal plate is lowered and the wiper again lays on metallic powders. Then, the next layer of metallic powder in the focusing level of the beam is ready for processing (Wehmöller *et al.*, 2005; Van Bael *et al.*, 2011). Parameters like laser power, spot size, scan speed and hatch distance must be adjusted carefully to improve porosity and surface quality, as well as to decrease residual stress in the parts produced by SLM (Meier and Haberland, 2008). Figure 18.1 shows a typical highly porous and well ordered microstructure, which has been produced by EBM.

Comparing laser and electron beams, the latter has higher energy density, which leads to reduced building times and consequently reduced manufacturing costs (Parthasarathy *et al.*, 2010). Arcam's EBM process is currently in use for low volume production of medical components in Europe and the USA (Ola *et al.*, 2008). Additive manufacturing allows the building of parts with very complex geometries without any sort of tools or fixtures, and without producing any waste material. Both methods enable manufacture of patient-specific orthopaedic implants directly from computer generated models. As an example, patient-specific total knee implants from biocompatible alloys such as Ti-6Al-4V produced via EBM and SLM were reported to have suitable mechanical, structural and frictional properties for use as bone implants (Heinl *et al.*, 2008; Murr *et al.*, 2011; Biemond *et al.*, 2011). Wehmöller *et al.*, (2005) shows that it is possible to produce extremely complex-shaped implants such as cortical lower jaw or lumbar vertebrae. Furthermore, titanium dental implants with a porosity gradient through screw cross section were also reported, which were produced by SLM (Traini *et al.*, 2008).

18.3.5 Self-propagating high-temperature synthesis of porous NiTi

Within the last few decades, the exothermic reaction between elemental Ni and Ti powders has been intensely investigated with the aim of manufacturing

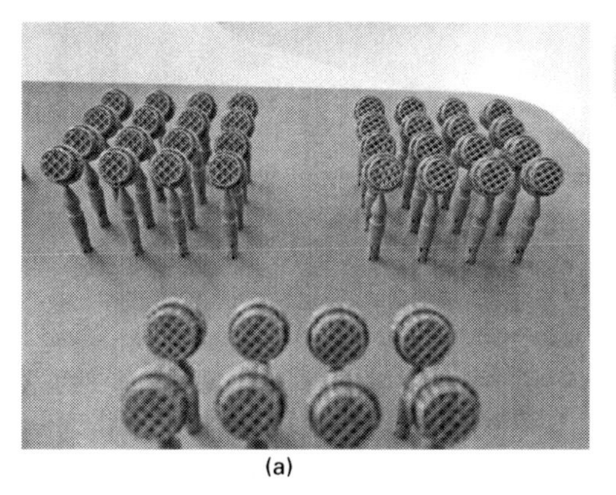

(a)

(b)

18.1 (a) Titanium mesh implants to treat bone defects (courtesy of Institute for Medical and Analytical Technologies, University of Applied Sciences Northwestern Switzerland) (b) Detail of titanium mesh, produced by EBM.

of highly porous NiTi parts to be used as scaffolds for bone tissue engineering (Li *et al.*, 2000; Biswas, 2005). The main course of the reaction is given by the following equation:

$$Ni + Ti \rightarrow NiTi + 67\ kJ\ mol^{-1} \qquad [18.2]$$

Depending on the processing conditions, the exothermic reaction may become self-sustaining (self-propagating high-temperature synthesis, SHS). Nickel and titanium elemental powders are mixed and consolidated into a green compact by uniaxial or isostatic pressing. Afterwards, the reaction is started by igniting the top of the sample, for example by induction heating followed by the formation of a reaction front running through the sample. The energy discharge during the exothermic reaction leads to a sudden temperature increase inside the compact caused by self-heating, which is coupled with the formation of high internal pressures. An explosive wave is generated for pressure release resulting in expansion of the sample and formation of interconnected and linear-aligned pore structures (Li *et al.*, 2000). The appearance of partial liquid phase supports pore formation by capillary forces. Typically, porosities in the range of 50–70 vol% and pore sizes in the range of 200–500 μm were achieved, which were found to be suitable for bone tissue ingrowth and vascularisation (Assad *et al.*, 2003). A drawback of the SHS method is the formation of secondary intermetallic phases like Ti_2Ni and Ni_3Ti, which increase the brittleness of the porous NiTi alloy and may cause deterioration of the biocompatibility by inducing corrosion processes in the physiological environment. Shaping must be done after completion of the SHS process. A porous spine implant, which was launched on the market in 2002, demonstrates the potential of the method (Assad *et al.*, 2003), even if the high nickel contents are thought to be critical for long-term implants (Shabalovskaya *et al.*, 2008).

18.3.6 Porous coatings

Introduction of classical PM technologies for biomedical applications started in the 1980s, when coarse titanium or CoCr beads were pressed and sintered on the surface of bone implants to improve their fixation by bone ingrowth (Clemow *et al.*, 1981). These kinds of implants are still on the market, but the risk of inflammation of the surrounding tissue in the case of loosening single beads under cyclic loading conditions has decreased their importance compared to other porous implants. Another drawback of the high temperature sintering process (temperatures >1200°C) required to achieve a stable coating is the coarsening of the implant microstructure, which might reduce the strength and fatigue resistance of the implant material clearly.

Another PM process for surface modification of biomedical implants is vacuum plasma spraying (VPS) of titanium powders. VPS coatings have been well established for the stem of hip and knee prostheses as well as for coating the outer shell of acetabular cups for several decades (Hahn and Palich, 1970). A detailed description of the VPS method is given by Fauchais (2004). Titanium powder is injected into the plasma gas stream produced by an electric arc. The powder is accelerated to high speed, melted or partially melted and impacted onto the implant surface. Adherence of the VPS layer to the implant surface can be improved by enhancing the surface roughness, for example by sand blasting. In contrast to sintering of beads, substrate materials retain their mechanical properties after the coating procedure owing to a low increase in substrate temperature during VPS coating. Varying degrees of porosity can be achieved if processing parameters are adjusted for this purpose. Special care must be taken to avoid the formation of impurity phases if oxidation sensitive powders like titanium are sprayed (Ryan *et al.*, 2006). Rather than producing open-cell porosity, the technique often delivers isolated or closed-cell pores. Although bone ingrowth into such a structure is not possible, attachment of the surrounding bone tissue to the rough implant surface ('ongrowth') is clearly improved compared to a non-treated surface. Stable anchoring of implants is demonstrated by *in vivo* tests.

18.4 Specific properties of powdered titanium and titanium alloy biomaterials

Generally, commercially pure titanium and its alloys combine good mechanical properties that avoid stress shielding, high corrosion and wear resistance, osseointegration, non-toxicity and do not cause any inflammatory or allergic reactions in the human body (Geetha *et al.*, 2009). Therefore, they are established materials for bone scaffold applications (Singh *et al.*, 2010). In the case of PM manufacturing of titanium and its alloys, for example by MIM, the properties are additionally influenced by the occurrence of residual porosity in the range of 2–4% after sintering. Usually, the pores are closed, isolated and typically smaller than 10 μm. In the case of mechanical loading, these pores may act as notches in the sintered microstructure causing deterioration especially of the tensile and fatigue properties (Ebel *et al.*, 2010). Thus, if maximum strength and ductility is required, for example in the case of bone plates, screws or other implants used for bone support after fracture, dense materials have to be targeted. To achieve this aim while still maintaining the advantages of MIM processing for net-shape-manufacturing, subsequent hot isostatic pressing (HIP) is recommended after sintering. Owing to the superposition of isostatic pressure during high temperature treatment, all pores that are not connected to the implant surface, will be completely eliminated. In Table 18.2 an overview is given of the mechanical properties achieved by MIM of Ti-6Al-4V powders. The values are taken from several specific studies (Ferri *et al.*, 2009, 2010; Obasi *et al.*, 2010), but reveal typical values,

Table 18.2 Typical mechanical data of MIM processed Ti-6Al-4V compared to ASTM standards

Material	Residual porosity (%)	Yield strength (MPa)	Ultimate tensile strength (MPa)	ε_f (%)	Endurance limit (MPa)
Ti-6Al-4V, sintered at 1350°C	3.6	720	824	14	450
Ti-6Al-4V, sintered at 1400°C	3.3	744	852	15	n.a.
Ti-6Al-4V +HIP	0.0	841	937	17	500
Ti-6Al-4V-0.5B	2.3	787	902	12	640
ASTM B348 grade 23	n.a.	> 760	> 825	> 10	n.a.
ASTM B348 grade 5	n.a.	> 828	> 895	> 10	n.a.
ASTM F2885-11, sintered	< 4.0	> 680	> 780	> 10	n.a.
ASTM F2885-11, densified	< 2.0	> 830	> 900	> 10	n.a.

if residual porosity and oxygen content are kept within a certain range, as discussed below. The experimental data are compared to standards for wrought and MIM processed Ti-6Al-4V. In ASTM F2885-11, which is valid for MIM processed titanium alloys, two cases are distinguished: the sintered state and the densified state. In the latter case, almost theoretical density is achieved, for example by conducting a HIP cycle after sintering.

It is obvious that the properties of samples fabricated by MIM do not fulfil the requirements of the ASTM B348 standard as long as porosity is present. However, after densification by HIP, the tensile properties become equal to those of wrought material. This dependence on the residual porosity is reflected too in the distinction between the sintered and densified state within the ASTM F2885 standard for MIM fabricated material. On the other hand, even in the as-sintered state, MIM components already show high strength and excellent ductility, although their microstructure differs from wrought material not only in terms of existing porosity, but also in grain morphology and grain size. In Fig. 18.2 typical micrographs of wrought and of MIM processed Ti-6Al-4V alloy are shown. The wrought material exhibits a globular structure consisting of equiaxed alpha grains with beta-phase at the grain boundary regions. In contrast, the MIM processed material exhibits a lamellar structure with alpha-beta colonies. This structure is formed during cooling after sintering when the temperature falls below the beta transus temperature. In addition, the grain size is significantly enlarged caused by sintering within the single phase beta-region. In the case of the wrought material, thermomechanical treatment including recrystallisation provides small grains in the final microstructure. These differences explain the deviations in the mechanical properties of sintered and wrought material. The lamellar structure is beneficial with regard to crack growth processes, but strength and ductility are generally improved by provision of a fine globular microstructure (Lütjering, 1998).

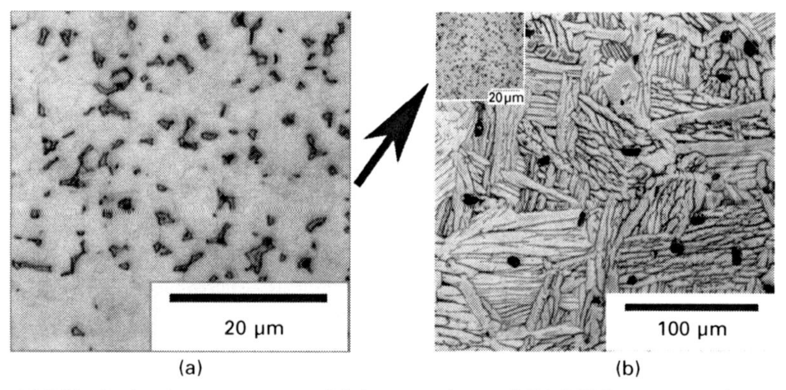

18.2 Typical microstructure of (a) wrought and (b) MIM processed Ti-6Al-4V.

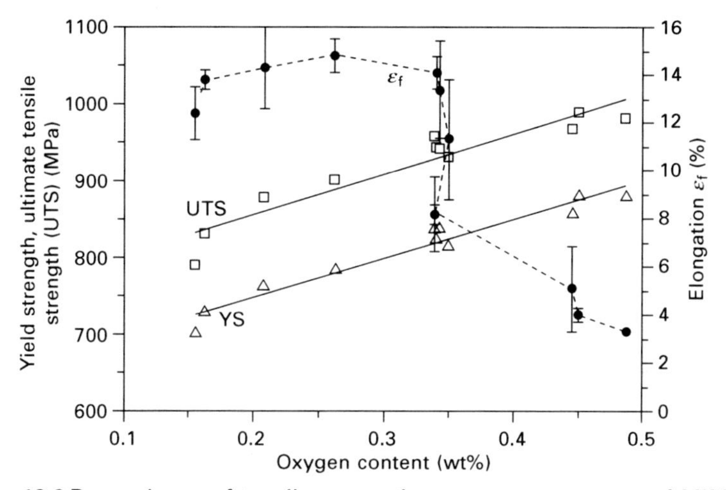

18.3 Dependence of tensile properties on oxygen content of MIM processed Ti-6Al-4V.

A further difference is the content of oxygen in the material. During MIM processing oxygen is taken up by the titanium and influences the mechanical properties. Figure 18.3 shows the dependence of the tensile properties on the oxygen content (Ebel *et al.*, 2011). In this study, samples were investigated which differed only in the oxygen content, ranging from 0.15–0.5 wt%. While the strength continuously increased with increasing oxygen content, a sharp decrease in the elongation was found at about 0.35 wt%. This limit must not be exceeded during processing. Interestingly, the maximum content stated in ASTM F2885-11 is only 0.2 wt%, equal to that of ASTM B348 for wrought material, although, at least for the tensile behaviour higher oxygen contents appear to be beneficial. The samples, whose experimental data are shown in Table 18.2, revealed oxygen contents between 0.19 wt% and 0.23 wt%.

Table 18.2 shows some samples, with an endurance limit at 10^7 cycles, under four-point bending test conditions in order to compare the fatigue properties of MIM processed Ti-6Al-4V in the as-sintered state, after densification by HIP and after alloying with small amounts with boron. A value of 400 MPa for as-sintered material is found, but wrought material shows typical fatigue strength in the range between 500 MPa and 800 MPa. Even closing the porosity by a HIP process does not improve the endurance limit by a satisfactory amount. Ferri *et al.* (2011) proved that the comparatively large grain size of MIM material mainly caused inferior fatigue behaviour. In order to achieve finer microstructures, elemental boron powder was added to the alloy powder during feedstock production. During heating to sintering temperature, boron reacts with titanium forming the intermetallic TiB phase. During sintering in the beta region the growth of the beta grains is impeded due to grain boundary pinning by these titanium boride particles. In addition, during cooling, the alpha phase nucleates at the particles. This leads to more and smaller alpha grains and the grain size is reduced from 130 μm to 18 μm, comparable to wrought material. Table 18.2 shows a strong effect on the endurance limit which is well within the range of wrought material after grain refinement. This proves the flexibility of powder metallurgy in optimising the material quite easily according to the specific demands of the required application.

The potential of PM technologies in biomedical implants is demonstrated by the fact that MIM of Ti-6Al-4V is already in use in the production of permanent implants, but limited to components stressed by low mechanical loads. In Section 18.6.1, a port system is shown as a case study, which has been implanted very successfully for several years without any problems including in terms of biocompatibility.

Highly porous titanium structures are promising candidates for bone implants, improving the implant fixation by bone ingrowth and reducing the risk of stress-shielding to a minimum owing to elastic properties that are well adapted to the human bone. On the other hand, it must be considered that the introduction of macropores also drastically reduces the mechanical strength. Therefore, this kind of implant should be preferentially used for implants, which primarily have to withstand compression loads. In the case of macroporous implants, the architecture and interconnectivity of the pores are crucial to allow vascularisation and for supplying nutrients to the developing tissue. Studies have led to the consensus that the optimal pore/interconnection radii for bone ingrowth is above 100 μm, but the maximum pore size required for stable implant fixation is still discussed controversially ranging from a few 100 μm up to several mm (Bobyn *et al.*, 1980; Whang *et al.*, 1999; Jones *et al.*, 2009).

Another important issue directly related to the size of the open pores is the roughness of the implant surface. It influences the friction between bone and implant, which is of great importance for the primary fixation of the implant after surgery. Figure 18.4 shows the microstructure of representative titanium foam typically used for biomedical implants. The foam was produced by pressing a titanium powder-space holder mixture (space holder urea, average size 400 μm, content 70 vol%), a solution of the space holder particles in water and subsequent sintering at 1200°C in argon (Tuncer and Arslan, 2009; Tuncer *et al.*, 2011a). X-ray microtomography imaging of the foam allows calculation of the amount of macro-and micropores as well as the size of pores and the interconnecting radii (Tuncer *et al.*, 2011a).

As mentioned before, contamination by interstitial elements, especially oxygen and carbon is a very important issue during processing. This is even more significant when porous biomaterials are considered, as they have a high specific surface area ready to capture interstitials during processing (Baril *et al.*, 2011; Tuncer *et al.*, 2011a, 2011b). The fact that the desired structural and mechanical properties of the implant strongly depend on substituted bone, the age and daily activity of the patient, necessitates a detailed understanding of the structure–property relationships in Ti foams. Porosity is the most effective foam property in mechanical behaviour (Figure 18.5). Strength and stiffness decrease as the porosity of the foam increases (Imwinkelried, 2007).

Elastic properties that depend on the amount of porosity can be calculated using a simple model proposed by Gibson and Ashby (1997), which is based on unit cells with a constant wall thickness. The scaling law in Equation (18.3) assumes a simple strut bending model applied to a cubic unit cell:

$$\frac{E}{E_s} = C_1\left(\frac{\rho}{\rho_s}\right)^{n_1} \qquad [18.3]$$

$$\frac{\sigma}{\sigma_s} = C_2\left(\frac{\rho}{\rho_s}\right)^{n_2} \qquad [18.4]$$

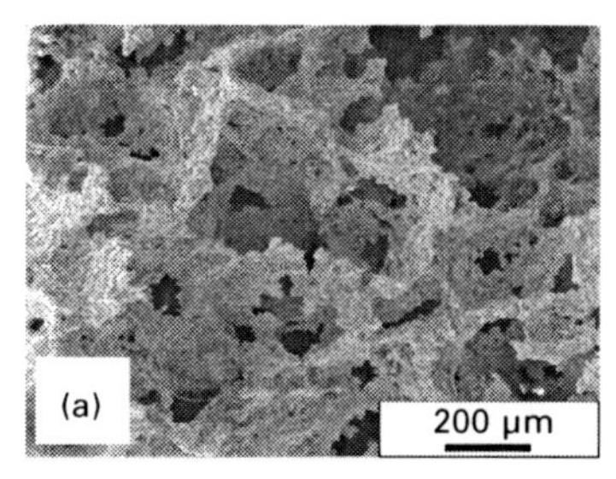

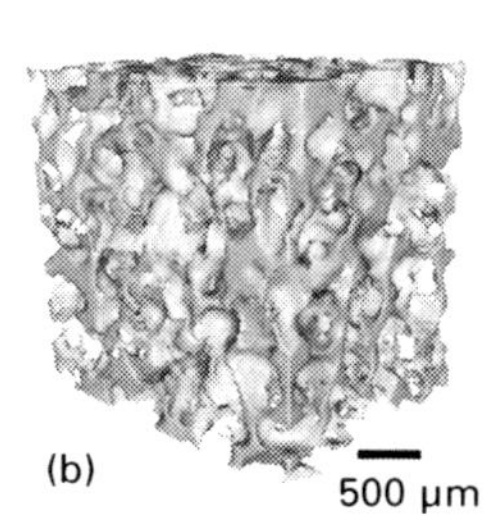

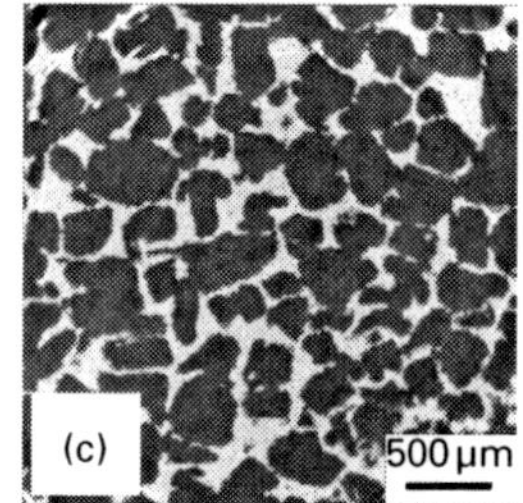

18.4 Microstructure of titanium foam suitable for bone implant applications produced with a 70 vol% space holder. (a) Fractured surface, (b) X-ray microtomography, (c) polished cross section.

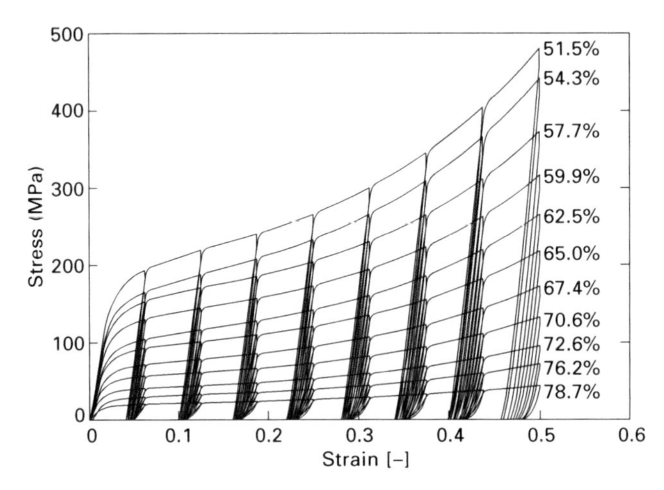

18.5 Stress–strain curves obtained by static compression of titanium foam samples (sample size ∅ 16 mm, h = 16 mm) depending on the porosity. At 1, 2, 3, 4, 5, 6 and 7 mm displacement, loading/unloading cycles were conducted seven times for each of the 11 samples at strains between 0.05 (displacement 1 mm) and 0.45 (displacement 7 mm). After maximum displacement of 8 mm (strain 0.5), all samples of the plastically deformed cylinders were unloaded and demounted. Percentages on the right hand side give the porosity of each individual sample. (courtesy of Synthes GmbH, Oberdorf, Switzerland).

In these equations, E, σ and ρ refer to elastic modulus, compressive strength and relative density of the foam, respectively, whereas subscript s stands for the bulk material property. C and n are constants depending on the geometrical features like the cell structure of the foam. Experimental evidence showed that n equals 2 for the elastic modulus and 1.5 for compressive strength. C is a constant of 1 for rigid polymers, elastomers, metals and glasses. However, there is a certain discrepancy between the predicted and experimental results which arises from underestimation of the real architectural effects (Tuncer *et al.*, 2011a, 2011b). Foam features such as pore aspect ratio, cell wall porosity, material homogeneity, pore size distribution and pore orientation have been shown clearly to affect the relationship between porosity and mechanical properties.

18.5 Specific properties of other powder metallurgy biomaterials

18.5.1 Magnesium and magnesium alloys

Magnesium and magnesium alloys have attracted great interest as materials for biodegradable implants in recent years. PM manufacturing of net-shaped magnesium implants, for example by MIM processing, is a promising method of establishing this material for biomedical applications. PM of magnesium is strongly influenced by its high affinity for oxygen, but in contrast to other metals with a high affinity for oxygen, for example titanium, the matrix does not show any solubility for oxygen atoms. As a consequence, oxygen uptake from the surrounding atmosphere leads to the formation of a thermodynamically stable MgO-layer with a thickness in the range 3–5 nm on the surface of the powder particles. This oxide layer drastically inhibits the diffusion processes required for densification of the material during sintering. To overcome this problem, several approaches were used (Wolff *et al.*, 2010). When pure magnesium powders were used as starting materials, improved sintering behaviour was found, when protecting magnesium powder compacts by embedding them in loose magnesium powders acting as the getter material. Additionally, a labyrinth-like crucible configuration was used. Under these conditions, residual porosity of approximately 15% was achieved, starting from commercial gas atomised magnesium powders and sintering at 630°C for 64 h in an argon atmosphere. The samples showed a compression stress of 170 MPa and a Young´s modulus of 8 GPa, approaching the mechanical properties of the cortical bone. Optionally, the interconnected porosity could serve for drug delivery or could be beneficial for cell attachment and bone ingrowth.

An improved sintering density was achieved when calcium was added as the alloying element. Considering the Mg–Ca phase diagram, a transient liquid phase is formed at low calcium levels (<1 wt%) when sintering is conducted at temperatures above 445°C (Massalski and Okamoto, 1990). In the compact, magnesium particle surfaces are wetted with this transient melt caused by the capillary action of the micropores. Then, MgO is removed by chemical interaction with calcium, resulting in a thermodynamically more stable CaO and intermetallic Mg_2Ca phase. After sintering in the temperature range 620–635°C for 16–64 h, both phases can be found as separated precipitates on the grain boundaries. Advantageously, these precipitates also pin the grain boundaries during sintering. If required, the metastable Mg_2Ca phase can be removed by an additional heat treatment with subsequent quenching. Calcium was added in the form of CaH_2 powders or Mg–Ca master alloys with calcium contents between 7 and 82 wt%. Alternatively, a pre-alloyed, gas atomised Mg–1Ca alloy powder was successfully used. Under the given conditions, the residual porosity of Mg–0.9Ca alloy was reduced to 3% after sintering. The average grain size was 23 μm. Ultimate compression strengths up to 325 MPa and yield strengths up to 90 MPa were achieved, demonstrating the potential of PM of magnesium alloys. An example of a related microstructure is given in Fig. 18.6(a).

If these results are transferred to MIM processing, interaction with the organic binder system will act as an additional source for the uptake of

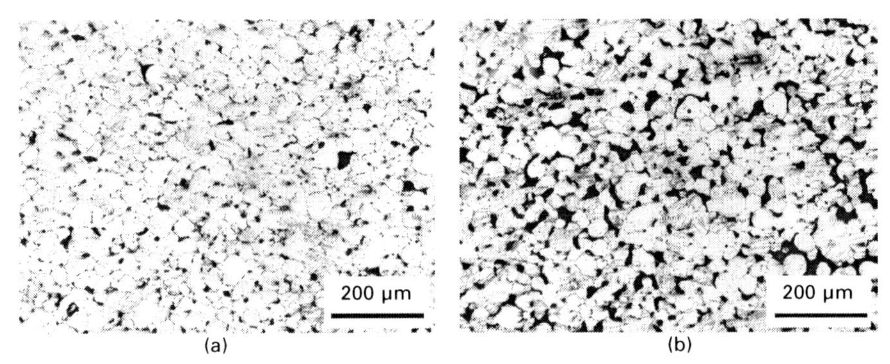

18.6 Microstructure of Mg–0.9 Ca parts. (a) Pressed powders without binder, porosity 5.3%. (b) MIM production, residual porosity 16.0%.

18.7 (a) Tensile samples made by MIM from Mg–0.9 Ca alloy powders, left: green state, right: after sintering. (b) Infiltration of porous magnesium with ethanol demonstrates the potential for drug delivery (infiltration time 5 s).

oxygen and carbon. Nevertheless, the first MIM parts could be produced from successfully processed Mg–0.9Ca alloy (Figure 18.7a) and their microstructure is shown in Fig. 18.6(b). The potential of porous magnesium for drug delivery applications is shown by the almost complete infiltration with ethanol within seconds. Similar to oxygen, magnesium does not show any solubility for carbon. Therefore, formation of magnesium or calcium carbides is expected. A detailed investigation of this phenomenon is in progress.

18.5.2 Nitinol

For net-shape manufacturing of PM Nitinol parts with improved mechanical properties, the application of pre-alloyed Nitinol powders is recommended. In this case, the formation of secondary phases, which do not show martensitic phase transformations, is reduced to a minimum. The best mechanical properties were achieved with Nitinol powders, which were prepared by electrode induction-melting gas atomisation (EIGA). The oxygen and carbon contents of the starting powders produced by this method were below 0.06 wt% and 0.03 wt%. Nevertheless, these contents already approach the maximum oxygen and carbon contents allowed in the ASTM standard F 2603-05 for wrought Nitinol alloys.

Figure 18.8 shows the microstructure, as well as the mechanical properties, of tensile specimens, which were net-shape manufactured by MIM technology. After debinding, the specimens were sintered at 1250°C for 5 h in vacuum. With the exception of electropolishing, no further secondary operations were applied (Mentz *et al.*, 2006). After sintering, the maximum impurity contents

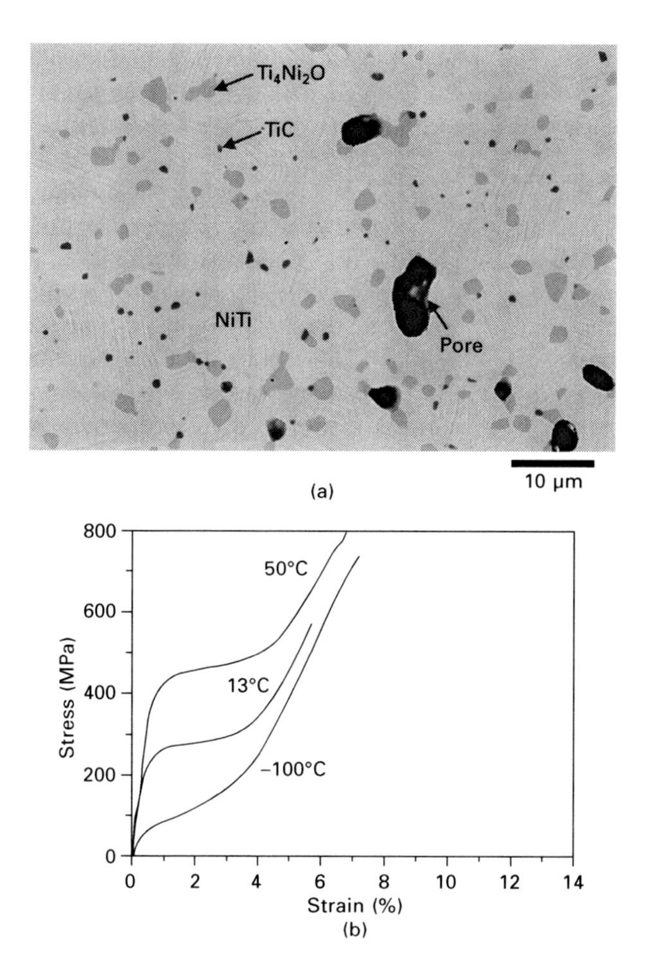

18.8 (a) Microstructure of a tensile specimen produced by MIM technology starting from pre-alloyed Nitinol powders. (b) Related stress–strain curve depending on testing temperature.

of samples were 0.12 wt% oxygen and 0.04 wt% carbon if EIGA powders were used. The porosity of the sintered specimen was below 3%. Owing to the negligible solubility of oxygen and carbon in the NiTi lattice, precipitation of homogeneously distributed oxygen-containing Ti_2Ni (light grey) and TiC phases (black) resulted (Fig. 18.8(a)). The related stress–strain curve showed that the maximum strain to failure was below 8% (Fig. 18.8(b)). The fracture strength lay between 600 MPa and 800 MPa, which is clearly lower than the fracture strength of wrought Nitinol. The $Ti_2Ni_4O_x$-phase was identified as initiating cracks already at low deformations (Mentz *et al.*, 2008). Owing to their spherical shape, residual pores were found to be less critical with regard to the mechanical properties.

As indicated by Fig. 18.8, formation of a fully pronounced martensitic plateau was also found for PM Nitinol. Probably because of the influence of TiC and $Ti_2Ni_4O_x$ phases on the formation and detwinning of the martensitic phase, a reduced plateau length coupled with a continuous stress increase in the plateau region was found if compared to Nitinol samples prepared by ingot metallurgy.

As a reference, dense NiTi samples were prepared by hot isostatic pressing of pre-alloyed NiTi powders in evacuated steel cans at temperature of 1050°C, pressure of 100 MPa and dwell time of 5 h (Mentz *et al.*, 2006, 2008). In this case, the influence of residual microporosity as well as an increase of oxygen and carbon contents caused by the contact with organic binder and sintering atmosphere was eliminated. Owing to the encapsulation of the powders, there is no increase in oxygen and carbon contents compared to the starting powder. The specimens exhibited increased strains to failure of up to 20% in the as-HIP state. An additional heat treatment of the HIP specimens at 1250°C for 10 h (comparable to the sintering conditions for the MIM specimens) was performed for further coarsening of the $Ti_2Ni_4O_x$ phase. Owing to the clear reduction in finely dispersed $Ti_2Ni_4O_x$-particles on the grain boundaries, strains to failure of up to 35% were achieved, which approach the ductility of samples prepared by conventional ingot metallurgy. The stress at fracture lay in the range 800–900 MPa.

To investigate the fatigue of the shape memory effect, 50 loading–unloading cycles were conducted with a maximum strain of 2% at a deformation rate of 0.1 mm min^{-1} (Krone 2005). Even if a continuous change in the curve was observed, especially within the first cycles, PM Nitinol remained almost completely pseudoelastic (Fig. 18.9). The shift of the curve with an increasing number of pseudoelastic cycles is already well known for wrought Nitinol alloys. It is thought to be related to the formation of dislocations at the austenite–martensite interface as well as by the interaction with non-transforming oxides and carbides.

The application of suitable spacer materials like NaCl enabled the manufacture of net-shaped porous Nitinol by MIM, which is highly attractive

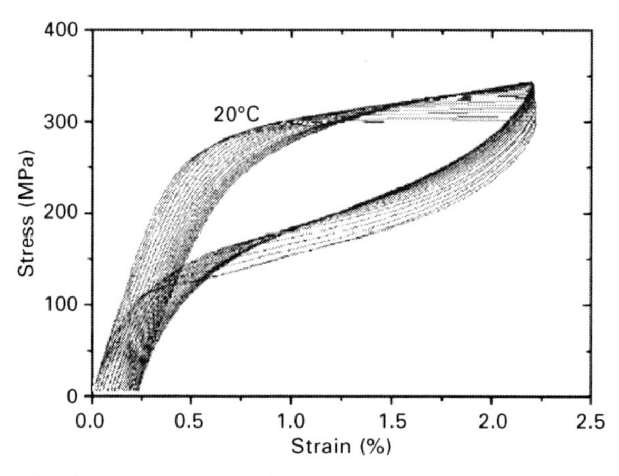

18.9 Cyclic stress–strain curves of pseudoelastic Nitinol produced by MIM.

for bone implants. Porosities in the range of 50–70 vol% in combination with fully pronounced pseudoelastic shape recovery lead to a perfect adaption of mechanical properties related to the bone tissue (Köhl *et al.*, 2009; Bram *et al.*, 2011). Therefore, the risk of stress-shielding is reduced to a minimum and fast bone ingrowth is expected. An overview of mechanical properties of porous Nitinol prepared by self-propagating high temperature synthesis is given elsewhere (Li *et al.*, 2002; Wen *et al.*, 2010).

18.6 Case studies

18.6.1 Port system

Port systems are used in the framework of cancer and pain therapy for repeated access to veins or epidural and intrathecal spaces for drug delivery. Commonly, the housing is made either from a biocompatible polymer or from a metal like titanium. The latter has the advantage of improved biocompatibility and ruggedness against accidental stitches by the syringe during application of the drug. However, an anatomical and ergonomic shape is hard to realise by conventional ingot metallurgy at reasonable cost. On the other hand, port systems made from polymers are already produced by injection moulding which offers almost no limits in terms of geometry, but suffer from inferior material properties.

Transferring the injection moulding process to titanium powders it becomes possible to combine the advantages and produce titanium port systems with optimised function ability at low cost. Figure 18.10(a) shows a commercial port system which is manufactured by a biomedical company in serial MIM production since 2005 (Oger *et al.*, 2006). The port is made from

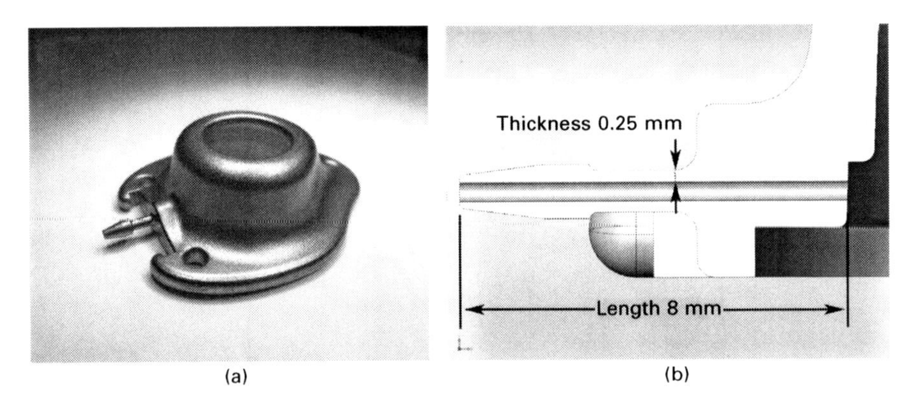

18.10 (a) Housing of a port system made by MIM from Ti-6Al-4V powder. (b) Detailed sketch of the port outlet (courtesy of Tricumed Medizintechnik GmbH, Kiel, Germany).

Ti-6Al-4V Grade 23 powder, approved as a long-term implant and successfully implanted in numbers of several hundreds.

Apart from the realisation of ergonomic geometry, MIM manufacturing allows cost savings. Owing to the high material utilisation in the case of the MIM, material loss during shaping was reduced by a factor of 0.9 compared to conventional machining of bulk titanium. Thus, even if powders are more expensive than ingot bars, in the end a reduction in material costs results. In addition, complex joining steps can be avoided as outlet and main housing are produced as one part. In Fig. 18.10(b) the cross section of the port outlet is given in detail. The fluid channel is about 8 mm long and has a diameter of only 0.45 mm. Minimum wall thickness is only 0.25 mm. The required length to diameter ratio of the outlet is very difficult to realise by common methods like drilling. In the as-sintered state, tight and fast attachment of surrounding tissue to the smooth surface was observed. Optionally, specific surface modifications like grinding, polishing or anodic oxidation are possible as well.

18.6.2 Bone screw

While the port system does not require high mechanical strengths, bone screws, as shown in Fig. 18.11, have to withstand strong forces. Thus, typically, titanium alloys like Ti-6Al-4V are used. The screw depicted in Fig. 18.11 was manufactured as a prototype demonstrating the potential of MIM starting from Ti-6Al-4V Grade 23 and from Ti-6Al-7Nb powders (Aust *et al.*, 2006). Its geometry is specialised for repairing fractures of the dense axis, which protrudes from the second neck vertebra and acts as the pivot for the head. A longitudinal boring enables the insertion of a guiding

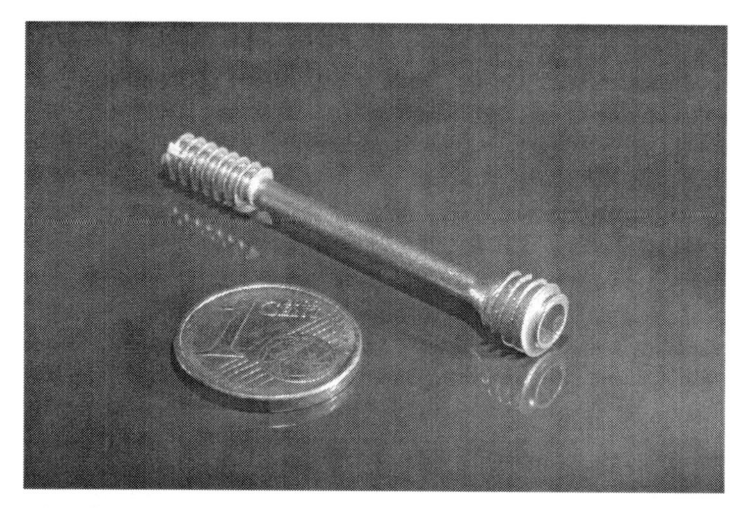

18.11 Prototype of a bone screw made by MIM from titanium alloy powder (design by Tricumed Medizintechnik GmbH, Kiel, Germany).

wire during operation. Furthermore, the screw features self-cutting edges, two threads with different pitches and a hexagonal socket.

The entire screw with all its features is made by a single injection moulding process. By a simple change in the core of the mould, which forms the boring, into a variant with a hexagonal cross-section instead of a circular one, the boring can be shaped such that a hexagonal screw driver can work even on the shaft. In this case, the force transmission on the screw is improved, the risk of fracture decreases and the chance of removing even a broken screw is clearly enhanced. Although this screw is not commercially available today, it reveals the great freedom related to geometry of the MIM process. In addition, it shows that more functionality by more complexity does not imply higher manufacturing costs.

18.6.3 Porous spine implants

Lumbar discs are the largest articulated non-vascularised tissue in humans; they are statically and dynamically extremely heavily loaded. Therefore, lumbar diseases are often reported in the human body (Wintermantel and Ha, 2002). When a complete lumbar disc replacement is needed, a common implant is the interbody fusion cage, which acts as a spacer between adjacent vertebral bones. It is usually designed to promote the ingrowth of bone tissue into the implant (osseointegration). Improved ingrowth behaviour is expected if the implant contains an open porous network. There are several methods for manufacturing such implants.

In 2007, a spine implant was launched on the market, which was manufactured by pressing titanium powder-space holder compacts with subsequent green machining and sintering (Imwinkelried, 2007). Figure 18.12 shows the geometry of this implant. This spinal implant is composed of a porous part (interconnected porosity in the range 60–65 vol%, pore sizes in the range of 100–700 μm) and a part with clearly reduced porosity, which is required to achieve sufficient stability for positioning the implant during surgery. The surface roughness of the porous part supports the primary fixation of the implant after surgery, while the interconnected porosity enables the bone ingrowth required for long-term stabilisation.

Another porous spine implant available on the market is made by self-propagating high-temperature synthesis starting from powder compacts made from elemental nickel and titanium powders (Li *et al.*, 2000; Assad *et al.*, 2003; Rhalmi *et al.*, 2007). This implant has a fully porous Nitinol structure with a porosity of 65 ± 10%. Another approach is the manufacture of the porous interbody fusion cage by selective laser melting (SLM) as described by Lin *et al.* (2007). A spine implant with a porosity of 52% was produced.

18.6.4 Bone staple with shape memory effect

The prototype of a biomedical foot staple was chosen to prove the suitability of the MIM process for manufacturing net-shaped Nitinol implants with a fully pronounced shape memory effect. Such foot staples are used to put a permanent load on bone fractures, stabilising the bone defect and supporting the bone regeneration (Esenwein *et al.*, 2004). The function principle of the foot staple is based on a one-way shape memory effect. Once the foot staple has been manufactured, it has an austenitic structure. The staple is cooled

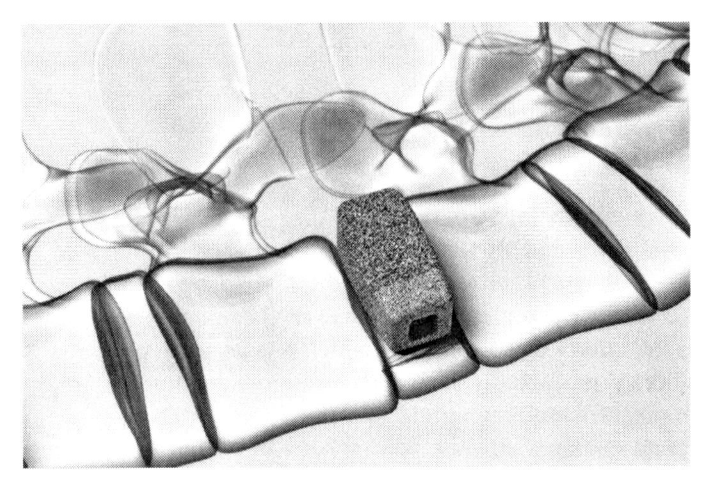

18.12 Spine implant (interbody fusion cage) with a stepwise gradient in porosity (design by Synthes GmbH, Oberdorf, Switzerland).

before the application and thus fully transformed into the martensitic state. In this state, it is opened with a special tool and stored below the austenitic start temperature (A_s) until surgery. Two small holes are drilled on both sides of the bone defect containing the ends of the foot staple. After positioning during surgery, the foot staple heats up to body temperature. At this temperature, phase transformation back to the austenite takes place, coupled with the shape recovery of the staple to its original shape. Since shape recovery is suppressed, the staple exerts a defined load on the bone defect supporting the healing process. Figure 18.13 demonstrates the function of a foot staple manufactured by MIM (Krone *et al.*, 2005).

18.7 Conclusions and future trends

PM technologies are innovative for the manufacture of biomedical implants if large quantities of complex-shaped implants or implants with functional porosity are required. Today, the potential of PM has been demonstrated by the marked launch of a port system made by metal injection moulding using Ti-6Al-4V powder. Furthermore, porous spine implants, as well as several

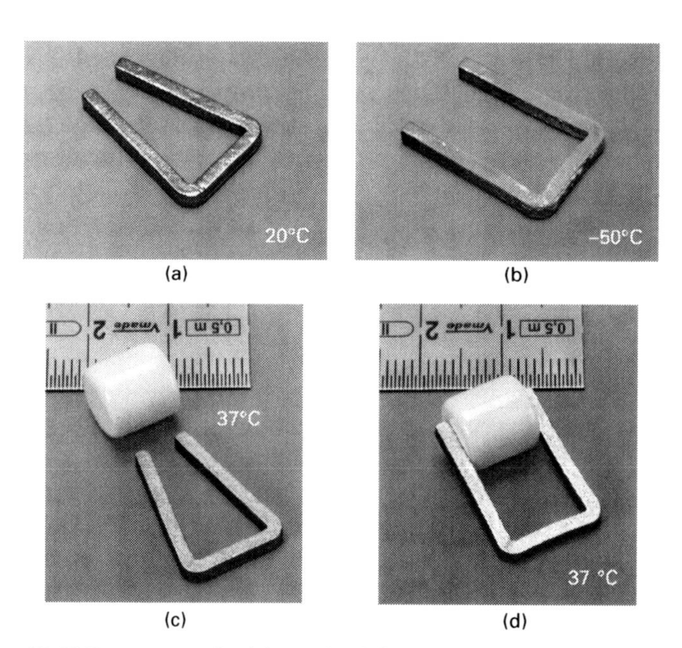

18.13 Prototype of a biomedical foot staple manufactured by MIM, demonstrating its function based on the one-way shape memory effect. (a) Initial sample, sintered and electropolished, (b) cooling to –50°C, mechanically opened, (c) heating to body temperature, full shape recovery and (d) applying a permanent load by suppression of shape recovery.

implant coatings with functional porosity made from titanium powders, are already in clinical use improving implant fixation by bone ingrowth. Furthermore, prospective additive manufacturing techniques like SLM and EBM offer the possibility for economic production of individual implants tailored precisely to the patient.

In the case of titanium and titanium alloys, components with excellent properties that fulfil the demands of common standards can be produced, if adequate powders and binder systems are used and if sintering is conducted considering the special requirements of titanium. Up to now, there is a lack of standard powders, feedstocks and facilities for PM processing of titanium powders aimed at biomedical applications. On the other hand, the fact that a new ASTM standard for MIM of titanium has been established reveals the industrial interest and shows that research is on the right track.

Magnesium-based alloys are promising candidates for biodegradable implants. PM manufacturing of such alloys has been investigated on a laboratory scale so far and the first promising results have been achieved. At present, more basic research is required before the potential of PM magnesium implants can be demonstrated on net-shaped prototypes.

The applicability of PM manufacturing for Nitinol implants was demonstrated by metal injection moulding of a foot staple, starting from pre-alloyed Nitinol powders. After sintering, high dimensional accuracy, as well as fully pronounced shape memory properties, have been shown. Nevertheless, the lack of standard Nitinol powders and the high affinity for oxygen and carbon uptake during processing makes it difficult to achieve ASTM-standard F2063-05, which specifies wrought Nitinol alloys for medical devices and surgical implants.

The ongoing progress in PM manufacturing of titanium and titanium alloys during the last few years has helped to introduce this technology for the production of medical devices and surgical implants on an industrial scale. Nevertheless, development of titanium alloys adapted to the specific requirements of PM manufacturing, as well as standardisation of each processing step, is required to make PM titanium an inherent part of the emerging biomedical market:

In the case of magnesium alloys and Nitinol, more basic research is required before PM technologies can be transferred to industry. The following tasks must be fulfilled before parts made from these alloys are ready to be introduced to the market.

- development of standards for PM manufacturing of magnesium and Nitinol
- supply of powders produced by established powder manufacturers following these standards
- further reduction of oxygen and carbon uptake during PM processing

- further improvement of sintering behaviour, e.g. by adding suitable sintering aids
- proof of controlled degradation of PM magnesium alloys by corrosion and *in vivo* tests
- proof of biocompatibility of PM Nitinol parts by *in vitro* and *in vivo* tests
- proof of fatigue resistance.

18.8 Further reading

In order to start with MIM of titanium it is helpful to read review articles, for example by German (2009), Baril (2010) and Ebel (2008). For a deeper understanding, references given in these articles should be consulted.

Review papers including the PM production of porous titanium have been written by Ryan *et al.* (2006), Dunand *et al.* (2004) and Singh *et al.* (2010). Bansiddhi *et al.* (2008) and Wen *et al.* (2010) summarise the state-of-the-art of porous Nitinol alloys.

Some general information about PM facility suppliers, case studies and basic properties of PM parts can be obtained via www.empa.com, www.mpif.org, www.mimaweb.org, www.jpma.gr.jp and www.mim-experten.de. However, the sections about titanium or porous implant materials are usually quite limited.

A visit to PM congresses is also recommended, mostly organised by European Powder Metallurgy Association (EPMA) or American Metal Powders Industries Federation (MPIF). Today, special sessions about PM of metals like titanium, magnesium or Nitinol as well as their application as biomaterials are often offered and the current activities and developments in this field can be followed. Much of the information given in this chapter is based on presentations and discussions at such congresses.

18.9 References

Aller A J and Losada A (1990), 'Rotating atomization processes of reactive and refractory alloys', *Metal Powder Rep*, **45**, 51–5.

Assad M, Jarzem P, Leroux MA, Coillard C, Chernyshov AV and Charette S (2003), 'Porous titanium-nickel for intervertebral fusion in a sheep model: part 1. Histophometric and radiological analysis 1', *J Biomed Mater Res*, **B64**, 107–20.

Aust E L, Gerling R, Oger B and Ebel T (2005), 'Herstellung einer komplexen Titan-Knochenschraube durch Metallpulverspritzguss', *Mat-wiss und Werkstofftechnik*, **36**, 423–8.

Aust E, Limberg W, Gerling R, Oger B and Ebel T (2006), 'Advanced TiAl6Nb7 bone screw implant fabricated by metal injection moulding', *Adv Eng Mater*, **8**, 365–70.

Bahrami Nasab M and Hassan M R (2010), 'Metallic biomaterials of knee and hip - A review', *Trends Biomat Artif Organs*, **24**, 69–82.

Bakan H I, Jumadi Y, Messer P F, Davies H A and Ellis B (1998), 'Study of processing parameters for MIM feedstock based on composite PEG-PMMA binder', *Powder Metall*, **41**, 289–91.

Bansiddhi A, Sargeant T D, Stupp S I and Dunand D C (2008), 'Porous NiTi for bone implants: A review', *Acta Biomater*, **4**, 773–82.

Baril E (2010), 'Titanium and titanium alloy powder injection moulding: Matching application requirements', *PIM Int*, **4**, 22–32.

Baril E, Lefebvre L P and Thomas Y (2011), 'Interstitials sources and control in titanium P/M processes', *Powder Metall*, **54**, 183–7.

Bidaux J E, Jochem A, Zufferey D and Carreno Morelli E (2007), 'Metal injection moulding of NiTi shape memory alloys', in *EURO PM 2007, Proceedings of the International Powder Metallurgy Congress and Exhibition*, European Powder Metallurgy Association EPMA, Shrewsbury, UK, Volume 2, 223–8.

Biemond J E, Aquarius R, Verdonschot N and Buma P (2011), 'Frictional and bone ingrowth properties of engineered surface topographies produced by electron beam technology', *Arch Orthop Trauma Surg*, **131**, 711–8.

Biswas A (2005), 'Porous NiTi by thermal explosion mode of SHS: processing, mechanism and generation of single phase microstructure', *Acta Mater*, **53**, 1415–25.

Bobyn J D, Pilliar R, Cameron W and Weatherly G C (1980), 'The optimum pore size for the fixation of porous-surfaced metal implants by ingrowth of bone', *Clin Orthop*, **150**, 263–70.

Bram M, Stiller C, Buchkremer H P, Stöver D and Baur H (2000), 'High-porosity titanium, stainless steel and superalloy parts', *Adv Eng Mater*, **2**, 196–9.

Bram M, Kempmann C, Laptev A, Stöver D and Weinert K (2003), 'Investigations on the machining of sintered titanium foams utilizing face milling and peripheral grinding', *Adv Eng Mater*, **5**, 441–7.

Bram M, Köhl M, Buchkremer H P and Stöver D (2011), 'Mechanical properties of highly porous NiTi alloys', *J Mater Eng Perform*, **20**, 522–8.

Clemow A J, Weinstein A M, Klawitter J J, Koneman J and Anderson J (1981), 'Interface mechanics of porous titanium implants', *J Biomed Mater Res*, **15**, 73–82.

Conrad H (1966), 'The rate controlling mechanism during yielding and flow of titanium at temperatures below 0.4 T_M', *Acta Metall*, **14**, 1631–3.

Cysne Barbosa A P, Köhl M, Bram M, Buchkremer H P and Stöver D (2009), 'Production of near-net-shape components with graded porosity by 2-C-MIM', in *Proceedings of 17th Plansee-Seminar*, Sigl L S, Rödhammer P, Wildner H (eds), Reutte, Austria, GT47/1–GT47/12.

Cysne Barbosa A P, Bram M, Buchkremer H P and Stöver D (2012), 'Fabrication of titanium implants with a gradient in porosity by 2-component powder injection moulding', *PIM Int*, **6**, 69–73.

Dunand D C (2004), 'Processing of titanium foams', *Adv Eng Mater*, **6**, 369–76.

Ebel T (2008), 'Titanium and titanium alloys for medical applications: opportunities and challenges', *PIM Int*, **2**, 21–30.

Ebel T, Akaichi H, Ferri O M and Dahms M (2010), 'MIM fabrication of porous Ti-6Al-4V components for biomedical applications', in *Proceedings of EURO-PM2010*, EPMA, Shrewsbury, UK, Volume 4, 797–804.

Ebel T, Ferri O M, Limberg W and Schimansky F-P (2011), 'Metal injection moulding of advanced titanium alloys', in *Proceedings of the 2011 International Conference on Powder Metallurgy and Particulate Materials, PowderMet 2011*, MPIF, San Francisco, California, USA, Volume 1, 45–57.

Esen Z and Bor S (2007), 'Processing of Ti foams using magnesium spacer particles', *Scr Mater*, **56**, 341–4.

Esenwein S A, Bodganski D, Krone L, Köller M, Epple M and Muhr G (2004), 'Zur Eignung von Formgedächtnislegierungen auf der Basis von NiTi (NITINOL®) als Implantatwerkstoff – Möglichkeiten zur klinischen Anwendung', *Biomed Technol*, **49**, 582–3.

Fauchais P (2004), 'Understanding of plasma spraying', *J Phy D: Appl Phy*, **37**, 86–108.

Ferri O M, Ebel T and Bormann R (2009), 'High cycle fatigue behaviour of Ti–6Al–4V fabricated by metal injection moulding technology', *Mater Sci Eng A*, **504**, 107–13.

Ferri O M, Ebel T and Bormann R (2010), 'Influence of surface quality and porosity on fatigue behaviour of Ti–6Al–4V components processed by MIM', *Mater Sci Eng A*, **527**, 1800–5.

Ferri O M, Ebel T and Bormann R (2011), 'The influence of a small boron addition on the microstructure and mechanical properties of Ti-6Al-4V fabricated by metal injection moulding', *Adv Eng Mater*, **13**, 436–47.

Frenzel J, George E P, Dlouhy A, Somsen C, Wagner M X F and Eggeler G (2010), 'Influence of Ni on martensitic transformations in NiTi shape memory alloys', *Acta Mater*, **58**, 3444–58.

Geetha M, Singh A K, Asokamani R and Gogia A K (2009), 'Ti based biomaterials, the ultimate choice for orthopaedic implants – A review', *Prog Mater Sci*, **54**, 397–425.

German R M (1997), *Injection Molding of Metals and Ceramics*, Metal Powder Industry Federation MPIF, Princeton, NJ, USA.

German R M (2009), 'Titanium powder injection moulding: A review of the current status of materials, processing, properties and applications', *PIM Int*, **3**, 21–37.

Gibson L J and Ashby M F (1997), *Cellular Solids*, second edition, Cambridge Solid State Science Series, Cambridge, UK.

Hahn H P and Palich W (1970), 'Preliminary evaluation of porous metal surface titanium for orthopedic implants', *J Biomed Mater Res*, **4**, 571–7.

Heinl P, Müller L, Körner C, Singer R F and Müller F A (2008), 'Cellular Ti–6Al–4V structures with interconnected macro porosity for bone implants fabricated by selective electron beam melting', *Acta Biomater*, **4**, 1536–44.

Hey J C and Jardine A P (1994), 'Shape memory TiNi synthesis from elemental powders', *Mater Sci Eng A*, **188**, 291–300.

Hohmann M and Jönsson S (1990), 'Modern systems for production of high quality metal alloy powder', *Vacuum*, **41**, 2173–6.

Hong T F, Guo Z X and Yang R (2008), 'Fabrication of porous titanium scaffold materials by a fugitive filler method', *J Mater Sci: Mater in Med*, **19**, 3489–95.

Igharo M and Wood J V (1985), 'Compaction and sintering phenomena in titanium-nickel shape memory alloys', *Powder Metall*, **28**, 131–9.

Imwinkelried T (2007), 'Mechanical properties of open-pore titanium foam', *J Biomed Mater Res, Part A*, **81**, 964–70.

Janning C, Willbold E, Vogt C, Nellesen J, Meyer-Lindenberg A, Windbergen H, Thorey F and Witte F (2010), 'Magnesium hydroxide temporarily enhancing osteoblast activity and decreasing the osteoclast number in peri-implant bone remodelling', *Acta Biomater*, **6**, 1861–8.

Jones A C, Arns C H, Hutmacher D W, Milthorpe B K, Sheppard A P and Knackstedt M A

(2009), 'The correlation of pore morphology, interconnectivity and physical properties of 3D ceramic scaffolds with bone ingrowth', *Biomaterials*, **30**, 1440–51.

Köhl M, Habijan T, Bram M, Buchkremer H P, Stöver D and Köller M (2009), 'Powder metallurgical near-net-shape fabrication of porous NiTi shape memory alloys for use as long term implants by the combination of the metal injection moulding process with the space holder technique', *Adv Eng Mater*, **11**, 959–68.

Köhl M, Bram M, Moser A, Beck T, Buchkremer H P and Stöver D (2011), 'Characterization of porous, net-shaped Ti and NiTi alloys regarding their damping and energy absorbing capacity', *Mater Sci Eng A*, **528**, 2452–62.

Krone L, Mentz J, Bram M, Buchkremer HP, Stöver D, Wagner M, Eggeler G, Christ D, Reese S, Bogdanski D, Köller M, Esenwein SA, Muhr G, Prymak O and Epple M (2005), 'The potential of powder metallurgy for the facrication of biomaterials on the basis of nickel-titanium: A case study with a staple showing shape memory behaviour', *Adv Eng Mater*, **7**, 613–9.

Laptev A, Bram M, Buchkremer H P and Stöver D (2004), 'Study of production route for Ti parts combining very high porosity and complex shape', *Powder Metall*, **47**, 85–92.

Leyens C and Peters M (Eds) (2003), *Titanium and Titanium Alloys*, Wiley-VCH, Weinheim, Germany.

Li B Y, Rong L J and Li Y Y (1998), 'Porous NiTi alloy prepared from elemental powder sintering', *J Mater Res*, **13**, 2847–51.

Li B Y, Rong L J, Li Y Y and Gjunter V E (2000), 'Synthesis of porous Ni-Ti-shape-memory alloys by self-propagating synthesis: reaction mechanism and anisotropy in pore structure', *Acta Mater*, **48**, 3895–904.

Li Y H, Rong L J and Li Y Y (2002), 'Compressive property of porous NiTi alloy synthesized by combustion synthesis', *J Alloys Comp*, **345**, 271–4.

Lin C Y, Wirtz T, LaMarca F and Hollister S J (2007), 'Structural and mechanical evaluations of a topology optimised titanium interbody fusion cage fabricated by selective laser melting process', *J Biomed Mater Res, Part A*, **83**, 272–9.

Lütjering G (1998), 'Influence of processing on microstructure and mechanical properties of ($\alpha + \beta$) titanium alloys', *Mater Sci Eng A*, **243**, 32–45.

Massalski T B and Okamoto H (1990), *Binary Alloys Phase Diagrams*, ASM International, Materials Park, Ohio, USA.

Massalski T B, Okamato H, Subramanian P R and Kacprzak L (1996), *Binary Alloys Phase Diagrams*, ASM International, second edition, Materials Park, Ohio, USA.

McCracken C G, Barbis D P and Deeter R C (2010), 'Key titanium powder chracteristics manufactured using the hydride-dehydride (HDH) process', in: *Proceedings of EURO-PM2010*, EPMA, Shrewsbury, UK, Volume 1, 71–7.

Meier H and Haberland C (2008), 'Experimental studies on selective laser melting of metallic parts', *Mat.-wiss. u. Werkstofftech*, **39**, 665–670.

Mentz J, Bram M, Buchkremer H P and Stöver D (2006), 'Improvement of mechanical properties of powder metallurgical NiTi shape memory alloys', *Adv Eng Mater*, **8**, 247–52.

Mentz J, Bram M, Buchkremer H P and Stöver D (2008), 'Influence of heat treatments on the mechanical properties of high-quality Ni-rich NiTi produced by powder metallurgical methods', *Mater Sci Eng A*, **481–482**, 630–4.

Murr L E, Amato K N, Li S J, Tian Y X, Cheng X Y, Gaytan S M, Martinez E, Shindo, P W, Medina F and Wicker R B (2011), 'Microstructure and mechanical properties of open-cellular biomaterials prototypes for total knee replacement implants fabricated by electron beam melting', *J Mech Behav Biomed Mater*, **4**, 1396–1411.

Obasi G C, Ferri O M, Ebel T and Bormann R (2010), 'Influence of processing parameters on mechanical properties of Ti–6Al–4V alloy fabricated by MIM', *Mater Sci Eng A*, **527**, 3929–35.

Oger B, Ebel T and Limberg W (2006), 'The manufacture of highly-ductile and geometrically complex MIM-parts based on TiAl6V4', in *Proceedings of EURO-PM2006*, Volume 2, 191–6.

Oh I H, Nomura N, Masahashi N and Hanada S (2003), 'Mechanical properties of porous titanium compacts prepared by powder sintering', *Scr Mater*, **49**, 1197–202.

Ola L, Harrysson A, Cansizoglu O, Marcellin-Little D J, Cormier D R and West H A (2008), 'Direct metal fabrication of titanium implants with tailored materials and mechanical properties using electron beam melting technology', *Mater Sci Eng C*, **28**, 366–73.

Panigrahi B B, Godkhindi M M, Das K, Mukunda P G and Ramakrishnan P (2005), 'Sintering kinetics of micrometric titanium powder', *Mater Sci Eng A*, **396**, 255–62.

Parthasarathy J, Starly, B, Shivakumar, R and Christensen A (2010), 'Mechanical evaluation of porous titanium (Ti6Al4V) structures with electron beam melting (EBM)', *J Mech Behav Biomed Mater*, **3**, 249–59.

Poumarat G and Squire P (1993), 'Comparison of mechanical properties of human, bovine bone and a new processed bone xenograft', *Biomaterials*, **14**, 337–49.

Quadbeck P, Stephani G, Kümmel K, Hauser R, Stahnke G, Adler J, Stephani G and Kieback B (2011), 'Structural and material design of open-cell powder metallurgical foams', *Adv Eng Mater*, **13**, 1024–30.

Rhalmi S, Charette S, Assad M, Coillard C and Rivard C H (2007), 'The spinal cord dura mater reaction to Nitinol and titanium alloy particles: a 1-year study in rabbits', *Europ Spine J*, **16**, 145–54.

Ryan G, Pandit A and Apatsidis D P (2006), 'Fabrication methods of porous metals for use in orthopaedic applications', *Biomaterials*, **27**, 2651–70.

Schiefer H, Bram M, Buchkremer H P and Stöver D (2009), 'Mechanical examinations on dental implants with porous titanium coating', *J Mater Sci: Mater in Med*, **20**, 1763–70.

Singh R, Lee P D, Dashwood R J and Lindley T C (2010), 'Titanium foams for biomedical applications: a review', *Mater Technol*, **25**, 127–36.

Shabalovskaya S, Anderegg J and Van Humbeeck J (2008), 'Critical overview of Nitinol surfaces and their modifications for medical applications', *Acta Biomater*, **4**, 447–67.

Shivashankar T S and German R M (1999), 'Effective length scale for predicting solvent-debinding times of components produced by powder injection molding', *J Am Ceram Soc*, **82**, 1146–52.

Staiger M P, Pietak A M, Huadmai J and Dias G (2006), 'Magnesium and its alloys as orthopedic biomaterials: A review', *Biomaterials*, **27**, 1728–34.

Thieme M, Wieters K P, Bergner F, Scharnweber D, Worch H, Ndop J, Kim T J and Grill W (2001), 'Titanium powder sintering for preparation of a porous functionally graded material destined for orthopaedic implants', *J Mater Sci*, **12**, 225–31.

Traini T, Mangano C, Sammons R L, Mangano F, Macchi A and Piattelli A (2008), 'Direct laser metal sintering as a new approach to fabrication of an isoelastic functionally graded material for manufacture of porous titanium dental implants', *Dental Mater*, **24**, 1525–33.

Tuncer N and Arslan G (2009), 'Designing compressive properties of titanium foams', *J Mater Sci*, **44**, 1477–84.

Tuncer N, Arslan G, Maire E and Salvo L (2011a), 'Investigation of PM parameters' effect on architecture of titanium foams using X-ray microtomography', *Mater Sci Eng A*, **530**, 633–42.

Tuncer N, Arslan G, Maire E and Salvo L (2011b), 'Influence of cell aspect ratio on architecture and compressive strength of titanium foams', *Mater Sci Eng A*, **528**, 7368–74.

Van Bael S, Kerckhofs G, Moesen M, Pyka G, Schrooten J and Kruth J P (2011), 'Micro-CT-based improvement of geometrical and mechanical controllability of selective laser melted Ti6Al4V porous structures', *Mater Sci Eng A*, **528**, 7423–31.

Wehmöller M, Warnke T P H, Zilian C and Eufinger H (2005), 'Implant design and production—a new approach by selective laser melting', *Int Congress Ser*, **1281**, 690–5.

Wen C E, Mabuchi M, Yamada Y, Shimojima K, Chino Y and Asahina, T (2001), Processing of biocompatible porous Ti and Mg, *Scr Mater*, **45**, 1147–53.

Wen C E, Xiong J Y, Li Y C and Hodgson P D (2010), 'Porous shape memory alloy scaffolds for biomedical applications: a review', *Phys Scr*, **T139**, 1–8.

Whang K, Healy K E, Elenz D R, Nam E K, Tsai D C, Thomas C H, Nuber G W, Glorieux F H, Travers R and Sprague S M (1999), 'Engineering bone regeneration with bioabsorbable scaffolds with novel microarchitecture', *Tissue Eng*, **5**, 35–51.

Wintermantel E and Ha S W (2002), *Medizintechnik mit biokompatiblen Werkstoffen und Verfahren*, Springer-Verlag, Berlin, Heidelberg, New York.

Witte F, Fischer J, Nellesen J, Crostack H A, Kraese V, Pisch A, Beckmann F and H. Windhagen H (2006), '*In vitro* and *in vivo* corrosion measurements of magnesium alloys', *Biomaterials*, **27**, 1013–18.

Witte F, Ulrich H, Rudert M and Willbold E (2007), 'Biodegradable magnesium scaffolds: Part I: Appropriate inflammatory response', *J Biomed Mater Res, Part A*, **81**, 748–56.

Wolff M, Dahms M and Ebel T (2010), 'Sintering of magnesium', *Adv Eng Mater*, **9**, 829–36.

Wu S K, Lin H C and Chen C C (1999), 'A study on the machinability of a Ti49.6Ni50.4 shape memory alloy', *Mater Letters*, **40**, 27–32.

Wu Y W R, Kwon Y S, Park S J and German R M (2006), 'Injection moulding of HDH titanium powder', *Int J Powder Metall*, **42**, 59–66.

19
Applications of powder metallurgy to cutting tools

J. S. KONSTANTY, AGH University of Science and Technology, Poland

DOI: 10.1533/9780857098900.4.555

Abstract: This chapter presents the role of powder metallurgy (PM) in the production of diamond impregnated tools and provides information on their applications. The chapter first reviews the historical background of PM diamond tools and provides the basics of tool design. The chapter then discusses tool fabrication roots, guides the end-users through the tool application conditions and shows the latest trends and developments.

Key words: diamond impregnated tool, hot pressing, metallic matrix, synthetic diamond.

19.1 Introduction

Modern applications of diamond tools extend back about 150 years. In 1862 the Swiss engineer J.R. Leschot of Geneva conceived the idea of making diamond drill bits (Hughes, 1980). The first diamond circular saw blades for cutting stone were developed by Felix Fromholt in France in 1885. A large diameter blade was first used in practice in the Euville stone quarries 13 years later.

The early tools used *carbonados*, a cryptocrystalline mass of microscopic crystals of diamond, graphite, amorphous carbon, and other accessory minerals locked in random directions. At that time the strong *carbonados* which were resistant to cleavage, were a valued material set around the tool periphery (Tolansky, 1967). These primitive blades were utilised to cut limestone and marble during the construction of large buildings in Paris in the 1900s (Hughes, 1980). Progress in the tool production routes that followed, by making use of powder metallurgy (PM) techniques, resulted in the development of diamond grit impregnated saw blades which were put into operation around 1940 (Hughes, 1980).

Further developments in the tool manufacturing technology may chiefly be attributed to the invention of synthetic diamond. Natural diamond has been used for centuries and efforts to manufacture synthetic crystals also date back at least several hundred years. They had remained fruitless until 1953, when positive and reproducible results were obtained by a team of researchers at ASEA (Lundblad, 1990). In 1955, quite independently and entirely without knowledge of what ASEA had been doing, General Electric announced its capability to manufacture synthetic diamonds on an industrial scale (Bundy *et al.*, 1955) and was first to apply for a patent (Hall *et al.*, 1960).

Permanent progress in diamond manufacturing technologies fostered the commercial importance of synthetic grits, which now account for almost 99% of all industrial diamonds consumed (USGS, 2011). Over the past six decades, modern production techniques based on diamond tooling have been implemented into evolving areas of industrial activity enabling the job to be done faster, more accurately and at less cost. These techniques have revolutionised machinery and processing in the stone and construction industries, road repair, petroleum exploration, production of glass, dense ceramics, and so on.

In the new millennium the market for diamond tools continues to grow rapidly. Recent figures indicate that the global production of synthetic diamond exceeded an impressive volume of 4.38 billion carats in 2010 (USGS, 2011), with China in the lead for both production and consumption. The marked decline in the price of industrial diamond makes it a commoditised product capable of competing, in terms of its price/performance ratio, with cemented carbides and conventional abrasives. Judging by the value of sales, by far the largest group of diamond tools comprises the so-called 'metal-bonded' diamond impregnated tools such as circular and frame saw blades, wire saws and core drills for cutting natural stone and construction materials, as well as core bits for drilling in medium to hard rock formations (Fig. 19.1).

19.1.1 Classification of diamond tools

The term 'diamond tools' has a broad meaning. The existing classifications of diamond tools (Fig. 19.2) are based on various criteria such as the quantity

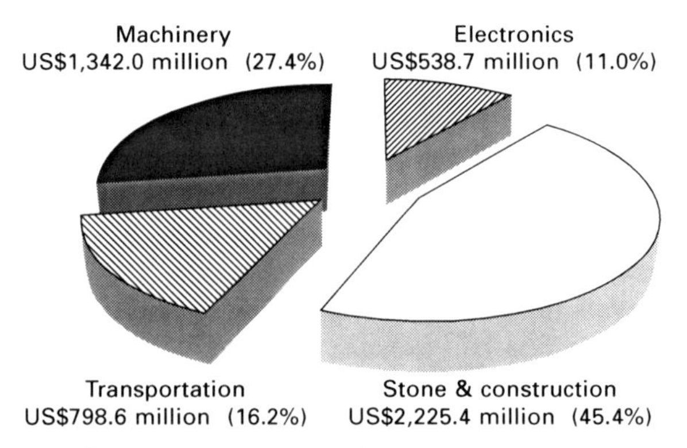

19.1 Global consumption of diamond tools by end-use market in 2007 (Dedalus Consulting, 2008).

of diamond involved and its origin, outward appearance and internal structure of the tool, its application, production route, and so on. For the purpose of this publication it is convenient to arrange the tool types into categories which are distinctive with respect to the various manufacturing methods involved (Fig. 19.2).

In production volume terms, metal-bonded tools (Fig. 19.3) account for around two-thirds of the whole bonded-grit diamond tool market. In the metallurgical sense the term metal bond should be applied to electroplated products. In diamond tools, however, it often refers to metallic (metal or cermet) matrix diamond impregnated composites manufactured by different PM techniques.

19.2 Tool design and composition

The design and composition of the diamond tool play an important role in the economic machining of materials where great demands are made on the tool life, its free cutting ability and the precision of the job involving the

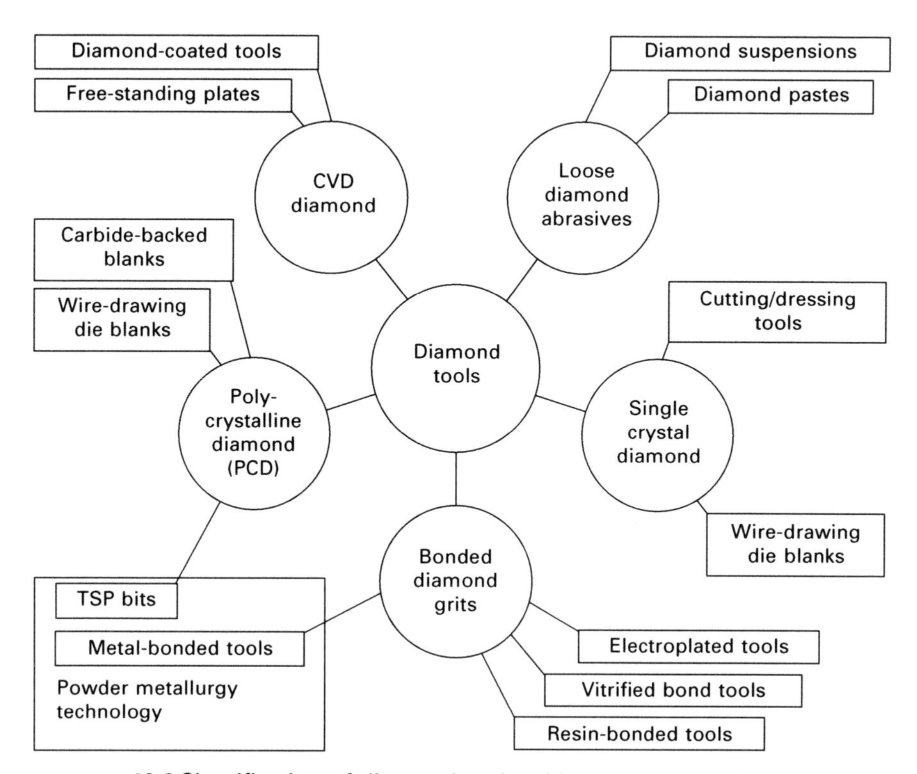

19.2 Classification of diamond tools with respect to production technology.

19.3 Selection of metal-bonded tools and diamond impregnated segments.

edge quality and surface finish of the workpiece. Tool design has been found to be of paramount importance especially in the case of circular sawing by circular saw blades (Konstanty, 2005, p 40). The proper choice of slot geometry, as well as the arrangement and shape of the diamond impregnated segments, renders improved tool performance in terms of quality of the cut, noise generated, abrasive wear and fatigue life of the steel centre, flow of coolant to the cutting zone, and reduced segment wear by abrasion. Industrial experience has resulted in the adoption of several saw blade shapes (Fig. 19.4), which have performed well in a wide range of applications.

Continuous rim blades (Type 1) are used when the quality of cut and perfect edge retention are of prime concern, for example in sawing of flat crystal glass, tiles and other types of hard and dense ceramics. Continuous rim blades are usually restricted to diameters ≤400 mm, whereas segmental saw blades are produced in sizes ranging between 100 mm and 3500 mm. Blades with narrow slots (Type 2) are used for sawing igneous types of stone and hard ceramics, where good edge retention and surface finish have to be achieved. By substituting each long segment by two, or more, shorter segments (Type 3), it is possible to decrease the abrasive action of the slurry by shortening the trailing part of the segment and by directing more coolant into the cutting zone. The application of a keyhole shaped slot (Type 4) hinders fatigue crack initiation at the slot base thus having a beneficial effect on the tool life under heavy duty conditions. The fatigue life of the steel centre becomes a critical consideration in primary sawing of stone blocks by large diameter saw blades. The largest steel centres are very expensive and have to be re-tipped many times to render the sawing process economically viable. Wide slots (Type 5) are best suited for these applications because they effectively prolong the fatigue life of the centre.

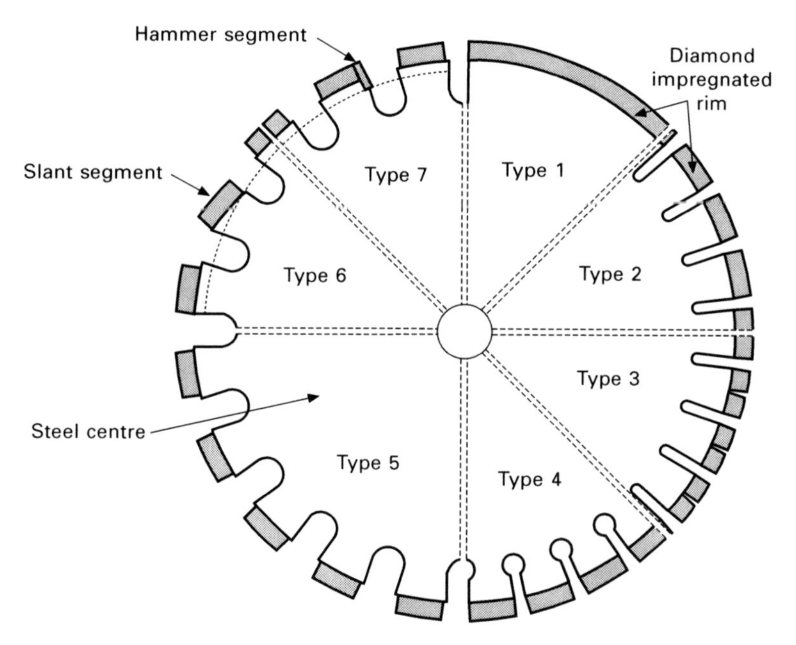

19.4 Typical periphery configurations of circular saw blades.

Wider slots allow more efficient flushing of the swarf from the cut and are therefore also used for fast sawing operations as well as in circular sawing of very abrasive materials, such as fresh concrete or asphalt. The presence of coarse and abrasive particles between the sides of the blade and the workpiece causes the slotted rim of the saw blade to wear laterally leading to premature loss of segments. To prevent this, a few wear-protective segments (Types 6 and 7) must be incorporated into the blade periphery.

There are also many ways in which diamond impregnated segments may be designed (Fig. 19.5). Uniform, rectangular segments are cheap and easy to manufacture; however, the blade fabrication or application requirements often justify selection of complex designs. The use of tapered segments, for example, decreases the power consumed for sawing by reducing the lateral friction of segments against the workpiece. The so-called 'sandwich' is a three-layer segment in which the outer layers differ from the inner one in their susceptibility to wear. Sandwiches are superior to the conventional segments owing to a desirable saddle-like wear profile, which imparts a self-guiding characteristic to the saw blade and prevents it from deviating from the cutting direction. For safety reasons in dry cutting blades the segments are laser welded to the steel centre. Therefore they have to contain a diamond-free base with suitable fusion characteristics, otherwise the carbon-containing material would embrittle the steel support after welding.

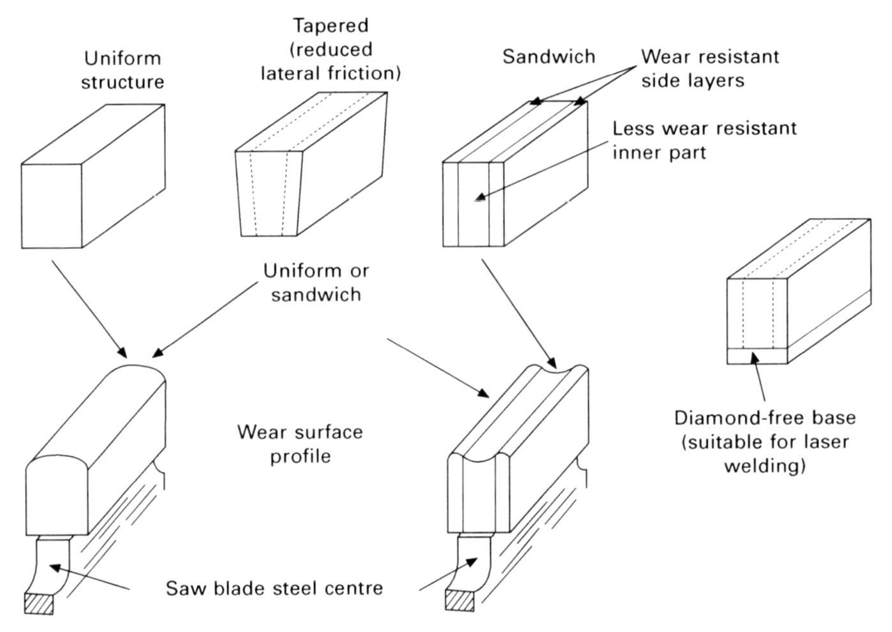

19.5 Basic saw blade segment designs.

19.2.1 Selection of metallic matrix

The two basic functions of the metallic matrix are to hold the diamonds tight and to wear at a rate compatible with the diamond breakdown. The key factors that determine the choice of a particular powder, used to form the matrix, include its chemical composition and degree of alloying, mean particle size and size distribution. The right combination of all these properties yields a material which does not degrade the diamond during sintering, can be consolidated to near full density at a moderate temperature (700–900°C) and has high as-sintered hardness, yield strength and toughness, and wear resistance ideally suited for the application conditions.

Cobalt

Historically, cobalt has been the most valued matrix metal, especially in professional tools, owing to its superior diamond retention properties and resistance to wear, which can be widely modified by forming alloys (with Cu, Sn, Ni, Fe, etc) or composites (WC, W_2C, W) (Konstanty, 2003). Unlike other metals, fine cobalt powder is available in a wide variety of grades (Fig. 19.6) which differ in sintering behaviour. The finer powders (Figs 19.6(a) and (b)) are most preferred because they can be cost effectively densified to virtually pore-free conditions by either hot pressing or pressureless sintering at temperatures ≤900°C.

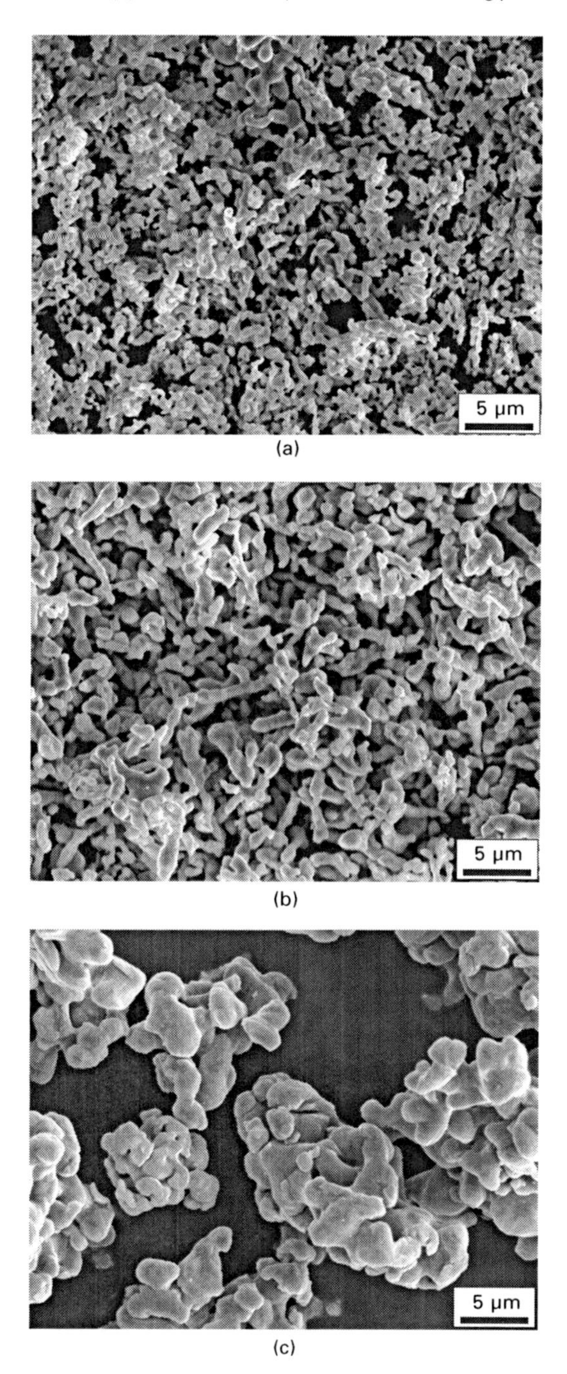

19.6 Examples of cobalt powders produced by Umicore (Olen, Belgium) for diamond tools: (a) sub micrometre size; (b) extra fine; (c) 400 mesh.

As the average price of cobalt is highly unstable (Fig. 19.7) and increasingly contributes to the overall tool cost, the recent trend is towards replacement of cobalt-based tools with low cobalt or, preferably, cobalt-free ones.

Cobalt substitutes

Until the early 1990s, the economical viability of the toolmaking process had been mostly affected by a high cost of diamond grits and, to a lesser extent, graphite moulds, whereas the contribution of cobalt and cobalt-based matrix powders had remained at an acceptable level. The forthcoming price cuts in the diamond supply side, tremendous increase in the service life of graphite moulds owing to broader application of nitrogen gas chambers in hot pressing equipment, and the price spikes of cobalt raised the spectre of substitution. The intensive search for cheaper matrix powders began in the mid-1990s and the first cobalt substitutes were marketed in 1997 and 1998 (Anon, 1997; Clark and Kamphuis, 2002).

To date, three families of fine copper-based and iron-based powders dedicated to fabrication of diamond impregnated tools have been developed and launched commercially under the brand names: *Cobalite* (Umicore, Belgium), *Next* and *Keen* (Eurotungstene Poudres, France). Their basic characteristics are given in Table 19.1.

These novel powders consist of a combination of at least two elements which have been co-precipitated in proprietary manufacturing processes (Standaert, 2001; Bonneau *et al.*, 2003; Kamphuis and Peersman, 2006)

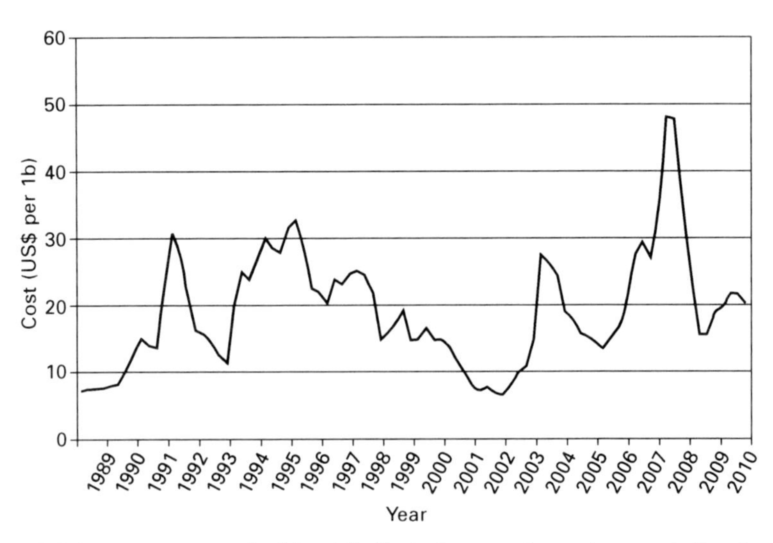

19.7 Average quarterly *Metal Bulletin* free market price quotation for 99.8% cobalt cathodes (Anon., 2007; Anon., 2011).

Table 19.1 Nominal chemical compositions and mean particle sizes of commercial cobalt substitutes

Designation	Chemical composition (wt%)				Fisher sub-sieve size (μm)	Producer (Source)
	Fe	Cu	Co	Others		
Next 100	29	46	25	–	0.8–1.5	Eurotungstene
Next 200	15	60	25	–	0.8–1.5	Poudres
Next 300	72	3	25	–	4	(Anon, 2005;
Next 400	50	35	15	–	2	Bonneau,
Next 900	80	20	–	–	3	2007)
Keen 10	58	17	25	–	2.5	
Keen 20	43	33	19	5Mo	3	
Cobalite 601	70	19.5	10.5	–	5.5	Umicore
Cobalite HDR	66	7	27	–	6.5	(Umicore,
Cobalite CNF	68.4	26	–	3Sn; 2W; $0.6Y_2O_3$	2.5	2011)
Cobalite XH	43	10	47	–	3	
Cobalite OLS	28	47	25	–	1.4	

to yield pre-alloyed agglomerates of sub-micrometre sized particles (Fig. 19.8).

The *Cobalite*, *Next* and *Keen* powders complement well the variety of fine cobalt powders as they have as-sintered hardness and strength properties similar to cobalt. With the exception of *Keen 20*, the other pre-alloyed powders show excellent consolidation behaviour achieving near full density after hot pressing at temperatures ranging from 650–800°C. Since the late 1990s novel cobalt substitute powders have continued to take an increasing share of the market reaching 25% in 2005 (Anon, 2005). Over this period the range of commercial grades have been markedly extended and a lot of effort has been made to refine their chemical composition and to modify the powder manufacturing process in order to couple the products' excellent mechanical properties and densification characteristics with field performance similar to cobalt.

19.2.2 Selection of diamond grit

To meet the requirements of given tool manufacturing techniques and application conditions it is essential to find the right combination of diamond type, size and concentration.

Diamond type

The workpiece material, for example stone, concrete, brickwork, ceramics, asphalt, and so on, primarily determines the type of diamond abrasive used

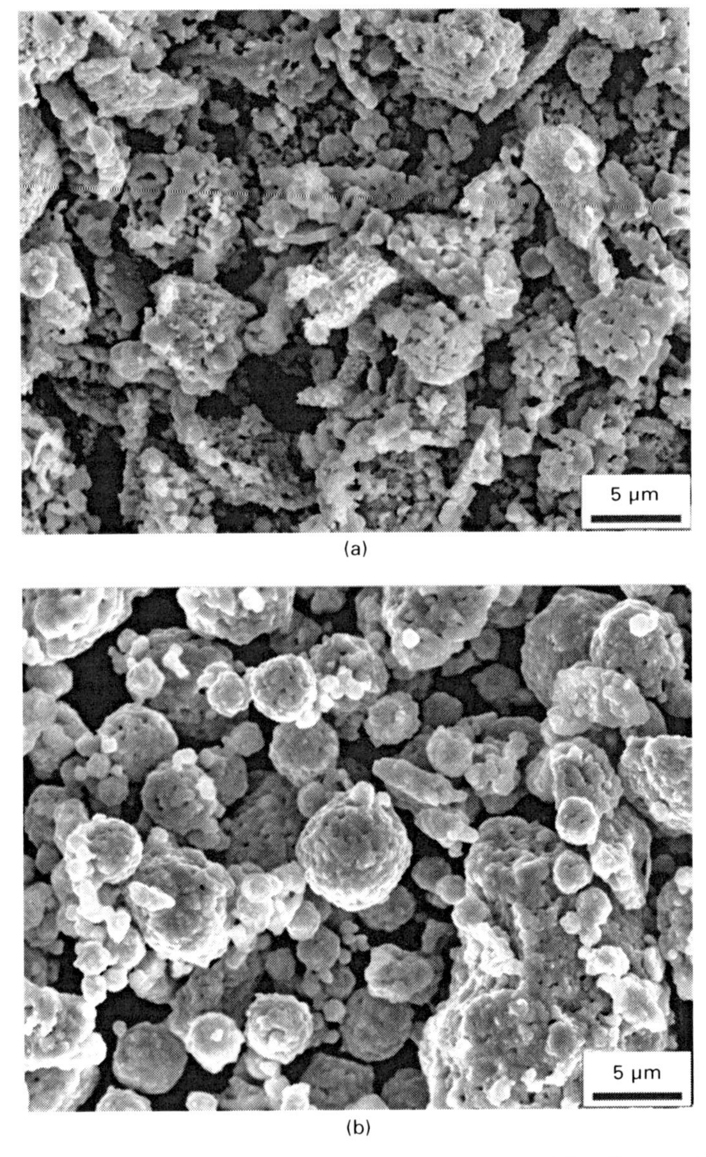

19.8 Examples of pre-alloyed powders dedicated for diamond tools: (a) copper-base *Next 200*; (b) iron-base *Cobalite 601*.

in the tool. A good general rule is that the tougher and more difficult it is to machine the workpiece, the stronger the diamond grit to be chosen.

Diamond manufacturers provide the toolmakers with abrasives showing a wide diversity of mechanical strength, thermal stability and matrix retention characteristics. Natural grits, made by crushing mined diamond boart, possess excellent bonding characteristics and therefore have been used for frame sawing of stone and light duty sawing and grinding applications (Konstanty, 2005, p. 56). As the diamond synthesis becomes more cost effective and prices of synthetic grits continue to decline, owing to increasing competition from low-cost producers in China, a growing tendency to replace natural diamond with economy synthetic grades is seen in the production of stone and construction tools. While most PM diamond tool manufacturers have moved to synthetic products, natural diamond still meets criteria for application in single point dressers, form dressers, water jets, dental tools, cutting tools for non-ferrous metals, drawing dies, and so on.

The main advantage of using synthetic diamond is that it can be designed and manufactured to satisfy virtually any specific application requirements. In response to the diversification of the demands of the market, two major diamond grit families with different particle characteristics have been developed. By using either a cobalt-based alloy or a nickel-based alloy as a catalyst in diamond synthesis, crystals with different internal structure can be obtained (Fig. 19.9). The cobalt grades are characterised by the

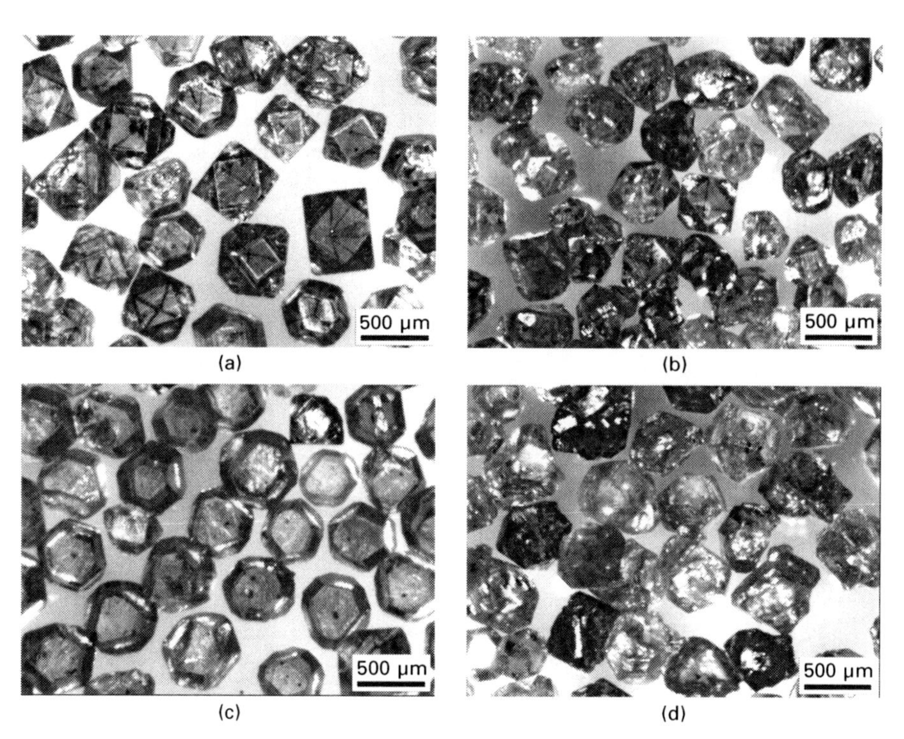

19.9 Synthetic diamond abrasives: (a) premium and (b) economy grade diamond grits manufactured using a cobalt-based alloy synthesis technique; (c) premium and (d) economy grade diamond grits manufactured using a nickel-based alloy synthesis technique.

presence of ordered arrays of metal inclusions within each particle (Figs 19.9(a) and (b)) and, therefore, they tend to fracture in an irregular manner which promotes a free cutting behaviour in the tool. In the nickel grades (Figs 19.9(c) and (d)) the impurities are uniformly distributed throughout the particle, resulting in excellent transparency, higher toughness and resistance to thermal degradation.

Grit size

Commercial diamond abrasives are classified into two broad bands of grit sizes. These are referred to as saw sizes and wheel sizes. Within each band, the grit sizes are commonly defined by two sieve aperture sizes, in the top and bottom defining sieves, conforming to the ASTM E11 standard. The saw diamond abrasives are coarser than 80 US mesh and find extensive application in a variety of sawing and drilling operations where fast material removal rates are essential. The general guidelines for diamond sizes used in the most typical tool applications are as follows:

- 20/30 US mesh: circular sawing and drilling of very abrasive sandstone, fresh concrete and asphalt, drilling of reinforced concrete;
- 30/40 US mesh: circular sawing and drilling of stone, concrete, reinforced concrete and asphalt, frame sawing of abrasive sandstone;
- 40/50 US mesh: circular sawing and drilling of less abrasive stone (e.g. granite, diorite, gabbro, fine-grained sandstone, limestone, dolomite, marble), concrete and refractory materials, wire sawing of stone and construction materials, calibrating of stone slabs;
- 50/60 US mesh: circular sawing and drilling of very tough fine-grained igneous stone (e.g. granite, granodiorite) and refractory materials, circular sawing, frame sawing and drilling of fine grained sedimentary and metamorphic stone (e.g. limestone, marble), calibrating of stone slabs;
- 60/80 US mesh: circular sawing, drilling and milling of glass, calibrating of ceramic tiles.

There is a general tendency to apply finer grits for slow, secondary sawing operations where perfect surface finish and edge definition are matters of great concern. In any brand of diamonds, the finer grits, being relatively stronger, are also recommended for sawing difficult to machine types of stone and ceramics. On the other hand, the main advantage of using coarser grit is a greater potential for faster cutting. This is because a coarse grit attains higher protrusions and hence produces greater clearance and enables easier swarf ejection from the cutting zone.

Diamond abrasives finer than 80 US mesh fall into the wheel sizes range and are generally used for grinding operations. Most commercial products are available as standard in a full complement of nine sieve sizes, from 80/100 down to 325/400 US mesh, to meet the requirements of the consecutive grinding steps.

Grit concentration

The amount of diamond in a segment is based on a scale in which 100 concentration is equivalent to 4.4 carats per cm^3 (25 vol%). All other concentrations are proportional. As shown in Table 19.2 the diamond concentration, in conjunction with the diamond particle size, governs the number of cutting points per unit area of the working surface of a segment. Which concentration to use depends on a number of factors. The material to be processed and its properties should be considered first when designing the tool composition. In general, the easier to cut and more abrasive the workpiece, the higher should be the diamond concentration.

Application of coated diamonds

As already mentioned, it is extremely important that the diamond crystals are firmly held in the matrix. It is, however, not easy to develop a matrix capable of effective utilisation of very strong synthetic grits characterised by a perfectly defined blocky shape and smooth surfaces. Therefore, the diamond itself has to cooperate with the matrix in order to minimise its excessive pullout. The most effective way of improving retention of high-grade diamond crystals in the matrix is to pre-coat them with a thin film of strong carbide former such as titanium, chromium or silicon (Konstanty, 2005, p. 62). The enhanced diamond retention arises from the coating which chemically bonds to the diamond and forms a metallurgical, diffusion attachment to the matrix. An additional nickel cladding (Fig. 19.10) may optionally be applied with the intention of aiding mechanical anchoring of the grits in bronze-based matrices (Kompella, 2005; Element Six, 2011).

The other benefits of coatings are protection of the diamond from surface

Table 19.2 Total number of diamonds and pullouts per cubic centimetre of the working face of a tool (GE Superabrasives, 1991)

US mesh size	Diamond concentration					
	15	20	25	30	35	40
25/35	14	19	24	29	34	38
30/40	22	30	37	45	52	60
35/40	26	34	43	51	60	68
40/50	38	51	63	76	88	101
50/60	65	87	109	131	153	174
60/80	85	114	142	170	199	227

19.10 Rough nickel cladding on Ti-coated synthetic diamond crystals.

graphitisation, oxidation and attack by aggressive matrix components and strengthening of flawed diamonds through healing of surface defects. Si-coated diamonds are particularly useful when effective protection of diamond crystals from destructive high temperature reactions is a matter of prime concern, for example when iron-based matrices are applied and hot pressing temperatures markedly higher than 900°C become necessary.

19.3 Diamond tool fabrication

A typical fabrication process consists of two separate stages. The diamond impregnated tool components are initially produced by the PM route (Fig. 19.11) and then they are attached to a suitable steel carrier which may require further processing. In certain cases, for example in the manufacture of continuous rim saw blades or infiltrated drill bits, these two stages are combined in a single process.

19.3.1 Powder metallurgy segment fabrication routes

Preparation of matrix powder–diamond mixture

This operation is usually carried out in a *Turbula* type mixer. The powder storage container is filled to about half its volume with diamonds, matrix powders and steel chains, used to aid mixing, and set into a chaotic motion of twisting and rotation for a certain period of time. Suitable organic binders and/or lubricants, for example monoethylene glycol, paraffin oil, zinc stearate,

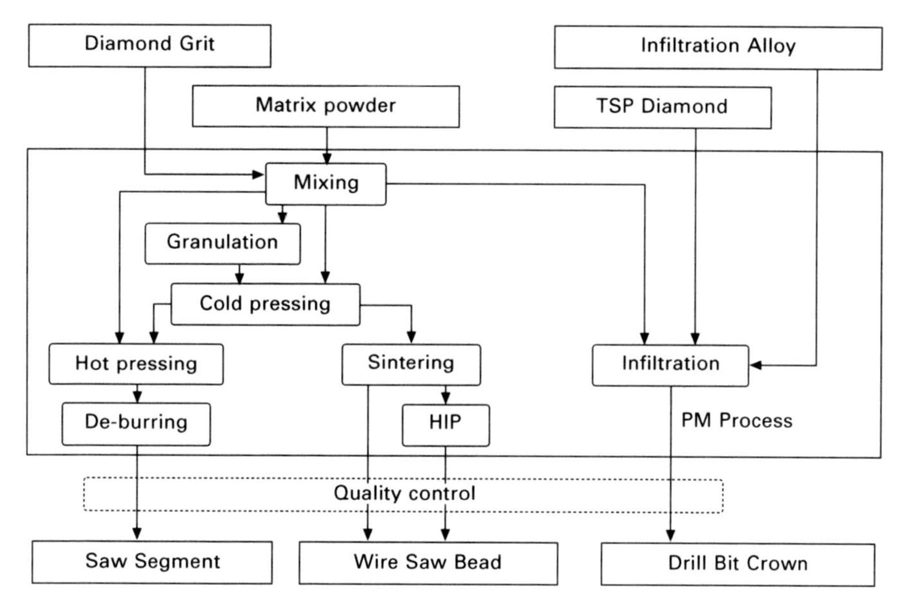

19.11 PM diamond tool production process.

are added to the container at various stages of mixing so as to prevent segregation and minimise steel die wear during subsequent cold pressing.

Granulation

When the mixture is to be compacted by a volumetric cold press, the powder has to be granulated to ensure the required flow and packing characteristics. Granulation may be carried out in a variety of ways (Burckhardt, 1997; Weber, 1999) but, practically, techniques based on either a high-speed mixing principle or mechanical rolling process have received widespread application in the diamond tool industry. Irrespective of the processing route, organic binders dissolved in suitable solvents are used to cement the mixture ingredients together, thereby imparting the desired mechanical strength to the granules. It is important that the binder has suitable thermal properties, which permit its complete removal from the material at the hot consolidation step. Otherwise, the segments have higher residual porosity, which may degrade their quality and create problems with brazing to the steel support.

Cold pressing

Cold pressing is always used before pressureless sintering but also prior to hot pressing of saw segments that have a layered structure. Cold pressing is performed in tool steel dies on double action presses at a moderate compaction pressure of up to 200 MPa. There are two types of machinery used in the diamond tool industry. The conventional presses are fitted with either vibratory or screw powder feeders and scales used to weigh out and fill the die with the correct amount of the diamond–matrix powder mixture. Alternatively, the machines may incorporate feed shoes operating on the volumetric filling principle. The gravimetric presses offer the higher flexibility necessary to manufacture segments in smaller quantities whereas the more efficient volumetric equipment is the preferred choice for mass production of tool components.

Hot pressing

The hot pressing technique, which has gained widespread use in the production of diamond impregnated segments, consists of a simultaneous application of heat and pressure in order to obtain a product nearly free from internal porosity. Compared to the conventional cold press/sinter route, hot pressing requires holding the powder for a short time, usually 2–3 min, at a substantially lower temperature to reach a markedly higher density level. The hot pressing process is typically realised in high resistance graphite moulds by passing electrical current directly through the mould. Industrial hot presses are fitted with chambers wherein the moulds are heated in vacuum and nitrogen and therefore their service life is markedly increased.

The principle of direct resistance heating has limitations with respect to the size and geometry of the tool components. Experience has demonstrated that the production of continuous rim saw blades and grinding wheels is slow and energy consuming owing to the large dimensions of these tools. Furthermore, difficulties are encountered in assuring uniform temperature distribution across the mould. Therefore the favoured equipment for this application is a furnace press in which several steel dies, piled one on top of the other, are simultaneously heated by radiation and convection.

Deburring

Saw blade segments, grinding segments and other hot-pressed diamond tool components require cleaning and removing edge residuals after consolidation. This is usually performed by tumbling the segments with coarse alumina or silicon carbide grit.

Sintering

Sintering is an indispensable step in fabrication of numerous PM structural parts and tool components. However, its application in the manufacture of diamond impregnated tools is limited by restrictions on the composition, mechanical strength and dimensional accuracy of the final product. An exception is the process for making wire saw beads (Fig. 19.12). In contrast to other techniques, the cost savings and higher rates of production achieved in this case by the use of the conventional cold press/sinter route apparently outweigh the shortcomings of the pressureless sintering operation.

Hot isostatic pressing

Hot isostatic pressing (HIP) finds limited application in the production of wire saw beads. The green beads are first sintered to a closed-pore condition or, alternatively, encapsulated in evacuated glass tubes (Fig. 19.12). The semi-finished parts are then loaded into the pressure vessel of a hot isostatic press which is evacuated and purged several times with a pressurising gas, typically argon, before the HIP cycle begins. The HIP units are capable of achieving pressures up to 200 MPa and allow a marked reduction in the consolidation temperature compared with the hot pressing technique.

Infiltration

Infiltration is ideally suited to make rotary surface set diamond bits, thermally stable polycrystalline diamond (TSP) bits and TSP hybrid bits used in drilling in earth formations. The process consists of filling the interconnected pores of loose powders, placed in a graphite mould, with a liquid infiltration alloy. It is common that TSP rectangles are incorporated in the working section of the drill bit with the intention of stabilising its lateral dimensions (Fig. 19.13).

19.12 Diamond wire saw and beads (glass-encapsulated for HIP and ready for mounting on a steel wire rope).

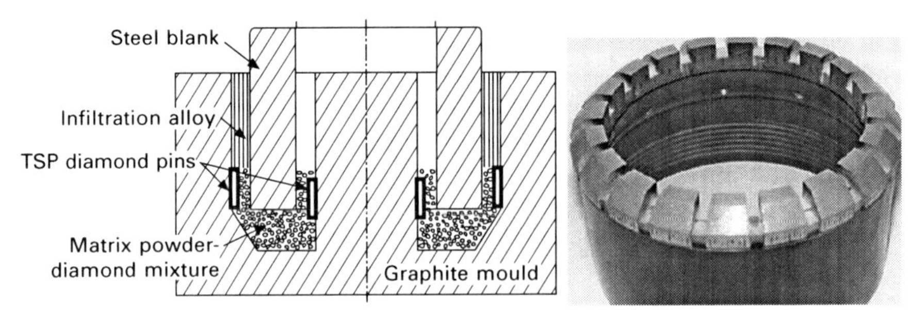

19.13 Sectional view of mould assembly and core head type TSP hybrid bit used for geological exploration.

The particle shape, size and size distribution of the matrix powders must be carefully selected to attain the required tap density and to avoid shrinkage during high temperature processing. To ease mould loading, the powdered material may be applied in a liquid hydrocarbon carrier which vaporises when heated. Once the powder has been placed in the mould, in such a way that it surrounds a certain portion of the steel blank, a suitable infiltration alloy, preferably in the form of coarse shot, is positioned beside the matrix powder. The whole mould assembly is then placed in a furnace and heated in a controlled atmosphere until the infiltration alloy melts and is drawn by capillary action into the powder mass bonding it to the steel blank.

Quality control

The quality control of diamond impregnated tool components is frequently limited to a hardness test. The Rockwell B test is the most widespread used technique owing to its simplicity and inexpensiveness. The matrix powder–diamond mixture consolidated to near full density acquires a narrow hardness range which mostly depends on the matrix composition and porosity. Incompletely densified segments have low toughness, which results in poor wear resistance and diamond retention. Therefore, if there is any doubt about the hardness readings, the evaluation of the as-consolidated density becomes another important quality check. In the case of infiltrated drill bits, visual examination is routinely used to assure correct diamond setting and protrusion but non-destructive techniques, such as radiography, may also be used to detect internal flaws and discontinuities in the bit crown.

19.3.2 Finishing operations

Diamond impregnated segments cannot be directly applied in a tool and need to be brazed or laser welded to a suitably shaped steel support to obtain a saw blade, core drill or grinding wheel. Brazing is used in the manufacture of tools for wet processing of natural stone, which can be re-tipped, whereas laser welding is used in the mass production of small diameter, dry cutting circular blades. In the latter case, the heat generated during sawing without coolant softens the conventional braze joints and hence there is a risk that a whole segment will break off the steel support during dry cutting operations. Since the bending strength of a laser-welded joint is from three to six times higher compared to the brazed one (Weber, 1991), laser welding in practice eliminates the possibility of segment detachment. As laser welding imposes restrictions in selection of the steel centre and segment matrix material, low-carbon steels with low hardenability are used to avoid formation of brittle martensite in the heat-affected zone owing to rapid cooling after the laser beam has passed. For this reason, the segment must also have a diamond-free base (Fig. 19.5) made of a suitable material that will yield a strong and flawless welded joint with the steel.

After brazing, or laser welding, circular saw blades must undergo truing to render the segments concentric with the bore of the tool body and to clean their sides and reduce lateral run-out. A subsequent sharpening operation is performed in order to remove the matrix from around the diamond particles, to produce sufficient protrusion and allow efficient cutting from the outset. In contrast to truing, sharpening is carried out in the down-grinding mode, that is the blade and grinding wheel turn in the same direction exerting less impact on diamond particles. Steel saw blade centres accumulate stresses during fabrication which often make them incapable of spinning without wobble. Therefore all circular blades are subjected to levelling, that is neutralising the unequal stresses to cause them to lie flat, followed by tensioning which consists of the addition of extra stress to stretch the centre section, counteracting the centrifugal force that tends to elongate the rim section of the rotating blade. Although the final checks of the blade's lateral run-out and stress condition have been fully mechanised and computerised, the arduous levelling and tensioning jobs consist of rolling and hammering by a skilful and experienced sawsmith.

In the production of diamond wire saws, sintered diamond impregnated rings are brazed onto solid cylindrical cores, which are subsequently bored through and threaded. The beads and spacers are then alternately mounted at regular intervals on a steel rope by means of simple assembly equipment or automatic machines (Anon, 2004). Injection-moulded plastic or vulcanised rubber is usually used to separate the individual beads and to provide a watertight protection to the rope against abrasion by aggressive slurries. The older type of diamond wire, with helical spring spacers and stopping rings, which can easily be assembled or repaired in the field, may still be employed for sawing marble and other non-abrasive materials. To avoid ovalisation of the working beads, the wire saw must be suitably pre-twisted around its longitudinal axis before the continuous loop is assembled by means of clamping pipes or threaded connecting sleeves.

19.4 Application of powder metallurgy diamond tools

Both diamond tool manufacturing techniques and application areas have undergone a spectacular evolution since the invention of synthetic diamond in the mid-1950s. Over this time different types of metal-bonded diamond tools have been developed and increasingly used to bring major productivity benefits in the extraction, sizing and final processing of natural stone, drilling and sawing of brickwork, concrete and reinforced concrete, controlled demolition of buildings, road repair, machining of glass and ceramics, exploration of gas and petroleum, and so on.

19.4.1 Sawing

Nowadays the PM diamond tool is the most effective means of sawing natural stone, concrete, reinforced concrete, asphalt, brickwork, glass and other ceramics. The main classification of the operations carried out on these materials comprises circular sawing, frame sawing and wire sawing.

Circular sawing

A diamond circular saw blade is a versatile tool that may be used on both portable and stationary machines. Continuous rim blades and segmental laser welded blades are primarily designed for dry cutting operations carried out by hand-held equipment although the latter are also recommended for wet sawing of asphalt and concrete by floor and road saws which may not ensure an adequate water supply.

Brazed segmental blades are commonly used for sawing walls and on stationary machines (Fig. 19.14), which secure sufficient amounts of coolant to prevent the brazed joint from overheating. These tools can be subjected to a number of re-tipping operations, which is economically justified in the case of saw blades 1000 mm in diameter and bigger. The contribution of the steel centres to the overall cost of the tool increases dramatically with the blade diameter and therefore is reused until its fatigue life is approached.

The general guidelines for operating parameters used in the most typical circular saw blade applications are provided in Table 19.3.

The easy to process, low power-consuming materials are generally sawn to full depth by the downward rotation of the blade, whereas the difficult to cut, high power-consuming materials, such as granite and dense ceramics, are

(a)

(b)

19.14 (a) Sawing a doorway with a remotely controlled wall saw and (b) production of modular granite tiles with a multi-blade saw (courtesy of Hilti Corporation and EHWA Diamond Ind. Co. Ltd).

always sawn in many passes by a reciprocating movement of the saw blade. On single-blade machines, a typical depth of cut ranges between 10 mm and 20 mm whereas the feed rate is chosen to comply with the recommended cutting rates given in Table 19.3. On multi-blade machines all blades must run freely and parallel to one another. Therefore to minimise forces acting on the blades a high feed rate of between 8 and 14 m min^{-1} is combined with a very shallow depth of cut ranging between 0.4 and 1.5 mm (Konstanty, 2005, p 134).

Table 19.3 General recommendations for peripheral speeds and cutting rates for use on single-blade and multi-blade machines (Konstanty, 2005, p 133)

Workpiece material	Blade peripheral speed (m s^{-1})	Cutting rate (cm^2 min^{-1})
Dry cutting with hand-held equipment	80–100	Adjusted to the saw blade and workpiece characteristics
Quartziferous granite	25–30	100–200
Low quartziferous granite	30–40	200–600
Marble	40–50	600–1200
Sandstone	40–65	300–1000
Ceramics	20–50	–
Concrete	35–50	–
Reinforced concrete	30–40	–
Asphalt	40–60	–

19.15 Frame saw fitted with twenty diamond blades (courtesy of BMTG Marek Gelbert).

Frame sawing

Frame saw systems have long been established as an efficient and low sawing cost means of mass production of big stone slabs. They are generally used for cutting marble, travertine, limestone, sandstone and agglomerates, however, syenite and the easiest to process types of granite may also be sawn. Frame saw blades are usually mounted horizontally on single or multi-blade (Fig. 19.15) machines, which are further classified as slow frames or fast frames depending on whether the product of the length of stroke and the flywheel rotational speed is lower or higher than 50 m min^{-1}.

The typical cutting rates used on slow and fast horizontal frames are listed in Table 19.4.

Table 19.4 Recommended cutting rates for sawing various types of stone on slow and fast frame saws (Konstanty, 2005, p 136)

Stone	Feed rate (cm h^{-1})[a]	
	Slow frame	Fast frame
Marble	10–18	20–35
Travertine	15–22	25–40
Limestone	12–25	20–30
Sandstone	20–30	30–40
Agglomerates	13–20	25–35
Easy to process types of granite and syenite	8–18	–

[a]increases with the linear speed of the frame and decreases with the number of saw blades

19.16 Multi-wire saw slabbing a stone block (courtesy of MC Diam Ltd).

Wire sawing

A diamond wire has become a standard stone quarrying tool which enables high production rates and increased output of blocks that are used for monumental purposes in areas where flawed or fragile stone is quarried. Owing to its adaptability to suit most sawing tasks, it has also made rapid progress in stoneyards, where both single-wire and multi-wire stationary machines are increasingly used for block division (Fig. 19.16), as well as for profiling of stone slabs. A typical wire saw contains 10–11 mm diameter diamond impregnated beads mounted at regular intervals on a flexible 5 mm diameter steel rope composed of many twisted together high strength stainless steel strands. The multi-wire machines utilise 6–8 mm beads on a 4 mm steel rope to minimise kerf widths and thus to maximise the yield of stone slabs per block.

The cutting action consists of pulling a properly pre-tensioned wire saw across the workpiece. The linear wire speeds and cutting rates achieved on stationary machines are similar to those applied in the quarry and depend on the stone type as shown in Table 19.5.

The versatility and economic advantages of the wire saw technology have also been recognised in the construction industry, where portable wire saw machines are used for various construction, renovation and controlled demolition purposes. The ability of the diamond impregnated wire to cut cleanly, quickly and accurately, with little noise and vibration, makes this tool an ideal alternative to blasting or jack hammering with flame cutting of the rebar, which were previously used for removal of thick sections of reinforced concrete or brickwork. The cutting rates achievable on construction materials may widely vary from 1–6 $m^2 h^{-1}$ on reinforced concrete, through 5–11 $m^2 h^{-1}$ on plain concrete, up to 10–18 $m^2 h^{-1}$ on masonry, depending on the type of concrete aggregates, percentage of steel reinforcing, brick composition, and so on.

It is essential for the tool performance that the diamond beads wear in a uniform manner over the whole working surface. In industrial practice, pre-twisting the wire, by applying one anti-clockwise twist per metre before a continuous loop is assembled, gives rise to its rotation in the kerf and consequently prevents bead ovalisation.

19.4.2 Drilling

Diamond core drills (Figs 19.17 and 19.18) are efficient tools used for cutting holes in ornamental stone, concrete, asphalt, brickwork, glass and other non-metallic materials, stitch-drilling doorways and vents in heavily reinforced concrete structures, as well as for exploratory drilling in earth formations.

In wet drilling it is common to inject water through the centre of the drill to cool its working part and flush the cuttings out. As a general rule, small

Table 19.5 Typical application parameters for quarrying stone by means of diamond impregnated wire saws (Konstanty, 2005, p 137)

Type of stone	No of beads per metre of wire length	Wire linear speed ($m s^{-1}$)	Cutting rate ($m^2 h^{-1}$)
Quartziferous granite	37–40	19–23	1–3
Low quartziferous granite	30–39	23–28	3–5
Marble	27–30	30–40	6–8
Limestone	27–30	28–40	7–8
Travertine	27–30	40–45	8–9

19.17 Selection of core drills for drilling in stone, concrete and ceramics.

19.18 Portable drill rig mounted for drilling through a concrete wall (courtesy of Hilti Corporation).

diameter drills, up to around 32 mm, have a continuous drill crown design with waterways incorporated in it, whereas bigger drills have diamond impregnated segments brazed to the end of the core barrel and suitably spaced to facilitate efficient evacuation of the slurry.

Drills used for dry coring by means of hand-held equipment are guided by a cemented carbide-tipped drill mounted axially on a short barrel encompassing lateral air vents. The central guiding drill projects out over the segments, which, for safety reasons, should be laser welded to the barrel. Typical core drilling parameters are listed in Table 19.6.

Drilling tools used in geological exploration and mining of mineral deposits, made by the infiltration route, include coring and non-coring surface set diamond bits, as well as diamond impregnated core bits. The surface set bits offer higher penetration rates but they are prone to damage while drilling through broken and fractured formations. Impregnated core bits, which are more robust and capable of achieving a longer bit life are used more frequently. Compared with surface set bits, they display better directional stability and drilling performance in the hardest rock formations, enabling extraction of clean core for laboratory examination of rock structures.

To optimise economy of drilling, the rotational speed and feed rate of an impregnated bit should be balanced to suit the rock type being drilled. These two parameters combine to yield a bit revolutions per centimetre of penetration (RPC) index that has been found to be an extremely useful guide to successful drilling. Ideally, the pressure on the bit must be kept below 15 MPa and adjusted to aim at 80–100 RPC to keep the bit sharp (Boart Longyear, 2009). If the load is insufficient, the diamonds will be polished and the bit will get blunt. If the load is too high, excessive and uneven wear can occur.

In most situations water is used as the coolant but cutting oils or lubricating fluids may alternatively be applied for drilling in hard rock formations (Boart Longyear, 2009).

19.4.3 Grinding and polishing

In addition to sawing and drilling, metal-bonded diamond impregnated tools have also been used with great success in a variety of grinding and polishing operations, offering the best combination of long tool life, fast stock removal and good surface finish. In stone working, calibrating, grinding and rough polishing steps are realised with metal-bonded diamond tools installed either

Table 19.6 Peripheral speeds and feed rates recommended for core drilling (Konstanty, 2005, p 141)

Workpiece material	Drill peripheral speed ($m\ s^{-1}$)	Feed rate ($cm\ min^{-1}$)
Very hard and difficult to cut ceramics	~1	1–2
Granite, igneous stones	1–2	3–4
Glass	1.5–2.5	3–4
Concrete, reinforced concrete	2–4	4–8
Marble, travertine, limestone	3–5	5–10
Sandstone, asphalt	6–8	10–20

on single-head machines or multi-head automatic surface polishing lines. The latter are universally used in the manufacture of stone tiles for flooring and wall cladding in public and private buildings. The grinding operations are carried out in several stages, mostly with metal-bonded wheels containing progressively finer grits.

Diamond-grinding techniques have also branched into civil engineering projects. Planing concrete floors (Fig. 19.19) and walls, slitting concrete and brickwork structures for wiring, grooving the surface of roads, bridges and airport runways, to reduce the risk of aquaplaning in wet weather, are typical examples of jobs carried out with metal-bonded diamond impregnated tools.

Another important application of diamond impregnated tools is in grinding glass and ceramic materials. Various shapes of metal-bonded cup wheels, peripheral wheels and pencil edge wheels are used here, to complement electroplated and resin-bonded tools, for bevelling of plate glass, shaping, edging and surfacing of architectural furniture, automotive and optical glass components, artistic decoration of glassware, and so on.

19.5 Latest trends and developments

The recent tremendous price cuts in the diamond supply side and increased competition with low cost labour countries have brought pressure on the leading edge of the diamond tool industry to increase the rate of innovation and to carve a market niche for non-commodity products. In practical terms, a great deal of effort is going into seeking cheap, high-performance matrix materials and optimum grit distribution in segments.

19.19 Levelling a concrete floor (courtesy of Hilti Corporation).

Although cobalt and its substitutes (see Section 19.3.1) are still used for professional tools, the trend is towards a broader application of premixed and mechanically alloyed iron-based powders. Good candidates are medium-carbon steels containing metastable austenite that readily transforms to hard martensite under tribological straining (Konstanty *et al.*, 2011). The strain-induced martensitic reaction imparts wear resistance to the matrix and, by generating compressive stress under the working face of the tool, favours retention of the most heavily loaded diamond crystals.

A more uniform diamond distribution in the working portion of the tool renders it more durable and enables free cutting. The general consensus on this matter has stimulated a major effort by toolmakers to find ways and means to control the grit distribution in diamond impregnated segments (Fig. 19.20). Among the various methods which have been developed to date, the technologies utilising either diamond placing by automated systems (Weber, 2010) or encapsulated diamonds (Egan *et al.*, 2010) are finding industrial applications.

In the former method the diamonds are both equally spaced and arrayed in a regular manner (Fig. 19.20(a)). To avoid narrowly spaced grits having the same height of protrusion and high fluctuations in the number of working

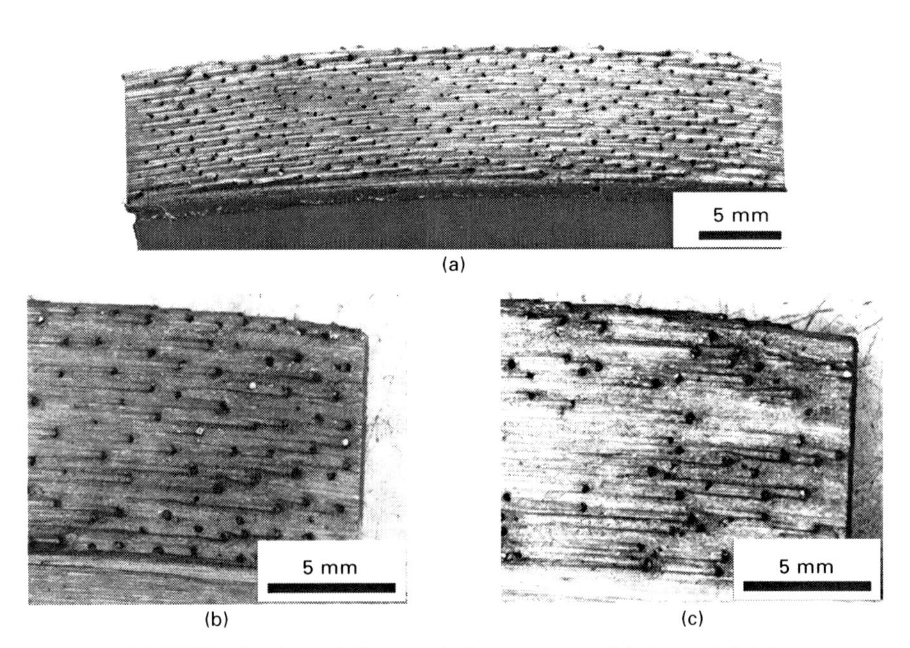

19.20 Distribution of diamonds in segments fabricated: (a) by a diamond placement technology (courtesy of EHWA Diamond Ind. Co. Ltd, Korea), (b) using encapsulated diamonds and (c) from a mixture of diamond grits and matrix powders (courtesy of Element Six Ltd., Ireland).

crystals, it is crucial that the rows of diamonds are properly angled versus the tool face (Lee *et al.*, 2008).

The latter method uses diamonds coated with a predetermined amount of matrix powder to yield the intended diamond concentration. To produce pellets containing exactly one diamond core (Fig. 19.21) a dual stage process has been developed (Gush *et al.*, 2009). In the initial stage the individual crystals are pre-coated in a shovel rotor apparatus, or alternatively in the fluidised bed, to attain a critical size without the risk that the pellets may contain more than one diamond and consequently show a significant size distribution. In the second stage, the pellets are built up to the final size by the rotating pan method which offers increased powder deposition rates.

19.21 Diamond pellets containing a single diamond crystal.

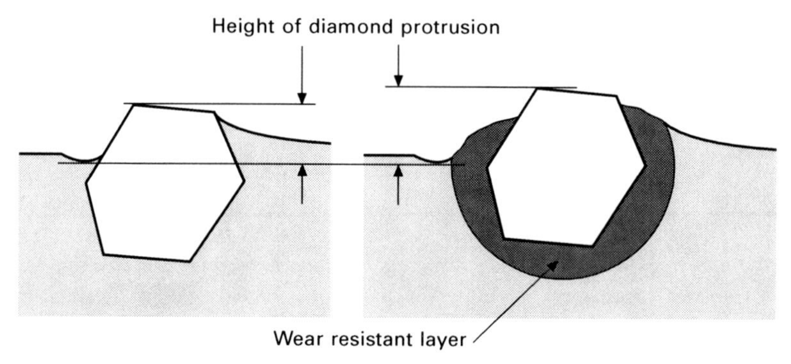

19.22 Increasing diamond protrusion by incorporating a wear resistant zone around each diamond.

Besides achieving uniform distribution of diamond grits in the tool (Fig. 19.20(b)) an additional benefit of increased diamond protrusion may be gained if the coating consists of two concentric layers of different composition where the inner layer attains higher resistance to wear upon segment consolidation (Fig. 19.22).

19.6 References

Anon (1997). 'A new generation of powders for the diamond tool industry'. *Marmomacchine International*, **18**, 156–7.

Anon (2004). 'Automatic threading and unthreading of diamond wires'. *Industrial Diamond Review*, **64**, 63–4.

Anon (2005). 'Keen – a new concept in prealloyed powders'. *Industrial Diamond Review*, **65**, 45–7.

Anon (2007). '2006 production statistics'. *Cobalt News*, [online] April. Available at: <http://www.thecdi.com/cdi/images/news_pdf/Cobalt_News_Apr07.pdf> [Accessed 15 July 2011].

Anon (2011). '2010 production statistics'. *Cobalt News*, [online] April. Available at: <http://www.thecdi.com/cdi/images/news_pdf/11-2_cobalt_news.pdf> [Accessed 15 July 2011].

Boart Longyear (2009). *Diamond Products Field Manual. Customer guide to the selection and field use of diamond coring products*. February 2009. [online] Available at: <http://www.mapek.com/uploads/default/Files/Boart_Longyear_Bit_Usage_Manual.pdf> [Accessed 29 September 2011].

Bonneau, M. (2007). personal communication, 9 March 2007.

Bonneau, M., Chabord, S. and Prost, G. (Eurotungstene Poudres) (2003). *Micronic Pre-alloyed Metal Powder based on Three-dimensional Transition Metal*. US patent 6,613,122 B1. September 2, 2003.

Bundy, E.P., Hall, H.T., Strong, H.M. and Wentorf, R.H. (1955). 'Man-made diamond'. *Nature*, **176**, 51–5.

Burckhardt, S. (1997). 'New technique for granulating diamond and metal powders'. *Industrial Diamond Review*, **57**, 121–2.

Clark, I.E. and Kamphuis, B.J. (2002). 'Cobalite HDR – a new prealloyed matrix powder for diamond construction tools'. *Industrial Diamond Review*, **62**, 177–82.

Dedalus Consulting (2008). *Global Markets, Applications & End-users: 2008–2013 analysis & forecasts*, Dedalus Consulting, New York, Section 2, p1.

Egan, D., Cormac, L. and Seamus, M. (2010). 'A study of the use of encapsulated diamond in saw-blades for the cutting of stone'. In: *EPMA (European Powder Metallurgy Association), Proceedings of PM2010 Powder Metallurgy World Congress and Exhibition*. Florence, Italy 10–14 October 2010, Volume 3, 559–66.

Element Six (2011). *SDBTN*. [online] Available at: <http://www.e6.com/en/media/e6/content/pdf/SDBTN_Data_sheet.pdf> [Accessed 24 July 2011].

GE Superabrasives (1991). *MBS Diamond Products for Sawing and Drilling Applications*. (product information brochure GES 91–966).

Gush, K., Munday, M.G., Schmock, P. and Kelly, S. (Element Six, Ltd.) (2009). *Method for Manufacturing Encapsulated Superhard Material*. WO 2009/101605 A1 (International Application PCT/IB2009/050626).

Hall, H.T., Strong, H.M. and Wentorf, R.H. (General Electric Company) (1960). *Method of Making Diamonds*. US patent 2,947,610. August 2, 1960.

Hughes, F.H. (1980). 'The early history of diamond tools'. *Industrial Diamond Review*, **40**, 405–7.

Kamphuis, B-J. and Peersman, J. (Umicore) (2006). *Pre-alloyed Bond Powders*. US patent 7,077,883 B2. July 18, 2006.

Kompella, J. (2005). 'Increased profitability through diamond coatings on tool-maker and end-user economics'. *Diamante Applicazioni & Tecnologia*, **42**, 83–9.

Konstanty, J. (2003). *Cobalt as a Matrix in Diamond Impregnated Tools for Stone Sawing Applications*. 2nd edition, AGH-UWND: Krakow.

Konstanty, J. (2005). *Powder Metallurgy Diamond Tools*. Elsevier, Oxford.

Konstanty, J., Stephenson, T.F. and Tyrala, D. (2011). 'Novel Fe–Ni–Cu–Sn matrix materials for the manufacture of diamond-impregnated tools'. *Diamond Tooling Journal*, **3**, 26–9.

Lee, H.W., Park, J.H., Lee, S.K. and Kim, D.G. (Shinhan Diamond Ind. Co. Ltd.) (2008). *Diamond Tool*. WO 2008/060018 A1 (International Application PCT/KR2007/001988).

Lundblad, E. (1990). 'Swedish synthetic diamond scooped the world 37 years ago'. *Indiaqua*, **55**, 17–23.

Standaert, R. (N. V. Union Miniere S.A.) (2001). *Pre-alloyed, Copper Containing Powder, and its Use in the Manufacture of Diamond Tools*. US patent 6,312,497. November 6, 2001.

Tolansky, S. (1967). 'Early historical uses of diamond tools'. In: *Science and Technology of Industrial Diamonds*. J. Burls, (ed.), Industrial Diamond Information Bureau, London, Volume 2, 341–9.

Umicore (2011). *Tool Materials*. [online] Available at: <http://www.toolmaterials.umicore.com/technicalData/csmTDS.pdf> [Accessed 16 July 2011].

USGS (2011). *Mineral Commodity Summaries 2011*, US Geological Survey, Reston, Virginia, 51.

Weber, G. (1991). 'Laser welding of diamond tools'. *Industrial Diamond Review*, **51**, 126–8.

Weber, G. (1999). 'Granulating: a new process for diamond tool producers'. In *EPMA (European Powder Metallurgy Association), Proceedings of International Workshop on Diamond Tool Production*. Turin, Italy 8–10 November 1999, 73–82.

Weber, G. (2010). 'DiaSet – now fully automatic setting of diamond arrays'. *Diamond Tooling Journal*, **2**(1), 38–40.

Index

CPSIA information can be obtained at www.ICGtesting.com
Printed in the USA
BVOW04*1419080414

349811BV00005BC/64/P

9 780857 094209